T0396422

Scimat
Anthology

Histophysics, Art
Philosophy, Science

LOVE
WISDOM
LOVE
FREEDOM

WHEN HERE DWELL POETICALLY

Science Matters Series

Lui Lam
Founder and Editor

 Scimat (Science Matters) is the new multidiscipline that treats all human matters as part of science, wherein, humans (the material system of *Homo sapiens*) are studied scientifically from the perspective of complex systems. That *Everything in Nature Is Part of Science* was well recognized by Aristotle and Da Vinci and many others. Yet, it is only recently, with the advent of modern science and experiences gathered in the study of evolutionary and cognitive sciences, neuroscience, statistical physics, complex systems, and other disciplines, that we know how the human-related disciplines can be studied scientifically. Science Matters Series covers new developments in all the topics in the humanities and social science from the scimat perspective, with emphasis on the humanities.

Published

1. *Science Matters: Humanities as Complex Systems*
   M. Burguete & L. Lam, editors
2. *Arts: A Science Matter*
   M. Burguete & L. Lam, editors
3. *About Science: Philosophy, History, Sociology & Communication*
   M. Burguete & L. Lam, editors
4. *Humanities, Science, Scimat: From Two Cultures to Bettering Humanity*
   L. Lam
5. *Scimat Anthology: Histophysics, Art, Philosophy, Science*
   L. Lam

Science Matters Series | No. 5

Scimat
Anthology

Histophysics, Art
Philosophy, Science

Lui Lam

San Jose State University, USA

World Scientific

EW JERSEY · LONDON · SINGAPORE · BEIJING · SHANGHAI · HONG KONG · TAIPEI · CHENNAI · TOKYO

Published by

World Scientific Publishing Co. Pte. Ltd.

5 Toh Tuck Link, Singapore 596224

USA office: 27 Warren Street, Suite 401-402, Hackensack, NJ 07601

UK office: 57 Shelton Street, Covent Garden, London WC2H 9HE

Library of Congress Control Number: 2024943400

British Library Cataloguing-in-Publication Data
A catalogue record for this book is available from the British Library.

Cover design: Lui Lam

Science Matters Series — Vol. 5
SCIMAT ANTHOLOGY
Histophysics, Art, Philosophy, Science

ISBN 978-981-12-9704-5 (hardcover)
ISBN 978-981-12-9705-2 (ebook for institutions)
ISBN 978-981-12-9706-9 (ebook for individuals)

For any available supplementary material, please visit
https://www.worldscientific.com/worldscibooks/10.1142/13956#t=suppl

Typeset by Stallion Press
Email: enquiries@stallionpress.com

Contents

About the Author

 Lui Lam, humanist and physicist, is professor emeritus at San Jose State University, California; recipient of SJSU's 2017 Distinguished Service Award; guest professor at Chinese Academy of Sciences and China Association for Science and Technology. Education: BS (First Class Honors), University of Hong Kong; MS, University of British Columbia; PhD, Columbia University, thesis mentor: Philip Platzman, Bell Labs.

Apart from the United States, Lam had worked in Europe (Belgium and West Germany) and Beijing (Institute of Physics, CAS, 1978-1983); worked 30 years in natural science (physics, chemistry, complex systems), then 20 years in the humanities (history, art, philosophy).

Lam invented Bowlics (1982), one of three existing types of liquid crystals; Active Walks (1992), a new paradigm in complex systems; and two new multidisciplines: Histophysics (2002) and Scimat (Science Matters, 2007). He published over 180 papers and 25 books, including *Nonlinear Physics for Beginners* (1998), *Arts* (2011), and *Humanities, Science, Scimat* (2024).

Lam is founder of International Liquid Crystal Society (1990); founder and editor of two book series: Science Matters (World Scientific) and Partially Ordered Systems (Springer); current research: philosophy, humanities-science synthesis, and innovation. (For more, see A2.)

Email: lui2002lam@icloud.com
Website: www.sjsu.edu/people/lui.lam/scimat
More: https://www.researchgate.net/profile/Lui-Lam/research

Preface

Scimat (science of human) is a new multidiscipline proposed by Lui Lam in 2007. Scimat treats all studies on human as a unified enterprise. In terms of content, Scimat = Humanities + Social Science + Medical Science. The rationale behind this is that these three categories of human-study are looking at human from different angles, interrelated to each other, while to understand any object thoroughly, human in particular, every angle is useful. In terms of research method, Scimat advocates the use of humanities-science synthesis, and collaboration between the humanists and natural scientists. The ultimate aim of Scimat is to better humanity by bettering the humanities.

In my research, I happened to spend the first 30 years on natural science (physics, chemistry, complex systems) and the last 20 years in the humanities, and so the humanities-science-synthesis approach is a natural for me. And I have studied history, art, science, and philosophy in turn, coming up with some interesting and important results—I believe or hopefully so.

For example, by combining physics with history, we have created the new discipline called *Histophysics* (physics of history) and found quantitative laws in the case of Chinese history. Looking back millions of years in evolutionary history, we came up with a new interpretation of art's origin and nature. Moreover, by examining the economic and societal conditions underpinning the ancient Greek and Chinese societies, we understand better why the philosophies in the West and East differ so much from each other. Finally, the distinctions between simple and complex systems gleaned from physics and beyond help us clarify many of the important issues that plague the philosophers and historians of science, such as the definition of science and the Needham Question.

Scimat Anthology collects 26 original articles in the humanities, 21 published and 5 unpublished, from 2000 to 2024, by the founder of Scimat, plus a related article by Xiao-Fan Luo, my student and friend. They are all reformatted, with typos corrected, references updated, and a few words replaced for clarity or unity; redundancies are kept when they are unavoidable. For ease of reading, the articles are arranged according to content, not the dates they appeared. The book consists of five parts: Scimat, Histophysics, Art, Philosophy, and Science. It ends with two appendices: Hawking and His Legacy, and Lui Lam's Academic Life.

For those who care about my publications, a few words of explanation are in order. The fact that in this collection, there are more articles published in books instead journals is simply due to the heavy teaching duty I encountered (two courses and three labs per week) and the many conferences I helped to organize around the world, including the biennial series of International Conference on Science Matters in Portugal. Historically, it was at this series' first conference that Scimat was first proposed. And there is some comfort knowing that some great works, like those of Copernicus and Newton, were published in book forms. Good work is good work no matter where it is published.

Finally, we thank the publishers and authors for their cooperation that makes this book possible, and the many friends and colleagues for helpful discussions in the last two decades.

This book is for students and scholars of any discipline, from the humanities to natural science, as well as artists and laypeople.

Lui Lam

April 26, 2024
San Jose, California

About Chinese Names

There is no perfect way to write Chinese names in English. The spelling and ordering conventions of a Chinese name's characters are different in different geological areas—mainland China, Hong Kong, Taiwan, and United States. The conventions adopted in this book are as follows.

All Chinese names in text and references are written with family name *first*, with first name's characters connected by a hyphen.

All Chinese names from mainland China are spelled out in pinyin.

For those who made their career in the US, whether they settled later in mainland China or not, their name's old spelling is adopted, i.e., *not* in pinyin. For example, Chen-Ning Yang (or Yang Chen-Ning) in this book would be Yang Zhen-Ning if he made his career in mainland China but not in the US.

Lui Lam made his career in both places, outside and inside China. The name Lui Lam appears the same as an author and in text while his pinyin name Lin Lei appears also in text and reference lists. (His family name, Lin in pinyin, is Lam in Cantonese.)

PART I

SCIMAT

The Scimat Story

Lui Lam

Scimat is a term I coined to represent a new multidiscipline that I proposed in 2007 [Lam 2008]. Scimat is about the scientific study of humans. In terms of content, scimat is a collective term for the humanities, social science, and medical science. Scimat can be understood as the "science of human"—the goal that the Enlightenment (1688-1789) wanted to do but did not achieve. Scimat advocates a rational discussion of any question about humans—all questions, big or small—from a scientific point of view that is based on scientific knowledge about humans. It aims to raise the scientific level of the humanities and broaden the field of study of "natural science" by encouraging collaboration between scholars from the two sides. The ultimate *aim* of Scimat is to better humanity by bettering the humanities. Here is the story of scimat—how it began, what had been done, and what lies ahead.

1 Motivation

I was interested in people when I was a child, at least with beautiful female classmates. Because I grew up with low-end people, I knew about human sufferings early on, and I thought about it from time to time. And when I grew up, my thoughts expanded into a concern for past and present world developments and the fate of humankind. Even before my physics training, I was curious in everything and loved to ask why about everything. All these led me to return to China and joined the "revolution" there, for the sake of a better tomorrow—the tomorrow of the humankind [Lam 2015].

After I invented the *Active Walk* for dealing with complex systems in 1992 [Lam et al. 1992] I began to dabble in some humanities topics (such as economic history). But it was not until 2002 when I proposed the new discipline of *Histophysics* (physis of history) [Lam 2002] that my formal study of human problems began—marking the ultimate combination of research direction and personal interest. The reason for turning from history to *Scimat* (science of human) [Lam 2008] is simple. It takes only one founding paper or a few to establish a new field of study, and that could be finished in a few years. And so, by 2006 [Lam 2006], I was ready to tackle the *ultimate* question concerning the entire humanities: Why the overall research level of the humanities has been stagnant for such a long time—2,400 years, in fact, since the time of Plato (427-347 BC). Frustration with this lack of progress in the humanities has been raised by some Western scholars [Gottschall 2008].

2 Preparation and First Conference

Beijing

In July 2005, I met Maria Burguete from Portugal at the 22nd International Conference on the History of Science in Beijing. She invited me to visit. Maria holds a PhD in History from Ludwig Maximilians University in Munich, Germany, and works at the Bento da Rocha Cabral Institute in Lisbon, the capital of Portugal.

Portugal

Although I had worked in Europe for two years (1975-1977) and visited almost all countries in Western Europe, I never went to Portugal because it is located at the southwestern tip of Europe, which is not on route to other European cities, and because I was unfamiliar with Portugal. I did not find a reason to go.

This time there was a reason. So, on March 25, 2006, taking advantage of the spring break of my college and after securing the air-ticket fee from the dean of the Faculty of Science, I flew to Lisbon. Maria took me around and finally we ended up at the Vila Galé Hotel (Fig. 1) in the seaside town of Ericeira, where we drank too much and wanted to do something together. After thinking about it, I concluded that we could organize a series of international conferences together. Since I have been thinking mostly about the science of human lately, I suggested this as the central theme of the conference series. The agreed division of labor: Maria manages the hardware (raising money, arranging venue), and I manage the software (conference content).

Fig. 1. Portuguese coast. *Left*: Foz da Arelho. *Right*: The Vila Galé Hotel in Ericeira.

Seoul

On May 17-19, 2006, I gave a presentation on "The two cultures and The Real World" at the 9th International Conference on Public Communication of Science and Technology in Seoul [Lam 2006]. The report points out two factual errors in C. P. Snow's book *The Two Cultures*, and provides the historical origins of the two cultures that Snow did not address as well as the remedy for bridging the gap between the two cultures. The remedy I recommend is *not* the patch-up, general-education approach that Snow suggested but rather the root-cured scimat approach, i.e., the integration of the humanities and (natural) science based on the fact that the two share the same roots.

The report's discourse on science and its conclusions—the humanities, like physics, are a branch of science—became the theoretical basis for Scimat a year later. See my tone-setting article in the first scimat book *Science Matters: Humanities as Complex Systems*, which, additionally, has the words "Let the Earth be peaceful forever!" near the end—a hint at the ultimate aim of Scimat [Lam 2008].

Paris

In order to make the meeting a success, I flew from Beijing to Paris on January 2, 2007, joined Maria and met Paul Caro at a restaurant. We asked him to support and attend the first scimat meeting. After three rounds of drinking, he readily agreed.

Paul is a Corresponding Member of the French Academy of Sciences and a Member of the French Academy of Technology, and before his retirement, was a research director of the CNRS (Centre National de la Recherche Scientifique) and a rare-earth expert. In the 1980s, he became interested in popular science. Until 2001, he was in charge of "scientific affairs" at the Cité des Sciences et de l'Industrie in Paris. He has been a consultant for DG Research European programs in the field of education and is also a scientific advisor to the Portuguese program Ciència Viva.

Of course, I also went to the Rodin Museum, which is Rodin's former home.

Ericeira

From 28 to 30 May 2007, the First International Conference on Science Matters was held in Ericeira, Portugal. Maria and I are co-chairs (Fig. 2). I gave the conference's keynote presentation, "Science Matters: From Aristotle to you and me," and the special-subject presentation, "Histophysics: integrating history and physics."

The slogan of the conference was: Everything in nature is part of science! This simple sentence means:

1. Problems relating to humans should and can be studied scientifically.

2. The study of humans—the humanities, social science, and medical science—is part of natural science. The disciplinary division is only a division of labor; the three are at different stages of scientific development depending on their respective difficulty.

3. In terms of their current *scientificity* (i.e., scientific level), medical science is the highest, followed by social science, and the humanities are the lowest.

To announce and advertise a conference, one has to make a poster. But I could not think of a suitable title for the conference. In a hurry, I used the term "Science Matters," meaning that everything that the conference wants to talk about, especially the humanities, is part of science. In the first scimat book *Science Matters: Humanities as Complex Systems* (2008), *Scimat*—a new word I coined—appeared as an abbreviation for Science Matters, which was not Chinese translated until 2013. In 2013, Chinese translation of the book was published by China Renmin University Press, in which the term 人科 (*Renke*), meaning Human Science, appears as the Chinese name of scimat and the title of this book (Fig. 3, second left). After that, Scimat and 人科 became the official names of this new multidiscipline.

3 Scimat

Regarding the content of science and the position of each discipline, scimat looks at it like this:

> Science = Natural Science
>
> = Nonhuman-system Science + Human-system Science

where Nonhuman-system Science = "Natural Science" —the usual usage of the term; Human-system science = Humanities + Social Science + Medical science = Scimat.

4 International Scimat Program

The 2007 conference marked the beginning of the International Scimat Program. It has 6 points and is carried out in 6 steps:

Step 1: We launched an international scimat **conference series** which is held every two years. Each conference focuses on a specific topic. The first one in 2007 set the tone for scimat. The second one in 2009 was about art. Since our focus is on the humanities, we do not talk about "natural science" per se, although at this one we did put art and science together.

The third one in 2011 was devoted to all aspects of science: the nature, philosophy, history, sociology, and science communication of science.

The sixth one in 2017 marks the 10th anniversary of scimat. In this one, we have finally revealed the ultimate goal of scimat: to build a better tomorrow. Scimat's ideal is a little grander than Plato's. Plato thinks about how to make a country better, but scimat thinks about how to make the whole world better. I was co-chair of the first three and the sixth conferences (Fig. 2), but I planned all the first to sixth ones.

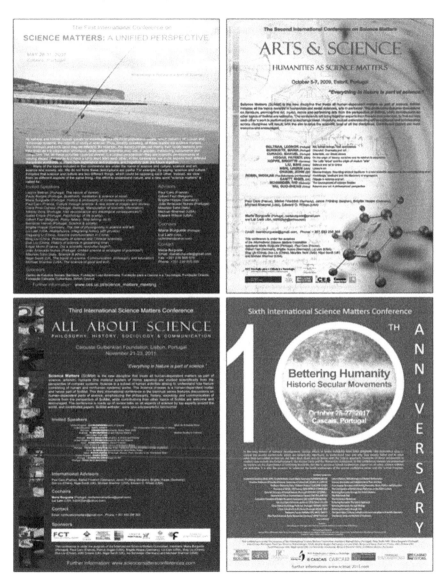

Fig. 2. Poster of the International Conference on Science Matters. From *left to right, up to down*: 2007, 2009, 2011, and 2017.

Step 2: we set up an International Scimat **Committee**. The committee was established on May 30, 2007, and grew from 9 to 18 members, including the 2005 Nobel Prize winner Robin Warren and the *Scientific American* columnist Michael Shermer. I also served as the committee's coordinator and liaison. The committee aims to promote the scimat philosophy and oversee the international scimat program.

Step 3: I established and edited an English scimat **book series**, published by World Scientific in Singapore. The first three books are *Science Matters* (2008), *Arts* (2011), and *All About Science* (2014) (Fig. 3). Only one-third of the articles in these books are selected from the conference reports; the rest were written by relevant experts from around the world, solicited after the conference.

At present, in addition to my series called Science Matters, there are two other book series called Scimat, published in the UK and Portugal.

Fig. 3. First three English scimat books published by World Scientific (*left to right*: 2008, 2011, and 2014) and a Chinese translation of the first book (*Science Matters*) published by China Renmin University Press (2013).

Step 4: We are in the process of establishing a number of Scimat **centers** (100 eventually) around the world, which would be independent from but collaborating with and reinforcing each other. The Center is:

1. To do fundraising to support the Center financially.
2. To organize international workshops/conferences and summer/ winter schools.
3. To communicate the Scimat ideas to the public.

4. To give out an Award every two years (in the donor's name perhaps) for an individual who contributes significantly in the advancement of scimat.

5. To host short-term visiting scholars who will give lectures/short courses, who will also collaborate with existing faculty members and students of any discipline, especially from the humanities.

6. To help match faculty members from the humanities and science departments, and give them release time to create new interdisciplinary courses (e.g., Physics of History).

7. To help promoting the new general-education course on "humanities + science" for undergrads of all majors (Fig. 4).

Note that the Center will *not* do research within itself, and so the maintenance fee is very minimal. With enough (outside) money, it can advance scimat by funding interdisciplinary research within a university. The scimat center will then be in a leading position academically in the most important multidiscipline of the 21st century.

Step 5: Establish an international Scimat **society**.

Step 6: Publish an international Scimat **journal**.

We are now working on step 4. Steps 5 and 6 are for the future, hopefully the near future. (For more see: www.sjsu.edu/people/ lui.lam/scimat.)

5 Conclusion

Scimat's motto is "Everything in nature is part of science." Scimat's key insight is that we have "One culture, two systems, three levels"—science culture, simple and complex systems, three research levels (empirical, phenomenological, and bottom-up). What we have presented here is the initial stage in the birth of a new multidiscipline—more precisely, a new paradigm—called *Scimat*.

1. It is similar to the case of History of Science (initiated by George Sarton early last century) and of Artificial Life (by Christopher Langton in 1986), but not quite. Scimat is much larger in scope since it incorporates the research of everything related to humans, and thus will be more far reaching in its influences.

Fig. 4. A scimat course at the 2015 International Summer School, China Renmin University. Course name: Humanities, Science, Scimat: An Interdisciplinary and Intercultural Experience (Number: SH1518). Lecturer: Lui Lam. *Top*: Class photo. *Bottom*: Course materials.

2. It provides a unified perspective for all the disciplines in the humanities, social science and "natural science."

3. It is a rally point to raise the *scientificity* of the humanities, making the world a better place. (The reason is that many large-scale human tragedies can be traced to the underdevelopment of the humanities

in the last 2,400 years since Plato. See "Bettering humanity: The Scimat approach," article 4 below in this book.)

4. It is the foundation behind the "synthesis" of the humanities and "science," solving the so-called two-culture problem at the foundational level.

5. It provides the basic rationale for general education and a route to make it successful.

6. It provides the broadest framework in interdisciplinary learning/ teaching, and science teaching [Matthews 2015].

Above all, Scimat is the most interesting and important multidiscipline in the 21st century.

In short, Scimat advocates the understanding of our world through science and rational thinking, whereas the humanities are recognized as part of science. Let us work together for a better humanity and make the world a peaceful place forever, for us and our children!

References

Gottschall, J. [2008]. *Literature, Science, and a New Humanities*. New York: Palgrave Macmillan.

Lam, L. [2002]. Histophysics: A new discipline. Modern Physics Letters B **16**: 1163-1176.

Lam, L. [2006]. The two cultures and The Real World. The Pantaneto Forum, Issue 24 (2006).

Lam, L. [2008]. Science matters: A unified perspective. *Science Matters: Humanities as Complex Systems*, Burguete, M. & Lam, L. (eds.). Singapore: World Scientific. pp 1-38.

Lam, L. [2015]. From physics to revolution and back. Science **348**: 1170.

Lam, L, Freimuth, R. D., Pon, M. K., Kayser, D. R., Fredrick, J. T. & Pochy, R. D. [1992]. Filamentary Patterns and Rough Surfaces. *Pattern Formation in Complex Dissipative Systems*, Kai, S.(ed.). Singapore: World Scientific.

Matthews, M. R. [2015]. *Science Teaching: The Contribution of History and Philosophy of Science*. New York: Routledge.

Published: Lam, L. [2022]. The Scimat Story. *China Complex*, Lam, L. San Jose: Yingshi Workshop. Original in Chinese; English translation here.

The Two Cultures and The Real World

Lui Lam

The "two cultures" refer to the scientific culture and the literary culture, pointed out by C.P. Snow in the 1950s. The former derives from the study of material systems from the natural sciences, while the latter comes from the understanding of humans. However, humans are *Homo sapiens*—a (biological) material system and is thus a part of the natural sciences since the latter is the study of all material systems. Consequently, science and the humanities are unified at the fundamental level. The apparent "gap" comes from the different levels of scientific development, the deficiency in the school curricula, and the unfortunate misconception reinforced by current science communications. To help close this gap, a general-education course—The Real World—was introduced and taught by the author at San Jose State University (SJSU). The idea is to introduce students to the unifying principles behind the humanities and "natural science" and the world of nonlinear and complex systems.

1 The Two Cultures: Snow's Lecture

Forty-seven years ago, on May 7, 1959, Charles Percy Snow (1905-1980) gave the lecture "The Two Cultures and the Scientific Revolution" at Cambridge University [Snow 1998]. The lecture essentially contains three themes: the distinction and non-communication between the scientific culture and the literary/humanistic culture in the West, the importance of the science revolution (defined by Snow to mean the application of the "atomic particles," presumably nuclear physics and quantum mechanics), and the urgency for the rich countries to help the poor countries. Very interesting, big themes—but nothing original, as admitted by Snow himself (see "The Two Cultures: A Second Look (1963)" in [Snow 1998]).

The lecture generated tremendous interest and discussion round the world, which helped to earn Snow twenty honorary degrees (mostly from universities outside of England) and carve his name in history. While the other two themes are definitely worth talking about, it is the "two cultures" theme that causes the most controversy and debates. This is not at all surprising. Many in the literary circle felt slighted by Snow in his lecture and had to defend themselves or their profession (see Stefan Collini's "Introduction" in [Snow 1998]). And, by definition, literary people are those who can write.

The purpose of this article is not to discuss Snow's lecture but to present the deep reasons behind the apparent gap between the two cultures, which are hardly touched upon by Snow himself. (After all, the existence and importance of the "gap" is not in doubt.) Furthermore, our effort in helping to close this gap is described.

But before we do that, for historical sake, Snow's errors concerning scientific matter in his lecture should be pointed out. Apparently, despite the worldwide fame of this lecture, no one has done this before.

Snow is wrong when he writes, "No, I mean the discovery at Columbia by Yang and Lee"[1] [Snow 1998: 15]. While T. D. Lee indeed worked at Columbia University, C. N. Yang's "permanent" address at that time was the Institute for Advanced Study at Princeton, New Jersey (see the address bylines in [Lee & Yang 1957]). In fact, Yang has never been associated with Columbia University. A few sentences later, still referring to the work of Yang and Lee, Snow makes another mistake in his sentence, "If there were any serious communication between the two cultures, this experiment would have been talked about at every High Table in Cambridge." The work of Lee and Yang is purely theoretical, which is to point out that there was no experimental evidence supporting or refuting parity conservation in weak interactions at that time. They went on to propose several experiments to settle this issue without predicting the outcome of these experiments. Parity nonconservation was indeed discovered in an experiment by C. S. Wu [Wu et al. 1957], a colleague of Lee at Columbia.

These two errors are minor by themselves and do not affect the rest of the lecture. Yet, they are factual errors that should be easily avoided. Lee and Yang received their Nobel Prize in December 1957, a mere 17 months before Snow delivered his lecture. With all the reporting of the parity nonconservation story in the newspapers and magazines, not to mention the more formal academic publications, only a careless writer like Snow could miss the basic facts. And this is not an isolated incident. In 1932, Snow had to recant publicly his "discovery" of how to produce Vitamin A artificially after his calculation was found faulty. Snow, a trained chemist, decided to leave scientific research completely after this incident and became a novelist [Snow 1998: xx]. He indeed made the correct career move, judging by later developments.

2 The Two Cultures: The Essence

2.1 Emergence of the two cultures

About 10,000 years or longer ago on earth, the early *Homo sapiens*, our ancestors, started to wonder about the things around them—things in their immediate surroundings and things in the sky. Curiosity serves not just human needs but for those who figure out how things work from their observations; it is a survival skill via the evolutionary mechanism according to Darwin.

Later, among these activities, literature is the description of humans' reflection on and understanding of nature. Here, nature includes all (human and non-human) material systems, such as falling leaves in autumn, the changing weather and seasons, effect of moonlights on lovers, the way humans treat each other in different spatial and temporal settings, and, quite often, thoughts in one's brain as a function of happenings inside or outside the person's body. When the authors write all these down, they are using their bodily sensors (sighting, touching, smell, hearing, etc.) as the main detectors and their brain as the major information processor. Apart from that, for latecomers, they do benefit from reading what previous writers wrote.

As time went by, the observation and understanding of certain kinds of phenomena progressed faster. For example, how things fall under the influence of gravity can be predicted and measured with high accuracy. This is achieved not because the falling object under study is simple, but because (1) we can approximate it by something simple, and (2) we use detectors and information processors other than those from our own bodies. For example, a human body falling from a tall building is the same complicated human body described in a piece of literature, but in physics we pretend that it is a point particle (i.e., an idealized particle with zero size) in our calculations. This is an approximation; it works because the size of the Earth, the major source of the gravitational force, is much greater than the size of the human body. Furthermore, we can record the positions of the falling body by digital cameras and compare them with our calculations, with the help of calculators or computers. (For smaller falling objects, low-tech devices are used to record the positions at regular time intervals. This is routinely done in freshmen physics labs.)

This branch of study is now called "natural science," which involves mostly nonliving systems even though living systems (such as humans in free fall and other simpler biological bodies) are not excluded. As we just pointed out, natural science presently succeeds because it chooses to deal only with a special subset of phenomena. And literature is stuck with the complicated aspects—such as pride and prejudice—of the complex system called humans.

As study deepened, specialization became essential and we were left with two distinctive groups of practitioners, the writers in the literature profession and what Snow called scientists for those working in natural science. Since writers use their own bodies as tools, only those with supreme bodily sensitivity and suitable hard wiring of neurons in their brains can become good writers, while scientists need other types of quality (such as supreme self confidence) to succeed. There is no overlap between these two groups of professionals, as Snow painfully found out for himself.

2.2 Why the gap

The fact that scientists can talk to each other is true only to a certain extent. There is not much to talk about between a particle theorist and a condensed matter physicist if the subject is the standard model of particles. But all scientists, be they physicists, biologists or chemists, do share some common knowledge such as the second law of thermodynamics, because this law is a required learning in the college education of these scientists.[2]

Professional activities require high concentration of attention and usually are time consuming, especially in the case of science, which involves very keen competition. Time is short, for the professionals. Many first-rate scientists do not read books, particularly science books, because what contained in books is usually not fresh enough. Instead, they read research papers that they think might be helpful to their (present or future) work. That is what the scientist had in mind when he, asked by Snow what books he read, replied, "Books? I prefer to use my books as tools." [Snow 1998: 3] Tools, here, mean something that will help him to do his research. There is in fact a fair chance that literary books will be read by scientists, for relaxing purpose, e.g., when they are in a plane after attending a conference. But these books are not Shakespeare's. The same goes for the literary people. Why should they read any science book if they cannot find anything there that would help them to do their job?

In short, the non-communication between these two groups is not due to the non-overlap of the people involved, but due to the absence of any common language or principle in their trades, at least in the 1950s when Snow delivered his lecture. This is no longer so. Since the 1980s, some general principles arise from the study of simple and complex systems, which are applicable to both the natural and social sciences, and to both living and nonliving objects. Here, we are referring to fractals, chaos and active walks [Lam 1998]. (See Sec. 3.2 for more.)

2.3 Why close the gap

We want the literary people to learn a little bit basic science and the scientists to read some good literature. The reason is not that we afraid they have nothing to talk to each other in a cocktail party. They can always talk about Ang Lee's *Brokeback Mountain* (2005) or his other movie, *Crouching Tiger, Hidden Dragon* (2000). The movie's storyline is as deep as Shakespeare's, and perhaps more entertaining.

And if the purpose is to make the literary people to appreciate the mental achievement of the humankind via the elegant theories established in science, then learning basic science is not the easiest way to do so and may not even be necessary.[3] In Snow's days, the television and telephone, and if not enough, the two atomic bombs in 1945 and the Sputnik in 1957 should convince every sensible person on earth the high achievement of science, without the need to know the theories behind them. These days, a cell phone will do.

Yes, knowing some basic science supposedly will help you to cast sensible votes as a citizen on scientific matter, such as laws regarding global warming. But we actually rely on the experts on their professional opinions on these matters. Our rudimentary scientific knowledge is not sufficient for us to make the judgments ourselves, even though it may help us to pick which expert to trust. Unfortunately, this is easier said than done. Sometimes you can't even trust the Nobel laureates. (For example, there is one who believes in astrology, big foot, and that we never went to the moon.) The situation is like picking which financial advisor to help managing your money. You go to the big institutions and also look at their track record. It helps if you have some financial knowledge, but that is not enough.

A good reason for ordinary people to learn some science is to increase their personal safety. Some science knowledge could help people to eat and live healthy and avoid accidents. It could also help them to recognize the crooks when they see one. For example, if someone told you that the earth had exploded three times in the past and it was he who repaired it (which is so against what we know in science), you could safely ignore what else he told you and should never give him a dime, nor your other valuable possessions.

But why we want the scientists to read some good literature?[4] This is less clear, not even Snow has anything meaningful to say about this. What is sure is that it is not a bad thing to do, unless you happen to be a young scientist who needs undivided attention to your research. It could help the scientists to meet more interesting people; many of them, at the time this article is written, know only literature but not science.

2.4 How to close the gap

The best time to get the literary people and the scientists to learn something from each other's trade is when they are still students in schools or colleges, when they are forced to attend classes. Apart from requiring the students to take some general education classes in both science and literature before they are allowed to graduate, as is the practice in most American universities, it would be wonderful to teach them something, if exist, that they could use for the rest of their life no matter what profession they end up with. Fortunately, these wonderful things do exist. They are the general principles governing many complex systems (see Sec. 3.2).

And the way to achieve this in the classroom is through educational reform; outside of the classroom, science communication [12]. But first, let us review the essentials of science.

3 Science: The Essentials

3.1 What science is about

Science is about the systematic understanding of nature. And nature, of course, includes all material systems. On the other hand, human beings are biological and thus material bodies called *Homo sapiens*. Consequently, any study related to human beings, literature in particular, should be a part of science [Lam 2002].

Since nature consists of everything in the universe, the two terms science and natural science are identical to each other.[6] In other words, in terms of the objects under study in science, we have

Science = Natural science

= Science of nonliving systems + Science of living systems (1)

whereas

Living systems = Nonhuman biological systems + Humans (2)

Since we human beings (and not, e.g., the ants) are the ones who do the study and control the research budgets, it is not surprising that a large part of the science activity is related to and is for the benefit of humans. In terms of the disciplines, these human-related studies fall into one of two categories, viz., social science and the humanities (and medical science). Social science consists of anthropology, business and management, economics, education, environmental science, geography, government policy, law, psychology, social welfare, sociology, and women's studies.[5] Philosophy, religions, languages, literature, art and music make up the humanities. History, by its very nature, should be part of social science, but is listed in the humanities at some universities.

The aim of literature, music and art in the humanities is to stimulate the human brain—through arrangement of words or colors, sound or speech, or shape of things—to achieve pleasure and beauty, or their opposites, via the neurons and their connecting patterns [Pinker 1997]. The brains, some sort of computer, of the creator and the receiver at the two ends of this process are heavily involved. The scientific development of these disciplines as complex systems is at a primitive level, and that is why they are separated from social science, which are at an intermediate level. Linguistic is the study of the tools involved in written words and speeches, supporting the three disciplines mentioned above.

In terms of the disciplines, Eq. (1) could be rewritten as:

Science = Natural Science

$$= \text{Physical science} + \text{Social Science} + \text{Humanities} \qquad (3)$$

whereas "physical science" includes not just physics, but biology, chemistry, etc.

3.2 Three general principles

There are three established principles that are able to unify many different phenomena found in nature, with examples taken from both the natural and social sciences, and even the humanities. They are [Lam 1998]:

1. **Fractals**—the principle of self-similarity. Self similar means that if you take a small part of an object and blow it up in proportion, it will look similar or identical to the original object. Self-similar objects are called fractals, which quite often have dimension not equal to an integer. A

famous example is the Sierpinski gasket [Lam 2004b]. Fractals are everywhere, ranging from the morphology of tree leaves, rock formations, human blood vessels, to the stock market indices and the structure of galaxies. Fractals are even relevant in the corporate culture [Warnecken 1993] and the arts [Barrow 1995].

2. **Chaos**—the common (but not universal) phenomenon that the behavior of many nonlinear systems depends sensitively on their initial conditions. Examples of chaos include leaking faucets, convective liquids, human heartbeats, planet motion in the solar system, etc. The concept is found applicable in psychology, life sciences and literature [Robertson & Combs 1995; Hayles 1991]. A general summary is available [Yorke & Grebogi 1996].

3. **Active walks**—a major principle that Mother Nature uses in self-organization. Active walk is a paradigm introduced by Lam in 1992 to handle complex systems [Lam 2005a, 2006a]. In an active walk, a particle (the walker) changes a deformable potential—the landscape—as it walks; its next step is influenced by the changed landscape. For example, ants are living active walkers. When an ant moves, it releases chemicals of a certain type and hence changes the spatial distribution of the chemical concentration. Its next step is moving towards positions of higher chemical concentration. In this case, the chemical distribution is the deformable landscape. Active walk has been applied successfully to a number of complex systems coming from the natural and social sciences. Examples include pattern formation in physical, chemical, and biological systems such as surface-reaction induced filaments and retinal neurons, the formation of fractal surfaces, anomalous ionic transport in glasses, granular matter, population dynamics, bacteria movements and pattern forming, food foraging of ants, spontaneous formation of human trails, oil recovery, river formation, city growth, economic systems, and, most recently, human history [Lam 2002, 2004, 2006a].

All three principles are an integral part of complex-system science, which is becoming important in the understanding of business, governments, and the media.[7]

4 Educational Reform: A Personal Journey

University educational reforms could involve three possible components: (1) the contents of the courses, (2) the way of teaching by the instructors, and (3) the learning method of the students. No matter how it is done, an unavoidable constraint that will crucially affect the success of the reform is usually not mentioned or ignored completely by the reformers; i.e., the reform should not increase the teaching load of the instructors. Also, the quality of the students taking a course—like the quality of a sample in a physical experiment or the raw material in a factory—is of primary importance; this factor is never emphasized enough. Obviously, with a defective sample, no good experimental result can be expected, no matter how skillful the experimentalist is. This last factor points to the need to start any educational reform from grade one on, or even better, from the kindergartens.

With the constraints understood and resources limited, I tried to do my best as a teacher. There is not much we can do about item 3. It is very hard for the students to change their learning habit after being wrongfully taught for 12 years before they show up in college, and this is not their fault. I therefore concentrated my effort in the first two items.

On item 2, I have tried something radically different. It is called "MultiTeaching MultiLearning" (MTML) [Lam 1999]. We note that in a physics class, the teacher usually does not have enough time to cover everything. The attention span of a student is supposed to be about 15 minutes. Students in a class have different learning styles. Some students are more advanced than others. Active learning and group learning are good for students. To overcome these problems in the teaching of two freshmen classes in Mechanics and Thermodynamics, around 1999, I have tried a zero-budget and low-tech approach. In these classes, we cover about one chapter per week, using *Physics* by Resnick, Halliday and Krane as the textbook. In each course, there are three sessions per week, each 50 minutes long. In the last session of every week, the class is broken up completely. Different "booths" like those in a country fair are set up in several rooms, manned by student volunteers from the class. The rest of the class is free to roam about, like in a real country fair, or like what the professional physicists will do in a large conference with multiple sessions. In this way, we are able to simultaneously offer homework problem solving, challenging tough problems for advanced students, computer exercises, web site visits, peer instruction, and one-to-one tutoring to the students. The students seemed to enjoy themselves and benefited from it. However, this approach was soon discontinued. It did require a little bit of extra preparation time from the instructor, but more importantly, it did not seem to raise significantly the grades of the students. The "inferior raw material" factor might be at work here.

The next thing I tried, with better luck this time, is to integrate popular science books into my physics classes [Lam 2000, 2005b, 2008]. This is done by giving extra credits to the students who would buy a popular science (PS) book, read it and write up a report [Lam 2000, 2001: 330-336]. The instructor does not actually teach the books, and hence will not find the teaching load increased. It is like a supplementary reading, a practice commonly used in the English classes but rarely adopted by science instructors. The aim of this practice is (1) to broaden the knowledge base of the students, (2) show them the availability and varieties of PS books in their local book stores, (3) encourage them to go on to buy and read at least one PS book per year for the rest of their life, and (4) become a science informed citizen—a voter and perhaps a legislator who is science friendly. It is about lifetime learning of science matters. This practice is quite successful and is still going on in my classes.

This PS book program is not trying to alter the course content per se. My first attempt in this direction, item 1 in educational reform, actually happened earlier. Soon after I started teaching at SJSU in 1987, I created two new graduate courses, Nonlinear Physics and Nonlinear Systems, respectively [Lam 1998]. But these two were for physics majors. In Spring 1997, I established a general-education course called The Real World, opened to upper-division (i.e., third and fourth year in college) students of any major. It results from my many years of research ranging from nonlinear physics to complex systems [Lam 2000]. There were only nine students, majoring in physics, music, philosophy and so on, plus two physics professors sitting in. It was fun. The course stopped after one semester due to nonacademic reasons, falling victim to the sociology of science education.

Five years later in Fall 2002, the course was resurrected with the same name but modified to suit incoming freshmen students. It is this general-education freshmen course that will be described in detail in the next session.

5 The Real World: A General Education Course

In 2001 we have a new provost in campus. This very energetic and ambitious man, Marshall Goodman, wanted to make SJSU distinctive from the other twenty plus campuses of the California State University system. Introducing international programs with a global outlook was his way of doing that. But perhaps more important, with lightning speed as administrative things went, he was able to push through the university senate and actually had 100 brand new freshmen general-education courses set up and running in about half-a-year's time. Each of these courses is limited to no more than 15 incoming freshmen students. The program starting in Fall 2002 is called the MUSE program (MUSE stands for Metropolitan University Scholar's Experience). "MUSE/Phys 10B (Section 3): The Real World" was one of the 100 courses. Some details of the course are given below (see [Lam 2008] for more):

Course Description: To understand how the real world works from the scientific point of view. The course will consist of two parallel parts. (1) The instructor will introduce some general paradigms governing complex systems—fractals, chaos, and active walks—with examples taken from the natural and social sciences, and the humanities. (2) The students will be asked to pick any topic from the newspapers or their daily lives, and investigate what had been done scientifically on that topic, with the help from the web, library, and experts around the world. Outside speakers and field trips are part of this course.

Learning Objectives and Activities for this Course: This course qualifies as an Area B1 (Physical Sciences) course in your General Education requirements. It is designed to enable you to achieve the following GE and

MUSE learning outcomes. By the end of this course, you should be able to:

- Use methods of science and knowledge derived from current scientific inquiry in life or physical science to question existing explanations;
- Demonstrate ways in which science influences and is influenced by complex societies, including political and moral issue;
- Recognize methods of science, in which quantitative, analytical reasoning techniques are used;
- Understand the learning process and your responsibility and role in it; and
- Know what it means to be a member of a metropolitan university community.

After successfully completing this course, stendents will:

- Realize that there are general paradigms—fractals, chaos, and active walks—governing the functioning of complex systems in the real world, physical and social systems alike.
- What nonlinearity is.
- How "dimension" is defined mathematically.
- The meaning of self-similarity and fractals.
- Recognize and able to evaluate data to show that any physical structure or pattern in the real world is a fractal or not.
- What a chaotic system is.
- Able to distinguish a chaotic behavior from a random behavior given the time series of a system.
- To realize that many complex systems in the real world can be described by Active Walk, and be familiar with a few examples.
- Recognize that there are multiple interpretations or points of view on some ongoing, forefront research topics, and that these interpretations can co-exist until the issue is settled when more accurate data and a good theory become available.

- Know the difference between science and pseudoscience, and the real meaning of the "scientific method."
- How scientific research is actually done.
- Able to find out the latest scientific knowledge about any topic of interest in the future.
- Have improved skills in communicating both orally and in writing.
- Have increased familiarity with information resources at SJSU and elsewhere.

Course Materials: The following two books are required:

1. Lui Lam, *Nonlinear Physics for Beginners: Fractals, Chaos, Solitons, Pattern Formation, Cellular Automata, and Complex Systems* (World Scientific, Singapore, 1998), paperback (list price: $28). Reading assignments from this book will be announced in class. Additional materials will be provided by the instructor. Other information could be found from the web, magazines, research journals and books from the library.

2. *A Spartan Scholar from the Start* (published by SJSU).

6 Conclusion: What Is to be Done?

The "two cultures" issues are clarified. What the literary people and the natural scientists do are similar to each other at the basic level. They both try to understand the world around them. What differentiates them is, roughly speaking, that literary people confine their investigation to using their body as the detector and their own brain as the information processor, while modern scientists use tools other than their own body to do their work (Sec. 2.1). All these activities could be viewed as parts of a big project—to understand nature (human and nonhuman systems) systematically, except that literature is still doing it empirically and is at a less developed level, scientifically speaking. But this is also the case with the study of many other complex systems, because the problem involved is much harder (Sec. 3).

The gap between the two cultures can never be completely closed, and there is no need to do so. What should be and could be done is to teach everybody the fact that our real world is governed by some unified principles, which are applicable to both the human and nonhuman systems and could be shared beneficially by people in the two cultures, and in fact, in any culture. And a course for this purpose has been designed and tried (Sec. 5), which should be taught at school of any level, the earlier the better. Furthermore, to reinforce the effect, lifelong learning through popular science books is strongly urged [Lam 2005b]. Beyond that, it would be good to make both natural science and literature writings an essential reading for college students. This can be easily done by incorporating a few popular science books—such as James Watson's *The Double Helix*—into the list of required readings in the general education of every student in every university.

It does not help if in science communication of any kind, we keep conveying to the public the wrong impression that natural science, social science, and the humanities are three very different things, without anything in common. A remedy to correct this in the science museums is to show the unifying themes of all natural and social phenomena before the museum exit [Lam 2004c].

Through appropriate effort, the merge of natural science and the humanities is possible. Examples include the merging of biology and sociology to form Sociobiology [Wilson 1975], economics and physics for Econophysics [Mantegna & Stanley 2000], sociology and physics for Sociophysics [Galam 2004], and more recently, the creation of a new discipline called *Histophysics* through the link up of history with physics [Lam 2002, 2004a, 2006].

Notes

1. In the famous paper that earned Lee and Yang the Nobel Prize in 1957, the authors' names appears as Lee and Yang [Lee & Yang 1957]. The ordering of the two names in this and other joint papers apparently is not a small matter; it plays an important role in the two men's subsequent total breakup of collaboration and friendship [Yang 1983; Lee 1986; Chiang 2002; Zi et al. 2004].

2. The second law of thermodynamics is the example used by Snow to test the scientific knowledge of the literary people in a gathering [Snow 1998: 15]. This is in fact quite unfair, because the second law is less universal and useful than people think. It applies only to closed systems and only to their thermodynamic

equilibrium states. It applies neither to humans—an open system and the interest of literary people—nor to the expanding "cosmos" as Snow wrongly claimed [Snow 1998: 74]. The reason is that our universe is ever expanding and is never in an equilibrium state [Lam 2004a]. See [Zhao 2003] for a detailed discussion.

3. Snow, a chemistry major, is mistaken when he writes that asking someone to define "acceleration" is "the scientific equivalent of saying, Can you read?" [Snow 1998: 15].) The definition of acceleration ($a \equiv dv/dt$) involves calculus and the concept of vectors [Halliday & Resnick 1988], and may even be found difficult by some students in a freshmen physics course.

4. In practice, as good literature is concerned, unlike the case in science, there is no unique choice suitable for everybody. Reading Shakespeare or Tang poem/Song prose will equally do.

5. http://www.sosig.ac.uk.

6. With this understanding, every possible enquiry undertaken would be about nature. The term "science" in its German sense of Wissenchaft—any systematic body of enquiry—and its use in the English language will coincide with each other.

7. See http://www.trafficforum.org/budapest for a description of an upcoming conference on "Potentials of complexity science for business, governments, and the media," Budapest, Aug. 3-5, 2006.

References

Barrow, J. D. [1995]. *The Artful Universe: The Cosmic Source of Human Creativity*. New York: Little, Brown and Co.

Chiang, Tsai-Chien (江才健), *Biography of Yang Chen-Ning: The Beauty of Gauge and Symmetry*. Taibei: Bookzone.

Galam, S. [2004]. Sociophysics: A personal testimony. Physica A **336**: 49-55.

Halliday, D. & Resnick, R. [1988]. *Fundamentals of Physics*. New York: Wiley.

Hayles, N. K. [1991]. *Order and Chaos: Complex Dynamics in Literature and Science*. Chicago: University of Chicago Press.

Lam, L. [1998]. *Nonlinear Physics for Beginners: Fractals, Chaos, Solitons, Pattern Formation, Cellular Automata and Complex Systems*. Singapore: World Scientific.

Lam, L. [1999]. MultiTeaching MultiLearning: A zero-budget low-tech reform in teaching freshmen physics. Bulletin of the American Physical Society **44**(1): 642.

Lam, L. [2000]. Integrating popular science books into college science teaching. Bulletin of the American Physical Society **45**(1): 117.

Lam, L. [2001]. Raising the scientific literacy of the population: A simple tactic and a global strategy. *Public Understanding of Science*, Editorial Committee (ed.). Hefei: University of Science and Technology of China U. P.

Lam, L. [2002]. Histophysics: A new discipline. Modern Physics Letters B **16**: 1163-1176.

Lam, L. [2004a]. *This Pale Blue Dot: Science, History, God.* Tamsui: Tamkang U. P.

Lam, L. [2004b]. A science-and-art interstellar message: The self-similar Sierpinski gasket. Leonardo **37**(1): 37-38.

Lam, L. [2004c]. New concepts for science and technology museums. The Pantaneto Forum, Issue 21, 2006 (http://www.pantaneto.co.uk). See also Lam, L. in *Proceedings of International Forum on Scientific Literacy*, Beijing, July 29-30, 2004.

Lam, L. [2005a]. Active walks: The first twelve years (Part I). Int. J. Bifurcation and Chaos **15**: 2317-2348.

Lam, L. [2005b]. Integrating popular science books into college science teaching. The Pantaneto Forum, Issue 19 (2005) (http://www.pantaneto. co.uk).

Lam, L. [2006a]. Active walks: The first twelve years (Part II). International Journal of Bifurcation and Chaos **16**: 239-268.

Lam, L. [2006b]. Science communication: What every scientist can do and a physicist's experience. Science Popularization, 2006, No. 2: 36-41. See also Lam, L. in *Proceedings of Beijing PCST Working Symposium*, Beijing, June 22-24, 2005.

Lam, L. [2008]. SciComm, PopSci and The Real World. *Science Matters: Humanities as Complex Systems*, Burguete, M. and Lam, L. (eds.). Singapore: World Scientific. pp 89-118.

Lee, T. D. [1986]. *Selected Papers, Vol. 3.* Boston: Birhauser Inc.

Lee, T. D. & C. N. Yang, C. N. [1957]. Question of parity conservation in weak interactions. Physical Review **104**: 254-258.

Mantegna, R. N. & Stanley, H. G. [2000]. *An Introduction to Econophysics.* New York: Cambridge U. P.

Pinker, S. [1997]. *How the Mind Works.* New York: Norton.

Robertson, R. & Combs, A. (eds.) [1995]. *Chaos Theory in Psychology and the Life Sciences.* Mahwah, NJ: Lawrence Erlbaum Associated.

Snow, C. P. [1998]. *The Two Cultures.* Cambridge: Cambridge U. P.

Warnecken, H. J. [1993]. *The Fractal Company: A Revolution in Corporate Culture.* New York: Springer.

Wilson, E. O. [1975]. *Sociobiology.* Cambridge, MA: Harvard University Press.

Wu, C. S., Ambler, E., Hayward, R. W., Hoppes, D. D. & Hudson, R. P. [1957]. Experimental test of parity conservation in beta decay. Physical Review **105**: 1413-1415.

Yang, C. N. [1983]. *Selected Papers 1945-1980 with Commentary*. New York: Freeman.

Yorke, J. A. & Grebogi, C. (eds.) [1996]. *The Impact of Chaos in Science and Society*. Tokyo: United Nation University Press.

Zi, Cheng (季承), Liu, Huai Zu. (柳怀祖) & Teng. Li (腾丽) (eds.) [2004]. *Solving the Puzzle of Competing Claims Surrounding the Discovery of Parity Nonconservation: T. D. Lee Answering Questions from Sciencetimes Reporter Yang Xu-Jie and Related Materials*. Lanzhou: Gansu Science and Technology Press.

Published: Lam, L. [2006]. The two cultures and The Real World. The Pantaneto Forum, Issue 24 (2006). Paper presented at *The 9th International Conference on Public Communication of Science and Technology*, Seoul, May 17-19, 2006.

Science Matters: A Unified Perspective

Lui Lam

What is science? The answer is that "Everything in nature is part of science." On the one hand, what we called "natural science" is actually the science of (mostly) simple systems; they are knowledge about nonhuman systems. On the other hand, the humanities/social science—knowledge about the human system—belong to the science of complex systems. Demarcation of nature according to human and nonhuman systems, and the recognition that complex systems are distinct from simple systems allow us to understand the world differently and profitably. For completeness, the nature of simple and complex systems is briefly presented. The origin of the "two cultures" (made famous by C. P. Snow), the humanities and "natural science" cultures, is traced and some confusing issues clarified. While a gap between humanists and "scientists" does exist due to historical reasons, there is no intrinsic gap between the humanities/social science and "natural science." If these disciplines look different from each other, it is because they are at various level of development, scientifically speaking.

To properly bridge the gap and to advance the search for human-system knowledge, a new discipline—Science Matters (Scimat)—is introduced. *Scimat* treats all human-system matters as part of science, wherein, humans are studied scientifically from the perspective of complex systems with the help of experiences gained in physics, neuroscience and other disciplines. Consequently, all the topics covered in the humanities and social science are included in Scimat. The motivation and concept of Scimat, and a successful example (*Histophysics*, the physics of human history) are presented and discussed. Four major implications of Scimat are described. In particular, a new answer to the Needham Question is offered for the first time. This chapter ends with discussion and conclusion.

1 Introduction

All earnest and honest human quests for knowledge are efforts to understand nature, which includes both human and nonhuman systems, the objects of study in science. Thus, broadly speaking, all these quests are in the science domain. The methods and tools used may be different; e.g., the literary people use mainly their bodily sensors and their brain as the information processor, while natural scientists may use, in addition, measuring instruments and computers. Yet, all these activities could be viewed in a unified perspective: They are scientific developments at varying stages of maturity and have a lot to learn from each other.

That "Everything in nature is part of science" (see Sec. 2) was well recognized by Aristotle and Da Vinci and many others. Yet, it is only recently, with the advent of modern science and experiences gathered in the study of statistical physics [Lam 1998; Paul & Baschnagel 1999], complex systems [Lam 1997, 1998] and other disciplines, that we know how the human-system disciplines can be studied scientifically.

Science Matters (Scimat) is the new multidiscipline that treats all the study of human matters as part of science. Scimat is about all human-system knowledge, wherein, human (the material system of *Homo sapiens*) are studied scientifically from the perspective of complex systems. Here, the term "complex systems" means simply "very complicated systems," in the sense adopted by common people; they may not be fractals or chaotic. After all, when fractal and chaotic systems are usually complex, not all complex systems are fractals or chaotic; and there exists no unique and satisfactory definition of complex systems, technically or otherwise [Lam 1998, 2000]. Scimat includes all the topics covered in the humanities and social science, with human history as a particular example [Lam 2008a].

This chapter is organized as follows. The very nature of science is revealed in Sec. 2, followed by an introduction and analysis of the "two cultures" in Sec. 3. Demarcation of everything in nature according to human and nonhuman systems is introduced in Sec. 4. Section 5 discusses simple and complex systems, including a brief introduction of the latter, one of which is the human system. The motivation, concept and an example (Histophysics) of Scimat is given in Sec. 6. Four major implications of Scimat, including a new answer to the Needham Question, are presented in Sec. 7. Finally, Sec. 8 concludes this chapter with discussion.

2 What Is Science?

About 2,600 years ago, Thales (c.624-c.546 BC) proposed the first Theory of Everything: Everything is made of water.[1] (See Fig. 1.) Subsequently, Aristotle (384-322 BC) studied various aspects of the universe—astronomy, physics, biology, botany, zoology, logic, ethics, politics, and so on—from the same platform [Llyod 1970]. In other words, he was interested in almost all the subjects of study existing in the universities today. This was not accidental.

Fig. 1. Two water-loving philosophers. Left: Thales (c.624-c.546 BC) from the West. Right: Guanzi (管子, c.723-645 BC) from the East.

The fragmentation of knowledge into different disciplines is a relatively recent phenomenon, starting only a few hundred years ago. It results more from management convenience than the intrinsic nature of knowledge itself. *Knowledge knows no separating boundaries.* After all, the highest degree conferred by a university is still called Doctor of Philosophy (not Doctor of Physics, for example), wherein, philosophy means "wisdom"—all kinds of wisdom. As will be explained below, there is a material basis underlying the unified, intrinsic nature of knowledge.

Knowledge about our universe or the world could be divided into two groups:

1. Those about nonhuman systems—called human-*independent* knowledge.
2. Those about the human system—called human-*dependent* knowledge.

For instance, Newton's three law of mechanics are human-independent. That is, if there were extraterrestrial intelligence (ET), sooner or later, these three laws could also be discovered by them, even though the laws might not be named after Newton. Examples of human-dependent knowledge (nonexistent if human never exists) are literature and dance. An ET might not dance like us, because it could have three, not two, legs.

Human-independent knowledge is commonly called "natural science"; human-dependent knowledge, the humanities and social science. However, this classification is inaccurate and inappropriate. On the one hand, humans are *Homo sapiens*, a material system consisting of atoms— the same atoms that make up the systems studied in "natural science." Consequently, all human-dependent knowledge is part of natural science, since the objects studied in natural science are *all* material systems.

On the other hand, science is about the study of nature and a means to understand it in a unified way. Nature consists of everything in the universe—all material systems, humans and nonhumans. The two terms science and natural science are thus identical to each other.[2] It then follows that there could be only one conclusion [Lam 2006a]:

Science = Natural Science

$$= \text{Physical science} + \text{Social Science} + \text{Humanities} \qquad (1)$$

where "physical science" includes not just physics, but chemistry, biology, and so on.[3] In other words: *Everything in nature is part of science*. This conclusion was known to the early Greeks. If some of our contemporaries do not know about this, it is because the word Science is either misunderstood or misused.

3 The Origin and Nature of the Two Cultures

Forty-nine years ago, on May 7, 1959, Charles Percy Snow gave the lecture "The two cultures and the scientific revolution" at Cambridge University [Snow 1998].[4] The lecture essentially contains three themes: the distinction and non-communication between the scientific culture and the literary culture in the West, the importance of the science revolution (defined by Snow to mean the application of the "atomic particles," presumably nuclear physics and quantum mechanics), and the urgency for the rich countries to help the poor countries. Very interesting, big themes—but nothing original, as admitted by Snow himself (see "The two cultures: a second look (1963)" in [Snow 1998]).

The lecture generated tremendous interest and much discussion around the world, which helped to earn Snow 20 honorary degrees (mostly from universities outside of England) and carve his name in history. While the other two themes are definitely worth talking about, it is the "two cultures" theme that causes the most controversy and debates. This is not at all surprising. Many in the literary circle felt slighted by Snow in his lecture and had to defend themselves or their profession (see Stefan Collini's "Introduction" in [Snow 1998]). And, by definition, literary people are those who can write. Now, the important question, not addressed by Snow himself in his lecture, is this: What is the origin of the two cultures?

3.1 Emergence of the two cultures

About ten thousand years or longer ago on earth, the early *Homo sapiens*, our ancestors, started to wonder about the things around them—things in their immediate surroundings and things in the sky. Curiosity serves not just human needs but for those who figure out how things work from their observations; it is a survival skill *via* the evolutionary mechanism according to Charles Darwin (1809-1882).

Later, among these activities, literature is the description of humans' reflection on and understanding of nature. Here, nature includes all (human and nonhuman) material systems, such as falling leaves in autumn, the changing weather and seasons, effect of moonlights on lovers, the way humans treat each other in different spatial and temporal settings, and, quite often, thoughts in one's brain as a function of happenings inside or outside the person's body. When the authors write down all these, they are using their bodily sensors (sighting, touching, smelling, hearing, and so on) as the main detectors and their brain as the major information processor. Apart from that, for latecomers, they could also input information by reading what other writers wrote.

As time went by, the observation and understanding of certain kinds of phenomena progressed faster. The speed-up process started in science with Galileo about 400 years ago. The success results from three crucial and cleaver steps:

1. We pick simple systems (such as a ball rolling down an inclined plane) to study.
2. We do big and daring approximations in constructing theories (e.g., approximating the ball by a point particle—an idealized particle with zero size).
3. We use detectors and information processors other than those from our own bodies.

Consequently, e.g., how things fall under the influence of gravity can be predicted and measured with high accuracy. Let us consider the case that the falling object is a human body. The human body falling from a tall building is the same complicated human body described in a piece of literature, but in physics we pretend that it is a point particle in our calculations. This is an approximation; it works because the size of the Earth, the major source of the gravitational force, is much greater than the size of the human body. Furthermore, we can record the positions of the falling body by digital cameras and compare them with our calculations, with the help of calculators or computers.[5]

This branch of study is now called "natural science," which involves mostly nonliving systems even though living systems (such as humans in free fall and other simpler biological bodies) are not excluded. However, the so-called "natural science" is actually "science of (mostly) simple systems," while all human-dependent studies (humanities/social science) are about complex systems since, in fact, a single human being is the most complex system in the universe.

As we just pointed out, "natural science" succeeds because it chooses to deal only with a special subset of phenomena. And literature is stuck with the complicated aspects—such as pride and prejudice—of the complex system called human.

As study deepened, specialization became essential and we were left with two distinct groups of practitioners: the writers in the literary profession and what Snow called "scientists" for those working in "natural science." Since writers use their own bodies as tools, only those with supreme bodily sensitivity and suitable hard wiring of neurons in their brains can become good writers, while scientists need other types of quality (such as supreme self-confidence) to succeed. There is no overlap between these two groups of professionals,[6] and we end up with "two cultures"—with a gap in between.[7]

3.2 The gap today

The method to close the gap between humanists and "natural scientists" as proposed apparently by Snow is to encourage the literary people to learn some freshmen physics, and the "scientists" to read some good literature.[8] This method is widely adopted in the universities and other places but, in fact, it is *problematic* and *ineffective*.

To understand this, we have to examine how the gap is formed *presently* in practice (while what described in Sec. 3.1 is how the gap was formed historically), and the very nature of the gap itself.

1. How the gap is formed today

As correctly pointed out by Snow [1998], the existence of today's gap is due to the design of our education system. Students in high schools and universities are directed too early towards either the humanities or "science." In response, the way to bridge the gap is to cancel this early division of students in high schools, and (after the gap is formed) force them to take general-education courses in the universities. These remedies are actually carried out in some countries and in most universities.

However, we are no longer in the early Greek days. Economically there is strong competitive pressure to arm our students early with special professional skills. It is impossible to get all countries, especially the developing countries, to agree on a slow-down schedule in their education systems. And so, to narrow the existing gap, the response is to increase the dose of general-education courses in the universities and the enhancement of popular science activities (reading popular science books in particular) in the society. (See [Lam 2008b].) But will this work in its present form? And is this effective and necessary?

2. Nature of the gap today

The gap today exists in the form of different knowledge contents picked up by the two camps of people, humanists and "scientists," during their schooling periods and beyond. And this is the rationale behind the proposal to encourage them to read something from the other camp. But (1) this is hard to achieve; (2) it is ineffective even achieved.

To illustrate Point 1, let us take the different groups within physics as an example. The fact that physicists can talk to each other is true only to a certain extent. There is not much to talk about between a particle theorist and a condensed matter physicist if the subject is the standard model of particles. But all scientists, be they physicists, biologists or chemists, do share some common knowledge such as the second law of thermodynamics, because this law is a required learning in the college education of these scientists.[9]

Professional activities require high concentration of attention and usually are time consuming, especially in the case of "natural science" which involves very keen competition. Time is short, for the professionals. Many first-rate scientists do not read books, particularly science books, because what contained in books is usually not fresh enough. Instead, they read research papers that they think might be helpful to their (present or future) work. That is what the scientist had in mind when he, asked by Snow what books he read, replied, "Books? I prefer to use my books as tools." [Snow 1998: 3] Tools, here, mean something that will help him to do his research. There is in fact a fair chance that literary books will be read by scientists, for relaxing purpose, e.g., when they are in an airplane after attending a conference. But these books are not Shakespeare's. The same goes for the people working in literature or the humanities. Why should they read any science book if they cannot find anything there that would help them to do their job? Time is short for them, too.

As for Point 2, let us assume for the moment that the humanist now knows something about basic physics and the "scientist" has read some Shakespeare or other great literature, and they meet in a cocktail party. If they ask each other what is new in the other's profession, they will not be able to go too far in their discussion because a sensible opinion in literature or "science" these days requires more knowledge than what is in their possession. Instead, for example, they can converse on Ang Lee's *Brokeback Mountain* (2005) or his other movie, *Crouching Tiger, Hidden Dragon* (2000). These movies' storylines are as deep as Shakespeare's, and perhaps more entertaining.

3. The proper way to bridge the gap

The gap can never be completely closed, nor should it be. What makes the world interesting is diversity; diversity requires some of us to be writers or artists and others physicists, and so on. What we can and should do is to *bridge* the gap. In our final analysis, the non-communication between the two camps is not due to the non-overlap of the people involved, but due to the *absence of any common language or principle* in their trades. Isn't it wonderful to teach every student something, if exist, when they are still in high schools or universities that they could use for the rest of their life no matter what profession they end up with, in the humanities or "natural science"? That would guarantee that everybody can communicate with anybody else, in a cocktail party or on the beach, say. Yet, this "something" did not exist in the 1950s when Snow delivered his lecture;[10] that is why Snow resorted to the *ineffective* remedy in bridging the gap— a remedy that is still being blindly adopted by others presently.

Today, fortunately, this "something" does exist. Since the late 1970s, some general principles applicable to almost all disciplines (and thus could serve as the common language mentioned above) have been discovered (see Sec.

5.2). Before these general principles can be introduced and appreciated properly, let us look more carefully at what the right-hand side of Eq. (1) actually means.

4 Demarcation According to Human and Nonhuman Systems

The "physical science" listed in Eq. (1), historically and as explained in Sec. 3, is mostly about the study of inanimate and simple systems. However, with the advancement of chaos theory[11] and the ubiquity of personal computers, and perhaps also due to the stagnation of research in particle physics (superstring or M theory notwithstanding) [Smolin 2007], quite a number of physical scientists have turn their attention the other direction, towards systems of larger and larger scales and "discover" complex system (such as those from Cells and up in Fig. 2) [Lam 1998].

Generally speaking, social science consists of anthropology, business management, economics, education, environmental science, geography, government policy, law, psychology, social welfare, sociology, and women's studies.[12]

Philosophy, culture, religion, language, literature, art, music, movie and performing arts make up the humanities, at least most of them. History, by its very nature, could be part of social science, but it is listed in the humanities at some universities. The aim of literature, music, and art in the humanities is to stimulate the human brain—through arrangement of words or colors, sound or speech, or shape of things—to achieve pleasure and beauty, or their opposites, *via* the neurons and their connecting patterns [Pinker 1997].[13] The brains, some sort of computer, of the creator and the receiver at the two ends of this process are heavily involved. Linguistic is the study of the tools involved in written words and speeches, supporting the three disciplines mentioned above.

The *scientific* development of these disciplines in the humanities is at a primitive level—the empirical level, using methods that are largely analytic, critical or speculative.[14] And that is why they are separated from social science, which is at an intermediate level (while physical science is at the highest level).

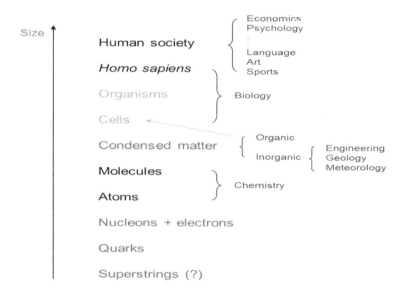

Fig. 2. Different systems (left column) studied in various disciplines (right column), listed from bottom to top as the system's size increases.

At this point, it becomes clear that the three items listed on the right-hand side of Eq. (1) are classified distinctly because of their scientific level in development, and *not* because of the nature of the objects under study in each category. Actually, the three items are arranged, from left to right, in decreasing level of scientific development.

This classification scheme may be convenient, but is definitely neither logical nor natural. To study something seriously and logically, nature in this case, one would like to group the objects under study according to their intrinsic nature and not how much we happen to understand them presently. For instance, at least in the beginning, if we want to study orange we will focus our attention on all kinds of oranges and put them in the same category, instead of starting from the category of oranges and clams, say—even though we may compare the two and benefit from the findings of orange and clam studies during the research process. Another example: If we are interested in electrons, we will not group them with rocks and study electrons and rocks together. Common sense, isn't it?

Consequently, to study nature a natural way is to categorize all objects in nature into two broad classes, viz., human system and nonhuman systems.[15] Equation (1) then becomes

Science = Natural Science

$$= \text{Nonhuman-system Science} + \text{Human-system Science} \qquad (2)$$

Here, "Nonhuman-system Science" means the study of inanimate and nonhuman-biological systems—what people usually call "natural science." "Human-system Science" consists of the humanities, social science, and medical science, whereas medical science includes neuroscience and genetics in particular. Apart from the obvious fact that medical science is about humans, the inclusion of medical science in human-system science is dictated by the *belief* and recent findings that many significant human characteristics and behavior (such as morality [Shermer 2004]) do have a biological basis, as revealed in neuroscience and genetics studies.

By grouping the humanities and social science together under one umbrella—human-system science, one can understand anew and more logically the connection between the constituent disciplines (Table 1). For the sake of convenience and with full respect for life—an interesting phenomenon in nature with yet an unknown origin, let us call a human being a "body." There are several basic facts about such a body [Lam 2002]:

1. Each body is macroscopic, about 40 centimeters to 200 centimeters long; it is a classical particle (i.e., quantum mechanics is irrelevant to these bodies).

2. Each body in their daily life moves very slowly compared to light; no need for Einstein's special relativity theory here.

3. The mass of each body is so small (compared to that of a planet, say) that Einstein's general theory of relativity can be forgotten, too.

4. Each body consists of layers and layers of structures (molecules, cells, organs, and so on) and many *internal states* (memory, thought, mood, and so on).

Table 1. Classification of the human system in a focused study according to the number of bodies involved, with examples and major relevant disciplines.

	One-body	Few-body	Many-body
Example	A Greek male, a Tang dynasty female, Einstein, Barbra Streisand, Hark Tsui, you, me	Romeo and Juliet, husband and wife, husband and wife living with mother-in-law, a person with two lovers, small-size family, the Beatles	Large physics class, tribe, city, country, Roman Empire, society, stock market, IBM
Discipline	Art, music, performing arts, language, literature, psychology, history (biography), neuroscience, genetics, medical science, law	Psychology, literature, performing arts, history, (family) law	Anthropology, (mass) psychology, philosophy, literature, culture, religion, history, business management, economics, education, environmental science, law, social welfare, sociology, women's study

5. All bodies are derived from the same ancestor (African Eve, say) only a few million years ago, and, according to Charles Darwin's evolutionary theory, human body and human nature take a long time to evolve and thus are practically unchanged over the last 6,000 years or so—the period in which human history is recorded.

6. Each body is an *open* system, inputting oxygen and food and outputting something else; the second law of thermodynamics does not apply here since the law is for closed systems (and equilibrium states) only [Lam 2004a].

7. Each body is under the influence of *external fields*, the most important of which is the community or society to which the body happens to belong.[16]

Keeping these facts and Table 1 in mind is important and advantageous when a human-system study is being undertaken. It allows you to pick the right tools and the right approximations (i.e., simplifying the problem by ignoring some irrelevant factors) to do the research. And it allows you to borrow or be inspired by some successful experience from other areas of study such as physics (wherein, same classification like that in Table 1 is used).

For example, in physics, a two-body problem interacting gravitationally with each other[17] is solvable, while such a three-body problem is chaotic and unsolvable [Stewart 2002]. Now you probably understand why it is so difficult living with your mother-in-law since it is a three-body problem. Just kidding! You cannot simple-mindedly take a result from physics and apply it without thinking (or with wrong thinking) to human affairs, because the interacting force between you and your mother-in-law does not obey the inverse-square law as in gravity.[18] Talking seriously, by ignoring all internal states of a body and treating it like a point particle, and using simple rules of interaction between the particles, computer modeling is able to explain and predict many human-group behaviors, ranging from pedestrian movement and traffic flow to voting processes, economic markets and war [Ball 2004].

As the classification of systems in nature is concerned, apart from dividing them into human and nonhuman systems as discussed in this Section, there exists in fact another way, i.e., dividing them into simple and complex systems according to the system's complexity. However, as we will show in the next Section, the latter approach, though very valuable in clarifying a lot of problems, is not that rigorous for demarcating systems in nature.

5 Simple and Complex Systems

Let us start by explaining what it means to be complex, followed by a brief introduction to complex systems. The reason that complex systems are relevant to our problem of understanding humans will then present itself.

5.1 What it means to be complex

There is a feeling that our world is very complex. According to the Merriam-Webster dictionary:

- *Complex*: composed of two or more parts...hard to separate, analyze, or solve.
- *Complex* suggests the unavoidable result of a necessary combining and does not imply a fault or failure.

- *Complicated* applies to what offers great difficulty in understanding, solving, or explaining.

Scientifically, the mechanisms driving the complexity of our world are more than what the dictionary suggests; it is not purely due to the existence of two or many parts [Lam 2004a]. And, indeed, complex systems are hard to analyze, but progress has been made in the last two decades or so [Lam 1998].

It would be nice if we can quantitatively define *complexity* and use it to compare different situations. Unfortunately, complexity is like love, you know it when you encounter it but it is hard to pin it down. In fact, people came up with more than 30 definitions of complexity, all different from each other. For our purpose, a working definition is enough: *Complexity is determined by the length of the shortest description of a system*; the longer the description, the more complex the system. For example, let us look at three sequences consisting of 1's and 0's.

A. 11111111111111111111111111111111111

B. 1100101000110000011001010110100110

C. 0010111001010111000000110101000111

Sequence A can be described by "All 1's." Sequence B is generated randomly, and so is described by "1's and 0's generated randomly." There is no pattern in sequence C,[19] and the description has to be a recitation of the whole sequence; sequence C is thus the most complex among the three.

Such a definition is, of course, problematic. First, if we are not informed that sequence B is generated randomly, it could be hard to know this by examining it supe, and we will assign it the same complexity as that for C. Second, the definition of the "shortest description" is also subjective. It could be we are not smart enough to identify the rhythm or pattern in C, or we may be able to identify it a month later. Well, let us keep these limitations in mind and move on.

5.2 Complex systems

Partly and largely due to the difficulty in defining complexity, there exists no rigorous and universal definition of complex systems. Generally speaking, a complex system usually consists of many interacting components; each component could have a few or many internal states and is adaptive in its behavior. The weakness of such a definition is easy to see. For example, a system may appear complex only because we do not understand it yet. Once understood, it becomes a simple system. Moreover, whether a system is complex or not may depend on what aspect of it we want to study. If we want to know the inner structure and formation mechanism of a piece of rock, the rock could be a complex system. But if we only want to know how the rock will move when given a kick, using Newtonian dynamics will do the job and the rock is simple. The lack of a technical definition and the ambiguity in the concept of simple/complex make it unsuitable to be used as a demarcation tool of anything.

For our purpose in this chapter, a *working* definition could be adopted: Almost all subjects covered in the universities, except those in the traditional curricula of physics, chemistry, and engineering departments, fall into the domain of complex systems [Lam 1998]. In other words, at the minimum, *biological systems including humans, and all topics covered in the humanities and social science belong to the domain of complex systems*. Most of the rest belong to simple systems.[20]

The importance of complex systems makes it worthwhile to know something more about them. Here is some basic knowledge about complex systems.

All material systems in nature are made up of "elementary" particles. Going from small to large in size, we have many layers of materials: quarks (or perhaps superstrings), nucleons (protons and neutrons), atoms (nucleons plus electrons), molecules, condensed matter (liquids and solids), cells, human organs, and human beings (Fig. 2). It is commonly known that at each layer of organization, there are many *emergent* properties; i.e., properties not easily guessed from the lower layer of constituents.

Life is such an emergent property: The fact that a human body is made up of cells and organs does not automatically lend itself to the expectation of life. Another example: The fluidity of water, a property not transparent from knowing that water is made up of H_2O molecules. To describe and understand the emergent property at each layer, one does not need to start from the very bottom level. For example, to describe the flow of water, one does not need to start from the quarks, not even from the molecular level. In fact, based on a few basic principles of symmetry, physicists are able to derive a coupled set of phenomenological equations describing water flows—the Navier-Stokes equations—which are still being used today. Similarly, to understand complex systems related to human phenomena, one can start from one of three levels: *empirical, phenomenological,* and *bottom-up* [Lam 2002]. And one has to study complex systems case by case.

Fortunately and surprisingly, since the late 1970s and through the extensive study of simple and complex systems, three general principles of organization in nature have been discovered. These three unifying principles can be applied to many—though not all—living and nonliving systems,[21] coming, in particular, from the humanities and social science. And we are referring to fractals, chaos, and active walks [Lam 1998].[22]

1. Fractals

Fractals were introduced by Benoît Mandelbrot in 1975 [Mandelbrot 1977]. A fractal is a self-similar (mathematical or real) object, possessing quite often a fractional dimension. Self-similar means that if you take a small part of an object and blow it up in proportion, it will look similar or identical to the original object. A famous example is the Sierpinski gasket [Lam 2004b]. Fractals are everywhere, ranging from the morphology of tree leaves, rock formations, human blood vessels, to the stock market indices, and the structure of galaxies. Fractals are even relevant in the corporate culture [Warnecken 1993] and the arts [Barrow 1995].

2. Chaos

Chaos has been investigated by Henri Poincaré at about the turn of the century and subsequently by a number of mathematicians. The modern period occurred in the late 1970s after Mitchell Feigenbaum discovered the "universality" properties of some simple maps, which was preceded by the important but obscure work of Edward Lorentz (1917-2008) related to weather predictions [1993]. Chaos is the phenomenon observed in some nonlinear systems, wherein, the system's behavior depends sensitively on their initial conditions.[23] Examples of chaos include leaking faucets, convective liquids, human heartbeats, and planetary motion in the solar system. The concept is also found applicable in psychology, life sciences, and literature [Robertson & Combs 1995; Hayles 1991]. A review of chaos for general readers is available [Yorke & Grebogi 1996].

3. Active walks

Active Walk (AW) is a major principle that Mother Nature uses in self-organization; it is a *generic origin of complexity* in the real world. Active walk is a paradigm introduced by Lui Lam [2005, 2006a] in 1992 to handle complex systems. In an AW, a particle (the walker) changes a deformable potential—the landscape—as it walks; its next step is influenced by the changed landscape.[24] Active walk has been applied successfully to a number of complex systems coming from the natural and social sciences. Examples include pattern formation in physical, chemical, and biological systems such as surface-reaction induced filaments and retinal neurons, formation of fractal surfaces, anomalous ionic transport in glasses, granular matter, population dynamics, bacteria movements and pattern forming, foraging of ants, spontaneous formation of human trails, oil recovery, river formation, city growth, economic systems, parameter-tuning networks [Han et al. 2008], and human history[25] [Lam 2002, 2006a, 2008a] (see also [Zhou et al. 2008]).

These three general principles are what we referred to at end of Sec. 3. All three principles are now an integral part of complex-system science, which is becoming important in the understanding of business, governments, and the media, among other things. But, of course, in the study of complex systems there remain a lot of virgin lands waiting to be explored.

6 Science Matters

The motivation and concept of Science Matters (Scimat) are given here, followed by an example (Histophysics). Implications of Scimat will be presented in the next two Sections.

6.1 Motivation

The discussion presented in Secs. 2-5 shows that there is *no* gap between the humanities/social science and "natural science"; they are all part of science. After all, there exists no natural dividing line among the items listed in Fig. 2; it is a continuum. The gap referred to by Snow *is* between humanists and "natural scientists," which was formed historically and is maintained by the education system; this gap is not intrinsic in nature. It is then *possible* to narrow or bridge this gap (between the two camps of people).

It was almost half-a-century ago that Snow gave his lecture on the two cultures; the world today is quite different. We now realize that the *real* reason to bridge this gap is not simply to let the two sides to have something to converse on in a cocktail party but [Lam 2006b]:

1. To have citizens who are better-informed on both the humanities/social science and "natural science" and thus can vote more sensibly on issues that could be scientific and/or ethical in nature (such as funding for stem-cell research).

Furthermore, to bridge the gap between humanists and "natural scientists" *properly* and *effectively* as well as, more importantly, to advance *knowledge* about the world and *humanity* (as explained below), we— humanists, social scientists and "natural scientists"—need to work together:

2. To raise the scientific level of the humanities.

While both aims are noble and important, the second one is dearer to us, epistemologically speaking.

6.2 Concept

New disciplines of study are born from time to time, like in the case of human babies, but less frequently; or, like new stars emerging in the sky, being suddenly noticed after a long period in the making.

Science Matters (Scimat), as a new multidiscipline, is created for the two aims listed above [Lin 2008]. Scimat is the scientific study of all things about the human system. Equation (2) is now rewritten as:

Science = Natural Science

= Nonhuman-system Science + Scimat   (3)

By naming "human-system science" in Eq. (2) as Scimat, we want to emphasize the fact that all human-system matters *are* part of science.

The *concept* and *method* of Scimat are: Following the good tradition of Aristotle and using the successful experience gained in physics (especially statistical physics), neuroscience and other disciplines, all human-system studies are treated as part of science and studied from the perspective of complex systems. The fact that there do exist general principles (see Sec. 5.2) that cut across all disciplines tells us that this approach is entirely possible.

Figure 3 illustrates what we discussed so far. Out of all the objects in nature, Scimat focuses on humans (the right box in the upper panel), the most complex system in the universe. Should be fun!

6.3 An example: Histophysics

History concerns itself with what happened to the *Homo sapiens* in the past [Stanford 1998].[26] The focus could be on an individual (such as Cleopatra, Alexander the Great, or Ava Gardner), a family or an empire; i.e., the system under study could be, respectively, one-body, few-body or many-body (Table 1). Traditional historians would collect historical records, analyze and put some order in the data or information at hand, then come up with some insights on why something happened and not merely how it happened, and perhaps offer some historical lessons; they stop there usually. No matter how convincing they are, these insights are frequently just educated guesses. As far as I know, no historians in the last few thousands of years had come up with any historical laws; most of them even doubt the existence of any historical laws.

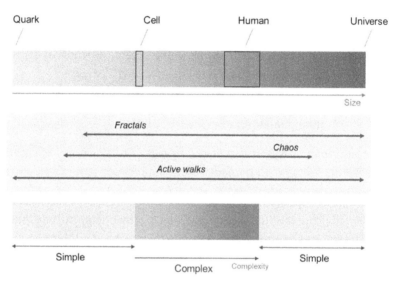

Fig. 3. *Upper panel*: Systems arranged in increasing size from left to right (not to scale). Humans are the object of study in Science Matters. *Middle panel*: Range of applicability of fractals, chaos, and active walks (from top to bottom). *Lower panel*: Complexity increases from cell to human (the middle region), the domain of complex systems; simple systems sit on the left and right regions, respectively, outside of this region.

Histophysics, the physics of human history, is a new discipline that views human history as the past dynamics of a complex system, from the perspective of Scimat [Lam 2002, 2008a]; i.e., there is a material basis underlying everything happened in human history. History is very complicated or complex, but could be discerned if one is lucky and the right kind of research tools are used. Techniques borrowed from physics and complex systems—such as statistical analysis, computer modeling, computer simulation, and the Zipf plot—have been successfully used to tackle problems in history. In particular, *quantitative* laws are found in the distribution of war casualties and of lifetimes of Chinese dynasties (from Qin to Qing, spanning 2,133 years). The latter are in fact laws in macrohistory, favored by the French Annales school [Burke 1990]. (See [Lam 2008a] for details.)

The success of Histophysics confirms the fruitfulness of the Scimat approach in studying human matters; it reinforces our confidence in the direction outlined in Scimat.

7 Implications of Science Matters

Calling the study of human-system matters by Science Matters is not merely a change in name. There are important implications. Four major implications of Scimat are presented here. More are given in the next and last Section.

7.1 Clearing up confusion in terminology

Let us designate by x the item included in the humanities/social science, where x could be art, literature, music, culture, society, and so on. And, for the convenience of discussion, we even allow x to represent "science" in its conventional usage, which actually means "natural science," the science of (mostly) simple systems. As we all know, there already exist studies of the scientific aspect of x, called the *"science of x,"* say; e.g., the "science of art." What is wrong with that?

Nothing, except that it is *confusing*. According to Scimat, art (art studies, not artworks) is already part of science [Lam 2011]. As we explained above that the present state of Art as we know it does not look like a science is simply because art, a very complex thing by itself, is at the early stage of its scientific development. Our brain is still the best processor in handling very complex things; that is why artists are still using their brain and not a supercomputer in creating art. Some day in the future, perhaps, when a super-supercomputer better than our brain is available, we will see artists using it to make a living—like the way that physicists are using their personal computer to solve a nonlinear equation these days.

Will it be depressing when this happens to art? Don't worry; the super-supercomputer still needs someone to input something and analyze the results, like the physicist has to decide what nonlinear equation to study, and how to interpret and use the computer results. Anyway, like in the case of physics, one would have more time to go fishing or go to an art museum and be happy.[27]

Why is art developing so slowly as its scientific level is concerned? There are two reasons: (1) Art is a very complex thing and hence it is very difficult to raise its scientific level—the reason just mentioned. (2) And this is very important as art (and in fact arts which include performing arts) goes: Humans have buying power; there are enough number of them willing to buy from the artists and thus helping to keep their discipline at a low scientific level. In other words, low-scientific-level art products sell so well already, there is no need to raise its scientific level. It is the market force at work here.

It is interesting to compare this to the case in physics. Physicists are able to control an electron and make it "dance"—a performing (electron) art. The electrons do *not* have buying power; no electron would come and buy a ticket to watch the show.[28] Consequently, the physicists have to raise the scientific level in their trade by doing two things: (1) They build a "superhighway" within a computer chip, which involves Nobel-prize-level breakthroughs. (2) They force the electrons to run like slaves in the superhighway inside the chip. Then they sell this computer chip, electrons included, to their fellow humans. And humans *do* have buying power— you see.

To avoid confusion, the correct way to say "science of art" is "art as a science matter" [Lam 2011]. The case of *"science and x"* is worse; it is *misleading*. For instance, let x be Culture here. "Science and Culture" is misleading because it implies that science and culture are two different things, which could even be opposing each other while, in fact, culture is part of science; culture is a many-body problem in Scimat (see Table 1). The use of the term "science and culture" is unfortunate because it endorses Snow's ineffective remedy of bridging the gap (see Sec. 3.2) and thus *prolongs* the gap between the humanists and "scientists." To properly bridge the gap, we should help both camps (and everybody else) to understand "culture as a science matter."

7.2 The Scimat Standard

Since everything is part of science—the fundamental basis of Scimat, there should be one and only one standard in validating the "correctness" of any theory in the humanities and social science, i.e., the one adopted in "science," established through thousands of years of painful trial and error and the sacrifice of numerous human lives. According to the American Physical Society (*APS News*, June 1999):

> The success and credibility of science anchored in the willingness of scientists to: (1) Expose their ideas and results to independent testing and replication by other scientists. This requires the complete and open exchange of data, procedures and materials. (2) Abandon or modify accepted conclusions when confronted with more complete and reliable experimental evidence. Adherence to these principles provides a mechanism for self-correction that is the foundation of the credibility of science.

With these words in mind, let us propose the following *Scimat Standard*:

1. We will be honest with the reader and ourselves and present our findings in clear writings, and will not try to hide our relevant thinking.
2. We will not quote anyone's writing to support our own argument.
3. We will not be ashamed to admit our own mistakes in our findings and correct them as soon as possible.
4. A conjecture[29] or *hypothesis* becomes a (temporary) *theory* only after it is confirmed by experiments or by practices in the real world.
5. We will abandon (or revise) the theory if it does not agree with confirmed and irrefutable evidences.

Explanations for these five rules are in order. **Rule 1** is the basic ethic of any honest researcher or knowledge seeker, but is not always practiced in non-physical disciplines. Yes, we understand that complex systems such as those studied in the humanities/social science are very complicated, and one does not always have a clear idea of what one's thought really is. If that is the case, please tell it to your reader which part is clear to you, which part is not, and mark your work as "work in progress." Better still, present your ideas like these in a seminar or cocktail party but not in a conference. If every paper was written clearly and findings/results were presented as "objectively" as possible, and if the paper *always* ended with a section of Discussion/Conclusion in which the author presented what lessons she/he had learned, perhaps the Sokal hoax [Sokal & Bricmont 1998] would never have to happen and the Science Wars [Labinger & Collins 2001] could be avoided.

Quoting others to support one's argument is a common practice in non-physical disciplines. But this is completely useless. For example, while Einstein was proven right in his many writings such as the two theories of relativity, he could not always be right and he did not [Kennefick 2005]. We all know about this; that is why **Rule 2**.

One should not be ashamed of making mistakes when doing complex-system studies since the job itself is so difficult. What one should do is simply admit their mistakes once recognized and correct them as soon as possible [Shermer 2001]. In this regard, good economists are real scientists who know their limits and act accordingly; they keep on adjusting their predictions of the stock index or the gross domestic product (GDP) and should be respected for doing that. That explains our **Rule 3**.

If anyone puts out an educated guess (what we mean by hypothesis), this guess has to be confirmed before it can be called a theory. Common sense, right?! **Rule 4** is copied from the practice in physics and other "natural sciences." We just want to unify our terminology in communicating to each other, since in Scimat we very likely are coming from different disciplines with different training and background.

Let me emphasize this: We do not mean that physical science is superior to the humanities/social science. It is not. In fact, the opposite could be true.[30] Humanists and social scientists are tackling very complex systems, while most physical scientists are dealing with simple systems. Those dealing with complex problems could be more courageous and should be respected. In fact, to be a *good* artist is more difficult than being a good physicist. In physics, there are rules to be obeyed and experience to follow and the choices in solving any physical problem are more restricted[31] than what is available to a painter who wants to create something new. The painter has infinite possibilities and really needs imagination and talent. That is why there are more good physicists than good painters in the world.

With Rule 4 in place, no social theory in the form of political ideology of *any* kind could be validated, since it is unethical to try experiments on living human beings, especially in large numbers.[32] Political leaders are advised to try their "experiments" with computer simulations and be prepared to adjust their policies frequent enough.

Rule 5 is obvious. Finally, it follows from the spirit of this Scimat standard that we will adopt a better standard if that becomes available.

7.3 There is always the reality check

There is something called the "reality check" as science matters are concerned. We accept Einstein's result, $E = mc^2$, not because it comes from Einstein but because it comes from the special theory of relativity which agrees with reliable experimental findings. Furthermore, the relativity theory gives a sequence of predictions that are later confirmed, even just last month; it helps our confidence in it. The fact that the atomic bomb, built according to $E = mc^2$, works is another plus. This is an example of reality check, even though reality check does not call for an atomic bomb every time, luckily for humans.

Now comes this old woman from Africa who tells us $E = mc$; the reason is that she does not like superscripts. And comes this philosopher from Europe who advocates $E = mc^{\sqrt{2}}$. His reason is that he wants to show his independence from Pythagoras (born between 580 and 572 BC, died between 500 and 490 BC) who abhorred irrational numbers [Lloyd 1970], and, besides, the philosopher honestly thinks the superscript $\sqrt{2}$ is more aesthetically appealing than 2. This phenomenon is called the *multicultural view of science* [Liu, B. 2008]. Who do you think a university will hire into her faculty?

We agree that each one of these three individuals should be fully respected by others for their honest attempts to understand nature, and be allowed to air their views (freedom of speech) or even publish their findings in a suitable journal/magazine/newspaper of some kind. These days, no opinion can be completely suppressed. At the minimum, there is always the Blog on the web that they can air their results, and it is free.

In practice, whether someone like you or me would spend our valuable time and listen to an individual's opinion (or read her/his article) on something, science included, does not depend entirely on the quality of that opinion, which we do not know ahead of time anyway. It depends on the *reputation* of that individual; in other words, it is a history dependent process. For instance, we are more inclined to read Einstein's article than that coming from the African woman or the European philosopher, because Einstein has an established reputation. It does not mean, however, everything uttered by Einstein is correct; but even wrong, his uttering could be inspiring. And that is why we make that choice; it is a matter of betting one's time. And we could miss something very valuable and important because that philosopher's writing could turn out to be very exciting and useful. We accept the risk, and do a catch up by Googling it after the philosopher's finding is reported in the newspaper, say. It is all a matter of allocating finite resources; it has nothing to do with disrespect for Africa or Europe, or *local knowledge* for that matter. The same goes for research funding. Misjudgments are made from time to time; the remedy is to open up more avenues of funding, like those coming from private wealthy individuals or private foundations, just in case.[33]

In fact, people *can* always ignore the reality check and hold on to whatever view of science they want and be happy—and we respect their right to be happy—as long as they do *not* try to put their "theory" to work. To build a cell phone you need quantum mechanics, not any kind of mechanics. But not everyone needs a cell phone, right? And the right of not wanting or producing cell phones should be respected. It is called *cultural diversity*. (See also [Liu, D. 2008].)

7.4 The Needham Question

In 1954, in his book *Science and Civilisation in China* Joseph Needham (1900-1995) [1954: 3-4] at Cambridge, UK, asked a question that goes something like this: Why did modern science develop only in the West after the 16th century (and not in China who, in the past, applied natural knowledge to practical technology and invention more efficiently than the West did)? [Liu, D. 2008]. There are many explanations offered [Liu 2000]. Some are obvious; others, not. Here is a new explanation which, I think, is right on the mark.

Remember Aristotle? Aristotle studied and pioneered a number of disciplines, in increasing order of complexity: physics, astronomy, biology and zoology, logic, ethics, government, politics, and so on. Today, his work on physics and astronomy are known to be completely wrong; his biology/zoology is partially wrong; but his logic and ethics studies are still found useful. The same smart Aristotle; how did this happen? The answer is that physics and astronomy are about (mostly) simple systems, biology/zoology is about complex systems, and the rest related to humans are extremely complex systems. It just shows that human-related matters as complex systems are very difficult to study, and we have not made much progress since Aristotle's times in these complex areas. Not Aristotle's fault; we still respect and admire him despite his failures.

The ancient Chinese—Confucius (551-479 BC) included, for whatever reason (which could be incidental or for financial reason), decided to start their enquiry of nature with complex systems—humans. They came up with some great insights but no clear conclusions [Wolpert 1993]. Worse, for unknown reason and unlike Aristotle, they did not or chose not to write down their findings in unambiguous language (i.e., they disobeyed Rule 1 in the Scimat Standard; see Sec. 7.2). That is like publishing a paper in a physics journal with writings that the reader can interpret in multiple ways; no way to make progress. In contrast, the Greeks did concern themselves with both simple and complex systems right from the beginning; e.g., Archimedes (c.287-c.212 BC) studied buoyancy of simple bodies and his own body, and was rewarded with the Archimedes Principle. Eventually, the Greek's successful results of simple systems got passed on in the West and ended up in the hands of Galileo, who started modern science. The ambiguous findings in complex systems from ancient Chinese passed on and kept *confusing* and *entertaining* the Chinese for more than two thousand years, even today. This is my answer to the Needham Question.

Here is the Lam Question: Why did modern science arise in Italy and not in other European countries?

8 Discussion and Conclusion

Here are ten points of interest:

1. There are always grey areas as demarcation of any kind is concerned. Some mathematicians find out about this and come up with a new mathematics called *fuzzy logic* [Klir & Yuan 1996]. Similarly, the division between humans and nonhumans, and that between simple and complex are not sharp divisions. For example, how many cells have to develop in an embryo before you will call the system a human being? In the grey areas, it is common that new and interesting phenomena and questions might pop out. Pay attention to the grey areas.

2. According to Scimat, Chinese traditional medicine (CTM) [Ma 2007] *is* science at the empirical level. We hope this will settle the debate on this topic once and for all (see [Liu, B. 2008]). The traditional "theory" of CTM offered in the old books or by its practitioners may seem strange to outsiders, but they could be some kind of phenomenological theory (or such a "theory" in the making, continuing for over two thousand years)— the next step beyond the empirical level in any scientific development— that works, partially or completely. The fact that the "theory" so far does not match anything in Western medicine implies one of three things: (1) the "theory" is on the wrong track and should be modified or abandoned in the future; (2) the "theory" is on the right track except that it will take time to connect it to that in Western medicine; (3) Western medicine is wrong or irrelevant. Case 3 is unlikely. Case 2 actually happened in the history of superconductivity: The Landau-Ginzburg phenomenological theory (1950) turned out to be correct and could be derived from the BCS microscopic theory (1957) after the latter was discovered [Tinkham 2004]. Whether that is also the case with CTM remains to be seen. But we all know this: The debate on CTM involves something more than prestige. In China, CTM is heavily funded (quite a number of hospitals and research

institutes in CTM are in place), but anything identified as pseudoscience would be banished.

3. No artists, writers or other humanists should feel threatened by Scimat. They could go on doing what they do best. The humanities are such a vast field that we need a lot of people working on it at the empirical level. Advancing the scientific level of the humanities needs to be done mostly by trained "scientists" with the help of or in collaboration with the humanists. We do hope that artists and writers will help.

4. There is no need to abandon the general-education courses in universities. But in all courses, general-education courses in particular, the instructor could start by introducing the concept of Scimat, explaining to the students why Eq. (1) should be replaced by Eq. (3), and so on. That would help tremendously in narrowing the cultural gap for our students, possibly our future writers and "scientists." Naturally, for the benefit of everybody, we would like to see a Scimat course like *The Real World* [Lam 2008b] be included in all universities as a *required* general-education course.

5. There could be a brand new theory about macroscopic humans waiting to be discovered, like quantum mechanics lurking there in the case of microscopic systems about 100 years ago. The *only* way to find out is to set out to look for it, *assuming* temporarily that the new theory indeed exists and is just hidden somewhere, like children searching for Easter eggs on the lawn of the White House each year. What is needed is the smoking gun, similar to the black-body radiation (as explained by Max Planck) or the double-slit experiment in the case of quantum mechanics.

6. Airplanes and humans, both complex systems, could be very different.

7. When the rule(s) of the Scimat Standard was broken, it often happened that it was humanity and not merely personal honor that suffered.

8. There is already a crowd out there doing Econophysics and Sociophysics [Chakrabarti et al. 2006; Ball 2006]; doing the humanities as complex systems could be more challenging and rewarding, more fun guaranteed.

9. Leonardo da Vinci (1452-1519) could be the last person in history who succeeded in mastering quite a number of topics from both "science"/technology and the arts. His failure to build many of his own designs in engineering, not to mention bringing them to the market, is due to insufficient funding and the absence of a large enough team, and also the non-existence of a suitable industry in society at his times. With the explosion of knowledge in modern times, perhaps no one could be as broad and deep as Da Vinci was any more. And there is no need to be. What we have to do is encourage people to be experts in two disciplines. With enough number of these bi-disciplinary scholars, all disciplines in the world will be able to link up with each other, directly or indirectly (Fig. 4). Here, we are talking about the flourishing of interdisciplinary education and scholarship, and the proper use of science communication [Lam 2008b], not just for Histophysics and Scimat but for all interdisciplinary studies.[34]

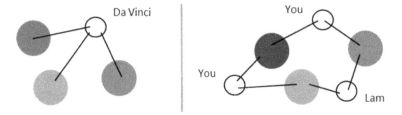

Fig. 4. One need not do everything like Da Vinci did (left), but does interdisciplinary work (right). The filled circles represent different disciplines; open circles, individuals.

10. It was for these reasons that an international conference on Scimat was held in Ericeira, Portugal, May 28-30, 2007 (Fig. 5) [Sanitt 2007]. We are looking forward to more conferences like this one, to provide an international platform for people to exchange ideas face to face. And, learning from the French: to drink, to eat and to sleep [Glover 2000]. Naturally, the purpose of doing all these is to reach the goal of "Let the Earth be peaceful forever!"

To conclude, Science Matters matter because science matters. But ultimately, Science Matters matter because humans matter!

Fig. 5. Poster of the First International Conference on Science Matters, Ericeira, Portugal, May 28-30, 2007.

Notes

1. The Chinese philosopher, Guanzi (c.723-645 BC), also favored water [Liu 2006: 272]. Among many water-related utterances, he said: "Human, water too. Vitality of male and female combine, and water flows and takes shape."

2. With this understanding, every possible enquiry undertaken would be about nature. The term "science" in its German sense, Wissenchaft (any systematic body of enquiry), and its use in the English language will coincide with each other.

3. In this chapter, "natural science" with quotation marks means the science of nonhuman systems, identical to that in the conventional usage of this term; the same goes for "natural scientist."

4. There are at least two factual errors in this famous article, apparently never pointed out by anyone before. (1) Snow is wrong when he writes, "No, I mean the discovery at Columbia by Yang and Lee" [Snow 1998:15]. In fact, in the famous paper that earned Lee and Yang the Nobel Prize in 1957, the authors' names appears as Lee and Yang [Lee & Yang 1957]. (The ordering of the two names in this and other joint papers by the two authors is not a small matter; it plays an important role in the two men's subsequent total breakup of collaboration and friendship [Yang 1983; Lee 1986; Chiang 2002; Zi et al. 2004].) While Lee indeed worked at Columbia University, Yang's address at that time was the Institute for Advanced Study at Princeton, New Jersey (see the address bylines in [Lee & Yang 1957]). The truth is that Yang has never been associated with Columbia University. (2) A few sentences later, still referring to the work of Yang and Lee, Snow makes another mistake in his sentence, "If there were any serious communication between the two cultures, this experiment would have been talked about at every High Table in Cambridge." In reality, the work of Lee and Yang is purely theoretical, which is to point out that there was no experimental evidence supporting or refuting parity conservation in weak interactions at that time. They went on to propose several experiments to settle this issue without predicting the outcome of these experiments. Parity nonconservation was discovered in an experiment by Chien-Shiung Wu (1912-1997) [Wu et al. 1957], a colleague of Lee at Columbia University.

5. For smaller falling objects, low-tech devices are used to record the positions at regular time intervals. This is routinely done in freshmen physics labs.

6. This was painfully experienced by Snow himself. In 1932, Snow had to recant publicly his "discovery" of how to produce Vitamin A artificially after his calculation was found faulty. Snow, a trained chemist, decided to leave scientific research completely after this incident and became a novelist [Snow 1998: xx]. He indeed made the correct move, judging by later developments in his career.

7. These days, the two separate groups in the two cultures are commonly understood to be humanists and "natural scientists," respectively.

8. As good literature is concerned, unlike the case in science, there is no unique choice suitable for everybody. Reading Shakespeare or Tang poem/Song prose will equally do.

9. The second law of thermodynamics is the example used by Snow to test the scientific knowledge of the literary people in a gathering [Snow 1998: 15]. This is in fact quite unfair, because the second law is less universal and useful than

people think. It applies only to closed systems and only to their thermodynamic equilibrium states. It applies neither to humans—an open system and the interest of literary people—nor to the expanding "cosmos" as Snow wrongly claimed [Snow 1998: 74]. The reason is that our universe is ever expanding and is never in an equilibrium state [Lam 2004a]. See [Zhao 2003] for a detailed discussion.

10. The powerful evolutionary theory of Charles Darwin does cut across all biological systems, but stops at inanimate systems.

11. See Sec. 5.2 for an introduction to chaos.

12. http://www.sosig.ac.uk.

13. And quite often, especially in the case of literature, to promote or stimulate a person's understanding of the world.

14. http://en.wikipedia.org/wiki/Humanities (July 16, 2008).

15. This is due to the fact that it is humans who do the study and control the research budget. If ants were in control, they would classify nature into the two groups of ants and nonants.

16. Other fields could be physical in nature, such as electromagnetic fields if cell phone is used; sunlight when the body is outdoor (guaranteed in summer in San Jose, California but not necessarily so in Beijing well before Olympics 2008); and unavoidably penetrating (harmless) neutrinos and non-penetrating cosmic rays (harmful in large dosage).

17. The gravitational interaction between two bodies is given by Newton's law of gravity, which states that the force on each body is inversely proportional to the square of the separation between the two bodies; the electric force between two charged bodies due to Coulomb's law is similar.

18. The uncritical application of physical results beyond physics is common, too common, among non-scientists, even among some non-physical scientists. The carefree misuse of chaos in human affairs published in numerous popular science/nonscience books is another example. A further example: Since Einstein's relativity theory tells us that mass and energy can be exchanged, $E = mc^2$, and since my body does have mass, therefore it gives me energy to dance. The fallacy is obvious: A piece of rock also has mass but it does not dance. All these pitfalls lie in the use of analogies without bounds, a symptom of parallelism [Scerri 1989].

19. Admittedly, this point is hard for the reader to see; to the naked eye, sequences B and C both look random. However, a technical test can differentiate the two: In the *difference map* (i.e., the plot of x_n vs. x_{n+1} where $\{x_n\}$ is the given sequence of numbers) a random sequence (with large number of elements) will spread out uniformly in the plot, while a non-random sequence will not. Let us say that such a test has been performed.

20. Note that "simple" does not imply "easy"; these are two very different concepts.

21. Some people call these principles "universal," a misnomer.

22. Self-organized criticality (SOC) proposed by Per Bak (1948-2002) et al. [1987] was advanced as another such general principle for complex systems [Bak 1996]. Unfortunately, SOC is at odd with many experimental findings in real systems, the ultimate judge in these kinds of things (see, e.g., [Cross & Hohenberg 1993]).

23. Chaos, a daily word, is used by scientists as a technical word with specific meanings. Before one can call a time sequence of numbers chaotic, several tests have to be performed, such as showing the Lyapunov exponent to be positive [Lam 1998]. The mere look of being random or chaotic is not enough—a pitfall committed by many laypeople.

24. For example, ants are living active walkers. When an ant moves, it releases chemicals of a certain type and hence changes the spatial distribution of the chemical concentration. Its next step is moving towards positions of higher chemical concentration. In this case, the chemical distribution is the deformable landscape.

25. In the active walk (AW) application in history, think of the walker as an active digger on a soft land. The digger could dig a round trough and keep himself moving in circles; he could dig himself a hole deeper and deeper and got himself trapped; or he could dig himself out of a hole and survive. These three situations, respectively, could be used to model what happened to some historical figures; or, when applied sequentially, three stages in the life of an individual. It all depends on the landscaping rule and the stepping rule involved, either or both of which could be time dependent; there are infinite possibilities. That is why AW is such a powerful modeling tool or metaphor in history and other studies.

26. This is the best book on historiography in my opinion.

27. It seems that people get happy by watching something more complex than what they are doing daily in their lab/office. You never see a physicist leaving her/his lab and go watching the swing of a pendulum. Galileo's days are long gone.

28. The physicist would be an ultra-billionaire if only 0.001% of the electron population within a 1-mm long copper wire showed up for the show by buying tickets at 0.001 dollars each.

29. A mathematical conjecture is a small theorem that is proved. We do not mean this kind of conjecture; the word conjecture used here simply means an educated guess.

30. My daughter is an artist and my hero.

31. Examples of the restrictions are: All the established laws have to be obeyed and confirmed experimental results respected. New theory cannot ignore or negate them; new theory could improve on them and/or find out where the validity boundary of the old theory is.

32. Of course, sadly, this has not prevented some historical figures from trying, with disastrous results. An example is what happened in Cambodia: From 1975 to 1979, two million Cambodians were killed under the Pol Pot regime because the leaders mistook a hypothesis in social science as a theory and applied it to their people; i.e., they broke Rule 4 of the Scimat Standard. This and other examples point to the urgent need of greatly improving the scientific training of political leaders, President George W. Bush included.

33. An example is the funding of extraterrestrial-intelligence (ET) searching in USA. The government funded it for a year and stopped; a rich man came along and continued the funding of SETI, the ET-searching institute in Mountain View, California. Everybody is happy; ET not found up to this moment.

34. In China, there is the journal *China Interdisciplinary Science* (Science Press, Beijing) which treats interdisciplinary studies seriously. It started in 2006 and has published two volumes so far.

References

Bak, P., Tang, C. & Wiesenfeld, K. [1987]. Self-organized criticality: An explanation of 1/f noise. Phys. Rev. Lett. **59**: 381-384.

Bak, P. [1996]. *How Nature Works: The Science of Self-Organized Criticality*. New York: Copernicus.

Ball, P. [2004]. *Critical Mass: How One Thing Leads to Another*. New York: Farrar, Straus and Giroux.

Ball, P. [2006]. Econophysics: Culture crash. Nature **441**: 686-688.

Barrow, J. D. [1995]. *The Artful Universe: The Cosmic Source of Human Creativity*. New York: Little, Brown and Co.

Burke, P. [1990]. *The French Historical Revolution: The Annales School 1929-1989*. Cambridge, UK: Cambridge University Press.

Chakrabarti, B. K., Chakraborti, A. & Chatterjee, A. (eds.) [2006]. *Econophysics and Sociophysics: Trends and Perspectives*. Berlin: Wiley-VCH.

Chiang, Tsai-Chien (江才健) [2002]. *Biography of Yang Chen-Ning: The Beauty of Gauge and Symmetry*. Taibei: Bookzone.

Glover, W. [2000]. *Cave Life in France: Eat, Drink, Sleep...* Lincoln, NE: Writer's Showcase.

Cross, M. C. & Hohenberg, P. [1993]. Pattern formation outside of equilibrium. Rev. Mod. Phys. **65**: 851-1112.

Han, X.-P., Hu, C.-D., Liu, Z.-M. & Wang, B.-H. [2008]. Parameter-tuning networks: Experiments and active-walk model. EPL **83**: 28003.

Hayles, N. K. [1991]. *Order and Chaos: Complex Dynamics in Literature and Science*. Chicago: University of Chicago Press.

Kennefick, D. [2005]. Einstein versus the Physical Review. Phys. Today, Sept. 2005: 43-48.

Klir, G. J. & Yuan, B. (eds.) [1996]. *Fuzzy Sets, Fuzzy Logic, and Fuzzy Systems: Selected Papers by Lotfi A. Zadeh*. Singapore: World Scientific.

Labinger, J. A. & Collins, H. [2001]. *The One Culture? A Conversation about Science*. Chicago: University of Chicago Press.

Lam, L (ed.) [1997]. *Introduction to Nonlinear Physics*. New York: Springer.

Lam, L. [1998]. *Nonlinear Physics for Beginners: Fractals, Chaos, Solitons, Pattern Formation, Cellular Automata and Complex Systems*. Singapore: World Scientific.

Lam, L. [2000]. How nature self-organizes: Active walks in complex systems. Skeptic **8**(3): 71-77.

Lam, L. [2002]. Histophysics: A new discipline. Mod. Phys. Lett. B **16**: 1163-1176.

Lam, L. [2004a]. *This Pale Blue Dot: Science, History, God*. Tamsui: Tamkang University Press.

Lam, L. [2004b]. A science-and-art interstellar message: The self-similar Sierpinski gasket. Leonardo **37**(1): 37-38.

Lam, L. [2005]. Active walks: The first twelve years (Part I). Int. J. Bifurcation and Chaos **15**: 2317-2348.

Lam, L. [2006a]. Active walks: The first twelve years (Part II). Int. J. Bifurcation and Chaos **16** ; 239-268.

Lam, L. [2006b]. The two cultures and The Real World. The Pantaneto Forum, Issue 24 (2006). [Paper presented at The 9th International Conference on Public Communication of Science and Technology, Seoul, May 17-19, 2006.]

Lam, L. [2008a]. Human history: A science matter. *Science Matters: Humanities as Complex Systems*, Burguete, M. & Lam, L. (eds.). Singapore: World Scientific. pp 234-254.

Lam, L. [2008b]. SciComm, PopSci and The Real World. *Science Matters: Humanities as Complex Systems*, Burguete, M. & Lam, L. (eds.). Singapore: World Scientific. pp 89-118.

Lam, L. [2011]. Arts : A science matter. *Arts: A Science Matter*, Burguete, M. & Lam, L. (eds.). Singapore: World Scientific. pp 1-32.

Lin, Lei (Lam, L.) [2008]. Science Matters: The newest and biggest interdiscipline. *China Interdisciplinary Science*, Vol. 2, Liu Zhong-Lin (刘仲林) (ed.). Beijing: Science Press. pp 1-7.

Lee, T. D. & Yang, C. N. [1957]. Question of parity conservation in weak interactions. Phys. Rev. **104**: 254-258.

Lee, T. D. [1986]. *Selected Papers*, Vol. 3. Boston: Birhauser.

Liu, Bing (刘兵) [2008]. Philosophy of science and Chinese sciences: The multicultural view of science and a unified ontological perspective. *Science Matters: Humanities as Complex Systems*, Burguete, M. & Lam, L. (eds.). Singapore: World Scientific.

Liu, Dun (刘钝) [2000]. A new survey of the Needham Question. Studies in the History of Natural Sciences **18**(4): 293-305.

Liu, Dun [2008]. History of science in globalizing time. *Science Matters: Humanities as Complex Systems*, Burguete, M. & Lam, L. (eds.). Singapore: World Scientific.

Liu, Da-Chun (刘大椿) [2006]. *Philosophy of Science*. Beijing: Renmin University of China Press.

Lloyd, G. E. R. [1970]. *Early Greek Science: Thales to Aristotle*. New York: Norton.

Lorentz, E. [1993]. *The Essence of Chaos*. Seattle: University of Washington Press.

Ma, X.-T. [2007]. Understanding traditional Chinese medicine in contemporary cultural context. *Science Defeated by Superstition?* Jiang, X.-Y. & Liu, B. (eds.). Shanghai: Huadong Normal University Press.

Mandelbrot, B. [1977]. *Fractals: Form, Chance and Dimension*. New York: Freeman.

Needham, J. [1954]. *Science and Civilisation in China*, Vol. 1. Cambridge, UK: Cambridge University Press.

Paul, W. & Baschnagel, J. [1999]. *Stochastic Processes: From Physics to Finance*. New York: Springer.

Pinker, S. [1997]. *How the Mind Works*. New York: Norton.

Robertson, R. & Combs, A. (eds.) [1995]. *Chaos Theory in Psychology and the Life Sciences*. Mahwah, NJ: Lawrence Erlbaum.

Sanltt, N. [2007]. The First International Conference on SCIENCE MATTERS: A unified perspective, May 28-30, 2007, Ericeira, Portugal. The Pantaneto Forum, Issue 28 (2007).

Scerri, E. R. [1989]. Eastern mysticism and the alleged parallels with physics. Am. J. Phys. **57**: 687-692.

Shermer, M. [2001]. I was wrong. Sci. Am., Oct. 2001.

Shermer, M. [2004]. *The Science of Good and Evil: Why People Cheat, Gossip, Care, Share, and Follow the Golden Rule*. New York: Henry Holt/Times Books.

Smolin, L. [2007]. *The Trouble with Physics: The Rise of String Theory, The Fall of a Science, and What Comes Next*. New York: First Mariner Books.

Snow, C. P. [1998]. *The Two Cultures*. Cambridge, UK: Cambridge University Press.

Sokal, A. & Bricmont, J. [1998]. *Fashionable Nonsense: Postmodern Intellectuals' Abuse of Science*. New York: Picador USA.

Stanford, M. [1998]. *An Introduction to the Philosophy of History*. Malden, MA: Blackwell.

Stewart, I. [2002]. *Does God Play Dice: The New Mathematics of Chaos*. Malden, MA: Blackwell.

Tinkham, M. [2004]. *Introduction to Superconductivity*. New York: Dover.

Warnecken, H.-J. [1993]. *The Fractal Company: A Revolution in Corporate Culture*. New York: Springer.

Wolpert, L. [1993]. *The Unnatural Nature of Science*. London: Faber and Faber.

Wu, C. S., Ambler, E., Hayward, R. W., Hoppes, D. D. & Hudson, R. P. [1957]. Experimental test of parity conservation in beta decay. Phys. Rev. **105**: 1413-1415.

Yang, C. N. [1983]. *Selected Papers 1945-1980 with Commentary*. New York: Freeman.

Yorke, J. A. & Grebogi, C. (eds.) [1996]. *The Impact of Chaos in Science and Society*. Tokyo: United Nation University Press.

Zhao, K.-H. [2003]. The end of the heat death theory. *Philosophical Debates in Modern Science*, Sun, X.-L. (ed.). Beijing: Peking University Press.

Zhou, T., Han X.-P. & Wang B.-H. [2008]. Towards the understanding of human dynamics. *Science Matters: Humanities as Complex Systems*, Burguete, M. & Lam, L. (eds.). Singapore: World Scientific.

Zi, Cheng (季承), Liu, Huai Zu. (柳怀祖) & Teng. Li (腾丽) (eds.) [2004]. *Solving the Puzzle of Competing Claims Surrounding the Discovery of Parity Nonconservation: T. D. Lee Answering Questions from Sciencetimes Reporter Yang Xu-Jie and Related Materials*. Lanzhou: Gansu Science and Technology Press.

Published: Lam, L. [2008]. Science Matters: A unified perspective. *Science Matters: Humanities as Complex Systems*, Burguete, M. & Lam, L. (eds.). Singapore: World Scientific. pp 1-38.

Bettering Humanity: The Scimat Approach

Lui Lam

1 Introduction

Large-scale tragedies in the world can roughly be divided into two types: natural or human-caused. The former includes volcano eruptions, earthquakes, and tsunamis; the latter, unnatural famines, wars, and genocides. Quantitatively speaking, human-caused tragedies could be more damaging than the natural ones. And being human-caused, they could be avoided or minimalized through human efforts.

Past efforts on bettering humanity through reduction of human tragedies can also be divided into two types: religion-based or secular. Important (non-political) secular movements of historical significance that cut across country boundaries include, in the recent past, the Enlightenment (1688-1789) and the Vienna Circle movement (early 20th century), and presently, the Humanism movement in UK and the USA [Lamont 1997]. Fair to say, in spite of some successes, we cannot claim that these movements were successful by looking at the world around us.

The Scimat program was started 10 years ago with the first international science matters conference in Portugal [Lam 2024: 419-428]. Scimat focuses on the science of human as a means of bettering humanity. In this talk, we will discuss what the Scimat approach is and how it agrees and differs from the previous movements. The emphasis is on educating the policy makers, present and future, to make sure that they are humble persons who will not knowingly sacrifice thousands or millions of human lives to achieve their political goals. Among other things, we will discuss why science and rational thinking are not enough in making humanity better, and why a secular movement coexisting with religion (the case of Scimat) stands the best chance of being successful.

2 Two Types of Tragedies

Quantitatively, numbers of deaths caused by human's wrong decision-makings outnumber by an order-of-magnitude of those due to natural disasters (Fig. 1).

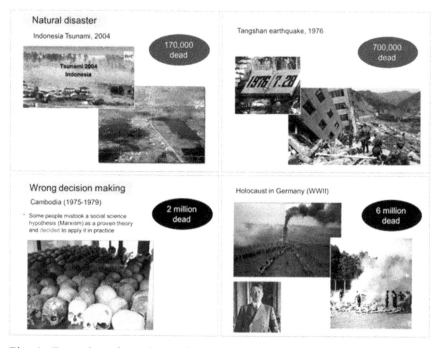

Fig. 1. Examples of numbers of deaths caused by natural and human-made disasters.

3 Is Humanity Worth Saving?

Humanity includes both abstract and physical aspects. For example, love belongs to the former and love-making, the latter. The former also includes sympathy and empathy for others, especially the weak; treasuring life (of human, other animals, and plants); fair play, honesty, and humbleness. The latter includes everything human-made. Look around yourself (Fig. 2); aren't they worth saving?

Fig. 2. Human-made things in Cascais, Portugal (2017). [photo/Lui Lam]

4 Two Types of Bettering Humanity

There are two approaches in bettering humanity: religion-based and secular. The former is built on faith (i.e., giving up independent thinking) but has the immediate effect of injecting a morality bottom-line to the believers, and, as long as the religion is peaceful, it does make the world a more peaceful place to live. The latter, in the last few centuries, show up as three major movements: Enlightenment, Vienna Circle, and Humanism (Fig. 3).

5 Three Secular Movements Before Us

5.1 The Enlightenment

The Enlightenment movement (1688-1789) in Europe, started one year after Newton's *Principia* was published, has the goal of creating a Science of Man but failed. It failed for two reasons: (1) The human system, unlike most of those covered in Newtonian mechanics, are complex systems— very complicated systems, which are difficult to study, scientifically speaking. (2) The tools to study it (such as mature probability theory and computer) were unavailable then [Lam 2014].

5.2 The Vienna Circle

The Vienna Circle, originated in Vienna (1922), tried to pick up where the Enlightenment failed. Their doctrine of *logical positivism* abandoned metaphysics (following the 1867 definition of Science [Lam 2024: 172]) and tried to constrict science into a jar, which failed, too, done in by Kurt

Gödel's incompleteness theorems. The Vienna Circle movement run in central Europe in the 1920s and 1930s, and spread to the USA when some of its members immigrated there as the Nazi Party came to power in Germany [Sigmund 2017].

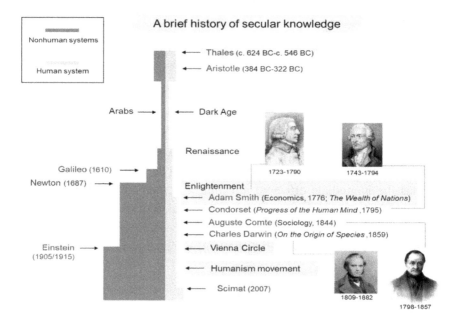

Fig. 3. A brief history of secular knowledge and the three modern secular movements in bettering humanity: Enlightenment, Vienna Circe, and Humanism.

5.3 The Humanism movement

In 1589, the word *Humanist* appeared, meaning non-theological studies and differing from its present meaning. Later, in the 19th century, *Humanism* the word appeared.

In the 20th century, in its current usage, Humanism means "an approach to life based on reason and our common humanity, recognizing that moral values are properly founded on human nature and experience alone." The humanist ideas are [Andrew & Grayling 2015]:

1. There are *no* supernatural beings.
2. The material universe is the *only* thing that exists.

3. Science provides the *only* reliable source of knowledge about this universe.

4. We only live this life—there is no after-life, and no such thing as reincarnation.

5. Human beings can live ethical and fulfilling lives *without* religious beliefs.

6. Human beings derive their moral code from the lessons of history, personal experience, and thought.

Humanists are well organized. There is this International Humanist and Ethical Union (IHEU), which was founded in 1952 and is an umbrella organization of humanist, atheist, rationalist, secular, skeptic, freethought and ethical culture organizations worldwide. There is also this British Humanist Association (Humanists UK), which "wants a tolerant world where rational thinking and kindness prevail," with the aim "to defend freedom of speech, support the elimination of world poverty, remove religious privileges, right to abortion."

In USA, the American Humanist Association (AHA) encompasses the following:

1. Humanism encompasses a variety of nontheistic views (atheism, agnosticism, rationalism, naturalism, secularism…).

2. Grounded in the philosophy of the Enlightenment, informed by scientific knowledge, and driven by a desire to meet the needs of people.

3. Works to protect the rights of humanists, atheists, and other nontheistic Americans.

4. Encourages individuals to live informed and meaningful lives *without* beliefs in any gods or other supernatural forces.

5. Membership numbers under 50,000, with over 575,000 followers on Facebook and over 42,000 followers on Twitter.

The AHA consists of local humanist chapters/affiliates (236 groups). It holds annual conferences (76th in 2017) and gives out Humanist of the Year Award (established in 1953); recent winners are: Adam Savage

(2017), Jared Diamond (2016), Lawrence Krauss (2015), Gloria Steinem (2012), Linus Pauling (1961), and Leo Szilard (1960). It also publishes a bimonthly magazine *The Humanist*, which claims to be a magazine of critical thinking and social concerns, with the slogan "Putting the Good in the 'Good without a God'" printed on its Sept/Oct 2017 issue.

In short, the common theme of the humanism movement is to *replace* (not coexist with) religion. The humanist organizations try to act as a counterpart to religious organizations in the West. They are not very effective for two reasons: (1) They spread too thin. (2) Something is lacking (because science and rational thinking are not enough; see below).

In contrast, the nonpartisan-nonprofit organization American Civil Liberty Union (ACLU), founded in 1920, is effective because it focuses on a single mission and is not anti-religion. Its mission: Defend and preserve the individual rights and liberties guaranteed to every person in this country by the Constitution and laws of the United States. With an annual budget of over $100 million, ACLU works through litigation and lobbying, with over 1,000,000 members and local affiliates active in almost all 50 states, the District of Columbia, and Puerto Rico.

6 Why Science Is Not Enough

Science is built on approximations. Every theory, a confirmed hypothesis, is an approximation of "reality"; there would be a better theory later. Approximations are unavoidable for the following reasons: (1) Even for an "exact" theory, it is rare to find exact solutions. (2) An equation (e.g., $F = ma$) is confirmed by measuring the quantities and showing that the left-hand side of the equation is equal to the right-hand side. But every measurement has uncertainty (called error) dictated by the apparatus used and so the equation can only be confirmed approximately.

Thus, it is a myth that "exact science" ever exists. Science never proves anything, rigorously speaking (in the mathematical sense of proof). In fact, science lives and thrives with approximations.

Moreover, science is very successful with simple systems but *not* that successful with complex systems, which include the human system. And that is why the humanities and social science are less developed when compared to natural science. People's impression that "science is very successful" is based on their neglect of the basic differences between simple and complex systems.

7 Why Rational Thinking Is Not Enough

Only simple systems in the classical world is certain (e.g., stone falling). The quantum world is inherently probabilistic (even though the equation is deterministic). The human world, though classical, is inherently probabilistic (due to unaccountable factors). We are thus living with uncertainly, no matter how much science we know or can know.

Living in an uncertain world, it is advisable for everyone to master some basic probability concepts (e.g., events with low probability could actually happen) and prepare for the "worse," and, for scientists, be humble.

Rational thinking is not enough for the following reasons:

1. Rational thinking is never complete (e.g., rational thinking becomes irrational when more factors are considered; vice versa).
2. Human system is probabilistic.
3. Any prediction about humans can only be given with probability.
4. Decision makings based on rational thinking thus always involve gambling.

Gambling means decision making (called betting) with inherit incomplete information. In this case, whether it is making political decisions or betting in the casino, other considerations are or should be involved (e.g., empathy for the underprivileged, affordability of total loss); otherwise, tragedies could result and human lives could be lost—sometimes in the millions (see Sec. 2).

8 Educational Level of Political Leaders and Good Decision Making

Since large-scale human tragedies are due to decision making of political leaders, it is worthwhile to pay attention to their educational levels (Table 1). Not surprisingly, those who caused many lives lost in history (like

Stalin and Hitler) have quite low educational levels—high school education at the most.

Table 1. Educational level of some major political leaders.

Name	Born-death	Country	Primary school	Middle school	High school	College	Highest degree
Joseph Stalin	1878-1953	USSR	Yes	Yes	Yes		
Adolf Hitler	1889-1845	Germany	Yes	Yes	Drop out		
Mao Ze-Dong	1893-1978	China	Yes	Yes	Yes		
Deng Xiao-Ping	1904-1997	China	Yes	Yes			
George W. Bush	1946-	USA	Yes	Yes	Yes	Yes	MBA

Even though it is true that the educational level of an individual could affect the person's grasp of scientific knowledge—an important consideration in decision making, there are other qualities that are equally important in making good decisions:

1. *Smartness*: Usually reflected in IQ, which is essentially born with, nothing much could be done with it through personal efforts.

2. *Personal background*: Parents' professions and the historical period (peaceful or not) a person went through could affect greatly his later decision makings.

3. *Good at learning*: Could be improved tremendously through personal effort like book-reading, which sometimes depends on the family background and the environment (like Newton who has a large private library).

4. *Critical thinking*: Usually gained through many successful experiences in solving problems (like fighting in wars) and good learning of failures—one's own or others'.

5. *Empathy*: Sympathy for the unfortunates, helping the poor, can be obtained by watching others nearby, and the activation of mirror neurons in the brain.

6. *Humbleness*: Could be gained through a proper understanding of human's short written history (thousands of years) compared to the long history of the universe (13.7 billion years).

While items 1 and 2 are beyond the influence of education, items 3-6, especially item 6, can be taught by the parents and schoolteachers. But what to teach? Below, we will concentrate on item 6 because, we believe, a humble leader will not *knowingly* sacrifice millions of lives to achieve a political goal, no matter how noble the goal is. To this end let's first introduce *Scimat*, the multidiscipline proposed by Lui Lam in 2007 [Lam 2008].

9 The Dao of Scimat

Scimat (science of human) aims to raise the scientific level of the humanities through the use of humanities-science synthesis and by encouraging interaction between humanists and natural scientists.

In *one* sentence: Everything in nature is part of science! That is, everything in nature, humans included, are legitimate objects of study in science.

Conceptually, Scimat represents the four tenets:

1. Science is human's effort to understand nature without bringing in God or any supernatural.

2. Science covers everything in nature.

3. Nature includes humans and all nonhuman systems.

4. All research on human matters, humanities in particular, are part of science.

Disciplinarily, Scimat represents the collection of research disciplines that deal with humans; i.e., Scimat = Humanities + Social Science + Medical Science.

The 1-2-3 *insight*: One culture, two systems, three levels!

1. There is only *one* culture—the scientific culture.

2. All systems are simple or complex systems; the *two* are quite different from each other.

3. There are always *three* research levels (empirical, phenomenological, bottom-up) in any discipline.

10 The Scimat Approach in Bettering Humanity

The Scimat approach in bettering humanity consists of three steps.

Step 1: Put 3 basic messages into elementary education

-   Message 1: *It all started with the big bang.* Everything in the universe originated from the big bang 13.7 billion years ago. Afterwards, atoms were formed, and part of it coalesced into the solar system 4.7 billion years ago, one of which was Earth. One billion years later, life appeared on Earth (Fig. 4).

-   Message 2: *We are one family.* All humans, dead or alive, share the same family tree; we are descendants of fish [Shubin 2009] (which could be the one called Microbrachius, 8 cm long and lived 0.4 billion years ago [Cressey 2014]).

-   Message 3: *We are stardust.* Everything on Earth, humans included, is made up of atoms. All atoms came from the stars (except H, He, and Li were formed soon after the big bang; Ag and Au, produced from neutron-star mergers). We are thus stardust. Every atom in our body is recycled from somewhere else (which could be other peoples' body, dead or alive, which you never know). We thus could be related to each other physically. We are recycled stardust.

Lesson: The recorded human history is only a few thousand years long, which, when put in the background of billions years of the universe, is less than a flash of time. All matters, personal or national, should be considered in this perspective. Moreover, we are descendants of fish and share the same family tree; our bodies are stardust, consisting of atoms which could be recycled from other's bodies. Live peacefully with others who shared this Earth. And be forgiving and humble.

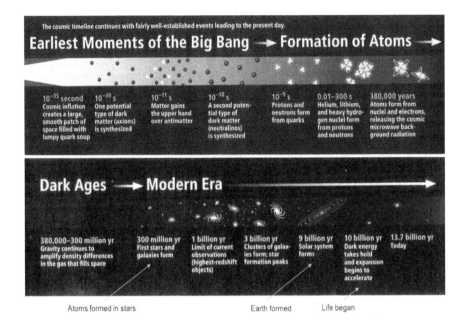

Fig. 4. Evolution of the universe. (Adapted from [Turner 2009].)

Step 2: Teach the HuSS general-education course in every university

The general-education course "Humanities, Science, Scimat" (HuSS) was created and test-taught by Lam in the International Summer School 2015 and 2016 at the Renmin University of China [Lam 2017; Luo 2017]. The course teaches undergrads of any major and any level the proper definition of science and the relationships between the humanities and "science" from the perspective of Scimat as well as recent developments in philosophy, arts, and history. In addition, the basics of probability and uncertainty deemed necessary knowledge of everybody are included (Figs. 5 and 6).

Step 3: Establish 100 Scimat centers worldwide

The Scimat Center is to organize international workshops/conferences and summer/winter schools, give out a Scimat Award every two years, host short-term visiting scholars, help match faculty members from the humanities and science departments to collaborate in research and creating new humanities-science courses (e.g., Histophysics), and help spread the new, ultimate general-education course HuSS for undergrads of all universities worldwide.

11 Discussion and Conclusion

Superstition probably existed millions years ago when people didn't know better about nature and human matters. Religions, as organized activity based on some kind of supernatural (e.g., God), are more recent, starting with Hinduism about 3,500 years ago. Yet, once started, it will never go away because religion does provide immediate answers to some deep and desperate questions that humans ask, like the meaning of life and where to go after death while science, basing itself on rational thinking, cannot do so. Moreover, *peaceful* religions do provide immediate bottom line in morality, a good thing for humanity anyway.

In any case, since 1867 when science consciously decoupled itself from religion, the two will forever coexist with each other. Any apparent conflict between the two is superficial and can be solved when both sides do not overclaim, with dialogues maintained from time to time [Lam 2014, 2024].

The number of religious people in the world outnumbers that of secular people. Presently, religious people is about 70% of the world population, and is projected to *increase* from 5.8 billion in 2010 to 8.1 billion in 2050 [Grim 2015]. Against this background, any secular movement explicitly positions itself against religion, like the Humanism movement, is bound to fail or capable of limited success only (Table 2), not to mention that science alone in bettering humanity is never enough (see Secs. 6 and 7).

Table 2. Comparison of four secular movements in bettering humanity.

Approach	Methodology	Religion
Enlightenment	Follow Newtonian mechanics	Against
Vienna Circle	Logical positivism	Yes/No
Humanism	Educational (outside system)	Against
Scimat	Educational (within system)	No position

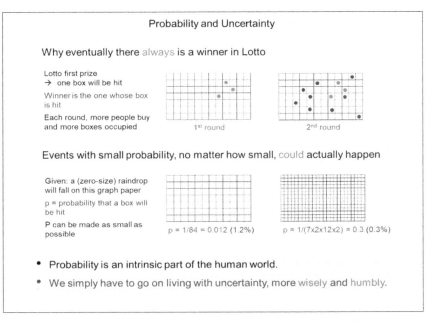

Fig. 5. Two basic knowledge about probability and uncertainty.

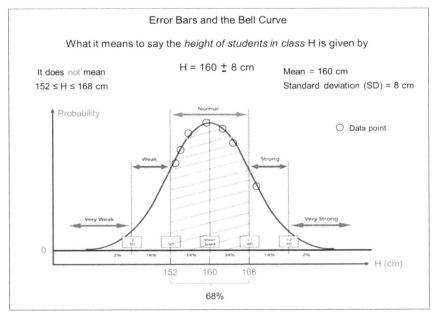

Fig. 6. Basic knowledge about error bars and the Bell curve.

In this regard, note that the Scimat movement is the only one which holds no particular position on religion (Table 2), knowing that science, concerning itself purely with things in the universe, has nothing to say about God—a supernatural who created the universe and is thus beyond the universe.

As history shows, past political leaders whose decisions caused large-scale human casualties are non-religious, secular people, usually with low educational level (see Sec. 8) and not good at self-learning. It thus seems obvious the religious approach in bettering humanity will not work with these kind of leaders, and we are left with the secular approach. And the practical way is to prevent this kind of leaders, present and future, to appear; the way to do that is to make sure everyone be humble enough so they will not do evils, through secular education.

But since we don't know who will be a major leader in the future and how many years of schooling they will go through, we suggest that we teach every school children from grade one on the three basic messages about humans, as explained in Sec. 10, and reinforce them year after year and at the university level through the mandatory HuSS general-education course.

In short, we believe the Scimat approach is a viable and practical way to better humanity, taken into account all the bloody historical lessons.

References

Andrew, C. & Grayling, A. C. (eds.) [2015]. *The Wiley Blackwell Handbook of Humanism*. Hoboken, NJ: Wiley.

Cressey, D. [2014]. Fossils rewrite history of penetrative sex. Nature (2014). https://doi.org/10.1038/nature.2014.16173 (Feb. 14, 2024).

Grim, B. J. [2015]. How religious will the world be in 2050? https://www.weforum.org/agenda/2015/10/how-religious-will-the-world-be-in-2050/ (Feb. 14, 2024)

Lam, L. [2008]. Science Matters: A unified perspective. *Science Matters: Humanities as Complex Systems*, Burguete, M. & Lam, L. (eds.). Singapore: World Scientific. pp 1-38.

Lam, L. [2014]. About science 1: Basics—knowledge, Nature, science and scimat. *All About Science: Philosophy, History, Sociology & Communication*, Burguete, M. & Lam, L. (eds.). Singapore: World Scientific. pp 1-49.

Lam, L. [2017]. Humanities, Science, Scimat: A new general-education course. *Interdisciplinarity and General Education in the 21st Century*, Burguete, M. & Connerade, J.-P. (eds.). Cascais, Portugal: Science Matters Press.

Lam, L. [2024]. *Humantities, Science, Scimat.* Singapore: World Scientific.

Lamont, C. [1997]. *The Philosophy of Humanism.* Amhurst, NY: Humanist Press.

Luo, Xiao-Fan [2017]. The HuSS course: A student's experience. *Interdisciplinarity and General Education in the 21st Century*, Burguete, M. & Connerade, J.-P. (eds.). Cascais, Portugal: Science Matters Press.

Shubin, N. [2009]. *Your Inner Fish: A Journey into the 3.5-Billion-Year History of the Human Body.* London: Pantheon/Allen Lane.

Sigmund, K. [2017]. *Exact Thinking in Demented Times: The Vienna Circle and the Epic Quest for the Foundations of Science.* New York: Basic Books.

Turner, M. S. [2009]. The universe. Sci. Am. Sept. 2009: 36-43.

Unpublished: Lam, L. [2017]. Bettering humanity: The Scimat approach. Based on my two talks given at the Sixth International Science Matters Conference on *Bettering Humanity: Historic Secular Movements*, October 25-27, 2017, Cascais, Portugal.

.

Humanities, Science, Scimat: A New General-Education Course

Lui Lam

My experience of teaching the new general-education course "Humanities, Science, Scimat" in the International Summer School 2015 and 2016 at the Renmin University of China is presented. The course teaches undergrads of any major and any level the proper definition of science and the relationships between the humanities and "science" from the perspective of Scimat as well as recent developments in philosophy, arts, and history. Concurrently, the research method is taught by guiding the students in real research, from picking topics to publishing papers. In 2015, all students were organized into research teams—the open-teaching mode. In 2016, part of the students formed research teams while the rest worked as individual researchers on topics taken from the textbook—the mixed-teaching mode. The textbook used in both years was *Humanities, Science, Scimat* written by the author. The conclusion is that this course can be taught by any instructor in any university in the world.

1 Introduction

General-education (GE) courses are offered in (almost) all universities in the United States and some other countries, as a means to bridge the gap between the humanities and "science,"[1] to prepare students to face the world after they graduate. Unfortunately, most the available GE courses are too narrow in their scope, which are confined either to the humanities or "science" and, occasionally, with some overlapping between the two. The GE course, "Humanities, Science, Scimat" (HuSS) created by Lam, offers something completely new. It is an interdisciplinary and cross-cultural introduction to the humanities and "science" from the unified perspective of *Scimat* (Science Matters), a new multidiscipline introduced by Lam in 2007 [Lam 2008a, 2008b].

Essentially, the course introduces students to the *proper* definition of science and its relation to the humanities—history, arts, and philosophy in particular—while the students are guided in *real* research from picking topics to writing and publishing papers. Here, I will present my experience of teaching this course in the International Summer School 2015 and 2016 at the Renmin University of China (RUC), Beijing.

2 General Education

The "general education" introduced in the 1930s in the United States, is an invention in American curriculum reform. While *liberal education* in the United States (starting late 19th century) looks back to the past, *general education* focuses at the present and eyes the future [Miller 1988]. There are two major books on the meaning and history of GE (Fig. 1): Earl McGrath's *General Education and the Plight of Modern Man* [1976] and Gary Miller's *The Meaning of General Education* [1988].[2] The latter contains more details and is more informative.

Ideally speaking, in the United States at least, the aim of GE courses [Miller 1988: 5] is to develop in individual students the (1) attitude of inquiry, (2) skills of problem solving, (3) individual and community values in association with a *democratic* society, and (4) knowledge needed to apply these attitudes, skills, and values—to maintain a lifetime learning process and function as self-fulfilled individuals and fully participants in society. Accordingly, the characteristics of GE courses are: (1) comprehensive in scope, (2) emphasize on specific and real problems faced by students/society, (3) concern with the future's needs, and (4) application of democratic principles in methods, procedure, goals of education.

Of course, when GE is adopted in other countries, the word "democratic" could be and has been substituted by other social values the countries see as suitable replacements (see, e.g., [Zhang 2017]).[3]

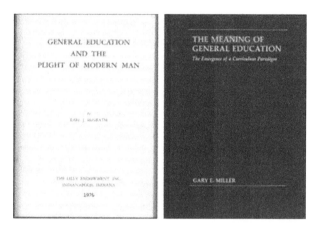

Fig. 1. Two general-education books by Earl J. McGrath (1976) and Gary E. Miller (1988), respectively.

3 Renmin University of China and Its Summer School

The Renmin University of China was founded by the Central Committee of the Communist Party of China in 1937, in Yan'an. Its present name started in 1949 with first batch of students admitted in October, 1950. RUC's curriculum concentrates on the humanities and social science, with small departments in the natural sciences. With no engineering school, RUC does have schools covering computer and information sciences. Right now, RUC has about 33,000 students whereas 45% are grad students.

Partly to promote GE (called "tongshi" or general-knowledge education in China), since 2009, an international summer school (ISS) with courses conducted entirely in English has been installed at RUC [Zhang & Xuan 2017]. Each undergrad has to take one ISS course (no more than two, though) before graduation even though grad students are allowed to take them, too. Thus, there is a steady pool of students for these summer courses. Tuition is free for RUC students but not for external students. The latter include a number of foreign students since there are many courses on Chinese language, philosophy, and culture.

The slogan of the ISS 2015 is "BEST": Broaden your views; Enjoy excellent education; Superb service; Touch the future. In particular, my new HuSS course was offered in 2015, and again in 2016 (Figs. 2 and 3). I was paid 200 USD per 45 min teaching, which amounts to 6,400 USD for the course. And, like other summer instructors, I was given a room in the campus hotel in the Huixian Building with free parking for the month of July. Since Scimat is part of the course title, a brief explanation is in order.

RUC International Summer School (course number: SH1518)
Humanities, Science, Scimat: A Trans-Disciplinary and Cross-Cultural Experience
Summer 2015

Lecture hours: Tue., Thur., 2:00 pm-5:30 pm
Prerequisite: None
Instructor: Lui Lam. Email: lui2002lam@yahoo.com. Phone: 1355 2008 xxx.
              Office hour: by appointment.
Course language: English

Fig. 2. Beginning of the course description (2015).

4 The Dao of Scimat

The new multidiscipline Scimat deals with the science of human—a biological system made up of atoms. Scimat's *aim* is to raise the *scientificity* (scientific level) of the humanities by encouraging interaction and collaboration between humanists and natural scientists. The essence of Scimat is:

1. <u>One sentence</u>. Everything in nature is part of science!

2. <u>Four tenets</u>. *Conceptually*, Scimat represents the four tenets:

   (1) Science is humans' effort to understand nature without bringing in God or any supernatural.

   (2) Science covers everything in nature.

   (3) Nature includes humans and all nonhuman systems.

   (4) All research on human matters, the humanities in particular, are part of science.

Disciplinarily, Scimat represents the collection of research disciplines that deal with humans, i.e.,

Scimat = Humanities + Social Science + Medical Science

3. <u>One insight</u>. The Scimat's *1-2-3 insight*:

One culture, two systems, three levels!

- There is only *one* culture—the scientific culture.
- All systems are simple or complex systems; the *two* are quite different.
- There are always *three* research levels (empirical, phenomenological, bottom-up) in any discipline.

4. <u>Two messages</u>.

- It all started with the big bang (everything on Earth is made up of atoms, mostly coming from the stars).
- We are one family, descendants of fish (Fig. 4).

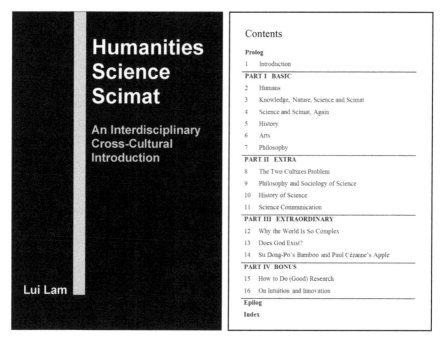

Fig. 3. Textbook adopted in the HuSS course (2015). Cover shown here is from v0.3.

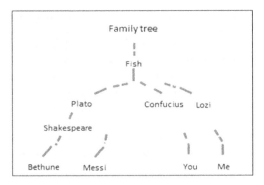

Fig. 4. Humans' family tree. Fish is our common ancestor.

For discussion on the birth of academic disciplines and science, supporting the claims above, see [Lam 2014]. For description of *The Scimat Program* concerning the international Scimat movement and an action plan to establish Scimat centers worldwide, see [Burguete & Lam 2016]. See also the Scimat website: www.sjsu.edu/people/lui.lam/scimat.

5 The HuSS Course in 2015: Open Teaching

The course description (like a contract) in Sec. 5.1 was finalized after discussion with students on Day 1. Note that the "no cheating" item in "Important Remarks" is a routine warning we issue to every class in any university. To be effective, it is important to make it clear early in class and put it in writing. How the class was actually conducted is outlined in Sec. 5.2 while the extra days relating to paper writing are described in Sec. 5.3. The important last day and feedback are presented, respectively, in Sec. 5.4 and Sec. 5.5. Section 5.6 describes how four student papers were published after summer.

5.1 The course description

This is a trans-disciplinary and cross-cultural course, taught according to Confucius' dictum of "Instruction knows no social distinction; teach according to the student's aptitude." The course consists of three parallel components:

1. The instructor will introduce the proper relationships between the humanities and (natural) science, from the perspective of Scimat.

2. The class (maximum 30 students) will be divided into five teams, with 5-6 persons per team. Each team will work on a (research) project of their choice and approved by the instructor, to investigate what had been done scientifically on that topic, with help from the web, library, and experts around the world. Students will present oral progress report in class, some with power-points. Each team will hand in a written report in English (in the form of a publishable paper) at end of course. Outside speakers could be used.

3. The teams will be treated and guided like research teams. They will learn how to do good research, make presentations and write research papers—both in English. (Some papers could be published in international research journals.)

Students are expected to attend every lecture.

Required textbook: *Humanities, Science, Scimat*, by Lui Lam (v0.2 will be provided free to students for their *personal* use).

Objective: At end of class, the students will learn:

1. The proper definition of science.
2. The proper relationships between the humanities and science.
3. The proper understanding of history, arts and philosophy.
4. The new multidiscipline called Scimat.
5. Use Excel to program, calculate and plot results of some stochastic systems (such as Random Walk).
6. How real research is done.
7. Teamwork.
8. Communicate efficiently and do *professional* presentations in MS power-point (ppt).
9. Write English papers in publishable form.

Schedule:

	Tuesday		Thursday
7/7	First lecture (form teams, discuss possible topics).	7/9	Finalize team projects; start research in teams.
7/14	Oral progress report from each team; continue research; teach Chaps 2-4.	7/16	Oral progress report from each team; continue research; teach Chap 5 (History).
7/21	Oral progress report from each team; continue research; teach Chap 6 (Arts).	7/23	Oral progress report from each team; continue research; teach Chap 7 (Philosophy).
7/28	Written **Exam** on Chaps 2-7 of textbook; oral presentation of **draft of paper** by each team; continue research; revise paper.	7/30	Oral presentation of **final paper** by each team; **submit** team paper; revise; submit final paper.

Exam: There will be *one* (multiple-choice + fill-in answer) exam on July 28, 2015, to check that the students have actually read Chaps 2-7. *No make-up exams will be given!*

Grades:

Item	Maximum grade
Oral presentation*	30%
Written Exam	20%
Final paper	40%
Participation	10%
total	100%

* Each student has to present *at least 3* oral presentations in class, with *at least one in ppt*; asking questions after other students' presentations is counted as an oral presentation. The top two grades of oral presentations will be chosen. You are welcome to give more than 3 if there is a chance to do so.

Important remarks:

- No cheating. If you cheat in any form, you will be dropped from class and receive a zero grade, and will be reported to your department.
- Extra credit will be given to students who volunteer and are chosen by the instructor to do extra work and present it in class.
- Debate and prepare to defend your own ideas! Be skeptical and critical to others' ideas!

- Be responsible! Be courteous!
- Starting July 14, presentation from each team at the *beginning of class* will have to be given with ppt.

5.2 How the class was conducted

Twenty nine students signed up for the course (maximum 30). Two came in one day and one week late, respectively, and were dropped immediately since each day's class is 3-hour long, amounting to 1/8 of the whole course. One dropped out at the end of the 2nd week because the chance to do volunteering work abroad came up, for which she applied before registering for the course. We ended up with 26 students (Fig. 5).

Fig. 5. Class portrait (July 30, 2015). The gentleman on the extreme left is Li Jian-Min, my teaching assistant and a first-year master's-degree student of the School of Philosophy, RUC. A student was absent in the last class, the day the picture was taken. Individual names of the 25 students in photo are given in Figs. 12-16.

On Day 1 (and every other day), we arranged the desks in a three-side rectangle so everyone could see the others' face (Fig. 6), and everyone was on a first-name basis. This is important to convey the idea that we are equal to each other, to encourage students to speak out freely—an essential part of good research. (It worked except they still called me Professor in class; only a few called me Lui after the course was finished.) To reinforce that, the instructor has to set the example by being willing to admit his own ignorance when that is the case (Fig. 7).

Fig. 6. Desks are arranged in a rectangle so everyone can see others' face.

Students were divided into five research teams by their own association. Each team has a "contact" person (not a leader) nominated by the team for me to contact. And the teams were treated as my own research teams working on five different projects under a big grant called "Understanding the World." Thus, there was no competition between the teams and we helped and learned from each other.

The students were all from the humanities (undergrads plus 4 grad students) except for 3 physics undergrads. No art students though, perhaps because art is considered not part of the humanities at RUC. I spread the 3 physics undergrads to three teams, to ensure those teams would have strong technical support. The teams then had group discussions (for about an hour), in classroom and in the corridor. Each team was asked to come up with a tentative team project and a team's name. They discussed enthusiastically among themselves and were very creative in generating names: Youth, Coal, TOB (Try Our Best), Pioneers, and Tornado—some sounding like sports teams.

Fig. 7. The instructor tries to figure out how to get the computer connected to the projector but fails, under the watchful eyes of students.

To round up the day, with time left, I decided on the spot to present my ppt on "Physics Innovation," a talk I gave at the Institute of Physics, Chinese Academy of Sciences, four days ago. (Only the first half of the talk was presented due to time limitation.) During the 3 and ½ hour class, we took brief breaks whenever I sensed the need for it. Throughout this course, copies of all ppt I presented in class were emailed to students and they were free to use them in any way they wanted.

Day 2, each team was given a copy of *The Beijing News* (Fig. 8), a daily selling for 1 yuan, and about 20 min to flip through it. They were asked to write down what could be potential research topics prompted by reading the daily, which covers local, national, international news as well as finance, entertainment and sports. The students were very smart; they divided the thick newspaper among the members. Each person's writing was then taped to the glass panels and doors (but not walls) in the corridor for everyone to read since this is the most efficient and cheap way to share information quickly (Fig. 9). Yet, after this exercise, no team wanted to change their research topic decided on Day 1.

Fig. 8. *The Beijing News*, July 9, 2015. Newspapers report on human happenings, thus a bountiful resource for inspiring research topics in the humanities—studies on humans.

During the 2nd and 3rd weeks, we started the class by hearing oral reports with ppt from *each* team. Every time the presenter was a different person. Yet, on July 28 when the draft of paper was presented, every team member had to get ready to do the presentation because I picked the speaker on the spot to present only one section of the paper until the whole paper was finished. And for the final presentation on July 30, only one person picked by the team presented the whole paper. After each oral presentation, questions from class followed; sometimes I called up individual students to do that (to ensure every student will speak in class at least 3 times in total). Each time the student's performance in giving presentation or asking questions, was graded with maximum 10 points. I showed them how to improve the quality of their ppt at the end, which included spelling, color, font, and style. As it turned out, ppt preparation was not a strong point of the students.

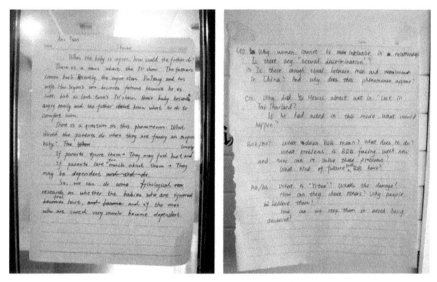

Fig. 9. Two samples of students' writing on questions that interested them, prompted by newspaper reading. Writing the name down is an option, which is the case on the left but not on the right display. Very interesting and insightful questions are raised.

I presented my Chapters 2-7 (see Fig. 3) with ppt to fill up the time, at the end of the day. Each presentation lasted 1 to 1 ½ hours. The ppt also served as examples of how good ppt could look like, which were made available to all students. The talks were not particularly long because the emphasis was not on the details (which students nowadays could find easily by themselves from the Internet) but on the proper understanding and relationship between the subjects covered. In fact, the only time these chapters were revisited was on the second last day when a 20-min written exam was administered. The exam consisted of 19 problems of multiple choices and fill-ins, which was just to check that the students had indeed read those 100 pages of Part I in textbook.

To show the students the capacity of Excel and so they should master it, and to introduce them to two simple but powerful stochastic models (with wide applications in the humanities and social science), viz., random walk and active walk [Lam 2005, 2006], volunteers were recruited to write Excel programs to demonstrate them. Each volunteer was given a few days to prepare before presenting it in class; both were given extra credits. Nathaniel Mitchell, an exchange math student from Warwick University, UK, took up the random walk model (4 extra points); Chen Xi-Zin from the physics department, the active walk model (5 extra points). The Excel programs were then distributed to all students with the hope that they will use or modify them in their future research.

Many students mistakenly think that they can find all the relevant literature from the Internet when researching on a specific topic and library visits are unnecessary. They are wrong for two reasons: (1) The Internet is never complete, but more importantly, (2) relevant articles may come from other topics, fields, or journals that are not directly related to the topic under study. To show this to the students, in the beginning of the 3rd week, 4 extra credits were given to each who would visit the university library for 3 or more consecutive hours and submit a 1-page report on what relevant journals and articles they found there. Eighteen students did just that. Selections from four reports:

1. Cao Shi-Peng, TOB team member working on car-sharing economy:

   Query paper [printed] periodical is more useful than electronic journal. When we are looking up the paper periodical, we found it is easy to find related articles. And also some articles cannot be found in the Internet. Before this experience, we thought related articles are very few, but after the experience, we found that there are a lot of related articles; these articles put forward all kinds of different points of view. In addition, [by] using the method of keyword search articles on the Internet is far from enough, because a lot of related articles can be used to analogy, which are hard to find for us through keyword search…

2. Chen Zi-Xin, TOB team member working on car-sharing economy:

   …The journal[s] I have read teach me a lot about how the theory of Internet plus work[s], and how those new companies use it. And in other articles, I learned a lot of information about the things we survey. I learned the related information in the field of laws. I learned the political boundary met by Uber. By those information, I think I can make our paper more focused and deeper.

3. Yuan Jie, TOB team member working on car-sharing economy (italics added):

   From this experience we truly collected many relevant information and also broaden our topic scope. And also from some Journals of law and policy we could know some new things and make them as example to consider our topic… Constrained by network…because [of] the keyword selection problem [maybe we] can't find much forefront academic information, while [in] some *magazines* and *newspaper[s]* [we] had found the key problems.

4. Wang Yu-Ting, Youth team member working on youth films:

   In RUC library, there are some journals…like *Movie World* or *World Screen*, [which] have more entertainment and are less technical… *Movie Art* is a serious journal…Then I only read *Movie Art* and learned somethings about Chinese cinema… [S]urprisingly, I found an essay which is so relevant to our research subject. The essay is "Creation Confusions behind the Big Bang of Youth Films" and the author is Fu Yu, a movie director and producer… This essay does make me think about our research. Firstly, it is proved that youth film is really one of focuses in cinema. It is not only because of this essay, but also [the] author mentions and quotes some other essays and papers about youth films. [Confirming to me] doing this research is valuable for us…

5.3 Extra days: research and paper writing

After two progress presentations, at the end of the second week, it became clear that the teams needed more guidance in their research. I offered to meet each group in campus during the weekend, July 17-19. Each meeting lasted 2-3 hours. Four teams met with me on Friday to Sunday at the Canaan Pizza and Steak Restaurant while the Pioneers team met at The Study Café on Sunday evening. The instructor paid for all the drinks, of course. Since it was a Sunday evening, when the session ended at 10 pm, I offered the Pioneers a tour in my Honda Fit. They picked four members to pack into my car and we drove counterclockwise the 3rd Ring (which took 1 hour), stopping at Hou Hai (Rear Sea). There we listened to music outside the bars and rode a motor tricycle by the lake. For some, that was the highlight of this course.

On July 28, after a 20-min written exam, a preview of the team papers was presented by each team. Every team member had to prepare for the whole paper since I called them up randomly to present one section each and the averaged grade was shared by the whole team. It was obvious that those papers were not good enough to get published. I therefore arranged a room, the conference room at the philosophy department, to be available for the whole day of July 29, Wednesday, from 8 am to midnight, and invited the teams to work there. That was the last push to get the papers finalized and ready to be presented the next day, the last day.

Everybody (except one) showed up, with one student coming in at 7:30 am and the last two left at 11 pm. Meals were called in from outside (each person paid her/his own). At the end, everyone was exhausted. We called that the "jam" session (Fig. 10). A major part of the jam session was to teach them to format the papers, including the proper way of inserting footnotes and diagrams and the proper style of references. I taught these to two teams, then asked them to teach it to the others. We helped each other and worked like a big research team, like a big family in fact. (The extra-day sessions were conducted in Chinese most of the time whenever everyone present could speak Chinese.)

At this point, the five papers were in publishable form. Mission accomplished. And it was time to decide whether to get them published in journals or not. The considerations were twofold: (1) Even though the papers were in publishable *form*, they still needed to be polished before submitting to journals. And that meant more work for me, the instructor. (2) To get the papers published and appeared in print in a few months' time, we have to submit them to journals that required page charges. That would amount to about 1,000 USD and who would pay for that?

Fig. 10. The jam session (July 29, 2015).

After much soul searching, and simply not to disappoint the eager students who worked *so* hard the whole month *and* to show the world that RUC students are world-class students, I decided to go ahead. To that end, each team was asked to provide a "publisher" who would work with me after the course. They were: Guo Yue (Youth), Yan Run-Yu (Coal), Chen Zi-Xin (TOB), Yan Xi (Pioneers), and Tian Yan (Tornado). The first three were chosen because they lived locally and could worked with me in person before I left Beijing later in August. This was an advantage since correcting a Chinese-authored manuscript without the author sitting by your side could be a nightmare, from my past experience. What happened afterward regarding paper publishing is described in Sec. 5.6.

5.4 Last day: paper presentation and party

On the last day, July 30, I offered to return the written exam to the students. They refused. They didn't want to see it again. The reason: Except for a few students who scored highly, the rest were just average (for lack of preparation time). We then started the final-paper presentations. Each paper was presented by one person of the team's choice, with the grade shared by the whole team (Fig. 11). They were professionally presented, followed by rigorous Q & A.

Fig. 11. Two final-paper presentations (July 30, 2015).

The papers' titles are:

1. *Youth*: The Youth Image in Chinese and American Youth Films.
2. *Coal*: Relationships between Mobile-Phone/Internet Usage and Socio-economic Development Level.
3. *TOB*: Sharing Economy Encountered Legal Quagmire: When Private Cars Entered the Taxi Market.
4. *Pioneers*: From Arranged Marriage to Autonomous Marriage: Marriage Liberalization in India, Ancient Rome, United Kingdom and China.
5. *Tornado*: Teachers' Awareness of Cross-Cultural Communication in Confucius Institute.

The authors (and members) from each team are identified in Figs. 12-16. Some papers' focus had been revised more than once after progress presentation and discussion. For example, for the sake of comparison, Didi was added to the Uber case in the TOB paper on car sharing.

As it turned out, the choice of topic for three papers had something to do with the major/background of the team members. For the TOB paper on the legality of car-sharing, one author was from the School of Law, another from the School of Economics. Two of the authors of the Tornado paper on teachers were majoring in education from the School of Liberal Arts. The Pioneers paper on marriage was more interesting. It was prompted by team discussion on marriage categories after a minority Tibet student talked about the phenomenon of one wife-two husbands she personally knew exists in the area of Tibet.[4]

It was time to party. My never-failing TA already had the drinks and food ready (the instructor paid, of course). And we had a ball (Fig. 17). Concurrently, I passed around a little black book and asked them to volunteer any feedback (see Sec. 5.5). Finally, it was time to say goodbye. No handshakes, no hugs—the Chinese style.

5.5 Feedback

I bought a little black book with the RUC logo from a small shop inside the campus the day before the last day (Fig. 18, left). The students wrote down their feedback there before their final grades were known—not the best way to get honest feedbacks, only second best. But it was practical. Still, every student wrote down something (some are in Chinese). From the way they wrote, one can easily pick out those that really came from the bottom of their hearts.

A selected few are presented in Figs. 18-20. A common theme of these remarks is that the instructor was "rigorous," very rigorous (Fig. 18, right). This reflects my insistence on every detail, from punctuations, spellings, to the font color in ppts and in debates (see Sec. 5.6 for more). Others were amazed that it was possible to learn so much in a month—this was because I was very efficient in managing time and drove them very hard, with extra days, say. Some students said the (forced) presentations helped them overcome their shyness in public speech. And a few said they would try to use Scimat in their future research. Additionally, a personal experience of the course was recounted in some detail by one of the students after the summer school [Luo 2017].

Fig. 12. The Youth team.

Fig. 13. The Coal team.

Fig. 14. The TOB team.

Fig. 15. The Pioneers team.

Fig. 16. The Tornado team.

Fig. 17. Party after paper presentation (July 30, 2015).

5.6 Post course: paper polishing and publishing

In August, I worked a couple of days with each of three publishers (Guo Yue, Chen Zi-Xin, and Yan Run-Yu) to refine their papers in Beijing. Papers nos. 1, 4 and 5 on youth films, marriage and teachers, respectively, were relatively simple and straightforward. Before submission, they needed only improvement on English, and footnote and reference stylings. Paper no. 3 on car sharing needed, in addition, clarifications and proper translation of legal terms. Paper no. 2 by the Coal team on phone usage was a different story. Yan Run-Yu, the publisher of this paper, and I spent a lot of time exchanging emails before mid-October to replot the diagrams (restoring dropped data points), ascertained the accuracy of information quoted, and found some new results in the process.

Eventually, papers nos. 1 and 2 were submitted to the *International Journal of Humanities and Social Science* (Fig. 21) in October; nos. 3 and 4, in November 2014. They were swiftly accepted and published by the journal (Figs. 22-25) [Guo et al. 2015; Jiao et al. 2015; Cao et al. 2016; Ciren et al. 2016]. The publication charge for each paper is 200 USD, paid by me first and reimbursed by RUC. Unfortunately, paper no. 5 (Appendix 1) was submitted in January 2016, which, due to delay of the journal, was never published. The reason was that we could not meet the technical deadline of claiming the publication charge from RUC, which was set to be mid-March 2016.

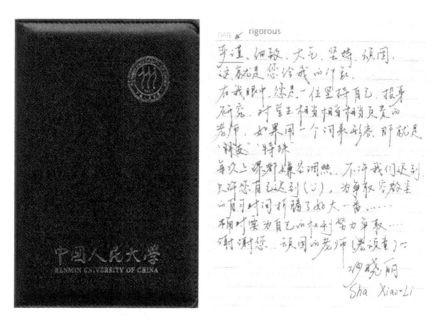

Fig. 18. Feedback 1 (July 30, 2015).

"With an apple, I will astonish Paris" To me, this class is like the apple which opens a new door during my college life. It's hard to imagine that I can learn so many things within just a few weeks. But it's true, I do have a special experience.

Firstly, I'm impressed with the new concept — "scimat". There are many misunderstandings about science. I think it's scimat that provides us with a new way and a correct way to define science and do research. I will try adapt some of the research methods mentioned in our class.

Secondly, my presentation skill is getting better thanks to our professor. For example, how to beautify my ppt and how to process my article on the computer.

Best wishes!

Xiao Fan

Thank you for your guidance and the new points you presented. The ideas of scimat and trans-disciplinary methods really inspired me a lot.

The experience and knowledge I have obtained will be useful in the future. And the idea of scimat interests me, perhaps it will be beneficial for my further study.

Best Wishes.

Yu-Xuan Xia

Fig. 19. Feedback 2 (July 30, 2015).

I haven't touched with physics for almost 5 years or more before this class. So thank you for give me another chance to experience the beauty of physics. And I also learned some skills about how to operate PPT and Word and Excel which I think it's helpful for my future study no matter in which field. Thank you!

Zhang Xi

July 30th 2015

Professor Lam:

Thank you very much for teaching me how to do professional research in English. When I first gave a presentation before all classmates, I was very nervous.

But today, when I stood there again, I felt good and confident. I think without this course, I will not have this amazing change.

Also, I think another thing I have learned in this class is on time. Doing research requires awareness of time. And I think through this process I have understood one of the significant thing in researching is keeping time sense.

Thank you for teaching me so much!

Tran Yan

July 30th. 2015

Fig. 20. Feedback 3 (July 30, 2015).

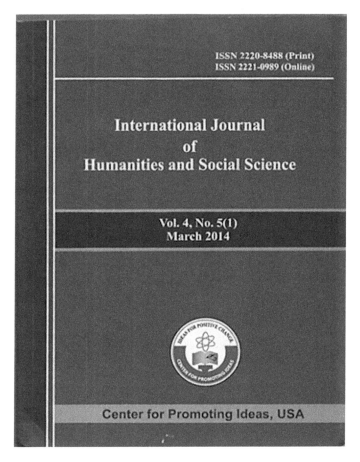

Fig. 21. The *International Journal of Humanities and Social Science*, which is published both online (ijhssnet.com) and in print, with ISSN numbers.

In October 2015, Guo Yue was invited to orally present her team's paper on youth films at the Fifth International Science Matters Conference on Interdisciplinary Education and Teaching in the 21st Century, October 28-30, 2015, Cascais, Portugal. Her expenses were paid by RUC. After the conference and when the four papers appeared online, the editor of the journal *The Pantaneto Forum* found them to be "exceptionally good" and reprinted them in their Issue 63, July 2016 (www.pantaneto.co.uk). Thus, in this area, the high quality of RUC students has been demonstrated and RUC has attained world-class level.

International Journal of Humanities and Social Science   *Vol. 5, No. 10; October 2015*

The Youth Image in Chinese and American Youth Films

Yue GUO
School of Journalism and Communication, Renmin University of China
Beijing 100872, China

Wen-Jing LIU
School of Sociology and Population Studies, Renmin University of China
Beijing 100872, China

Hong-Xian NIU
School of Science, Renmin University of China
Beijing 100872, China

Tong-Hui QU
School of Public Administration and Policy, Renmin University of China
Beijing 100872, China

Yu-Ting WANG
School of Labor and Personnel, Renmin University of China
Beijing 100872, China

Abstract

Youth film is a kind of film that is very popular all over the world, which has developed very fast in China in recent years. The audience, especially teenagers, set off a wave to watch this kind of movies. But after watching high box-office youth films screened in the last two or three years, we find that Chinese youth films have a tendency to be homogenous, and they have an obvious style different from that from the United States. Films can reflect culture, and youth culture is an important part of any culture. Consequently, it is extremely important to know what the differences between Chinese and American youth film are and why is that. We examine this problem from a specific angle—the youth image in youth films. Significant differences are found in this comparative study.

Keywords: Youth film, Youth image, Culture and society, China, USA

1. Introduction

After nearly a century of development, although not on a par with some developed countries in the film industry, Chinese films have received awards in important film festivals over the world in recent years. A group of highly ambitious filmmakers appeared, contributing strongly to the development of Chinese film industry. Youth culture can reflect the essence of social change, and the youth films are an important part of the youth culture. Thus, youth images in youth films can reflect the culture and the society.

Since the "reform-and-opening up" movement starting in 1978, Chinese society has changed a lot. People have experienced the rapid changes of the country and were influenced by a mixture of native and foreign cultures; they wavered between abandonment and acceptance of the changes. In recent years, Chinese youth-film directors have focused on the life of the broad masses of youth, which could have the effect of lessening the young audience's growth pain caused by society's upheavals. The films focus on and record growing-up confusion, contradictions and choices experienced by the youths.

According to statistics, China's current movie audiences 'average age has declined from 25.7 year in 2009 to 21.7 year in 2013[Liu&Liu,2013]. In recent years, youth films have attracted a large audience and been popular in China gradually but noticeably. American youth films have a much longer history than that in China and have a larger influence since it is distributed worldwide. In comparison with the American case, Chinese youth films are more homogenous both in the plots and in the youth images. It would be interesting to find out why.

100

Fig. 22. The 1st student paper published in *International Journal of Humanities and Social Science*, in October 2015. It is from the Youth team. The authors' names are arranged alphabetically. Note the various schools with which the students are affiliated.

International Journal of Humanities and Social Science _____ _Vol. 5, No. 10(1); October 2015_

Relationships between Mobile-Phone/Internet Usage and Socioeconomic Development Level

Tianyu JIAO
School of Finance, Renmin University of China
Beijing 100872, China

Nathaniel MITCHELL
Mathematics Institute, University of Warwick, Coventry
United Kingdom

Runyu YAN
School of Philosophy, Renmin University of China
Beijing 100872, China

Xin YANG
School of Information, Renmin University of China
Beijing 100872, China

Jingyi ZHONG
College of Art and Science, University of Washington, Seattle
Washington, USA

Abstract

With the rapid development of the smartphone market and smartphone technology, the smartphone has become a necessity for many people nowadays. More and more people develop a dependency on their smartphone. Potentially it could have a negative impact on many people's lives. Having noticed this apparent trend of people's addiction to the smartphone, we decided to explore the reasons why people use their phones so much, and it turned out that the overall socioeconomic development level (locally and nationally) has a huge impact on how people use their smartphones. In particular, in China's provinces, there is a positive (linear) dependence of the Internet usage on disposable income. For developed countries, a power law is found in the Zipf plot of smartphone usage while there is also a negative dependence of smartphone usage on the country's development level. These results are presented after a brief survey of previous studies. Finally, the positive effects of smartphone use in developing countries are noted.

Keywords: Mobile device, Economic development, Smartphone, Development level, Comparative research

1. Introduction

In recent years, in addition to traditional communication services like sending text messages and making phone calls, smartphones allow users to participate in popular online social networks like Facebook, WeChat, Line, and so on. With a smartphone, people are able to play a wide variety of games both online and offline, to stream movies and live sports, to take and share personal photos and videos, and to browse the Internet. The rapid growth of the Internet and mobile communication devices has caused significant effects, both positive and negative, on individual consumers and on the entire society. As more fancy functions are developed, it has become a global phenomenon that many people are finding it more and more difficult to stay away from their smartphone.

Researches including surveys (some have been posted on the web) have produced statistics and results concerning differences in smartphone usage patterns among various demographic groups (age, gender, and socioeconomic level)[Forgays et al, 2014; Sarraute et al, 2014; van Deursen et al, 2015; Pourrazavi et al, 2014; Hsiao & Chen, 2015; Li, 2014].

94

Fig. 23. The 2nd student paper published in _International Journal of Humanities and Social Science_, in October 2015. It is from the Coal team.

International Journal of Humanities and Social Science   *Vol. 6, No. 1; January 2016*

Sharing Economy Encountered Legal Quagmire: When Private Cars Entered the Taxi Market

Shi-Peng CAO
School of Law, Renmin University of China, Beijing 100872, China

Zi-Xin CHEN
Department of Physics, Renmin University of China, Beijing 100872, China

Rong-Ze MA
School of Economics, Renmin University of China, Beijing 100872, China

He YAO
School of Administration and Policy, Renmin University of China, Beijing 100872, China

Jie YUAN
School of Sociology and Population, Renmin University of China, Beijing 100872, China

Abstract

As society continues to advance, technology has spearheaded the transformation of the way we work, live and play. Sharing economy has been developed and spread gradually and then rapidly, but many problems have happened at the same time. The legal quagmire is an important issue; we will take the private cars entering the taxi market as an example to analyze this problem. By comparing Uber and Didi—a kind of "special car"(zhuanche in Chinese) service in Beijing, China—three major legal problems are analyzed: (1) the legal responsibility of the driver, car owner and the company; (2) the labor relationship between the company and the driver ;and (3) insurance issues. Moreover, changes of the Chinese laws governing the sharing economy of private cars in the market are proposed. Similarly, in anticipation of the law changes, modifications for the calling-car apps aiming to avoid the legal quagmire are also proposed.

Keyword: Sharing economy, Uber, Didi, Private car, Legal quagmire, Operation mode

Introduction

Sharing economy as a business model has a long history. In 1978, "collaborative consumption", a term coined by Felson and Spaeth [1978], emerged. In the mid-2000s, it evolved to "sharing economy", a new business mode inspired by enabling social technologies (e.g., Facebook and WeChat), global population growth and resource depletion. For example, one inspiration was thata privately owned car when shared with others will increase its value and decrease the waste of resources. That is, everyone can be both the owner and the consumer in a sharing economy.[1]

The sharing economy refers to economic and social systems that enable shared access to goods, services, data and talent[Felson & Spaeth, 1978] (Fig. 1). These systems take a variety of forms but all leverage information technology to empower individuals, corporations, non-profits and government with information that enables distribution, sharing and reuse of excess in goods and services. A common premise is that when information about goods is shared, the value of those goods increases, for the business, for individuals, and for the community.

---

[1] https://en.wikipedia.org/wiki/Sharing_economy#cite_note-11 (July 7, 2015).

145

Fig. 24. The 3rd student paper published in *International Journal of Humanities and Social Science*, in January 2016. It is from the TOB team.

118   *Lui Lam*

International Journal of Humanities and Social Science   *Vol. 6, No. 1; January 2016*

From Arranged Marriage to Autonomous Marriage: Marriage Liberalization in India, Ancient Rome, United Kingdom and China

Cuo-Mu CIREN
School of Business, Renmin University of China, Beijing 100872, China

Dan-Dan LIANG
School of History, Renmin University of China, Beijing 100872, China

Xiao-Fan LUO
School of Labour and Human Resources, Renmin University of China, Beijing 100872, China

Yu-Xuan XIA
School of Business, Renmin University of China, Beijing 100872, China

Xi YAN
School of History, Renmin University of China, Beijing 100872, China

Yu-Guang YANG
School of Economics, Renmin University of China, Beijing 100872, China

Abstract

The change of arranged marriage to autonomous marriage is analyzed and compared in four different countries: India, ancient Rome, China and United Kingdom. For India, the historical period covered is from the Aryan times to present; for China, from the Shang Dynasty to present; for ancient Rome, from 8th century BC to the Rome Empire; for United Kingdom, from 10th to 19th century. The kind of arranged marriage and evolutionary path are different in the four countries. Three major questions raised in our study are: (1) Why arranged marriage in four countries all existed in the early age in history; (2) why autonomous marriage (partially in India) came into being later on in all the countries; (3) why the speed of this process differs from country to country. The answer to the first two questions is found to be this: It is mainly economical. However, the speed of this process is influenced mainly by cultural and political factors. By not limiting our study to fixed historical periods, unlike the case in previous studies, and by comparing four different countries from Asia and Europe, our study looks at marriage change globally which involves the change of marital laws and sociological factors.

Keyword: Arranged marriage, Autonomous marriage, Marriage liberalization, Marital law.

Introduction

Marriages of different countries or ethnic groups had been studied separately as the showcase of the society and family life with limited scope (see below). And economic, cultural and political factors contributed to the development of marriages. In this paper, we focus on these factors to find the different mechanisms governing marriage changes in different countries. Specifically, we study the change from arranged marriage to autonomous marriage: their different patterns and characteristics in different countries.

According to O'Brien [2008, pp. 40-42], arranged marriage means a type of marital union where the bride and groom are selected by a third party rather than by each other, which, in most cases, are their parents. Our paper is limited to the case that a young person's spouse is selected by the person's parents or relatives rather than by themselves. Autonomous marriage means young people can choose their own wives/husbands by their own will, mostly based on love. The motivation behind our study is described in the following.

114

Fig. 25. The 4th student paper published in *International Journal of Humanities and Social Science*, in January 2016. It is from the Pioneers team.

6 The HuSS Course in 2016: Mixed Teaching

In July of 2016, the HuSS course was taught again at RUC, with the name changed slightly to "Humanities, Art, Science." The idea was to attract students from the arts but it was not successful. Out of the 26 students who showed up in the first class, there was no art student. A major difference between HuSS in 2016 and that of last year is that in 2016, all students are first and second year undergrads while those in 2015 include some seniors and grad students. Some knew about the fame of the course from their classmates who attended last year.

6.1 Course and schedule

After the first class and knowing the demands of the course, some students dropped out. The class ended up with 20 students (Fig. 26). In contrast with the case of 2015, students were given two choices: (1) they could form research teams to work on a paper, or (2) they could each pick a section from the textbook, teach it in class and write a paper on that topic. After a warning that the first choice would be very time consuming, nine students decided to form two research teams and the rest opted for sectional teaching—the mixed-teaching mode. Accordingly, the class schedule was designed differently (Table 1).

The "Aim High" team of 4 students worked on "Climate change: Decision making under intrinsic uncertainty." The "High Five" team of 5 students worked on "Dream," a survey of dream analysis from the angle of humanity, art, physiology, and psychology. The quality of the two team papers was not as high as that of last years' because the students this time were first and second year undergrads only. Note that for each chapter, even with students teaching some of the sessions, the instructor still had to teach the introductory and concluding sessions, giving an overview and picking up where the students left off.

6.2 "Under the tree" office hours

This time, office hours that run one to two hours starting 3 pm on every Wednesday and Friday were introduced, which were conducted in Chinese since all students spoke the language. They were used regularly by the two teams and also by other students. I offered each one who showed up 2 extra credits for each visit. But that turned out to be unnecessary since they were motivated. RUC, like most Chinese universities, does not provide professors with individual offices. For lack of a proper place, I conducted the office hours *outdoor* near the Teaching Building #4 where our classroom was located. They were thus called "under the tree" office hours (Figs. 27 and 28). We even held it with an umbrella in hand when it rained. A lot of fun.

Table 1. Class schedule of the summer course "Humanities, Art, Science" (2016). Note that the written exam are split into two this time.

Tuesday		Thursday	
7/5	First lecture (some form teams, discuss possible topics; others pick topics). Instructor reviews HuSS course of last summer.	7/7	**Teams**: oral progress report with ppt; finalize team projects. **All** start research. Instructor talks on research and innovation. **TA** teaches Chap 2.
7/12	**Teams**: oral progress report with ppt. **All** continue research. Instructor teaches Chaps 3-4.	7/14	Written **Exam 1** (Chaps 2-4). **Teams**: oral progress report with ppt. **All** continue research. **Instructor/3 students** teach Chap 5 (History).
7/19	**Teams**: oral progress report with ppt. **All** continue research. **Instructor/4 students** teach Chap 6 (Arts).	7/21	**Teams**: oral progress report with ppt. **All** continue research. **Instructor/3 students** teach Chap 7 (Philosophy).
7/26	**Teams**: oral report with ppt. on **draft of paper. All** continue research; revise paper. **Instructor/2 students** teach Chap 7 (Philosophy).	7/28	Written **Exam 2** (Chaps 5-7). **Teams**: oral report with ppt. on final paper. **All:** submit final paper, ppt and collected e-materials in an e-folder.

In addition to regular office hours, I still met several times on the weekends with the Aim High team since their topic on climate change was quite complex and was of direct interest to me (see [Lam 2014: 82-83]). On two of these occasions, I tried to pass to them more systematically some of my personal experiences in doing research. They took very good notes so I asked them to write it up which was then distributed to the whole class (Appendices 2 and 3).

On July 27, Wednesday, the whole day before the last day, like last year, a jam session was held in a conference room of the philosophy department.

The two teams and a few other students came in to get help in improving the writing of their papers.

Fig. 26. Class portrait (July 28, 2016). The photo of 17 students was taken at the end of the farewell dinner at the campus' Huixian Restaurant. Three students could not attend the dinner and are absent here.

6.3 Take home message

On July 28 the last day, after a brief written Exam 2, each team presented their final paper orally by one person (about 40 min). Qi Wei-Jie, a physics undergrad and volunteer, gave an introduction of random walk with Excel programs. Then every team/individual presenter submitted to me their final paper, ppt and collected e-materials in an e-folder. These materials will help the instructor in his next teaching of the course. I ended the course with 3 slides of "take home message" (Fig. 29). Instead of a farewell party, we had a farewell dinner at the Huixian Restaurant in campus. The instructor paid, of course. No hugs, no handshakes, just a photo (Fig. 26). Later, a few students joined me in a car ride for a counterclockwise round on the 3rd Ring. It rained and we didn't stop at Hou Hai.

Fig. 27. An "under the tree" office-hour visit (July 13, 2016). The left 3 students belonged to the High Five team (the two team members absent here are Wei Ying-Ying and Xu Ying-Qi). Not shown here is the instructor's small, blue, plastic, folding low stool, located in front of the students and bought with 40 yuan. We named the nearest tree behind students "the Scimat tree."

7 Discussion

In principle, I could teach the course like a traditional course, covering the core Part I and supplemented it with some topics from the rest of the textbook—what I call the "closed-teaching" mode. The course would still be interesting and worthwhile for the students. I decided not to do this on these two occasions for two reasons:

1. A GE course should not be about details but more on the basic concepts and connections between different topics/disciplines so that the students (from all disciplines) could apply them to their

future studies. If so, I would be left with plenty of time in the course since it would not take long to teach the core Part I; then I would have to fill up the time by teaching the less-important topics from the rest of the textbook (see Fig. 3).

Fig. 28. The Aim High team "under the tree" (July 17, 2016).

2. More importantly, educators have been urging undergrads to participate in research as early in their careers as possible, to find out how real research looks like and to be trained in handling real-world problems. For a small number of students, this is achieved by joining the professors' research part time and/or full time in summer, in their upper-division years. Even for these lucky ones (more in natural science than the humanities), writing a paper is usually not part of the training. Therefore, while teaching the core Part I, using the HuSS course to train undergrads to do real research, presentation, and paper writing becomes a viable and excellent

option, especially in places where research and English paper writing in the humanities are less developed.

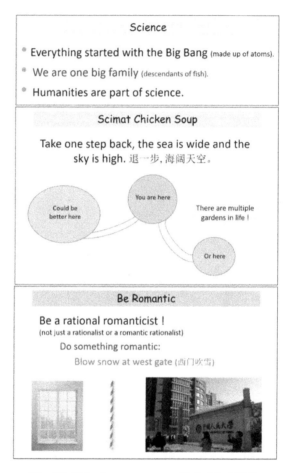

Fig. 29. Take home message (July 28, 2016).

While the open-teaching mode (Sec. 5) of conducting the HuSS course, with all students working in research teams, is most exciting and rewarding, for both the students and instructor, it is also very exhausting for both, especially when the course is conducted in the short span of one month. In comparison, the mixed-teaching mode (Sec. 6), with fewer research teams, is more relaxing. The drawback is that only a few motivated ones, not all students, would gain the experience of going through the rigorous training of doing research.

Finally, it should be noted that there exists a fourth mode of teaching—the "all-teaching mode," in which *each* student will teach a section and write up a paper, no research teams. This is most relaxing for the instructor, of course, while students do pick up something in the research method.

8 Conclusion

This HuSS course is important for everyone on earth for the following reasons. For *students*, through the teaching of the core Part I,

1. We try to impart to them the two basic scientific knowledge (big bang and common fish ancestor) which are essential in shaping their worldviews, leading, hopefully, to a fruitful and happy life.

2. By introducing the correct definition of science, we hope to clarify to them the perplexing issues on science and religion, science and art, etc.

3. By clarifying the relationships between the humanities (history, philosophy, and arts, in particular) and science, we hope to help them in choosing courses and majors in their student years that really suit their interests.

For *teachers* and everyone else, we note that

1. This is the only GE course that clarifies the connection between all the disciplines in a university, from a historical and unified perspective.

2. This is the only course that systematically teaches all undergrads to do research by really doing it, from picking topics to publishing papers.

3. The course provides the basic understanding about where we come from and why we are what we are, which should be shared by every living human being (in particular, the political leaders, present and future) in this world.

4. This course is cross-cultural and interdisciplinary, taught with everything recommended by the experts on how it should be done.

5. With a textbook available (from the author), this GE course can be taught by *any* instructor in *any* university worldwide.

Appendix 1: Abstract of the Unpublished Student Paper

This unpublished paper is from the Tornado team of 2015.

Teachers' Awareness of Cross-Cultural Communication in Confucius Institute

Ze HU [1], *Xiao-Li SHA* [2], *Yan TIAN* [3], *Lei YANG* [4] *and Xi ZHANG* [4]

[1] School of Science, Renmin University of China, Beijing 100872, China
[2] School of International Studies, Renmin University of China, Beijing 100872, China
[3] School of Philosophy, Renmin University of China, Beijing 100872, China
[4] School of Liberal Arts, Renmin University of China,

Abstract

Confucius Institute (CI) is a non-profit educational organization affiliated with the Ministry of Education of the People's Republic of China. The Institute's aim is to promote Chinese language and culture, support local Chinese teaching internationally, and facilitate cultural exchanges. Needless to say, Chinese teachers play an important role in the CI. Accordingly, this paper focuses on the cross-cultural awareness of the Chinese language teachers in the CI when they are first in touch with another culture. It is well known that difficulty in communication could occur between any two persons, even if they grow up in the same family and share the same culture for a long time. Thus, it is not hard to imagine what difficult times a Chinese teacher would go through when this person teaches Chinese language and culture in the CI abroad in an unfamiliar environment far from home. In this paper, we investigate what kinds of problem these Chinese teachers will encounter, why they experience these problems, and how they can overcome them. Our results are obtained by analyzing returned questionnaires we prepared ourselves and through interviews with the CI teachers who have worked abroad. Suggestions for the future are also given.

Appendix 2: On Literature Search and English Learning

Literature Search and English Learning

Aim High

On Saturday 7/9/16, three of us (Kang, Liu, Wu) met with Lui, our instructor, under the tree. Here is what we learned.

Literature search

Search engines

Two search engines are most helpful in literature search: Google Scholar and Bing Academic. It would be better if we were able to log in to Google Scholar. If we can't, Bing is a "good" substitute for it. The former is product of Google and the latter is from Microsoft, both are typical search engines. Baidu Scholar is inappropriate for our search because it's not scientific, rigorous and comprehensive enough.

Type in "bing" on a common search engine, and we will find the entry of Bing. What we need is English literature, so we should type in key words in English and choose Academic. For example, our topic is "decision making under uncertainty" and a typical situation is climate change, so we type in "decision making under uncertainty" and search for relevant information in Bing first. Furthermore, we add "climate change" into the key words for more information about this case. The search results in Bing cover all of the key words, while that in Baidu sometimes just cover some of the key words and are mixed up with some less relevant information. That's why Baidu is not recommended.

Choice of literature

Scanning the pages of search results, we can easily obtain a number of relevant documents. (If we cannot download for lack of authorization, we may go to the library for help, which will be discussed below.) After we preliminarily decide on our topic, what we need to do is read as much relevant literature as possible to grasp core concepts and make our research approach clear. For example, our topic is "decision making under uncertainty," so we need to know how many types of uncertainty and how many types of decision making people come up with and cover them when we are giving a review. Specifically, the latest *PhD theses* and *reviews* (both from Europe or America) are good material helping us to have a basic understanding of research status in this area and get much useful information about this topic. Besides, latest papers contain latest research progress and may inspire our thinking. The publisher of a journal and the author's affiliation are important reference factors. For example, "uncertainty of climate change" is more of an atmosphere problem than a geography problem, which is a reference factor in literature choosing, and English literature is more valuable than Chinese literature generally. *Wikipedia* in English is also a good start (it has same level of accuracy compared to *Encyclopedia*).

Review

When doing a literature review, the most important thing is reviewing what has been done about the topic. For example, our topic is "decision making under uncertainty" and climate change is a perfect example we decide to use; we are able to get a lot of literature including these key words. As noted above, while doing review, we should tell how many types of uncertainty and how many types of decision making under uncertainty are known. Moreover, when it comes to special cases, such as climate change, we should give more description of it and give specific explanation in terms of the topic. Those works will influence and guide our further research.

Library search

After discussion "under the tree," the three of us followed Lui to the library to search the resource we need. Here is what we did.

First of all, after picking out our topic in the library's computer, we put the key words of it into Bing and get many scholar papers. Then we looked through the papers to find out which journals they came from. And the journals would be our main aims.

Second, we consulted the database of school library for the journals. Unfortunately, the result is none. Therefore, there was nothing else to do but walking and looking though the shelves in the area of Chinese and foreign journals.

Once we saw and thought the title/contents of the shelf is relevant to our topic, we checked carefully of it. Looking though the contents of the journals, we found some related articles at last.

The last thing we need to do is to take a picture of the related journals and articles so that we can find more information of them online and print the papers. You can choose to print it in the library, too. Ask for help from the helpdesk if you need it.

English improvement

English learning can be divided into four parts: listening, speaking, reading, and writing. Lui thinks that there is no rush for us to make the first two parts perfect, as we will only need them badly when we study abroad. The best way to improve one's speaking skill is communicating with native-speakers all the time, and this is the reason why Lui asks us not to spend a lot of time hanging around with Chinese when we live abroad and have a chance to talk to foreigners.

As to reading and writing, different from the popular belief, Lui said that it was not so helpful for us to read too much. In other words, intensive reading is more important than extensive reading, especially when you want to improve your writing skill or the grammar. We were suggested to buy the thinnest grammar book. Then he recommended several newspapers to us: *New York Times*, *Los Angeles Times*, *Washington Post*, etc. *New York Times* is the best, and we should focus on the editorial, read slowly to learn how their words are organized, and more importantly, to learn the critical thinking by noticing how the writers try to convince you. The news can be ignored because they are usually written in a hurry. Our aim is not the content but the style of writing. When reading an article, we ask ourselves: What and how I would write if I was the writer? This is actually a process of learning. And we can learn how to write the last sentence of a piece which should be brief and forceful to express attitude and feeling.

Two or three paragraphs per day will be enough for us, and the point is perseverance. I remembered that Lui once said: "English is always a second language for us". So the key to perseverance, personally, might be the confidence and self-encouragement. We may not be as good as a native speaker, but we can become better.

Advice on life

Towards the end of the talking, Lui mentioned that we students should value our time, try to live simply and focus on study. To be more specific, he advised that we should avoid getting addicted to computer games (as well as TV entertainments and so on) and spend less time buying things on Taobao or Tianmao. The philosophy of Lui Lam is "Never buy anything unless you have to." And try to save time by reducing the need for decisions. For example, he has ten pairs of socks which are exactly the same, so he doesn't have to worry about choosing which socks to put on daily. Then he added: By the way, it is different if you are a girl who wants to find a boyfriend.

Note added by Lam (Sept. 24, 2016). As literature search is concerned, it turns out that sometimes, the latest research is found in *magazine* (rather than journal) articles. An example is the BBC article "What we learnt from reading people's dreams" by Chris Baraniuk [2016], which contains important, new developments on dream research not noted by the High Five team in their extensive literature search.

Appendix 3: On Preparing Notes before Writing Papers

Preparing Complete Notes before Writing a Paper

Aim High

On Sunday 7/17/16, the whole team met with Lui, our instructor, under the tree. Here is what we learned.

Here is an important step in *doing* a paper before you sit down and want to *write* the draft of a paper. Before writing the paper, you'd better prepare a complete "notes" detailing everything you have been doing. Of course, this step could and should be done when you are near the end of the research period. The notes don't ask for complete sentences (better not, just like in a ppt) but should *contain all details*. Write the notes in the format of a Master or PhD thesis, without complete sentences though, putting all details in appendices. (You write complete sentences in draft of the paper, of course.)

For writing up a paper, the notes are made up of five main sections:

1. The 1st section is Introduction. You should introduce briefly what people have published, why you choose the topic, what the paper talks about, and the meaning of your research.

2. The 2nd section is previous studies in some detail, i.e., background—with your own comments. You may use a table to summarize them since it is just notes.

3. The 3rd section is the Scimat perspective if you are doing an individual paper. For a team, you start with your team's research. This section can have subsections or followed by other sections on your research results.

4. The 4th section is discussion and conclusion, ending with take home message. Consult the four published student papers of last year.

5. Last but not least, the reference part. Once you quote others' ideas, words, data, experiment results and anything belonging to others, you need to give references *in text*. Otherwise, the reader will understand it to be your own contribution and you just commit the "crime" (worth than crime, in fact, because you are polluting the literature in print once published) of *plagiarism*. So, check carefully that every reference quoted in text is in the reference list, vice versa. The reference must be complete, including the title, author, and journal/book name (in former, volume and issue number, page numbers from beginning to end, too; in latter, publisher and city). Whenever possible, always try to quote the first paper that proposed it—giving proper credit to the originator; otherwise, quote a paper/book that mentions it and say so (e.g., …as reported in…).

Note that every researcher who publishes has a *professional reputation* that follow her/him for the rest of this person's career/life. People won't tell you in the face, but your colleagues know and will affect your employment and funding prospects, too.

You should *keep the notes for the rest of your life*. Reasons: You will need it to further the work or when challenged 30 years later or sooner by your fellow professionals.

Notes

1. "Science" (with quotation marks) means "natural science," the study of nonhuman systems.

2. Earl McGrath was commissioner of education under Presidents Truman and Eisenhower, and the executive director of the Institute of Higher Education at Columbia University. Gary Miller was executive director of the International University Consortium at University of Maryland.

3. In the text, unless specified otherwise, Chinese names are written with family name first.

4. This kind of marriage is depicted in the recent movie *De Lan* which won the Golden Goblet Award at the Shanghai International Film Festival 2016 [Zhu 2016].

References

Baraniuk, C. [2016]. What we learnt from reading people's dreams. http://www.bbc.com/future/story/20160727-what-we-learnt-from-reading-peoples-dreams (Aug. 9, 2016).

Burguete, M. & Lam, L. [2016]. The Scimat Program: Towards a better humanity. *Humanities as Science Matters: History, Philosophy and Arts*, Burguete, M. & Riesch, H. (eds.). Luton, UK: Pantaneto Press.

Cao, S.-P., Chen, Z.-X., Ma, R.-Z., Yao, H. & Yuan, J. [2016]. Sharing economy encountered legal quagmire: When private cars entered the taxi market. International Journal of Humanities and Social Science 6(1): 145-154.

Ciren, C.-M., Liang, D.-D., Luo, X.-F., Xia, Y.-X., Yan, X. & Yang, Y.-G. [2016]. From arranged marriage to autonomous marriage: Marriage liberalization in India, Ancient Rome, United Kingdom and China. International Journal of Humanities and Social Science 6(1): 114-120.

Guo, Y., Liu, W.-J., Niu, H.-X., Qu, T.-H. & Wang, Y.-T. [2015]. The youth image in Chinese and American youth films. International Journal of Humanities and Social Science 5(10): 100-105.

Jiao, T., Mitchell, N., Yan, R., Yang, X. & Zhong, J,-Y. [2015]. Relationships between mobile-phone/Internet usage and socioeconomic development level. International Journal of Humanities and Social Science **5**(10(1)): 94-103.

Lam, L. [2008a]. Science Matters: A unified perspective. *Science Matters: Humanities as Complex Systems*, Burguete, M. & Lam, L. (eds.). Singapore: World Scientific. pp 1-38.

Lam, L. [2008b]. Science Matters: The newest and biggest transdicipline. *China Interdisciplinary Science*, Vol. 2, Liu Zhong-Lin (刘仲林) (ed.). Beijing: Science Press. pp 1-7.

Lam, L. [2014]. About science 1: Basics—knowledge, Nature, science and scimat. *All About Science: Philosophy, History, Sociology & Communication*, Burguete, M. & Lam, L. (eds.). Singapore: World Scientific. pp 1-49.

Luo, Xiao-Fan [2017]. The HuSS course: A student's experience. *Interdisciplinarity and General Education in the 21st Century*, Burguete, M. & Connerade, J.-P. (eds.). Cascais, Portugal: Science Matters Press.

McGrath, E. J. [1976]. *General Education and the Plight of Modern Ma*n. Indianapolis: The Lilly Endowment.

Miller, G. E. [1988]. *The Meaning of General Education: The Emergence of a Curriculum Paradigm*. New York: Teachers College, Columbia University.

Zhang, Dong-Hui [2017]. Tongshi education in China: A new bottle with old wine or an old bottle with new wine? *Interdisciplinarity and General Education in the 21st Century*, Burguete, M. & Connerade, J.-P. (eds.). Cascais, Portugal: Science Matters Press.

Zhang, Dong-Hui & Xuan Tian-Ying [2017]. Cultivating innovative talents among the next-generation Chinese students: A case study of Renmin University of China. *Interdisciplinarity and General Education in the 21st Century*, Burguete, M. & Connerade, J.-P. (eds.). Cascais, Portugal: Science Matters Press.

Zhu, Xiao-Jie [2016]. Don't dare to simply imagine how they live: Golden Goblet Award movie *De Lan* films the Tibetan area. *Southern Weekly*, June 23, p F28.

Published: Lam, L. [2017]. Humanitles, Science, Scimat: A new general-education course. *Interdisciplinarity and General Education in the 21st Century*, Burguete, M. & Connerade, J.-P. (eds.). Cascais, Portugal: Science Matters Press.

The HuSS Course: A Student's Experience

Xiao-Fan Luo

As a freshman from Renmin University of China, I enrolled in the "Humanities, Science, Scimat" (HuSS) course taught by Professor Lui Lam in the International Summer School 2015. The course had 26 students and total class time of 24 hours, run in July. Introduced from the perspective of Scimat, the course opened the door of a trans-disciplinary and cross-cultural academic world for me. It also influenced me a lot on my ways towards pursuing knowledge and doing research. Here, my learning experience in the course is recounted in the form of a diary.

1 Introduction

The HuSS course was conducted by Professor Lui Lam using the "open-teaching" mode, which consisted of two major parts: lectures and research projects. In addition, seminars, progress reports, and English-writing training all served as important methods of instruction. Students of any major at any level could benefit from this teaching mode and participate actively in the class [Lam 2017].

The combination of lecture and project was very educational because the instructor and the students were connected so closely that the knowledge obtained from this class were well adapted to research process. For one thing, the instructor would introduce the proper relationships between the humanities and science in his lectures. For another, we students had to form teams to work on a certain project and do research under the guidance of the instructor.

If you are a beginner in learning Scimat [Lam 2014], I wish my comments about the class could help you better understand what you could obtain from this course. Here, I must make it clear that all of my comments are based on personal recognition on this course and it's you, the reader, who will decide whether these comments could be a useful reference or not. Another thing I have to emphasis on is that my comments are determined by the learning environment that I live in. In China, we speak Chinese as our native language. Although English is a compulsory course at school, most students' English skills still stay at the medium level after graduation. As for me, English writing and speaking are all challenging tasks. Luckily, in the HuSS class, we were required to use only English when doing presentation, writing the essay and so on. Therefore, apart from learning the Scimat perspective, I improved my English language skills after four-week training at the same time.

2 Course Dairy

The regular class sessions are on Tuesdays and Thursdays. For our team, three extra days (July 19, 20 and 29) relating to the writing of the team essay were added in, which resulted in a paper published after the summer school [Ciren et al. 2016].

Day 1 (July 7, Tuesday)

Today is my first day at the summer school. After finding myself a seat, I keep watching around like an excited rabbit until my instructor—Professor Lam—enters the classroom. I am surprised by the professor's neat and simple wearing style which gives me a deep impression.

Class starts. Professor gives us a personal introduction. He is quite an experienced scholar and his education background at Columbia University impresses me. Then, he talks about the HuSS course and the teaching plan.

From 7/7 to 7/30, lectures about HuSS, project study, English oral presentation, paper writing. What a challenging task! However, I love it and I realize that it will be a "full" summer.

To conduct the project study, all the students are required to form research teams and work together. We have 10 min to find team members. The whole class was booming during the team forming process. We talk a lot and try to know fellow students in order to seek the best match and pull together a strong research team. Luckily, I find 5 nice persons. They are Yan Xi and Liang Dan-Dan from the School of History, Xia Yu-Xuan and Ciren Cuo-Mu from the School of Business, Yang Yu-Guang from the School of Economics.[1] Well, a combination of history, business, economy and labor science. Interesting!

All the teams are formed. It's time to choose your research topic for your team! Our team brainstorms for about an hour and we think about various issues such economy, politics, culture and so on. Then Xia Yu-Xuan talks about the nationwide legalization of same-sex marriage in the USA, which leads to our discussion about marriage. Ciren Cuo-Mu then tells us about polyandry, a rare form of marriage in the world, prevalent for a very long history period in her hometown—Tibet. It is very interesting for polyandry means one wife could have many husbands while there is an opposite marriage type—polygyny and it is well-known for Mormonism. We find that the topic of marriage is quite worth of our research and we could dig deep to touch the rare-touched field. Therefore, after the brainstorm we add marriage into our tentative list.

All back to the classroom. Each team announces its potential research topics and professor gives his remarks. Professor evaluates the topics we are interested in and emphasize on relevance, timeliness and forward-looking. It seems that he agrees that we might find something special in studying marriage. He speaks highly of the topic about "Uber" from another team. Uber generates a hot debate among Chinese society and its combat with Didi, a local Chinese company who also provides tailored taxi service, which is causing serious concern. But professor announces that we could wait until next class to finalize the research topic.

In the first half of the class, we learn about the class, the teacher and the students. In the second half of the class, professor gives a lecture on innovation. He points out the importance of innovation at the very beginning and talks about barriers on China's way towards creative works, which leads to my worry about China's future of scientific innovation. I have to admit the whole course is overwhelming and I understand that from now on, I will have a busy and booming summer!

Day 2 (July 9, Thursday)

Class starts. We are shocked when professor hands out the fresh, today's *The Beijing News*. Newspaper? A pile of papers. Yes, that's the secret to topic choosing. "Find out issues you're interested in and share your ideas 15 minutes later." Fifteen minutes later, after going through the newspaper at hand, we put up what we write about the interesting issues and our remarks at the corridor for everyone to read and discuss.

Back to classroom. Our team has a thorough discussion and agrees that marriage will be our team project. It marks that we now truly start a research project! Every team put forward their team project in the class. And no team wants to change their project after reading the newspaper. Oh, it's really interesting for we have a colorful bunch of projects all together, ranging from shared economy to culture differences.

Professor then gives us some basic knowledge about (research) paper writing. What should the structure of a paper be like? Actually, I am afraid of paper writing for we will have to hand in a final paper written in English while I have never written that in English before. However, I relief myself and tell myself that you can do it!

Why should we avoid plagiarism? It shocks me when he emphasizes on the importance of properly quoting original works. The reason is that in China, most people are not aware of some quoting rules in their writing, and end up committing plagiarism in reality.

After that, professor leaves us enough time to do some research and choose our team's coordinator, or the "contact" person. We make full use of the time and talk about what we should do next. We reach an agreement that since we are at the beginning, we have to do enough reading to find information and get inspired. And we elect Cuo-Mu as our coordinator. Cuo-Mu is a nice girl from Tibet and is willing to be our coordinator.

Day 3 (July 14, Tuesday)

It's the second week and today each team should give an oral presentation. Yu-Guang is going to present our progress. Before that, our team met on July 12 and had a job division. Since we want to have a broad view of marriage internationally, our team is divided into 5 sub-teams studying four different regions. Yu-Guang will focus on marriage system in India. Yu-Xuan will pay attention to marriage in Ancient Rome. Dan-Dan and Xi work on that in China. Cuo-Mu does research mainly about Tibet. I will focus on that in England.

Yu-Guang does a presentation about our work and introduces this division of labor. I think he is a good presenter. Every time someone finishes his or her presentation we have a chance to challenge this person and have a discussion in the class. After all the teams finish the presentation, I am impressed by all of them and I can tell that some teams are really strong.

At the second half of the class, professor gives us a meaningful lecture which changes my recognition on science. Before attending this course, I regarded science as natural science, and it should include math, physics and some other disciplines studying nature. Professor changes my opinion and helps me realize that I was wrong. By starting with misunderstandings on the humanities and science, he points out a proper definition of science is needed. Then he looks back to the birth of disciplines and birth of science and finally, he puts forward the definition of science—science is humans' pursuit of knowledge about things in nature without bringing in God or any supernatural. In the world of Scimat, we study the humanities, social science and medical science.

Also, professor introduces his Scimat Program. This is really a thought-provoking lesson.

Day 4 (July 16, Thursday)

Continue researching! The topic marriage is too broad for us to study so the next step is to narrow down the subject and make our research down to earth. In order to fulfil this goal, our team begins to do literature review and want to go further after collecting enough information. We hope to find something worthy researching when going through the history of marriage systems.

Well, still, oral presentation. We have Cuo-Mu to introduce her part of research. She is expert at marriage system in Tibet for she is from Tibet and she shows some first-hand information. All the teams seem working well. One team even shows us the statistics they collected. We have to speed up.

And, today, the lecture is about history. I love this part. History is my favorite subject in school and I love reading history books in my spare time. Again, professor changes my view of history. History research can be done very scientifically! In the lecture, professor shows how physics can help history research. For example, we can make use of random walk and active walk. Then he gives us cases of three research levels: empirical level, phenomenological level and bottom-up level. Just as he says, there is much more in history than pure narratives.

Day 5 (July 19, Sunday)

We meet with professor at a café on July 19. We talk about our progress face to face; professor gives us some advice. He mentions that we can see marriage from the view of biology. There could be some biological reason behind marriage—the intimate bond between a couple. After that, he invites us to have a ride in his car touring Beijing city. That is really a wonderful experience. We enjoy the beautiful city light, music of Hou Hai and deep, profound talk.

Day 6 (July 20, Monday)

Our team meets again. And at this meeting, I put forward my new findings. When I was reading papers and surfing online, arranged marriage appealed to me. Arranged marriage means a type of marital union where the bride and groom are selected by a third party rather than by each other [O'Brien 2008: 40-42]. It first came to me when I was reading something about the marriage system in British history. That reminded me of the arranged marriage in ancient China. Arranged marriage had a long history in Chinese feudal history and marriage was not liberalized until the Opium War. Years of war introduced new production modes and Western ideas, especially individualism, which crushed the traditional big family of patriarchy, and arranged marriage began to collapse [Xu 1961].

Similarly, arranged marriage was once a common thing in the whole British society. During the 16th century to 19th century, British society had an obvious stratification phenomenon. In the upper class, if one wanted to inherit his or her parents' wealth and status, then this individual must obey them in choosing a spouse [Stone 1977]. It stroked me when I found that such two different countries, China and England, both playing an important role in the world history, could have similarities in the marriage system.

Why do these two countries once have arranged marriage? Do other countries in the world have arranged marriage in their history, too? Why does arranged marriage collapse and autonomous marriage emerge? I put forward all of these questions and propose further studying marriage liberalization to my fellow team members. It turns out that all of us are interested in this subject so we decide to focus on analyzing marriage liberalization—the development from arranged marriage to autonomous marriage.

Day 7 (July 21, Tuesday)

It's my turn to do oral presentation. I present my part of research on England and cite some important facts from the book *The Family, Sex and Marriage in England* [Stone 1977]. I am glad that I overcome my nervousness in front of the whole class and trying to inform and persuade my fellow students really gives me a feeling of achievement.

Professor Lam is an expert at using power point and teaches us how to make our ppt concise and informative. For example, he insists on format unification and likes to make his ppt look normative and impressive. It matters that you are a good speaker. It also matters that you have a good ppt to make your audience follow what you are talking about.

Today, the lecture is about arts. Art has a long history and appeared one million years ago. Professor divides arts into applied arts and pure arts. Pure arts is to kill time gently and passively [Lam 2011]. He points out that like science, art production is also a creative process and can reflect nature. Some great artworks are shown in the class and now we can see arts from a different view.

Day 8 (July 23, Thursday)

Things go on well. According to our research till now, Ancient Rome, China, England and India all have arranged marriage in the early age of history and autonomous marriage comes into being in later times. But we can see differences in the speed of change: For example, that of China lasts a long time while arranged marriage quickly changes into autonomous marriage so we want to find out the reason for this phenomenon. And today, it's Yu-Xuan who gives oral presentation. He is really a reliable teammate and is an expert of history in Ancient Rome and also good at English language. The presentation is excellent and the speaker is excellent too.

The lecture is about philosophy today. Professor points out that philosophical concept is built upon the best available scientific knowledge and contemporary philosophers have to be aware of the current scientific results. He also compares Philosophy in the Western world and Philosophy in China. Especially, he shares his opinion on Laozi with us. From his view, ancient Chinese philosophies concentrate in maintaining societal harmony and stability which, with help from the authority, has been kept alive for thousands of years.

Day 9 (July 28, Tuesday)

It's the written exam day. We have an examination on Chapter 2-7 of the textbook. The examination includes multiple choice and gap filling. I think it's a good way to check what we have commanded after 3 weeks' learning and most of the questions are about some important points the professor mentioned before which reflect the idea of Scimat.

The oral presentation have a different rule today. Each team gives oral presentation of their draft of paper and the presentation is finished by 5 or 6 persons of the team, which means everyone should know every part very well.

I think our presentation is done smoothly and fluently. We work together for 3 weeks and we now show our results in front of the whole class. This is a fantastic experience. As for me, I enjoy presenting what we have done and I have a sense of achievement. Also, our team is competitive and I love everyone's performance. By the way, all of the 5 teams in the class do a good job. And I am impressed with some particular teams for their presentation.

Day 10 (July 29, Wednesday)

It's the jam session. All teams try to have their papers perfected within these 24 hours. Tomorrow, we will hand in our final edition to the professor. Today, bonded to each other, we must make full use of the time left to revise it, reflect upon it and rectify it.

We have a meeting room used as our studio and everyone along with the air conditioner is running so fast to ensure the efficiency and the productivity. Yan Xi, the editor of our team during the jam session, takes up the responsibility of having everything in the orbit. The rest of us carefully check the whole paper again and again and correct the mistakes we find. Every team works so hard and I witness some students staying in front of the laptop and their eyes focusing on the screen for a quite long time. I believe that everyone has tried his or her best to make the contribution worthwhile. It's really an unforgettable jam session.

Day 11 (July 30, Thursday)

It's the last day. Yesterday, we have a tough time revising our research paper. Today, we present orally our final paper and submit the written version. And what I want to mention, is the party time. What we have at the party? Cakes and music! Professor prepares the party for us. At the party, we enjoy the cake and cheer for our achievement.

We have a happy time at the party but we also have sadness for saying goodbye. To be honest, I have never imagined that we can do so well in the end. I am glad that I make full use of my summertime and meet such a great teacher, great classmates and great teammates. So, when it's time to say goodbye, I am so enchanted to this class and I linger on without any thought of leaving. We have the Scimat and our work. We have science and the humanities. Paraphrasing Paul Cézanne, we have an apple and we want to astonish the world.

3 Conclusion

Here are what I have learned or improved after attending the HuSS course: the perspective of Scimat; the relationships between the humanities and science; new developments in history, arts, philosophy; team spirits especially for a research team; critical thinking; power-point presentation ability; academic research skills; English writing skills; English speaking skills.

As a female Chinese, the HuSS course introduces me to a new world of doing research and advancing knowledge. One month, six team members and one project, I get through it and I gain a lot. Apart from the Scimat perspective of viewing the world, the professor also teaches us the importance of respecting science and respecting reality. And I'm proud of my team for having our essay published after the summer school. The professor helps us a lot in improving the manuscript and we polish up the essay under his advice. We are grateful to our university for financial support by paying up the page charge. It's a summer I will never forget.

Notes

1. In this article, except that of Lui Lam, Chinese names are written with family name first.

References

Ciren, C.-M., Liang, D.-D., Luo, X.-F., Xia, Y.-X., Yan, X. & Yang, Y.-G. [2016]. From arranged marriage to autonomous marriage: Marriage liberalization in India, Ancient Rome, United Kingdom and China. International Journal of Humanities and Social Science **6**(1): 114-120.

Lam, L. [2011]. Arts: A science matter. *Arts: A Science Matter*, Burguete, M. & Lam, L. (eds.). Singapore: World Scientific. pp 1-32.

Lam, L. [2014]. About science 1: Basics—knowledge, nature, science and Scimat. *All About Science: Philosophy, History, Sociology & Communication*, Burguete, M. & Lam, L. (eds.). Singapore: World Scientific. pp 1-49.

Lam, L. [2017]. Humanities, Science, Scimat: A new general-education course. *Interdisciplinarity and General Education in the 21st Century*, Burguete, M. & Connerade, J.-P. (eds.). Cascais, Portugal: Science Matters Press.

O'Brien, J. [2008]. *Encyclopedia of Gender and Society*, Vol. 1. New York: SAGE.

Stone, L. [1977]. *The Family, Sex and Marriage in England, 1500-1800*. New

York: Harper & Row.

Xu, Jian-Sheng [1961]. Discussion about modern Chinese marriage and family change trend. Modern Chinese History Studies **3**: 139-167.

Xiao-Fan Luo is pursuing the bachelor's degree at School of Labour and Human Resources, Renmin University of China. Her current research is in labor and social welfare. She attended the Huss class in 2015 after her freshman year. Her team's research essay on marriage was published after class.

Published: Luo, Xiao-Fan [2017]. The HuSS course: A student's experience. *Interdisciplinarity and General Education in the 21st Century*, Burguete, M. & Connerade, J.-P. (eds.). Cascais, Portugal: Science Matters Press.

PART II

HISTOPHYSICS

The Histophysics Story

Lui Lam

Histophysics is a new discipline proposed by Lui Lam in 2002 with a paper published in *Modern Physics Letters B*. Histophysics means physics of history, which uses physical methods to tackle human history problems. From 2002 to 2010, three journal papers and two reviews are published. This article details how the idea of Histophysics came about and how Histophysics was developed and published. In short, this is the Histophysics story, told here for the first time.

1 Separation and Integration of the Two Cultures

The two cultures refer to the science culture and the humanities culture. In the era of Mozi (c.476-c.290 BC) and Aristotle (384-322 BC), "science" and the humanities were not separated, and there was no question of two cultures. During the Renaissance following the Middle Ages, Leonardo da Vinci (1452-1519) was probably the last person to be fluent in the two cultures; for him, the two cultures was not a problem either. In 1918, Cai Yuan-Pei (1868-1940), the president of Peking University and the founder and first president of the Chinese Academy of Sciences, initiated *General Education* requirement for students, ahead of the rest of the world. That is, before graduating, humanities students should take some science courses, and vice versa. This incident indicates that the separation of the two cultures had already appeared at that time. Later in 1956, Charles Percy Snow (1905-1980) formally raised the issue of the serious gap between the two cultures in society [Snow 1998; Lam 2008a].

Recently, someone pointed out: "The culture of the future will be a culture based on humanities-science synthesis, and a culture of continuous synthesis of different cultures. Future scholastic masters will be produced in the field of cross-integration of the humanities and science, and of different cultures." [Hu 2005]

Histophysics—using physical methods to study human history—is exactly such a discipline that integrates the humanities and science (Fig. 1) [Lam 2002]. This article gives the background and developmental history of this new discipline and uses it as an example to illustrate the process of academic and disciplinary innovation. The experience we have gained from this is relevant to the current development of science in China, and may also inspire the answer to the so-called Needham Question.

Fig. 1. *Left*: First page of Histophysics' founding paper [Lam 2002]. *Right*: The author's Chinese article "The humanities-science synthesized Histophysics: The unique role of physicists in developing human history," published in *Sciencetimes*, Aug. 29, 2003, as shown in ScienceNet.cn [Lin 2003a]. ScienceNet.cn is a Chinese online portal launched in 2007, co-sponsored by Chinese Academy of Sciences (CAS), Chinese Academy of Engineering (CAE), National Natural Science Foundation of China (NSFC), and China Association for Science and Technology (CAST).

2 The Founding Process of Histophysics

1998

In 1982, I invented a new type of liquid crystals—*Bowlics* [Lin 1982; Lam 1994]. In 1992, I proposed a new paradigm for dealing with complex systems—*Active Walk* [Lam et al. 1992; Lam 2005, 2006]. Liquid crystals is a branch of condensed matter physics; complex systems are broader than liquid crystals and include many systems in the natural and social sciences.

In 1998, my book *Nonlinear Physics for Beginners* was published [Lam 1998] (Fig. 2), in which Chapter 7 on complex systems talks about active walk, with particular reference to a new phenomenon discovered by my Nonlinear Physics Group in 1995: *Intrinsic abnormal growth* [Lam et al. 1995]. This phenomenon shows that in probabilistic complex systems (such as human history) composed of chance and necessity, if the control parameters are appropriate, the effect of chance will increase, resulting in abnormal system evolution. The "intrinsic" character explains the unpredictability of some historical phenomena, the abnormal growth of some plants, the irreproducibility of some medical experiments (not due to falsification), and rejects Stephen Jay Gould's (1941-2002) assertion that evolutionary history is *necessarily* unrepeatable in his book *Wonderful Life* [Gould 1989]. Our conclusion on this issue is different: If the evolution of life on earth can be "repeated," it is *possible* that humans will still appear on Earth.

After that, I wanted to do a broader and more important work, to combine Active Walk with a branch of the humanities or social science to create a *new* discipline. However, I did not know anything about these disciplines, and at that time I had considered literature, marketing, and political science. Then, by chance in 2000, I picked **history**.

2000

In **May** 2000, I invited Michael Shermer—a historian, and founder and editor-in-chief of *Skeptic* (Fig. 3)—to my college to be the speaker for my newly established "God, Science, Scientist" public lecture series. I asked him what the main journal of history is, and his answer was *History and Theory*.[1]

Fig. 2. My two books on nonlinear physics. *Left*: *Introduction to Nonlinear Physics* (Springer, 1997). *Right*: *Nonlinear Physics for Beginners* (World Scientific, 1998).

Fig. 3. *Left*: Cover of *Skeptic* magazine, Volume 8, Issue 3, 2000. Inside there is the author's article [Lam 2000]. *Right*: Michael Shermer (2003). [photo/Lui Lam]

In 2000, Shermer invited me to write a popular science article in the quarterly journal *Skeptic*. The article is titled "How nature self-organizes: Active walks in complex systems," near the end of which is the section: Modeling History [Lam 2000]. This is the first time I have written directly on **history**. The first paragraph of this section:

When historians write narrative histories, they use such phrases as "leave a mark in history" or "follow the giant's footsteps." But to model these scientifically, one needs a deformable landscape so that marks and footsteps can be left on it. Moreover, one needs a walker to do these things. As noted by the Chinese, through the action of individuals, bad things can sometimes be turned into good things, and vice versa. The walker should thus be an *active* walker. The active walk model, then, fits naturally into historical analyses. With prudent application, it could be the first step in achieving a quantitative theory of history.

This idea was fully reflected in the founding article of Histophysics two years later [Lam 2002].

2001

In 2001, I found the journal *History and Theory* in the college library. But after reading a few issues, I still did not know much about the current state of the history profession.

On **August** 12-16, 2001, I gave a keynote presentation at a futurology conference in Seattle, "Modeling history and predicting the future: The Active Walk approach" [Lam 2003a] (Fig. 4).[2] It is an extension of the "Modeling History" section of the 2000 *Skeptic* article. The future can be predicted because the active walk model can simulate the history of the past, so when time is moved forward, it is about the history of the future.

2002

On **January** 9, 2002, my luck came. I entered the University of Toronto bookstore in Canada and was shocked but delighted to see the General History bookshelf standing there, which is full of history books (Fig. 5). This happened because there is a well-known history professor at the university. I bought Richard Evans' *In Defence of History* [Evans 1997]. After reading it, I realized that history was in crisis of being oppressed by postmodernism, and the crisis represented an opportunity for me to enter the discipline as an outsider.

Fig. 4. The "Future of Humanity: Seminar 3" hosted by the Foundation For the Future, Seattle, Aug. 12-16, 2001. (Source: futurefoundation.org/programs/ hum_sem3 /html)

On **April** 4, 2002, I found a key book in the Stanford University bookstore: *History and Historians: A Historiographical Introduction* (2000) by Mark Gilderhus. The book was thin (140 pages) and at once taught me the traditional methods of historiography. My conclusion is that, in general, historians do not have an accurate understanding of the nature of science, and the *scientificity* (scientific level) of historical research is obviously not high, and that is where physicists can contribute and make breakthroughs. (I also bought the 2000 American edition of Evans' book: *In Defense of History*. Note that Defense is spelled differently from the 1997 version of Defence.)

Fig. 5. Bookstore at the University of Toronto, Canada. The four bookshelves in the second row are all general-history books (2002). [photo/Lui Lam]

On **April** 25 of the same year, I gave my first report on **Histophysics** in my physics department, entitled "Histophysics: A new discipline,"[3] which was well received.[4] The talk was reported by the student-run newspaper *Spartan Daily*. I started drafting an article based on the talk in San Jose in May and finished it in Beijing in June.

On **June** 18 of the same year, I gave the same talk "Histophysics: A new discipline" at the International Symposium on Frontier Science, held by Tsinghua University in Beijing for Chen-Ning Yang's 80th birthday.[5] The transcript of the talk was submitted in September and published in the proceedings by World Scientific of Singapore in 2003 [Lam 2003b].

In **July** of the same year, I found two good books in the University of Toronto bookstore: *An Introduction to the Philosophy of History* (1998) by Michael Stanford and *Dynasties of the World* (2002) by John Morby. The former tells me about various aspects of history (including the relationship between history and philosophy/social science as well as the themes, causes, and interpretations of history). The latter, published by Oxford University Press, has data on the years of emperors and dynasties in all countries of the world [Morby 2002]. And, of course, data is necessary for quantitative analysis—the basic skill of physicists. But I put this "dry" book on hold and did not touch it for almost two years.

2004

On **March** 13, 2004, a few days before I went to Montreal, Canada, to attend the March Meeting (March 22-26, 2004) organized by the American Physical Society, I opened the book *Dynasties of the World*, traced two curves based on the data on the Chinese dynasties, and found two *quantitative* laws about Chinese history and a *quantitative* prediction [Lam 2006]. These results were presented at the Montreal meeting. Finally, one of these laws developed into the broader *Bilinear Effect* [Lam et al. 2010].

2010

I and collaborators created a three-layer network model that mimics the structure of the ancient Chinese state; numerical results show that the Zipt plot of the dynasty lifetimes consists of two straight lines [Lam et al. 2010]. However, the mechanism and necessary conditions for the emergence of the bilinear effect remain unknown.

In short, I did Histophysics for ten years, from 2000 to 2010. In the meantime, since 2007, my research focus has shifted to *Scimat*—a multidiscipline that I pioneered in 2007 [Lam 2008a, 2014].

3 The Publication History of Histophysics

2002

My two academic innovations, bowlic liquid crystals [Lin 1982] and active walk [Lam et al. 1992], were published in 1982 and 1992, respectively, ten years apart. For the sake of perfect symmetry, I hope that the first article on Histophysics will be published in 2002. Therefore, in **June** 2002, I wrote an article entitled "Physics in history" and submitted it to the journal *Wuli* (Physics), the official publication of the Chinese Physical Society. In **November**, two referee reports returned, one saying it could be published, the other saying that the article involved social science and should be cautious. The following year, in February 2003, the editor-in-chief of *Wuli* decided to reject the manuscript. *Wuli* thus missed the opportunity to publish the first article in a new discipline.

In **December** 2002, World Scientific, without informing me, published my article "Histophysics: A new discipline" in their *Modern Physics Letters B* journal [Lam 2002].

2003

On **January** 14, 2003, while I was lamenting that the publication symmetry had been broken and that the first article in Histophysics could not appear in 2002, I suddenly received an email from Alexandre Wang from France, asking me to send him a copy of the Histophysics article. I scratched my head and looked it up on the Internet, only to know that there was indeed such an article published [Lam 2002]. Thankfully, the publication symmetry is not broken.

On **August** 29, 2003, I published in *Sciencetimes* (a newspaper published by the Chinese Academy of Sciences) a short Chinese article entitled "The humanities-science synthesized Histophysics: The unique role of physicists in developing human history" (Fig. 1), with my email address put at the end of the article [Lin 2003a]. It was reprinted by several websites in China, and some readers contacted me. I ended up connected with one of them, Han Xiao-Pu (韩筱璞), and after a few email exchanges, developed a cooperative research relationship.

In **December** of the same year, my long Chinese article "The humanities-science synthesized Histophysics" appeared in the book *Emerging Interdisciplinary Disciplines* published by Tsinghua University Press in Beijing [Lin 2003b].

In 2003, my article "Modeling history and predicting the future: The Active Walk approach" was published [Lam 2003a]. That was my presentation at the Foundation For the Future's workshop in Seattle in 2001.

2004

In **December** 2004, "How to model history and predict the future" [Lam 2004a] appeared in my first popular science book *This Pale Blue Dot: Science, History, God*, published by Tamkang University [Lam 2004b]. This book is a collection of three papers from the Tamkang Chair Lectures (淡江讲座) I gave at Tamkang University on December 9-11, 2003.[6]

2006

In **February** 2006, quantitative results on Chinese dynasty history were first published in Sec. 5.2 of my review article on Active Walk [Lam 2006], entitled "Two quantitative laws and a quantitative prediction in Chinese history." The review was published in the International Journal of Bifurcation and Chaos, a World Scientific journal.

2008

The first book in the Science Matters series, *Science Matters: Humanities as Complex Systems*, which I founded and edited, was published by World Scientific. Chapter 13 is "Human history: A science matter" [Lam 2008b].

2010

The paper "Bilinear effect in complex systems," with me as the first author, was published in *EPL* (Europhysics Letters) [Lam et al. 2010]. One of the effect's three examples shown there is the distribution of Chinese dynasty lifetimes, which I first reported in my 2006 review of active walks [Lam 2006]. In addition, in the paper, the term **Histophysics** appears in reference 29.

2013

The book *Renke: Humanities as Complex Systems* was published by China Renmin University Press, Beijing. *Renke* (人科) is Scimat's official Chinese name. Chapter 13 is my article "Research in human **history**: An example of Scimat" [Lin 2013]. The book is a Chinese translation of the English book *Science Matters*, published in 2008 by World Scientific.

In summary, with regard to Histophysics,[7,8] I published 3 journal papers [Lam 2002, 2006; Lam et al. 2010] as well as 2 English reviews [Lam 2003a, 2008b], 1 Chinese review [Lin 2013], and 2 major Chinese popularization articles [Lin 2003a, 2003b].

4 Discussion and Conclusion

1. The creation of the new discipline Histophysics, using the triple-jump method in track and field, took two years. The three-step jumps: (1) *Skeptic* article of 2000; (2) Seattle presentation 2001; (3) San Jose seminar 2002. (In those two years I also did granular physics research and organized two international conferences.)

Many times, my work habit is to set the time and topic of a talk first, and then spend one or more months forcing myself to come up with some content for the talk, rather than making a presentation after I have the content. This requires quick thinking and academic skills. The creation of Histophysics took only two years; the key is *not* reading those books. Engaging in innovation without reading books is the secret of innovation that only a few people who have successfully innovated know. For more discussion, see "Two types of reading" in the book *Research and Innovation* [Lam 2022a].

2. The material systems of nature vary in size but have continuity. Looking down at the small scale, there are molecules, atoms, nucleons and so on. Looking up at the large scale, there are molecules, condensed matter, cells, biological tissues, human and so on. Therefore, there is coherence between all disciplines, the so-called "knowing one, know them all." This is the reason that one can cross disciplines in doing research.

3. Some people think that science includes both the natural and social sciences but not the humanities. This view is wrong, because natural science is the study of all material systems in nature, including the biological system of human, and the humanities are also about the study of human, so the humanities, like social science, should be part of natural science. History, on the other hand, studies the history of humans; i.e., what has happened to individuals and societies. So, history is part of the

humanities or social science, and hence is part of natural science. The conclusion is that history can be studied in physics.

4. Scientific innovation or the birth of a new discipline requires three conditions:

(1) Sometimes it only takes a far-sighted and determined person who can use all the resources. For example, Einstein, who established the general theory of relativity.

(2) There is an academic environment without the pressure of publishing papers. For example, Einstein was an employee of the patent office then and had no pressure to publish papers; I am a tenured professor, with the freedom not to apply for research grants and not to publish papers—doing research and publishing papers is purely a personal interest. Unfortunately, tenure no longer exists in China.

(3) The person must be courageous and dare to challenge authority and popular opinion of the time.

For scholars to innovate and cross disciplines, the basic conditions are an "iron rice bowl" (guaranteed income) with a tenure system (no assessment, no form filling) and a "one-part salary" system (no need to apply for projects or doing administrative works).[9]

5. The so-called Needham Question asks: Why did modern science not arise in China? [Wang 2003]. Part of the answer is that the third condition above—those who dare to challenge authority—did not exist or were not numerous in old China, because of Confucianism and the suffocating feudal social system.

6. Finally, in China, there is a lack of good bookstores like that at Stanford and Toronto universities. It is these bookstores that allow outsiders in a discipline to know the ins and outs of the discipline and the latest progress in a short period of time. This lack may be one of the reasons why Chinese scholars tend to stick to a discipline, while academic innovation is the opposite, which quite often involves crossing disciplinary boundaries [Lam 2022a].

Notes

1. In December 1999, I created a public science lecture series "God, Science, Scientist" at San Jose State University, where I teach. The first speaker in this series was Michael Shermer, who delivered a lecture on "How people believe: Seeking God in the age of science." In November of the same year, he and I went to Beijing to attend the International Conference on Science Communications. After returning to the United States, Shermer wrote several articles on his visit to China as a columnist in the monthly magazine *Scientific American* [Shermer 2001a, 2001b].

2. In 2000, Sesh Velamoor, vice president of the Foundation For the Future in Bellevue, Washington, noticed my *Skeptic* article, particularly the section on "Modeling History" [Lam 2000]. He invited me to give a keynote presentation at the Foundation's Humanity 3000 Program, "The Future of Humanity: Workshop 3," Seattle, August 12-14, 2001.

3. The English name Histophysics for physics of history, is a new word I coined in 2002. (I have coined many other new words, such as Bowlic, Active walk, Bilinear effect, and Scimat.) At that time, my consideration was that Westerners often drew inspiration from Greek mythology or Latin when making up new words while most people in the world are not familiar with these ancient Western cultures, but were more familiar with English. Therefore, I merged the two English words History and Physics into Histophysics—the privilege of pioneers in naming their new-born "baby."

4. Actually, my colleagues in the physics department knew that Histophysics was a new thing but did not understand why it was important, because they did not understand History. Colleagues in the history department reacted emotionally, believing that physicists had invaded their turf, and even more, because they did not understand Physics. They mistakenly think that physics is only about nonhuman systems, and they do not know that there are three levels of research in any discipline: empirical, phenomenological, and bottom up. Starting with historical figures is research at the bottom-up level, which is the traditional approach of historians, while methods such as Zipf plot and active walk belong to the empirical and phenomenological levels. In general, the scientific training of historians is lacking or insufficient.

5. Chen-Ning Yang's 80th birthday international symposium was held at Tsinghua University in Beijing from June 17 to 19, 2002, with a large group of distinguished people, including 14 Nobel laureates and one mathematician who won the Fields Medal. Before going to Tsinghua, I prepared dozens of preprints of "Histophysics: A new discipline" and gave them out when I met the right people, one of whom was Murray Gell-Mann (1929-2019), the doctoral mentor of Philip Platzman (1935-2012)—my doctoral mentor at Bell Labs. Gell-Mann's parents immigrated from Austria to New York after World War I, and his father was a language teacher, so Gell-Mann had a deep understanding of various languages and a very

good memory. He took my preprint, looked at the title and the author's name Lui Lam. His first sentence was: "You are from Hong Kong." I said: "Yes." His second sentence was: "Histo- stands for tissue." This is also true; e.g., Histology stands for (cellular) tissues in pathology. But I responded super quickly, and I said: "How about Histogram?" Gell-Mann was speechless for a moment, and walked away silently with my preprint. I do not know if he later read the preprint.

6. Clement Chang (1929-2018) founded a futurology institute at Tamkang University in Tamsui, when he was president there. He attended the Foundation For the Future's "Future of Humanity: Seminar 3" workshop in Seattle in August 2001, listened to my talk on predicting the future, thought I was an expert in futurology, and invited me to give the Tamkang Chair Lectures at his school.

7. On January 8, 2011, I accidentally discovered a new online journal *Cliodynamics*, run by the University of California. (Clio is the goddess of history in Greek mythology.) After checking the Internet, I learned that the journal's editor Peter Turchin had published a book of the same name in 2003, so I sent him an email telling him that I had proposed the term and concept of Histophysics in 2002, a year before him, and pointed out that according to the content, Cliodynamics should be part of Histophysics, and Histophysics is part of Scimat. Turchin replied to my email five days later, not disagreeing with my claim, saying that he did not know about my work on Histophysics, but would read the literature I had sent him.

On April 4, 2014, I emailed Turchin and told him that I wanted to organize a session on "The Science of Human History" at the AAAS Annual Meeting in San Jose, February 12-16, 2015. And I proposed that he gives a review on Cliodynamics and I would talk about Histophysics. He replied two days later, saying thank you for the invitation but did not have time to attend. In the end, the session did not work out.

8. On July 19, 2022, someone posted a link to a paper on the physics of history on a WeChat group that I belong. The paper is entitled "Cliophysics: A scientific analysis of recurrent historical events" with 9 co-authors [Arukal et al. 2022]. This paper is published in the same journal *EPL* as my article 12 years ago on Bilinear Effect, in which Histophysics is mentioned [Lam et al. 2010]. The Cliophysics paper does not mention the word Histophysics or Cliodynamics or any works related to them. I emailed the nine authors the same day to tell them about it.

Only two authors responded to my emails. Bertrand Roehner of France replied one day later (July 20), saying that his Chinese co-author at Beijing Normal University mentioned my name and work to him after the article was published, which was too late to mention it in their paper but will mention my Histophysics works in their future papers.

Another author, Peter Richmond of Ireland, replied two days later (July 21) that the three words Histophysics, Cliodynamics, and Cliophysics, will compete in the academic world for peers to decide. He also promised to mention my work in future papers.

In this Cliophysics paper, 4 of the 9 authors are Chinese (Chen Xiaosong, Di Zengru, Wang Qing-hai, Yang Yang); 3 of them work in mainland China (Chen, Di, Yang are all at Beijing Normal University); 4 of them are in physics (Kim, Roehner, Richmond, Wang). The unusual thing is that all these 9 authors failed to notice the many Histophysics articles I published in physics journals and books, in English and Chinese, from 2002 to 2010 (see Sec. 3). At the same time, somehow and unfortunately, the long academic tradition of "who publishes first will name it" is broken by latecomers who show up 20 years later.

9. China presently uses the "three-part salary" system in universities/research institutes: basic salary, administrative salary, and performance salary. That is, one's salary will be drastically reduced if you are not leader of a group/department/college and if your *yearly* assessment performance is not good enough.

References

Arukal, Y., Baaquie, B., Chen Xiaosong, Di Zengru, Kim, B., Richmond, P., Roehner, B. M., Wang Qing-hai & Yang Yang [2022]. Cliophysics: A scientific analysis of recurrent historical events. EPL **138**: 22004.

Evans, R. [1997]. *In Defence of History*. London: Granta Books.

Gould, S. J. [1989]. *Wonderful Life: The Burgess Shale and the Nature of History*. New York: Norton.

Hu, Xian-Zhang (胡显章) [2005]. Preface to Tsinghua Humanities book series. *Misconduct of Scientists*, Shigeaki Yamazaki (山崎茂明). Beijing: Tsinghua University Press.

Lam, L. [1994]. Bowlics. *Liquid Crystalline and Mesomorphic Liquid Crystals*, Shibaev, V. P. & Lam, L. (eds.). New York: Springer. pp 324-353.

Lam, L. [1998]. *Nonlinear Physics for Beginners: Fractals, Chaos, Solitons, Pattern Formation, Cellular Automata and Complex Systems*. Singapore: World Scientific.

Lam, L. [2000]. How nature self-organizes: Active walks in complex systems. Skeptic **8**(3): 71-77.

Lam, L. [2002]. Histophysics: A new discipline. Modern Physics Letters B **16**: 1163-1176.

Lam, L. [2003a]. Modeling history and predicting the future: The Active Walk approach. *Humanity 3000, Future of Humanity: Seminar 3 Proceedings*. Bellevue, Washington: Foundation For the Future. pp 109-117.

Lam, L. [2003b]. Histophysics: A new discipline. *Proceedings of the International Symposium on Frontier of Science, 2002, Beijing: In*

Celebration of the 80ᵗʰ Birthday of C. N. Yang, Nieh, H. T. (ed.). Singapore: World Scientific. pp 456-471.

Lam, L. [2004a]. How to model history and predict the future. *This Pale Blue Dot: Science, History, God*, Lam, L. Tamsui: Tamkang University Press. pp 15-34.

Lam, L. [2004b]. *This Pale Blue Dot: Science, History, God.* Tamsui: Tamkang University Press.

Lam, L. [2005]. Active Walks: The first twelve years (Part I). International Journal of Bifurcation and Chaos **15**: 2317-2348.

Lam, L. [2006]. Active Walks: The first twelve years (Part II). International Journal of Bifurcation and Chaos **16**: 239-268.

Lam, L. [2008a]. Science Matters: A unified perspective. *Science Matters: Humanities as Complex Systems*, Burguete, M. & Lam, L. (eds.). Singapore: World Scientific. pp 1-38.

Lam, L. [2008b]. Human history: A Science Matter. *Science Matters: Humanities as Complex Systems*, Burguete, M. & Lam, L. (eds.) Singapore: World Scientific. pp 234-254.

Lam, L. [2014]. About science 1: Basics—knowledge, nature, science and Scimat. *All About Science: Philosophy, History, Sociology & Communication*, Burguete, M. & Lam, L. (eds.). Singapore: World Scientific. pp 1-49.

Lam, L. (林磊) [2022a]. *Research and Innovation.* San Jose: Yingshi Workshop.

Lam, L. [2022b]. *Science and Scientist.* San Jose: Yingshi Workshop.

Lam, L, Freimuth, R. D., Pon, M. K., Kayser, D. R., Fredrick, J. T. & Pochy, R. D. [1992]. *Filamentary Patterns and Rough Surfaces. Pattern Formation in Complex Dissipative Systems*, Kai, S. (ed.). Singapore: World Scientific.

Lam, L., Veinott, M. C. & Pochy, R. D. [1995]. Abnormal Spatio-Temporal Growths. *Spatiotemporal Patterns in Nonequilibrium Complex Systems*, Cladis, P. E. & Palffy-Muhoray, P. (eds.). Redwood City: Addison-Wesley.

Lam, L,. Bellavia, David C., Han Xiao-Pu, Liu Chih-Hui A., Shu Chang-Qing, Wei Zhengjin, Zhou Tao & Zhu Jichen [2010]. Bilinear effect in complex systems. EPL **91**: 68004.

Lin, Lei (Lam, L.) [1982]. Liquid crystal phases and "dimensionality" of molecules. Wuli (Physics) **11**(3): 171-178.

Lin, Lei [2003a]. The humanities-science synthesized Histophysics: The unique role of physicists in developing human history. Sciencetimes, Aug. 29, 2003.

Lin, Lei [2003b]. The humanities-science synthesized Histophysics. *Emerging Interdisciplinary Disciplines*, Liu Guo-Kui (ed.). Beijing: Tsinghua University Press.

Lin, Lei [2013]. Research in human history: An example of Scimat. *Renke: Humanities as Complex Systems*, Burguete, M. & Lam, L. (eds.). Beijing: China Renmin University Press. pp 232-251.

Morby, J. E. [2002]. *Dynasties of the World.* Oxford: Oxford University Press.

Wang, Qian-Guo-Zhong (王钱国忠) (ed.) [2003]. *The Bridge between East and West in Scientific Culture: A Study of Joseph Needham*. Beijing: Science Press.

Shermer, M. [2001a]. Starbucks in the Forbidden City. Sci. Am., July 1, 2001. 10.1038/scientificamerican0701-34.

Shermer, M. [2001b]. I Was Wrong. Sci. Am., Oct. 1, 2001. 10.1038/scientificamerican1001-30.

Snow, C. P. [1998]. *The Two Cultures*. Cambridge, UK: Cambridge University Press.

Published: Lam, L. [2022]. History of Histophysics. *China Complex*, Lam, L. San Jose: Yingshi Workshop. Original in Chinese; English translation here (with title modified).

How Nature Self-organizes: Active Walks in Complex Systems

Lui Lam

Ants do it. Birds do it. Humans do it.

1 Self-organization

When ants go out and look for food, they initially do not know where the food is. Somehow, they find it, carry it home, and recruit more ants from the nest to join them. Some kinds of trails are formed spontaneously, as shown in Fig. 1. There is no central planning, no central command. The ants just self-organize and get the job done efficiently. How do they do it?

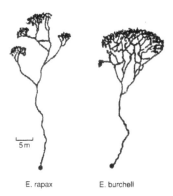

E. rapax   E. burchell

Fig. 1. Some real ant trails.

On power wires or building ledges we often see birds sitting side by side, chatting, or so it seems. Do they have a chat group? Or even a chat group leader? No. They just *self-organize*.

In early human history people self-organized to form groups, and groups self-organized to form societies. More recently, prompted by an external event—the World War II—countries self-organized to form the United Nations. On a smaller scale, in the city of San Jose (where I teach), everyone (or almost everyone) gets his or her hair cut, in one way or another. The number and locations of the barbershops are not decided by any organization or committee. It just happens, seemingly by a miracle. But it is not a miracle; it is *self-organization*.

Self-organization does not happen in a vacuum. It depends on the nature of the individuals, the interaction between the participants, and the environment. All three of these components mutually influence each other. We have a good understanding of self-organization from physics, e.g., thermal convection in a glass of beer.

However, for complex systems like ants and human societies, we know much less. The aim of scientists is to find a unified description of self-organization that covers all complex systems. Exciting headway has been made in the last eight years in describing complex self-organization through a theory called *Active Walk*.

2 Universality

In the long pursuit of an understanding of the universe, and especially of human societies, the emphasis has always been on the universal characteristics or behavior that are shared by all systems. As noted by Erwin Schrödinger in the preface of his 1967 classic *What Is Life?*, this emphasis is reflected in the choice of the word "university" in naming our institutes of highest learning. In this regard, one may also note that the highest degree of learning conferred by a university is the Doctor of Philosophy (PhD), irrespective of the subjects studied. Incidentally, philosophy is a Greek word meaning the "love of wisdom"; it is "wisdom" as the object, not any particular subject.

It is interesting to recall that Aristotle (384-322 BC), a Greek scholar, did not focus his attention on just one or two branches of knowledge but studied and contributed significantly to all fields, including biology, psychology, physics, and literary theory, as well as inventing formal logic as a system of thought. The fragmentation of learning and the compartmentalization of knowledge into different disciplines such as physics, chemistry, and economics is a rather recent phenomenon, occurring only in the last few centuries. It is suspected that this trend is due more to administrative convenience than to the nature of science itself.

Recent attempts to recoup a unified approach to knowledge with a view to finding a common description of both natural and social systems, include cybernetics and general systems theory from the 1940s to the 1960s, Ilya Prigogine's *dissipative structures* and Hermann Haken's *synergetics* in the 1970s, and the blossoming study of complex systems since the 1980s.

3 Complex Systems

A complex system is one that consists of a large number of simple elements, or "intelligent" agents, interacting with each other and the environment. The elements/agents may or may not evolve in time, and the behavior of the system cannot be learned by the reduction method, meaning that knowledge of the parts is not enough to predict the behavior of the whole system (unless the interactions between them are also known). But such a definition is not without problems. For example, a system may appear complex only because we do not understand it yet (by the reduction method or not). Once understood, it becomes a simple system.

Moreover, whether a system is complex or not may depend on the particular aspect of it that one happens to study. For example, if we wants to know about the inner structure and formation mechanism of a piece of rock, the rock could be a complex system. But if we want only to know how the rock will move when given a kick, then the classical Newtonian dynamics will do, and the rock is considered simple.

A precise definition of a complex system is thus difficult to devise, as is frequently the case in the early stages of a new research field. However, this difficulty has not hindered the study of complex systems much. In practice, it is safe to say that almost the subjects covered in the various departments of a university—except for those in the conventional curriculums of physics, chemistry, and engineering departments—are in the realm of complex systems. The topics studied in complex systems span a wide spectrum, including human languages, the origin of life, DNA and information, evolutionary biology, economics, psychology, ecology, earthquake geology, immunology, cellular automata, neural networks, and the self-organization of nonequilibrium systems. In other words, the study of complex systems is a study of the real world. The recent surge in the study of complex systems could be attributed to three developments.

First, in the early 1980s the science of *Chaos* was better understood, and time series obtained in almost every discipline—from a dripping faucet to a heartbeat to the stock market—were subjected to the same analyses as inspired by chaos theory. The importance of this development was that chaos theory seemed to offer scientists a handle (or an excuse if one was needed), to tackle problems from almost any field. A psychological barrier was broken; no complex system was too complex to be touched.

Second, a crucial influence that helps to propel and sustain complex systems as a viable research discipline is the prevalence of personal *computers* and the availability of powerful computational tools, such as parallel processors. While theoretical study of complex systems is usually quite difficult and sometimes appears impossible, one can always resort to some form of computer simulation (or computer experiment, as some like to call it.)

Finally, in the last 20 years the study of basic physics by going to smaller and smaller scales as practiced in the discipline of particle physics has hit an *impasse*. The forefront of particle physics is superstring theory—called the Theory of Everything in some quarters—which is very complicated mathematically and could not be tested by experiments for the time being. The impasse encouraged many physicists to turn their attention elsewhere, looking at systems of increasing scales and squarely facing the issue of complexity characterizing these large systems. It should be noted that biologists, engineers, paleontologists and many others have studied complex systems daily. What makes the present period of the study of complex systems distinct and exciting is the involvement of a large number of physicists who, working with computer scientists and others, bring with them new tools and concepts that give new hope in unifying the field.

4 Paradigms

If there were a universal law governing all complex systems, what would it look like? How would we recognize it? In my opinion, there are two signs of a successful universal law applicable to all complex systems:

1. The law should be simple enough to be stated in one or two reasonably short sentences.

2. The law should conform to our daily experience.

The first one follows from the wide applicability of the law. The second one is due to the fact that each one of us is a complex system by herself or himself. (As an example, the second law of thermodynamics that states "heat cannot flow on its own from cold to hot bodies" easily satisfies these two requirements.)

Unfortunately, such a simple, universal law for complex systems has not yet been found. Instead, in the last 20 years or so there emerged *four* general paradigms, each of which was found to be applicable to a large number of complex systems, if not all of them. The first is **Fractals**, founded by Benoit Mandelbrot in the early 1980s. Many complex systems exhibit the self-similar characteristic that a small part of itself, when enlarged in scale, resembles the whole; these are called fractals. Examples of fractals include the random walk, blood vessels, trees, rivers, clouds, stock market fluctuations, and so forth. Ubiquitous as they are, not all complex systems are fractals; e.g., the ant trails in Fig. 1 are not fractals.

The second paradigm is **Chaos**. Chaos was studied by the French mathematician Henri Poincaré at the turn of the last century, but was resurrected in the late 1970s through the works of Edward Lorenz and Mitchell Feigenbaum. In the realm of science, chaos is a technical word characterizing the behavior of some *deterministic* nonlinear systems that are sensitively dependent on initial conditions. This usage of the word chaos obviously differs from that adopted in our daily lives, in which chaos is synonymous with a state of utter confusion. When the distinction between these two usages is ignored—consciously by a few scientists, unconsciously by most laypeople and some science writers—a state of utter confusion does arise! And we are left with a bundle of dubious books like *The Tao of Chaos* and *Seven Life Lessons of Chaos*. (It is hard to claim that life is a deterministic system, or that life always depends sensitively on what you do in your early life, even though it may occasionally happen that way.)

As noted, it is obvious that chaos and fractals are two very different things. The connection between the two lies in the fact that for dissipative chaotic systems, in the mathematical *phase space* there exists one or more structures called *strange attractors*; and it happens that these strange attractors are fractals. (An example of a dissipative system is a simple pendulum immersed in water—energy kept drained away by friction and the pendulum will eventually come to a stop. This is not a chaotic system, however; but a pendulum with the hanging point in oscillation is.) This confusion between chaos and fractals led Martin Gardner to remark that chaos theory "is fashionable and interesting, but...it is mostly fractal geometry...more of a temporary fad like catastrophe theory" (see *Skeptic*, Vol. 5, No. 2, 1997, p 61). The truth is that chaos theory has been tested and confirmed in many real systems, from dripping faucets, to lasers, to the planet Pluto. Chaos theory is regarded as a great revolution in 20th-century physics, along with relativity and quantum mechanics. (See, e.g., my 1997 book *Introduction to Nonlinear Physics* and Abraham Pais' 2000 book *The Genius of Physics: A Portrait Gallery*.)

Another unnecessary confusion arises from the occasionally interchangeable use of the words "nonlinear dynamics" and "chaos theory." Chaotic systems are nonlinear systems, but not vice versa. In other words, not every nonlinear system (such as a simple pendulum) is chaotic. More recently, the word "chaoplexity" has crept into the literature. This is unfortunate and must be clarified. While a chaotic system does appear complex, it does not follow that every complex system is chaotic in origin. Chaos is just one of many routes that a system can become complex. A swarm of ants is very complex, but it is not chaotic. Each one of us is a very complex system—perhaps the most complex system in the universe—but most of us, most of the time, are not chaotic.

The third paradigm of complex systems is **Self-organized Criticality**, first proposed in 1987 by Per Bak, Chao Tang, and Kurt Wiesenfeld. Using the model of a sandpile, it asserts that large dynamical systems tend to drive themselves to a critical state with no characteristic spatial and temporal scales. While self-organized criticality, in the incarnation of power laws, shows up frequently in many computer models, it is less established in controllable, testable real systems, in contrast to the cases of fractals and chaos. Nevertheless, Vice President Al Gore apparently forgot what he learned as a kid playing on the beach, when he claimed that he found self-organized criticality "irresistible as a metaphor," and that it helped him to understand change in his own life.

The fourth paradigm is **Active Walk**, initiated by myself and my students in 1992. We proposed that the elements/agents in a complex system communicate indirectly with each other through their interaction with the deformable landscape they share. Each element is an active walker in the sense that it changes the landscape when it moves on its surface, and is influenced by the changed landscape in choosing its next step (Fig. 2). Complex behavior of the system could result from very simple rules governing the walker's interaction with the landscape.

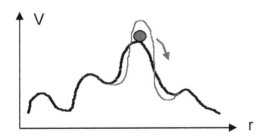

Fig. 2. Sketch of an active walk. The walker, represented by the solid dot, changes the landscape (from the black line to the blue line) around itself according to a landscaping rule.

For example, an ant is a living active walker. It releases a chemical as it walks, and it moves toward regions of high chemical concentration. In this case, the landscape is the spatial distribution of the chemical concentration, the so-called chemical concentration "field" or "potential." A swarm of such ants can efficiently self-organize to do a number of nontrivial chores, such as foraging for food, without any central direction from the queen ant. Active walks have been applied successfully to a variety of problems in both the natural and social sciences.

5 Active Walk

In 1990, I took an active walk myself, in the metaphorical sense. In January, I organized a Winter School in Nonlinear Physics at San Jose State University. In March, I chaired the American Physical Society Symposium on Instabilities and Propagating Patterns in Soft Matter Physics in Anaheim, California. In June, I was the Director of a NATO Advanced Research Workshop on Nonlinear Dynamical Structures in Simple and Complex Liquids, in Los Alamos. This was followed in July by a meeting in Vancouver, British Columbia, Canada, where I established the International Liquid Crystal Society during the 13th International Liquid Crystal Conference.

It was a few days before this conference in Vancouver that we rushed through a simple experiment in our Nonlinear Physics Laboratory at San Jose State University, in the basement of the Natural Science Building. The experiment was so simple that it could be finished in less than one second. What we did was take a liquid crystal cell—a thin layer of liquid crystal between two conductive, transparent glass plates (the same one you would find in any digital watch or calculator), replaced the liquid crystal by oil, and applied a high enough voltage across the cell. After a flash of light, the experiment was finished, and we found a complicated filamentary pattern left on the inner surfaces of the coated glass plates. An example is shown in Fig. 3, left. We presented these results in Vancouver.

The physical and chemical processes giving rise to these filaments are rather complicated, and are still being investigated. Essentially, the filaments are the locations that a series of dielectric breakdown—like lightning in the sky—takes place in the cell. After this experiment, without knowledge of the physical mechanisms, we set out to do a computer modeling of this filamentary pattern. We soon realized that a growing filament could be identified as the track of a walker. To grow the filament, we only have to tell the walker how to walk, and specify how the walker changes the local environment as it walks. One of our computer results is presented in Fig. 3, right, which agrees fairly well with the experiment. When I wrote up the paper for my 1992 book *Modeling Complex Phenomena*, I named it the "Active Walker Model," and subsequently called the process an "Active Walk."

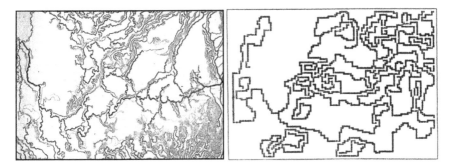

Fig. 3. *Left*: An experimental filamentary pattern. *Right*: An active walk computer simulation.

The use of a walker in modeling physical and other phenomenon runs a long history. The most well-known is a random walker, which has been used in mimicking the motion of a completely drunk person, the Brownian motion of a particle suspended in a liquid, or the fluctuations in a financial market. A random walker does not change anything in its environment and is what we would call a *passive* walker. In contrast, an active walker changes the landscape as it walks and is an *active* walker.

The description of an active walk thus involves two interacting components: the location of the walker as a function of time, and the deformable landscape as a function of time and space. The dynamics of an active walk are determined by three constituent rules:

1. The landscaping rule, which specifies how the walker changes the landscape as it walks.

2. The stepping rule, which tells how the walker chooses its next step.

3. The landscape's self-evolving rule, which specifies any change of the landscape due to factors unrelated to the walker, such as, in the case of ants, chemical evaporation or blowing wind.

These first two rules may evolve in time for "intelligent" active walkers. The details of these three rules should depend on the system under study.

The track of the active walker forms a filamentary pattern, while the landscape usually becomes a rough fractal surface after some time. Of course, any number of active walkers may coexist, and they communicate with each other indirectly through their individual interaction with the shared landscape.

The landscape could be one of two types. The first type are physically existing surfaces; the second type are abstract mathematical artifacts. Examples of physical landscapes include the following cases:

1. A woman walking on a sand dune; the sand dune surface is the landscape.

2. Percolation in soft materials; when water flows through a porous, deformable medium, the shape and distribution of the pores would be changed, and the porous medium itself is the (three-dimensional) landscape.

3. For ants, as already noted, the chemical distribution is the landscape. Frank Schweitzer, Kenneth Lao, and Fereydoon Family have successfully constructed an active walk model of ants. Ant trails like those in Fig. 1 are recovered. Furthermore, simulation of ants in the presence of randomly located food sources shows that the ants essentially attack and exhaust the food sources one at a time, a behavior observed in real ants.

4. Bacteria or fishes tend to move to regions with higher nutrients; as they move, they consume and reduce the nutrient concentration—the nutrient distribution thus constitutes the deformable landscape.

Abstract landscapes can be illustrated by two examples:

1. Urban growth can be modeled by the aggregation of active walkers. Initially, a "value" can be assigned to every piece of vacant land lot according to its location. For example, a lot on the flatland will have a higher value than one located on the hill; a river nearby can increase the value of the lot. The value could be taken to be the probability that the lot will be developed, by having a house or factory built on it. Once this happens, the value of the lands nearby will be increased and a new house will be added somewhere. The process is repeated. In this case, the spatial distribution of the value is the abstract landscape; the house is an active "walker," which does not walk though. Such a model is very realistic but has not yet been tried.

2. The second example is the fitness landscape employed in evolutionary biology. Every species is in coevolution with other species. The presence of species A (Fig. 4a) affects the fitness landscape of species B (Fig. 4b), which in turn changes the fitness landscape of A (Fig. 4c); the changed landscape of A then determines how A will move. If we want to describe this evolutionary process by a simple model involving A alone, we will go directly from Fig. 4a to Fig. 4c, with Fig. 4b hidden. It then seems that A deforms its own fitness landscape at every step of its movement; i.e., A acts like an active walker.

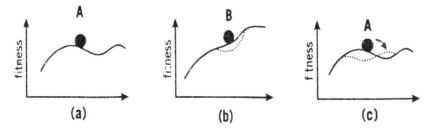

Fig. 4. Sketch of coevolution of two species A and B (see text).

A representative simulation from the active walk model is shown in Fig. 5, which compares very well with three real results from physical and biological systems. More amazingly, we produced our simulation without the experimental results in mind and before one of them was created in the laboratory. It shows that active walk is really in the bag of tricks of Mother Nature when she wants to produce complicated patterns with minimal effort.

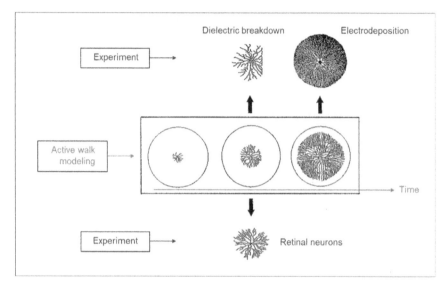

Fig. 5. Comparison of active walk computer simulation with experimental results. *Middle*: Time development of a computer generated pattern; time increases from left to right. *Top left*: Chemical reaction pattern induced by a series of dielectric breakdown in a thin layer of oil. *Top right*: An experimental electrodeposit pattern. *Bottom*: A retinal neuron pattern.

6 Applications in Economics

Increasing returns refers to self-reinforcing mechanisms in the economy, whereby chance events can tilt the competitive balance of technological products. For example, the Beta video system lost out to the VHS system, despite the slightly technical superiority of the Beta version, because more people happened to buy the VHS version in the beginning. (A student told me that this was because videotapes of X-rated movies were recorded in the VHS format at that time—this claim remains to be checked.) Increasing returns can be easily modeled by having an active walker jumping among the sites on a fitness landscape of the product types (Fig. 6).

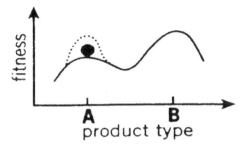

Fig. 6. The active walk description of competition between two different products A and B. The solid dot represents the active walker. When the dot is on site A, it means product A is bought by a new customer. The fitness of A is then raised by a constant amount. The same goes for B.

Fitness is a measure of how desirable the product is. The probability that a product would be bought by the next customer could be taken to be an increasing function of its fitness; i.e., the higher the fitness, the more likely the product would be bought. The uncertainty, in the form of the probability, is due to the customer being only partially rational or having only partial information about the products, or both. A typical numerical result is shown in Fig. 7.

We assume that the two products are equally good and desirable in the beginning. In Fig. 7a, when information is completely lacking, customers choose the products randomly and both products will survive and coexist.

But when some information is available or the customers are only partially rational, after a period of coexistence, one of the products will clearly win out (Fig. 7b)—like in the case of the competition between PC (personal computer) and Apple computer. However, which product will win is purely by chance and cannot be predicted.

Recently, Daniel Friedman and Joel Yellin at the University of California, Santa Cruz, have applied the active walk model to two problems in economics, viz., population games and the dynamics of rank-dependent consumption, respectively. Robert Savit and coworkers at the University of Michigan, Ann Arbor, succeeded in using active walk to model adaptive competition and market efficiency. These works also have implications for biological systems.

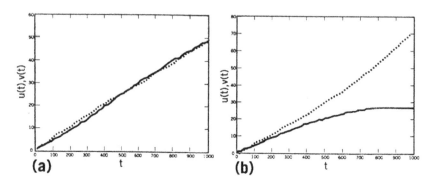

Fig. 7. Numerical results for two competing products. Here $u(t)$ and $v(t)$ are the fitness of products A and B at time t, respectively; each is proportional to the total number of that product sold. We assume the two products are equally good and desirable in the beginning. (a) The two products coexist in the market, when customers' buying is completely random. (b) One product wins out, by chance (see text).

7 Modeling History

When historians write narrative histories they use such phrases as "leave a mark in history" or "follow the giant's footsteps." But to model these scientifically, one needs a deformable landscape so that marks and footsteps can be left on it. Moreover, one needs a walker to do these things. As noted by the Chinese, through the action of individuals, bad things can sometimes be turned into good things, and vice versa. Thus, the walker should be an active walker. The active walk model, then, fits naturally into historical analyses. With prudent application, it could be the first step in achieving a quantitative theory of history.

For example, the "model of contingent-necessity" of Michael Shermer (see the final chapter of his 2000 book *How We Believe*), which essentially says that contingencies are important in the early stages of a historical sequence but less so in the later stages, could be understood as follows. In the beginning, the walkers are close to each other at the center of the landscape plane. There is less space for the walkers to navigate and, in a crowded situation, each walker is easily affected by the action of others through the changing landscape they share. In other words, each step counts. Later in the sequence of steps, when the walkers are more separated from each other and away from the center, each step counts less in influencing others (assuming that the change of the landscape by each walker is short ranged).

In another example, through the study of active walk models, we find that the relative importance of chance—versus necessity—could depend on where in the parameter space that the system belongs (Fig. 8). If our world happened to sit in the *sensitive zone*, then history could indeed be very different if life's "tape" is replayed, as advocated by Stephen Jay Gould in his 1989 book *Wonderful Life*; otherwise, history could be repeated, more or less. The problem is to know where our world sits.

Mingjun Lu, a translation scholar formerly at the Nanjing University (and later at University of Toronto), observes that there is more than one way to translate a sentence from one language to another. However, the choice is narrowed and affected by the particular translation of the previous sentence. Lu suggests that the translation process can be modeled by an active walk, a very innovative idea in the theory of translation. In fact, any decision-making process—translation and management in particular—is a "history-making" process and thus could be modeled by active walks.

8 What Does All This Mean?

There are more successful active walk applications: Migration of workers and economic agglomeration, collective opinion formation, and pedestrian traffic studied by Frank Schweitzer in Germany; cold production in heavy oil by Jian-Yang Yuan in Canada; tumor growth by Thomas Deisboeck of the Harvard Medical School; and surface filamentary pattern formation by Ru-Pin Pan in Taiwan.

Active walk is a (classical) field theory of complex systems. The field—the landscape—could be either physical or mathematical. In physics, we know that elementary particles interact with each other through a field they share. It is the same between elements and agents in a complex system. They are all active walk models.

Space limitation prevents us from spelling out all the "Life Lessons of Active Walk" here. But let me tell you the first one: Each one of us is an active walker. By changing the surrounding landscape with each action, every person has the potential to make history!

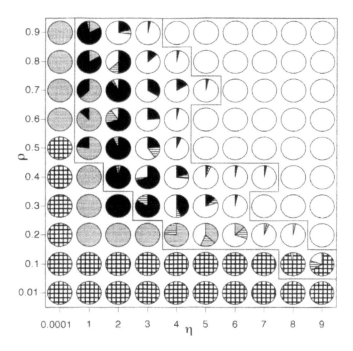

Fig. 8. A summary of the outcomes from a probabilistic active walk model. The two control parameters in the model are recorded on the horizontal and vertical axes. The pie chart at each point represents the percentage of each type of five outcomes obtained from 30 runs of the algorithm. Within the "sensitive zone" in the middle region—where chance plays a dominant role—the outcome is unpredictable.

Further Reading

Lam, L. [1998]. *Nonlinear Physics for Beginners: Fractals, Chaos, Solitons, Pattern Formation, Cellular Automata and Complex Systems.* Singapore: World Scientific.

Original: Lam, L. [2000]. How nature self-organizes: Active walks in complex systems. Skeptic **8**(3): 71-77.

Histophysics: A New Discipline

Lui Lam

History is the most important discipline of study. The system investigated in history is a many-body system consisting of biological material bodies, *Homo sapiens*, and hence can be studied scientifically. The unique role physicists can play in advancing the science of human history is presented. We will discuss the methods of study in history; worldviews; modeling history as a complex, dynamical system; predicting the future and retrodicting the past; and artificial history. In particular, active walk is shown to provide the foundation for a new worldview, and found to be widely applicable in modeling history, as illustrated by three examples from economic, evolutionary, and social histories, respectively.

1 Introduction

New disciplines of study are born from time to time, like in the case of human babies, but less frequently. Or, for that matter, like new stars emerging in the sky, being suddenly noticed after a long period in the making.

Historically, when physics is combined with other natural sciences, new disciplines are created and we have Astrophysics, Biophysics, Geophysics, and so on. More recently, *Econophysics* was born when physicists venture into economics, a branch of the social sciences [Mantegna & Stanley 2000]. (Similarly in the field of biology, in 1975, *Sociobiology* was created when biology was merged with sociology by Edward O. Wilson [1975].) In this article physics is linked to history, giving birth to a new discipline—*Histophysics* (physics of history).

Since the 19th century, history has been treated as a science intermittently through the efforts of Condorcet, Comte, Buckle, Taine, Adams, and others [Breisach 1994; Stanford 1998]. The progress has been uneven. And there are even doubts on whether this endeavor is at all possible, given the complexity and the irreproducible nature of historical processes. As argued in this article, the system under study in history is actually a classical, stochastic system. Yes, it is a very complicated, complex system; but like other complex systems, it can be studied scientifically.

In the following, the unique role physicists can play in advancing the science of human history is presented. Specifically, we will discuss the methods of study in history; worldviews; modeling history as a complex, dynamical system; predicting the future and retrodicting the past; and artificial history. In particular, active walk [Lam 2000, 2005, 2006] is shown to be the foundation for a new worldview and widely applicable in modeling history, as illustrated by three examples taken from economic, evolutionary, and social histories, respectively.

2 What Is History About?

The system under study in history is a many-body system. In this system, each "body" is a human being, called a "particle" or an "agent" in this article. Each constituent agent is a classical object (not quantum mechanical) and is distinguishable (i.e., each agent in the system can be identified individually). This many-body system is a heterogeneous system, due to the different sizes, ages, races…of the agents.

The interest of the historians is about anything happened in the *past* that is related to these agents (which exist presently or in the past). The system being investigated by the historians is thus a system of *material bodies*, and thus can be studied *scientifically*. But how?

For this purpose, let us examine the constituents more closely. For comparison, a well-studied many-body system in the physical sciences—water—will be used.

1. In water, the constituents are molecules, under the name of hydrogen oxide and represented by the symbol H_2O. In history, the

constituents are biological bodies, named *Homo sapiens* and represented by the symbol ☺ (or, with more details, by ♀ and ♂).

2. In water, each molecule is itself a composite system made up of two hydrogen atoms and one oxygen atom; each atom is made up of electrons and a nucleus; each nucleus is made up of smaller particles, protons and neutrons; each proton or neutron is made up of quarks (or superstrings). In comparison, in history, each biological body is made up of organs (brain, heart, stomach, and so on—the brain is considered by some historians to be the most important organ of them all, but other historians at one time or other considered the heart or the stomach as more important), blood and other liquids; each organ is made up of cells (a total of about 5 trillion cells in a body); each cell is made up of (mostly organic) molecules, such as DNA; each molecule is made up of atoms—and from this point on, there is no distinction between the two cases of water and history.

3. In both these two systems, there are layers and layers of structures, and many *internal states* in each constituent. (Internal states in a human body include memories, thoughts, and so on.) Even though a water molecule is several layers less in structure compared to a biological human body, it is actually quite a complicated system and possesses, in fact, as in the case of a human body, an infinite number of internal states.

4. The atoms inside a water molecule and a human body were (mostly) created in the stars long time ago, and have a long history behind them. The property of a water molecule remains the same since its creation, which makes a scientific study of water "easier" than the case if their property changed with time. (In the latter case, a scientific study is more difficult but not impossible.) Similarly, because all humans are derived from the same ancestor (the African Eve, say) only a few millions of years ago and, according to Charles Darwin's evolutionary theory, human body and human nature take a long time to evolve, they have practically unchanged over the last 5,000 years or so—the period in which human history is recorded—

it is "easy" for present historians to second guess the acts of past historical players.

5. The size of a water molecule is very small, about 10^{-8} centimeter; it is a quantum mechanical particle. In contrast, a human body is about 40 centimeter to 200 centimeter; it is a classical particle (i.e., quantum mechanics is irrelevant to these particles).

6. The fact that the historian herself or himself is one of, or one similar to, the particles under study in history should make the historian's job easier, not more difficult, when compared to the study of water.

The reasons that people are interested in history are slightly different from those in studying water. It is true that humans are curious-prone animals and hence should be interested in almost anything, water or history. But in history, there are more. We are interested to know what our parents, and great, great...great parents were doing. We want to know where we came from and why we end up like this. Furthermore, from the practical point of view, knowing the past may help us to understand the present and predict the future. The Chinese say it well:

Reflection from a mirror enables us to tidy up ourselves; reflection on history enables us to know the ups and downs of our time.

3 Motivation

There are several reasons behind the undertaking of Histophysics:

1. *The human system is the ultimate complex system in the universe.* Complex systems [Lam 1998, 2000] are those systems consisting of a large number of particles, agents or components, which usually cannot be studied by the reduction method. Examples include the economic system, ant colonies, the brain, traffic, humans, and extraterrestrial intelligences (if they exist) [Lam 2004]. The study of complex systems is about the real world. Anyone interested in complex systems should be fascinated by humans, the ultimate complex system in the universe.

2. *History is the most important discipline of study.* History as a research discipline, like physics, may be divided into two

categories: basic and applied. For example, Karl Marx is in basic research while Vladimir Lenin, applied. In either history or physics, the greatest impact shows up in the applied domain. The beneficial aspect of applied physics goes without saying. To name just a few, basic physics invented and applied physics developed semiconductors and lasers, which made our lives more convenient and improved dramatically our living standards. The impact of basic history is less clear. However, it is in the applied domain that history outstrips the combined impact of basic physics and applied physics. This can be seen through their negative effects. The one negative incident due to applied physics is the dropping of the two atomic bombs in Hiroshima and Nagasaki in 1945, resulting in a total of about 270,000 deaths counting the subsequent years. In contrast, from 1975 to 1979, two millions Cambodians were killed under the Pol Pot regime (ruled by applied historians inspired by Lenin), and this is just one of the many massacres recorded in history. In this sense, history is a far more important discipline than physics; and in fact, history is the most important counting all the disciplines.

3. *The time is ripe.* Important advances have been achieved in the study of complex systems and the biology of humans in recent years. Powerful computers are available to do simulations; chaos are understood; the human genome is decoded; real-time scans of the brain are performed in neuroscience experiments, and so on. The time is ripe to tackle human science from the physical and biological perspectives, as recently emphasized by Edward O. Wilson [1999]. In his words: "The human condition is the most important frontier of the natural sciences."

4. *Physicists, in collaboration with historians, can help turn history into a science.* (a) Physicists know how to deal with many-body problems. In fact, they have invented statistical mechanics to tackle these systems one hundred years ago and used it successfully ever since. (b) Physicists are "licensed" to do history. "Physics is what physicists do," says *Physics Today*, the magazine of the American Physical Society [Lubkin 1998]. (c) Physicists are urged to do history. "The task of physics is not only to understand the hydrogen

atom, but to understand the world," advocated Arthur Schawlow, the Nobel laureate [Wolf 1994: 53]. The world, of course, includes history. (d) Physicists can read research journals in history, but historians usually cannot read physics journals due partly to the jargons used there. (e) Physicists have good track records in interdisciplinary studies, as witnessed by the many x-physics new disciplines mentioned above. (Here, x stands for astro, bio, etc.) Consequently, in collaboration with historians, physicists can help raise the scientific level of history studies.

4 Methods of Study in History

In any science discipline, the "scientific method" consists of the following steps: (1) Starting from raw data or existing theory, a new hypothesis is proposed. (2) Consequences from the new hypothesis are checked with data. (3) In agreement, the hypothesis is confirmed and turns into a new theory—new laws are discovered, perhaps; otherwise, go back to step (1). (4) New findings from the theory are published in peer-reviewed, research journals. In exceptional cases, (5) popular science books are written, by the scientists themselves or, mostly, by others.

In the profession of history, these steps are fragmentized [Munslow 1997]. The proposal of a new hypothesis, step 1 above, is called constructionism. The discovery of new "laws" or trends, if exist, step 3, is called reconstructionism. And, many research results in history, expressed in the narrative form, are published as popular books [Stanford 1998], step 5, without going through step 4, peer-reviewed journals. It is not surprising then, that history is not yet a scientific discipline.

Some representative research methods in history [Evans 1997; Gilderhus 2000] are noted here:

1. The "internal states" of the brain are emphasized by Robin Collingwood when he insists that historians have to reenact the thoughts of historical figures in understanding history [Collingwood 1946]. This approach is called "method acting" in the movie industry; it is very difficult and can only be mastered by a few gifted persons.

2. History—like archaeology, paleontology and a large part of astronomy—is about things happened in the past. Its relationship to things happening in the present is obvious since there is no discontinuous break between past and present in the flow of history. Much resource is devoted to the understanding of the present state (and welfare) of *single* particles—humans; resulting in various disciplines such as medicine, psychology, art, and music. The present state of *many* particles is studied in social science, such as sociology and economics. The importance of social science in history study is advocated by the *Annales* school in Paris, associated with the names of Lucien Febvre, Marc Bloch, and Fernand Brandel [Burke 1990].

3. The objectivity in, and the possibility of reconstructing past "reality" through history narratives, are ruled out by Hayden White [1973] and Jacques Derrida [1976], and other deconstructionists because, they argue, the meaning of any writing is undecidable. These and other postmodern attacks lead to a crisis in the history profession in the 1990s [Evans 1997; Gilderhus 2000].

However, as we shall demonstrate below, research in history can benefit from the inclusion of some scientific approaches developed in physics.

5 Worldviews

Worldviews, through the action of powerful political leaders and governments, have tremendous consequence in applied history. The study of worldviews is important to historians [McNeill 1998].

All worldviews, in fact, follow from new advances in basic physics. For example, Newton's clockwork universe leads to the deterministic worldview, which seems appropriate since Newton's laws of motion govern classical particles and humans are classical particles (except that the human system is intrinsically probabilistic while inanimate particles are not).

What are inappropriate are the worldviews that "nothing is certain" and "an observer changes what's observed," supposedly inspired by quantum mechanics; and the worldview that "everything is relative," following from the success of Einstein's relativity theory. Some misunderstanding is involved here. Quantum mechanics is important only in the microscopic world (with the exception of two macroscopic phenomena: superfluidity and superconductivity) while humans are macroscopic bodies which are not known to be superfluid or superconduct. And special relativity is important only if the particle's velocity is close to that of light, while in general relativity the mass of the object has to be huge. Both these conditions do not apply to humans or human activities. Consequently, historians can safely forget quantum mechanics and relativity; they are irrelevant to the development of human history.

More recently, chaos theory inspires the worldview that "slight changes in initial conditions can lead to very different outcomes in history." Books on "virtual history" [Ferguson 1999] and "what...if" [Cowley & Ambrose 2000] appeared, but these attempts focus only on the *qualitative* aspects of chaos and thus has nothing to do with chaos per se. A *quantitative* analysis of real historical data to detect chaos has been tried [Shermer 1995; Pigliucci 2000] but the result is indeterminate due to insufficient data points in the time sequence.

Self-organized criticality, after its introduction in physics [Bak et al. 1987], has been invoked to be the metaphor for history [Bak 1996; Buchanan 2000]. The claim is that history works like sandpiles: It builds up to a critical state, crumbles, builds up, crumbles, ... While there is no denial that someone's life may work like this, we know for sure not everyone's life and not everything in history goes through this repetitive and depressing route. Think Bill Gates, for example.

History, resulting from a combination of contingency and necessity, is a stochastic process with many possibilities. Some aspects of it may appear periodic, sometimes; other aspects, moving in a spiral or chaotically. Yes, it may even occasionally build up and crumble like in a sandpile. But it may also go through weird paths in a strange landscape. Active walk, a paradigm introduced by us in 1992 to handle complex systems [Lam 1997, 1998, 2000, 2005, 2006], provides exactly the needed foundation for such a worldview. In an active walk, a particle (the *walker*) changes a deformable potential (the *landscape*) as it walks; its next step is influenced by the changed landscape. (In contrast, in a random walk, the particle changes nothing of its environment, and is thus a passive walker.) For example, ants are living active walkers. When an ant moves, it releases chemicals of a certain type and hence changes the spatial distribution of the chemical concentration. Its next step is moving towards positions of higher chemical concentration. In this case, the chemical distribution is the deformable landscape.

Active walk has been applied successfully to a number of complex systems coming from the natural and social sciences. Examples include pattern formation in physical, chemical, and biological systems such as surface-reaction induced filaments and retinal neurons, the formation of fractal surfaces, anomalous ionic transport in glasses [Lam 1997: 359-399], granular matter [Baldassarri et al. 2001], population dynamics, bacteria movements and pattern forming, food foraging of ants [Schweitzer et al. 1997; Helbing et al. 1997b], spontaneous formation of human trails [Helbing et al. 1997a, 1997b] and cart tracks [Huang et al. 2002], oil recovery [Yuan et al. 1999], and economic systems [Lam 1997, 1998, 2000; Lam & Pochy 1993; Schweitzer 2003].

In the last two years, the close connection between active walk and history is suggested [Lam 2000, 2003]. The connection comes naturally. When historians write narrative histories they use such phrases as "leave a mark in history" or "follow in a giant's footsteps." It follows that to model these scientifically, one needs a deformable landscape so that marks and footsteps can be left on it. Moreover, one needs a walker to do these things. As noted by the Chinese, through the action of individuals, bad things can sometimes be turned into good things, and vice versa; the walker should be an active walker. The active walk model, then, fits naturally into historical analyses.

6 Modeling History as a Complex Dynamical System

The first step in the scientific study of any subject is the collection of empirical data. This step is followed in both physics and history. The next step is to summarize the data, which leads usually to some *empirical* laws.

For example, for water, we have the Bernoulli's equation, which states that the sum of pressure, kinetic energy per unit volume, and gravitational potential energy per unit volume remains constant along a streamline of an ideal fluid. In history, there are fewer attempts at empirical laws, the existence of which is doubted by some historians. An example of such laws is Michael Shermer's "model of contingent-necessity," which states essentially that history results from the combination of contingency and necessity, with the former being more important in the early stages of a historical sequence [Shermer 1995].

6.1 Three research levels

Beyond the empirical level, research in history and physics differ from each other considerably. For example, physicists go on to study water at three different levels. The first is the *phenomenological* level. Based on a few very general symmetry laws, a set of coupled phenomenological equations, the Navier-Stokes equations, are written down. The equations do not even require that water is made up of molecules, and are still being used today to understand fluid flows and in the design of airplanes.

After the establishment of the molecular nature of water and the availability of powerful computers, water is studied at a deeper level—the *bottom-up* level. In molecular dynamics simulations, the motion of a few hundred number of water molecules are tracked at every time step, with calculations based on Newton's second law of motion and assumed pairwise molecular interactions. Here simplifications on the molecular interaction are made, and the number of molecules used is far below that in the real case, which is about 10^{23} molecules in one cubic-centimeter volume. In other words, physicists "cheat." The real details are not needed or used—that is how progress in physics is made and made possible.

The third level is the *artificial* level. In lattice gas automata, very artificial molecules move on a triangular lattice, with oversimplified rules of movement and collision. For example, when two molecules collide head-on horizontally, they move away from each other at 60^0 from the horizontal, either to the right or to the left; and they simply move one step forward if they encounter no other molecules. There are no such molecules in the world, but the result from this method is the same as that obtained from the other two levels. This is ultimate cheating on nature, and, again, this is how good physics is done.

In contrast, there is no study in history that matches any of these three levels. The problem is that historians tend to believe that to gain a good understanding of history, detail knowledge about the events or human interactions are required, the more the better. In other words, they aim immediately at the bottom-up level, and that is very difficult in human affairs. The lessons from physics are: (1) You don't have to know the whole thing in details. (2) Progress can be made by simplifications that still grasp the few essential driving factors. In what follows, history will be studied at the phenomenological level using active walks, and the artificial level using artificial societies.

6.2 Three stochastic systems

As stated above, historical processes are stochastic, resulting from a combination of chance and necessity. Chance is another word for contingency; and necessity, rules. The two combine to give rise to historical laws. The situation is like that in a chess or soccer game. There are a few basic rules that the players have to obey, but because of contingency, the detail play-by-play of each game is different, and the outcome of the game could be easily predicted, say, when the two competing persons or teams differ very much in strength. In principle, the rules governing historical processes, like those in a chess or soccer game, can be guessed by someone with sufficient skills and patience.

To gain insights into stochastic processes in general, let us look at three stochastic, physical examples. The first example is balls running through a triangular lattice of pegs on an inclined plane, one at a time. The first ball is released at the center of the upper horizontal line. It hits a peg and has equal chance of going to the left or right peg one row below. Hitting the second peg, it has equal chance again of going to the left of right peg on the row below, etc. This is a purely random process; the exact path of the ball cannot be predicted. Furthermore, if a second ball is released from the same spot as before, knowing the path of the first ball does not help us to predict the path of the second ball. On the other hand, if a large number of balls are released one at a time, like the first two balls, we can predict the number of balls landing at a particular location on the bottom line. It is simply a Gaussian distribution, also called the normal distribution. The lesson from this example is that even when it is futile to predict the exact path of each ball, it is still possible to say something about its landing statistically. But history is not a random process; the similarity stops here.

The second example is balls running down a deformable inclined plane (without the pegs), one at a time. Each ball is an active walker. In this case, assuming that the imprints of each ball rolling on the plane are retained long enough and are accumulative, the path of the first ball will influence that of the second ball, and so forth. Knowing the tracks of the first few balls could lead to prediction of those of later balls. It is a history-dependent stochastic process. And history is like this, an active walk process.

Our third example is a random walk on a horizontal plane. Being a random walk, its exact path cannot be predicted. But we can ask other questions. We can ask, for example, What is the morphology of the tracks? What is the dimension of this filamentary track pattern? Is it a fractal? How does its end-to-end distance vary with time? Physicists, in fact, have asked these questions and found answer to them. The same set of questions can be asked about any track patterns, including those left by active walkers. The lesson here for history is that knowing the historical track, usually a difficult task by itself, should not be the only thing obsessed by historians.

Here three examples, all involving the use of active walks, are presented to illustrate this point.

6.3 Three modeling applications

Example 1. *Modeling economic history: Why an initially disadvantageous product can catch up and win out in the market?* A Florence cathedral clock, designed in 1443, has hands that move "counterclockwise" around its dial [Arthur 1990]. Consequently, two types of clocks, with hands moving in different directions, must be in the market in those early years. However, since sundials, the timepiece before the invention of clocks, in the north hemisphere have the pointer's shadow moving clockwise, people in Europe are more comfortable with clocks having hands moving clockwise, too. Those clocks, like the Florence cathedral clock, are thus "inferior" products and they are soon run out of the market. That is why we are now left with only one type of clocks, those having hands running clockwise.

This is the case of an inferior product losing out, which is not at all surprising. What is surprising is the case of an inferior product winning out. The QWERTY keyboard, the type we are using today, is such an example [David 1986]. Invented in 1867, this keyboard is designed to slow down our typing so the mechanical parts will not be jammed that easily. Other superior designs, such as the Maltron keyboard—with 91% of the letters used frequently in English on the "home row" compared to 51% for the QWERTY design—coexist in the market but all lose out. Why?

To understand this, an active walk model with a single particle jumping between two sites is constructed. Each site, representing one product, is given a "fitness" height V_i, $i = 1, 2$. When the particle lands on a site, that corresponding product is bought by a customer, and the height of the site is increased by a fixed amount a. The probability that site i is chosen by the particle is proportional to $f(V_i)$, a given monotonic increasing function of V_i. (Note that the particle may stay in the same site.) A particular choice is $f(V_i) = \exp(\beta V_i)$, where the "inverse temperature" β varies from zero to infinity. The parameter β represents the combined effect of the rationality of the customers and how effective information is passed among the customers (through personal contacts or advertisements). A zero β corresponds to the two products being chosen randomly; an infinite β corresponds to the same product chosen all the time.

This two-site active walk model can be mapped to a one-dimensional position-dependent probabilistic walk and is solvable. Solutions of the model depend on the parameter β and the initial height difference x_0. Our analytic and numerical results show that (1) for zero β and $x_0 = 0$, the two products coexist in the market. (2) For infinite β, the product first picked by a customer always wins out if $x_0 = 0$, and the one with initial advantage wins out if $x_0 \neq 0$. (3) For $x_0 = 0$ and finite β, each product has equal chance of winning, but which product actually wins out is unpredictable. (4) For finite x_0 and β, the product with an initial advantage has a higher chance of winning, but the other product has a non-zero chance of catching up and win, too. And there is an optimal time that this chance of catching up becomes a maximum, implying that the initially disadvantageous product should stay in the market and not give up at least until this optimal time.

Of course, the initially disadvantageous product can change the rule of the game by increasing its own a, for example, by improving its quality or starting an advertisement campaign, or both. With sufficient real data, our model can be fitted to describe the competition between real products.

Example 2. *Modeling evolutionary history: Rewinding life's "tape," or how important is contingency in survival?* In 1989, Stephen Jay Gould published the book *Wonderful Life* [Gould 1989]. From the fossil record found outside of Vancouver, Canada, it seems that some "advance" organisms (with many legs, say) that should survive are wiped out suddenly. From this *one* data point, Gould concludes that contingency is extremely important, i.e., not the fittest will survive, contrary to what Darwin's evolutionary theory asserts. He asks: If life's tape is replayed, will history repeats itself and humans can still be found on Earth? His own answer is no. Debates go on but nothing is done seriously and scientifically. Worse yet, there is no second data point forthcoming. Our active walk model throws light on this debate [Lam 1998]. We find that the relative importance of contingency, versus necessity, could depend on where in the parameter space the system belongs. If our world happened to sit in the *sensitive zone*, in which chance is very important, then history could indeed be very different if life's tape was replayed. Otherwise, history could be repeated, more or less. The problem is to know where our world sits.

Example 3. *Modeling social history: Will all societies end up as liberal democratic societies?* Francis Fukuyama, considered one of fifty key thinkers on history [Hughes-Warrington 2000], published in 1989 an article, "End of history?". He asserts that every human being needs two satisfactions, viz., economic wellbeing and "recognition," with the latter meaning respect by others. He argues that since the liberal democratic society is the only one that can satisfactory its citizens on these two basic needs, consequently, given enough time, all societies will end up as liberal democratic societies. And that will be the end of history if history is understood to be directional change in societal forms [Fukuyama 1992]. Misunderstandings of Fukuyama's thesis ensure and debates go on in the history profession. Nothing is done scientifically to settle the issue.

In our view, the two human needs suggested by Fukuyama should be generalized to "body satisfaction" and "soul satisfaction." After all, body and soul (or spirit) comprise the whole of a human being. And we know for sure, for example, when someone joins a revolution to change the society, the person may give up her or his life before the revolution succeeds, if at all—and recognition is not in the person's mind. The degrees of satisfaction of "body" and "soul" in each society can be quantified by an index, obtained from a survey of its citizens.

To test Fukuyama's thesis, one can represent each society as an active walker, a particle, moving in the two-dimensional space of "body" and "soul" indices. At each point in this space, a fitness potential can be defined. The movement of each particle (usually, but not always, up the scales) will change the fitness landscape and influence the movement of other particles. The problem will be to find out, under what circumstances, all the particles will cluster together at the location corresponding to a liberal democratic society. It is thus a problem of clustering of active walkers in a two-dimensional deformable landscape. (The model can be generalized to include the possibility that two particles may combine into one, and some particles may split into two or three—some kind of chemical reactions—corresponding to the case in history that countries may get unified or fragmented in time.)

Such a problem has been studied before in physics in another context, and clustering of active walkers indeed occurs [Schweitzer & Schimansky-Geier 1994]. The corresponding investigation in history as outlined above will bring Fukuyama's historical study one level up, to the scientific level, and serve as an example in other cases.

7 Predicting the Future and Retrodicting the Past

In physics, the future of a complex system can sometimes be predicted without knowing the nature or mechanism of the system itself. Time series analysis is such a method, which has been applied in predicting the stock market, in particular.

Another possible avenue is to turn the data into tracks in some suitable space and, assuming that these tracks are generated by active walkers, one can then figure out how the landscape is modified and go on to predict the future or retrodicting the past. This approach is encouraged by the success in using active walk to reproduce the filamentary patterns in dielectric breakdown sequences [Lam 1997].

It is sometimes believed that due to the intrinsic complexity and the stochastic nature of history, one can never regenerate the historical "tracks." This is not true. Take the human trail formation for example [Helbing 1997a]. Once the present track pattern is reproduced successfully, through an active walk model in this case, the computer program can be rerun from time zero to show how the tracks are formed in the past, and how it will further evolve in the future. The same applies to any successful simulations in other systems.

8 Artificial History

Artificial Life, as a discipline, is created by Chris Langton in 1989 [Langton 1989]. Subsequently, artificial societies are investigated [Epstein & Axtell 1996]. Since human history, by definition, cannot be reproduced, it will be useful to treat the development in an artificial society as some kind of *artificial history*. Ask historians to study it seriously, pretending that they are real, and come up with their summaries and findings, preferably in the form of historical laws. The findings or laws can then be tested against data collected from the many reruns of the artificial society. Lessons learned from artificial history should help in the understanding of human history, in the same sense that artificial life has contributed to the progress in the study of real life forms.

A place to start is to use commercially available computer games such as *The Sims*, as an artificial society. Time evolution of data for the characteristics of the players in this community is collected. We are in the process of turning these data into tracks and using active walk model to figure out its history.

Needless to say, the predicting capacity coming from the study of artificial history is relevant in many real applications, such as the writing of better games, movie plots, novels, and military game plays with computers.

9 Conclusion

Historians have contributed significantly in preserving history. But history is too important to be left to the historians alone. Physicists can and should go into it. The complexity and stochastic nature of historical processes should not be a deterrent in studying them; we have experience and tools gained from studying other complex systems. It may be difficult for practicing historians to learn advanced mathematics and programming, but the teaming up between historians and physicists should be possible and fruitful. On the other hand, it may be time to make mathematics and programming required skills in the training of historian students in the universities, so the future generations of historians are better prepared scientifically. Asking history majors to take a physics minor will help, too. Similarly, it is recommended that physics majors should take a course on historiography if they want to work on Histophysics.

History is much, much more than story telling. Narrative history is not enough. Active walk is found useful and natural in the modeling of history. More real data are helpful for sharpening the models and comparing them with the real world. Paraphrasing a famous saying, here is my prediction for the future:

> The future history of science is the story of physics hijacking topics from social sciences.

This is only the beginning.

References

Arthur, B. [1990]. Positive feedbacks in the economy. Sci. Am., Feb. 1990: 92.

196 *Lui Lam*

Bak, P. [1996]. *How Nature Works*. New York: Copernicus.

Bak, P., Tang, C. & Wiesenfeld, K. [1987]. Self-organized criticality: An explanation of 1/f noise. Phys. Rev. Lett. **59**: 381-384.

Baldassarri, A., Krishnamurthy, S., Loreto, V. & Roux, S. [2001]. Coarsening and slow dynamics in granular compaction. Physical Review Letters **87**: 224301.

Breisach, E. [1994]. *Historiography: Ancient, Medieval, and Modern*. Chicago: University of Chicago Press.

Buchanan, M. [2000]. *Ubiquity: The Science of History...or Why the World is Simpler Than We Think*. London: Weidenfeld & Nichols.

Burke, P. [1990]. *The French Historical Revolution: The Annales School 1929-1989*. Cambridge: Cambridge University Press.

Cowley, R. & Ambrose, S. E. (eds.) [2000]. *What If?: The World's Foremost Military Historians Imagine What Might Have Been*. Berkeley: Berkeley Publishing Group.

Collingwood, R. G. [1946]. *The Idea of History*. New York: Oxford University Press.

David, P. A. [1986]. Understanding the economics of QWERTY: the necessity of history. *Economic History and the Modern Economist*, Parker, W. N. (ed.). New York: Blackwell.

Derrida, J. [1976]. *Of Grammatology*. Baltimore: Johns Hopkins University Press.

Epstein, J. M. & Axtell, R. [1996]. *Growing Artificial Societies: Social Science from the Bottom Up*. Cambridge, MA: MIT Press.

Evans, R. J. [1997]. *In Defence of History*. London: Granta Books.

Ferguson, N. (ed.) [1999]. *Virtual History: Alternatives and Counterfactuals*. New York: Basic Books.

Fukuyama, F. [1992]. *The End of History and the Last Man*. New York: Avon Book.

Gilderhus, M. T. [2000]. *History and Historians: A Historiographical Introduction*. Upper Saddle River, NJ: Prentice Hall.

Gould, S. J. [1989]. *Wonderful Life: The Burgess Shale and the Nature of History*. New York: Norton.

Helbing, D., Keltsch, J. & Molnar, P. [1997a]. Modelling the evolution of human trail systems. Nature **388**: 47.

Helbing, D., Schweitzer, F., Keltsch, J. & Molnar, P. [1997b]. Active walker model for the formation of human and animal trail systems. Physical Review E **56**: 2527.

Huang, S.-Y., Zou, X-W., Zhang, W.-B. & Jin, Z.-Z. [2002]. Random walks on a (2+1)-dimensional deformable medium. Physical Review Letters **88**: 056102.

Hughes-Warrington, M. [2000]. *Fifty Key Thinkers on History*. New York: Routledge.

Lam, L. [1997]. Active walks: Pattern formation, self-organization and complex systems. *Introduction to Nonlinear Physics*, Lam, L. (ed.). New York: Springer-Verlag.

Lam, L. [1998]. *Nonlinear Physics for Beginners: Fractals, Chaos, Solitons, Pattern Formation, Cellular Automata and Complex Systems*. River Edge, NJ: World Scientific Press.

Lam, L. [2000]. How nature self-organizes: Active walks in complex systems. Skeptic **8**(3): 71.

Lam, L. [2003]. Modeling history and predicting the future: The active walk approach. *Humanity 3000, Seminar No. 3 Proceedings*. Bellevue, WA: Foundation For the Future. pp 109-117.

Lam, L. [2004]. A science-and-art interstellar message: The self-similar Sierpinski gasket. Leonardo **37**(1): 37-38.

Lam, L. [2005]. Active walks: The first twelve years (Part I). Int. J. Bifurcation and Chaos **15**: 2317-2348.

Lam, L. [2006]. Active walks: The first twelve years (Part II). Int. J. Bifurcation and Chaos **16**: 239-268.

Lam, L. & Pochy, R. D. [1993]. Active walker models: Growth and form in nonequilibrium systems. Computers in Physics **7**: 534.

Langton, C. G. [1989]. *Artificial Life*. Menlo Park, CA: Addison-Wesley.

Lubkin, G. [1998]. A personal look back at *Physics Today*. Physics Today, May 1998: 24.

Mantegna, R. N. & Stanley, H. G. [2000]. *An Introduction to Econophysics*. New York: Cambridge University Press.

McNeill, W. H. [1998]. History and the scientific worldview. History and Theory **37**: 1

Munslow, A. [1997]. *Deconstructing History*. New York: Routledge.

Pigliucci, M. [2000]. Chaos and complexity: Should we be skeptical? Skeptic **8**(3): 62.

Schweitzer, F. [2003]. *Brownian Agents and Active Particles*. New York: Springer Verlag.

Schweitzer, F., Lao, K. & Family, F. [1997]. Active random walkers simulate trunk trail formation by ants. BioSystems **41**: 153.

Schweitzer, F. & Schimansky-Geier, L. [1994]. Clustering of active walkers in a two-component system. Physica A **206**: 359.

Shermer, M. [1995]. Exorcising Laplace's demon: Chaos and antichaos, history and metahistory. History and Theory **34**: 59.

Stanford, M. [1998]. *An Introduction to the Philosophy of History*. Malden, MA: Blackwell.

White, H. [1973]. *Metaphysics: The Historical Imagination in Nineteenth-Century Europe*. Baltimore: Johns Hopkins University Press.

Wilson, E. O. [1975]. *Sociobiology: The New Synthesis*. Cambridge, MA: Harvard University Press.

Wilson, E. O. [1999]. *Consilience: The Unity of Knowledge*. New York: Vintage Books.

Wolf, W. P. [1994]. Is physis education adapting to a changing world? Phys. Today, Oct. 1994: 48-55.

Yuan, J.-Y., B. Tremblay, B. & Babchin, A. [1999]. A wormhole network model of cold production in heavy oil. SPE 54097, Society of Petroleum Engineers (SPE) International Thermal Operations and Heavy Oil Symposium, Bakersfield, CA, March 17-19, 1999.

Published: Lam, L. [2002]. Histophysics: A new discipline. Modern Physics Letters B **16**: 1163-1176.

History: Present and Future

Lui Lam

1 What Really Is History?

History is the most important discipline of study [Lam 2002]. Yet, the link between history and science is underdeveloped.

Science is the study of nature and to understand it in a unified way. Nature, of course, includes all material systems. The system investigated in history is a (biological) material system consisting of *Homo sapiens*. Consequently, history is a legitimate branch of science, like physics, biology, paleontology and so on. In other words, history is not a subject that is beyond the domain of science. History can be studied scientifically [Lam 2002].

By definition, history is about past events and is irreproducible. In this regard, it is like the other historical sciences such as cosmology, astronomy, paleontology, and archeology. The way historical sciences advance is by linking them to systems presently exist, which are amenable to tests. For example, in astronomy, the color spectra of light emitted in the past from the stars and received on earth can be compared with those observed in the laboratory; the identity of the elements existing in stars is then identified. Similarly, the psychology, thoughts and behaviors of historical players can be inferred from those of living human beings, which can be learned by observations, experimentations, and neurophysiological probes [Feder 2005].

The system under study in history is a many-body system. In this system, each "body" is a human being, called a "particle" here; these particles have internal states (due to thinking, memory, etc.) which sometimes can be ignored. Each constituent particle is a (non-quantum mechanical) classical object and is distinguishable; i.e., each particle in the system can be identified individually. This many-body system is a heterogeneous system, due to the different sizes, ages, races, etc. of the particles.

A historical process, expressed in the physics language, is the time development of a subset or the whole system of *Homo sapiens* that happened during a time period of interest in the past. History is therefore the study of the past dynamics of this system. Historical processes are stochastic, resulting from a combination of contingency and necessity. In modeling, contingency shows up as probability and necessity is represented by rules in the model. The situation is like that in a chess or soccer game. There are a few basic rules that the players must obey, but because of contingency, the detail play-by-play of each game is different. In principle, someone with sufficient skills and patience can guess the rules governing historical processes, like those in a chess or soccer game.

In some cases, these two ingredients of contingency and necessity, through self-organization, may combine to give rise to discernable historical trends or laws. In other cases, either no laws exist at all, or the laws are not recognized by whoever studying them. Whether there actually exist historical laws cannot be settled by speculations or debates, no matter how good these speculations or debates are. A historical law exists only when it is found and confirmed. Furthermore, any historical law—like that in physics—has its own range of validity, which may cover only a limited domain of space and time.

Yet most people, including many historians, do not believe that any historical law could exist [Gardiner 1959]. They are wrong. Figure 1a shows a historical law; it exists. This power law on the statistics of war deaths is due to Lewis Richardson [1960]. Similar power laws are found in the distribution of earthquake intensities, called Gutenberg-Richter law (Fig. 1b), in the ranking of city populations, and in many other systems [Zipf 1949]. The fact that human events like wars obey the same statistical law as inanimate systems indicates that the human system does belong to a larger class of dynamical systems in nature, beyond the control of human intentions and actions, individually or collectively. More historical laws are given below.

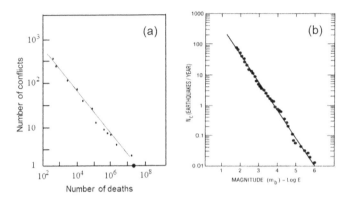

Fig. 1. (a) Statistical distribution of war intensities. Eighty-two wars from 1820 to 1929 are included; the dot on the horizontal axis comes from World War I. The graph is a log-log plot; a straight line indicates a power law. (b) Distribution of earthquake sizes in the New Madrid zone in the United States from 1974 to 1983 [Johnston & Nava 1985]. The points show the number of earthquakes with magnitude larger than a given magnitude m.

2 Two Quantitative Laws and a Quantitative Prediction in Chinese History

China has a long, unbroken history, which is probably the best documented [Huang 1997]. The dynasties from Qin to Qing range from 221 BC to 1912, with 31 dynasties and 231 regimes spanning a total of 2,133 years [Morby 2002]. (A regime is the reign of one emperor; a dynasty may consist of several regimes.) Some of these dynasties overlap with each other in time.

Let τ_R be the *regime* lifetime, and τ_D the *dynasty* lifetime; both are integers measured in years. The histogram of τ_R is found to obey a power law (Fig. 2), with an exponent equal to -1.3. This result implies that the dynamics governing regime changes is not completely up to the emperors, statistically speaking, but share some common traits with other complex systems such as those displayed in Fig. 1. This is the first *quantitative* law about Chinese history.

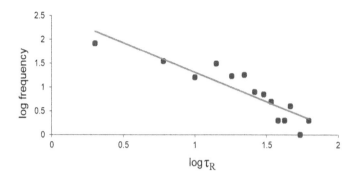

Fig. 2. Log-log plot of the histogram of regime lifetime τ_R, from 221 BC to 1912 in the Chinese history. The exponent of the straight-line fit is −1.3. (Bin width of the histogram is 4 years.)

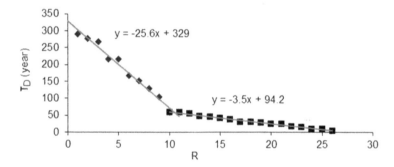

Fig. 3. The Bilinear Effect: Data points in a Zipf plot fall on two straight lines, with a transition point at the crossing. The data here are the lifetime of Chinese dynasties τ_D, from Qin to Qing. The transition occurs at around $\tau_D = 57$ years. The longest $\tau_D = 289$ years, the lifetime of Tang dynasty. R represents the "rank."

The second *quantitative* law is shown in Fig. 3, where the lifetime of 26 dynasties is arranged in a monotonic decreasing order. Figure 3 is a Zipf plot, with the largest τ_D assigned rank 1 (R = 1), the second largest rank 2, etc. The same rank is assigned to separate τ_D that is identical to each other, to ensure that the curve is monotonic decreasing, i.e., no horizontal parts, and hence there are less than 31 data point in Fig. 3. They fall on two straight lines, which I name the *Bilinear Effect*. What it means is that

1. The "curse of history," as Chinese dynasties are concerned, does exist.

2. A dynasty can survive every 3.5 years if it lasts 57 years or less; beyond that, every 25.6 years, meaning dynasty lifetime is discrete, or "quantized."

3. There is a transition point, at around $\tau_D = 57$ years, separating these two different behaviors.

This is the phenomenon that a human entity, a dynasty in this case, becomes stronger after existing for a period of time. The mere fact of survival reinforces its strength, through adaptive learning, restructuring or other means. Similar behavior is known to exist in the lifetime of corporations or biological species. What is surprising here is the presence of two linear lines and a sharp transition point.

The bilinear effect, exemplified in a Zipf plot, is a general phenomenon found also in other human affairs and complex systems. It constitutes a new class of behavior in Zipf plots, apart from the two well-known classes of power laws [Zipf 1949] and stretched exponents [Laherrère & Sornette 1998].

Here is a quantitative *prediction* derived from Fig. 3: Any dynasty after Qing, if exists, either

1. Will last 303 years or less and fall more or less on the two lines in Fig. 3, or

2. Will end definitely and exactly in its year 329.

Note that Fig. 3, in contrast to Fig. 2, is *not* a statistical plot, and this prediction is not a statistical prediction. As far as I know, no other quantitative non-statistical historical laws and predictions are known, to the historians or others. Note that these two laws and the prediction concerning Chinese history are model independent.

3 Modeling History by Active Walks

An important step towards the scientific study of any subject is to pick the right tools to tackle it. Historical processes are stochastic (i.e., with probability involved somewhere), resulting from necessity and contingency. The kind of physics suitable for handling many-body systems ingrained with contingency is statistical physics. Furthermore, the historical system is an open system with constant exchange of energy and materials with the environment and is never in equilibrium. Thus, for history, the appropriate tool is the stochastic methods developed in the statistical physics of nonequilibrium systems [Lam 1998; Paul & Baschnagel 1999; Sornette 2000]. In particular, active walk can be used to model history [Lam et al. 1992; Lam 2005].

In fact, a common metaphor for history is that it is like a river flowing; people talk about the "river of history." This metaphor is not so off mark if the water flowing in the river is able to reshape the landscape as it flows, and the river is allowed to branch from time to time under certain conditions. Active walk (AW) is a natural in matching such a metaphor. It is then no surprise that a whole class of probabilistic AW models are found to be relevant in studying history [Lam 2002]. For example,

1. The two-site AW model (see Sec. 4 in [Lam 2005]) is able to explain the real case in economic history that an inferior product such as the QWERTY keyboard [David 1986: 30-49] can actually win out in the market [Lam 2002]. Other examples are the competition between Apple computers and PCs, as well as VHS and Beta videotapes.

2. The active-walk aggregation (AWA) model is able to shed light on the debate in evolutionary history, initiated by Stephen Jay Gould [1989]. The question raised by Gould is that if life's "tape" is replayed, will history repeat itself and humans still be found on Earth? Gould's answer is "no"; the AWA model says "maybe" [Lam 1998]. It is "maybe" because if the world lies in the *sensitive zone* of the *morphogram* (the space of control parameters) then the growth outcome may not be repeatable; otherwise, it is repeatable.

3. An intriguing prediction in social history was given by Fukuyama [1989], which asserts that every human being needs two kinds of satisfaction, viz., economic well-being and "recognition," with the latter meaning respect by others. Since only liberal democratic

society can provide these two satisfactions, argued Fukuyama, all societies will end up as liberal democratic societies, given enough time. And that will be the end of history if history is understood to be directional change in societal forms. A phenomenological AW model with multiwalkers can be used to test this prediction [Lam 2002]. In this model, each country is represented as an active walker, a particle, moving on a common deformable landscape in a two-dimensional (2D) space. The problem will be to find out, under what conditions, if any, all the particles will cluster together at the location corresponding to a liberal democratic society (Fig. 4). It is thus a problem of clustering of active walkers on a 2D landscape.

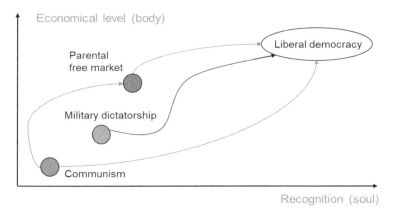

Fig. 4. Sketch of an active walk model for the evolution of political systems.

4 History in the Future

The importance of history can be seen, e.g., through its negative impact on human lives. Powerful political leaders could mistake an unproven historical hypothesis as firm theory, apply it to a confined population and cause millions of deaths in a few short years [Lam 2002]. Another example is the recent massive protests in China due to different interpretation of past history involving two countries (Fig. 5). Yet, in spite of its importance, the physical basis of history is unrecognized by most historians. For instance, in the historiography textbook *The New Nature of History*

[Marwick 2001: 248] the alleged "fundamental" differences between history and the sciences are listed:

1. Fundamental difference in the subject of study: Natural sciences concern natural world and physical world; history concerns human beings and human society, very different in character.

2. No controlled experiments by historians.

3. Historians develop theories and theses, but not concerned with developing laws and theories like that in sciences.

4. History studies do not have predicting power.

5. Relations and interactions in history studies are not expressed mathematically.

6. Historians report their findings in prose (articles or books), not in terse research articles.

Fig. 5. Protests in China, April 2005. (a) "Face Up to History" is the slogan on the left placard. (b) "Protect Diaoyudao" is a historical issue also raised in the protests. Diaoyudao, or Diaoyutai, is a group of tiny islands in the East China Sea. The protect-Diaoyutai movement was started by overseas Chinese students in the United States at the end of 1970 [The Seventies Monthly 1971].

Unfortunately, all six points are *wrong*, for the following reasons:

1. As explained Sec. 1, humans and thus human society are material systems, which are part of natural science. Human society share

same characteristics as other inanimate complex systems, as demonstrated in Figs. 1-3.

2. Some physical disciplines like astronomy and archeology also do not have controlled experiments.

3. It is untrue that all historians are not concerned with developing laws and theories in history. Some tried, not very successfully, partly due to their inadequate training in using scientific tools. Historical laws do exist, as shown in Sec. 2 and Figs. 1-3

4. History studies, like that in Sec. 2, do have predicting power.

5. Relations and interactions in historical studies can be expressed mathematically. An example is the landscape theory of Axelrod and Bennett [1993] to show how and why 17 European nations in World War II aligned themselves into two large groups. The pairwise propensities between nations are assigned numerical values, and the configuration energy in the (fixed) landscape is given in equations.[1]

6. Historians do report their findings in research articles, terse or not. That is why there exist quite a number of history journals, such as *History and Theory* and *American Historical Review*. It is true that many historians still skip the peer-reviewed journals and directly report their findings in books—not a healthy thing for the history discipline, epistemologically speaking [Lam 2002]. These are actually popular history books, like the popular science books written by physicists.

In the case of the history profession, there are at least three reasons behind this practice. (1) Many research results in history are still at the data gathering and empirical analysis stage, not very technical and can be presented in narratives. (2) There is enough number of readers out there who is willing to pay to find out what happened to their ancestors or their own kind in the past. In contrast, not that many will pay to learn what happened to the electrons. Bad for physics. (3) Historically, before history became a professional discipline in the universities in the second half of the 19th century, historians had to earn their living by writing books that are readable and salable to the public [Stanford 1998: 228]. In other words, writing popular history books was a survival need for historians, a tradition carried over to today.

These errors are due to misunderstanding of the nature of science, and the neglect of the material basis of the historical system itself. There are two reasons that historians, past and present, failed to find historical laws:

1. Historians kept on working at the bottom-up level, one of three levels (empirical, phenomenological, and bottom-up) in research of any discipline [Lam 2002]. That is, they try to understand history from the knowledge of the historical figures. This is very difficult and not the place that historical laws been found so far, which are found at the empirical level as shown in Figs. 1-3.

2. The inadequate scientific training received by historians. For example, the Chinese dynasty data have been lying there for 93 years; the plots in Figs. 2 and 3 could be carried out by hand without computers, and even by high school students. But unless one knows about power laws and the existence of the Zipf plot, there is no motivation to do so. And these are current topics in the study of complex systems. Ironically, the Zipf plot was first done by George Zipf (1902-1950), a Harvard linguist, with data from the humanities and social science.

5 Conclusion

Recently and surprisingly, while the importance of history is well recognized in Hong Kong, the history department of the University of Hong Kong is threatened with closure because it fails to attract enough number of students [Xie 2005: A15]. Anyway, it is time for all history departments to revamp their curriculum by taking, e.g., the following steps:

1. Increase the mathematical skill of their students.

2. Go beyond storytelling and make history research more technical and scientific.

3. Create a course on the physics of history (or *Histophysics* [Lam 2002, 2004]).

This revamp will help current students to become better historians after they graduate and may appeal to a new class of incoming students: those with a technical background but feel more attracted to the humanities than to the traditional science or engineering courses.

Notes

1. See [Galam 1998] for a comment on this work, and the following response by the original authors.

References

Axelrod, R. & Bennett, D. S. [1993]. A landscape theory of aggregation. British J. Political Sci. **23**: 211-233.

David, P. A. [1986]. Understanding the economics of QWERTY: The necessity of history. *Economic History and the Modern Economist*, Parker, W. N. (ed.) New York: Blackwell.

Feder, T. [2005]. Lab webs brain research and physics. Phys. Today, April 2005: 26-27.

Fukuyama, F. [1989]. The end of history? The National Interest **16** (summer): 3-18.

Galam, S. [1998]. Comment on "A landscape theory of aggregation." British J. Political Sci. **28**: 411-412.

Gardiner, P. (ed.) [1959]. *Theories of History*. Glencoe, IL: The Free Press.

Gould, S. J. [1989]. *Wonderful Life: The Burgess Shale and the Nature of History*. New York: Norton.

Huang, R. [1997]. *China: A Macro History*. Armonk, NY: M.E. Sharpe.

Laherrère, J. & Sornette, D. [1998]. Stretched exponential distributions in nature and economy: "fat tails" with characteristic scales. Eur. Phys. J. B **2**: 525-539.

Lam, L. [1998]. *Nonlinear Physics for Beginners: Fractals, Chaos, Solitons, Pattern Formation, Cellular Automata and Complex Systems*. Singapore: World Scientific.

Lam, L. [2002]. Histophysics: A new discipline. Mod. Phys. Lett. B **16**: 1163-1176.

Lam, L. [2004]. *This Pale Blue Dot: Science, History, God*. Tamsui: Tamkang University Press.

Lam, L. [2005]. Active walks: The first twelve years (Part I). Int. J. Bifurcation and Chaos **15**: 2317-2348.

Lam, L, Freimuth, R. D., Pon, M. K., Kayser, D. R., Fredrick, J. T. & Pochy, R. D. [1992]. Filamentary Patterns and Rough Surfaces. *Pattern Formation in Complex Dissipative Systems*, Kai, S. (ed.). Singapore: World Scientific.

Marwick, A. [2001]. *The New Nature of History*. Chicago: Lyceum.

Morby, J. E. [2002]. *Dynasties of the World*. Oxford: Oxford University Press.

Paul, W. & Baschnagel, J. [1999]. *Stochastic Processes: From Physics to Finance*. New York: Springer.

Richardson, L. F. [1960]. *Statistics of Deadly Quarrels*. Pittsburgh, PA: Boxwood.

Sornette, D. [2000]. *Critical Phenomena in Natural Sciences*. New York: Springer.

Stanford, M. [1998]. *An Introduction to the Philosophy of History*. Malden, MA: Blackwell.

The Seventies Monthly (ed.) [1971]. *Truth Behind the Diaoyutai Incident*. Hong Kong: The Seventies Monthly.

Xie, Xi [2005]. Historical feel regained. World Journal (Millbrae, CA), April 16, 2005.

Zipf, G. K. [1949]. *Human Behavior and the Principle of Least Effort*. Cambridge, MA: Addison-Wesley.

Published: Lam, L. [2006]. Active walks: The first twelve years (Part II). International Journal of Bifurcation and Chaos **16**: 239-268. Excerpt here.

History: A Science Matter

Lui Lam

Human history is the most important discipline of study. The complex system under study in history is a many-body system consisting of *Homo sapiens*—a (biological) material system. Consequently, history is a legitimate branch of science, since science is the study of nature which includes *all* material systems. A historical process, expressed in the physics language, is the time development of a subset or the whole system of *Homo sapiens* that happened during a time period of interest in the past. History is therefore the study of the past dynamics of this system. Historical processes are stochastic, resulting from a combination of contingency and necessity. Here, the nature of history is discussed from the perspective of complex systems. Human history is presented as an example of Science Matters (Scimat). Examples of various scientific techniques in analyzing history are given. In particular, two unsuspected *quantitative* laws in Chinese history are shown. Applications of active walk to history are summarized. The "differences" between history and the natural sciences erroneously expressed in some history textbooks are clarified. The future of history, as a discipline in the universities, is discussed; recommendations are provided.

1 What Is History?

Human history is the most important discipline of study [Lam 2002]. Yet, human history, or history in general, as a science is rarely discussed [Lam 2002; Krakauer 2007].

Science is the study of nature and to understand it in a unified way. Nature, of course, includes all material systems. The system investigated in history is a (biological) material system consisting of *Homo sapiens*. Consequently, history is a legitimate branch of science, like physics, biology, and paleontology, and so on. In other words, history is not a subject that is beyond the domain of science. History can be studied scientifically, as shown in a new discipline called *Histophysics* [Lam 2002].

By definition, history is about past events and is irreproducible. In this regard, it is like the other "historical" sciences such as cosmology, astronomy, paleontology, and archeology. The way historical sciences advance is by linking them to systems presently exist, which are amenable to tests. For example, in astronomy, the color spectra of light emitted in the past from the stars and received on earth can be compared with those observed in the laboratory; the elements existing in stars is then identified. Similarly, the psychology, thoughts, and behaviors of historical players can be inferred from those of living human beings, which can be learned by observations, experimentations, and neuro-physiologic probes [Feder 2005].

The system under study in history is a many-body system. In this system, each "body" is a human being, called a "particle" here; these particles have internal states (due to thinking, memory, mood and so on) which sometimes can be ignored. Each constituent particle is a (non-quantum mechanical) classical object and is distinguishable; i.e., each particle in the system can be identified individually. This many-body system is a heterogeneous system, due to the different sizes, ages, races...of the particles.

A historical process, expressed in the physics language, is the time development of a subset or the whole system of *Homo sapiens* that happened during a time period of interest in the past. History is therefore the study of the past dynamics of this system. *Historical processes are stochastic, resulting from a combination of contingency and necessity.*[1] Here, necessity is an assumption, which could only be confirmed by results showing that it really exists; contingence is due to the many other factors not included in the system under study—as usually is the case in many complicated situations —and could be represented as noise in the study of stochastic systems.

In modeling, contingency shows up as probability and necessity is represented by rules in the model. The situation is like that in a chess or soccer game. There are a few basic rules that the players have to obey, but because of contingency, the detail play-by-play of each game is different. In principle, someone with sufficient skills and patience can guess the rules governing historical processes, like those in a chess or soccer game.

In some cases, these two ingredients of contingency and necessity, through self-organization, may combine to give rise to discernable historical trends or laws. In other cases, either no laws exist at all or the laws are not recognized by whoever studying them. Whether there actually exist historical laws cannot be settled by speculations or debates, no matter how good these speculations or debates are. *A historical law exists only when it is found and confirmed* (as indeed is the case as shown in Secs. 2.1 and 2.4). Furthermore, any historical law—like that in physics—has its own range of validity, which may cover only a limited domain of space and time. Yet most people, including many historians, do not believe that any historical law could exist [Gardiner 1959]. They are wrong.

2 Methods to Study History

An important step towards the scientific study of any subject is to pick the right tools to tackle it. Historical processes are stochastic. The kind of physics suitable for handling many-body systems ingrained with contingency is statistical physics. Furthermore, the historical system is an *open* system with constant exchange of energy and materials with the environment and is never in equilibrium. Thus, for history, an appropriate tool is the stochastic methods developed in the statistical physics of nonequilibrium systems [Lam 1998; Paul & Baschnagel 1999; Sornette 2000].

However, there are other tools, too. In fact, there are at least *four* different approaches (see Secs. 2.1-2.4) applicable in understanding history. Examples are given below, with each reflecting either the *empirical*, *phenomenological* or *bottom-up* level commonly found in the scientific development of any discipline [Lam 2002].[2]

2.1 Statistical analysis

Statistical analyses of data are at the *empirical* level, without knowing the mechanism of the processes involved. Two examples are given here.

1. Power law in the distribution of war intensities

Figure 1a shows a historical law of statistical nature; historical laws do exist. The statistical distribution of war intensities obeys a power law[3] (Fig. 1a), first discovered by Richardson [1941] and confirmed by Levy [1983] using a different data set covering 119 wars from 1495-1975. In this new study, war intensity is defined by the ratio of battle deaths to the population of Europe at the time of the war. (Europe is used because for the earlier wars, estimates of the world population are unreliable.) More recently, this conclusion is interpreted by Roberts and Turcotte [1998] in terms of a forest-fire model. Similar power law is found in the distribution of earthquake intensities, called Gutenberg-Richter law (Fig. 1b), in the ranking of city populations, and in many other systems [Zipf 1949]. The fact that human events like wars obey the same statistical law as inanimate systems indicates that the human system does belong to a large class of dynamical systems in nature, beyond the control of human intentions and actions, individually or collectively.

2. Power law in the distribution of Chinese regime lifetimes

Another example is provided by the case in Chinese history. China has a long, unbroken history, which is probably the best documented [Huang 1997]. The dynasties from Qin to Qing range from 221 BC to 1912, with 31 dynasties and 231 regimes spanning a total of 2,133 years [Morby 2002]. (A regime is the reign of one emperor; a dynasty may consist of several regimes.) Some of these dynasties overlap with each other in time.

Let τ_R be the regime lifetime, an integer measured in years. The histogram of τ_R is found to obey a power law (Fig. 2), with an exponent equal to -1.3 ± 0.5 [Lam 2006a, 2006b]. This result implies that the dynamics governing regime changes is not completely up to the emperors, statistically speaking, but share some common traits with other complex systems such as those displayed in Fig. 1. To the best of our knowledge, this is the *first quantitative law concerning Chinese history.*

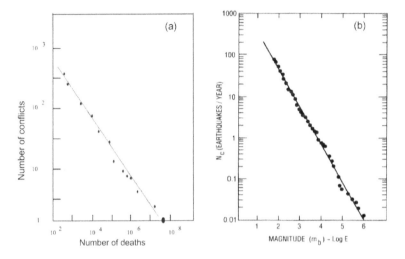

Fig. 1. (a) Statistical distribution of war intensities. Eighty-two wars from 1820 to 1929 are included; the dot on the horizontal axis comes from World War I. (b) Distribution of earthquake sizes in the New Madrid zone in the United States from 1974 to 1983 [Johnston & Nava 1985]. The points show the number of earthquakes with magnitude larger than a given magnitude *m*. The graphs in (a) and (b) are log-log plots; a straight line indicates a power law.

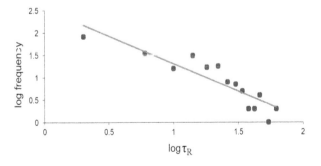

Fig. 2. Log-log plot of the histogram of Chinese regime lifetime τ_R, covering the years from 221 BC to 1912. (Bin width of the histogram is 4 years.)

2.2 Computer modeling

A common metaphor for history is that it is like a river flowing; people talk about the "river of history." This metaphor is not so off mark if the water flowing in the river is able to reshape the landscape as it flows, and the river is allowed to branch from time to time under certain conditions. Active walk (AW) [Lam 2005a, 2006b] is a natural in matching such a metaphor. It is then no surprise that a whole class of probabilistic AW models are found to be relevant in studying history [Lam 2002].

Active walk is a paradigm and method introduced by Lam in 1992 to handle self-organization and pattern formation in simple and complex systems. In AW, a particle (the walker) changes a deformable potential—the landscape—as it walks; its next step is influenced by the changed landscape. For example, ants are living active walkers. When an ant moves, it releases chemicals of a certain type and hence changes the spatial distribution of the chemical concentration. Its next step is moving towards positions of higher chemical concentration. In this case, the chemical distribution is the deformable landscape. Active walk has been applied successfully to a number of complex systems in "natural science"[4] and social science. Examples include pattern formation in physical, chemical and biological systems such as surface-reaction induced filaments and retinal neurons, the formation of fractal surfaces, ionic transport in glasses [Lam 1997], granular matter, population dynamics, bacteria movements and pattern forming, food foraging of ants, spontaneous formation of human trails, oil recovery, river formation, city growth, economic systems, parameter networks [Han et al. 2008] and, recently, human history [Lam 2002, 2004, 2006b]. Here are some examples of application of AW in history—modeling at the *phenomenological* level.

1. Modeling economic history: Why an initially disadvantageous product can catch up and win out in the market?

A Florence cathedral clock, built in 1443, has hands that move *counterclockwise* around its dial [Arthur 1990]. Consequently, two types of clocks, with hands moving in different directions, could be in the market in those early years. However, since sundials, the timepiece before the invention of clocks, in the north hemisphere have the pointer's shadow moving clockwise, people in Europe are more comfortable with clocks having hands moving clockwise, too. Those "counterclockwise" clocks, like the Florence cathedral clock, are thus "inferior" products and they are soon run out of the market. That is why we are now left with only one type of clocks, those having hands running clockwise. This is the case of an inferior product losing out, which is not at all surprising. What is surprising is the case of an inferior product winning out. The QWERTY keyboard, the type we are using today, is such an example [David 1986]. Invented in 1867, this keyboard is designed to slow down our typing so the mechanical parts will not be jammed that easily. Other superior designs, such as the Maltron keyboard—with 91% of the letters used frequently in English on the "home row" compared to 51% for the QWERTY design— coexist in the market but all lose out. Why?

The two-site AW model [Lam 2005a] is able to explain this and other real cases[5] in economic history that an inferior product can actually win out in the market [Lam 2002].

2. <u>Modeling evolutionary history: Rewinding life's "tape," or how important is contingency in survival?</u>

In 1989, Stephen Jay Gould (1941-2002) [1989] published the book *Wonderful Life*. From the fossil record found outside of Vancouver, Canada, it seems that some "advance" organisms (with many legs, say) that should survive are wiped out suddenly. From this *one* data point, Gould concludes that contingency is extremely important, i.e., not the fittest will survive, contrary to what Darwin's evolutionary theory asserts. He asks: If life's "tape" is replayed, will history repeat itself and humans can still be found on Earth? His own answer is "no." Debates go on but nothing is done seriously and scientifically. Worse yet, there is no second data point forthcoming. Our active-walk aggregation (AWA) model [Lam & Pochy 1993; Lam 2005a] is able to shed light on this debate. The AWA model says "maybe" in answering Gould's question [Lam 1998]. It is "maybe" because if the world lies in the *sensitive zone*,[6] then the growth outcome may not be repeatable; otherwise, it is repeatable, more or less. The problem is to know where our world sits.

3. <u>Modeling social history: will all societies end up as liberal democratic societies?</u>

Francis Fukuyama, considered one of fifty key thinkers on history, published in 1989 an article, "End of History?" [Fukuyama 1989]. He asserts that every human being needs two satisfactions, viz., economic wellbeing and "recognition," with the latter meaning respect by others. He argues that since the liberal democratic society is the only one that can satisfy its citizens on these two basic needs, consequently, given enough time, all societies will end up as liberal democratic societies. And that will be the end of history if history is understood to be the directional change in societal forms. Misunderstandings of Fukuyama's thesis ensures and debates go on in the history profession. Nothing is done scientifically to settle the issue.

In our view, the two human needs suggested by Fukuyama should be generalized to "body satisfaction" and "soul satisfaction." After all, body and "soul" (or spirit) comprise the whole of a human being. And we know for sure, e.g., when someone joins a revolution to change the society, the person may give up her or his life before the revolution succeeds, if at all—and recognition is not in the person's mind. The degrees of satisfaction of "body" and "soul" in each society can be quantified by two indices, obtained from a survey of its citizens. To test Fukuyama's thesis, one can represent each society as an active walker, a particle, moving in the two-dimensional space of "body" and "soul" indices (Fig. 3) [Lam 2002]. At each point in this space, a "fitness" potential can be defined. The movement of each particle (usually, but not always, up the scales) will change the fitness landscape and influence the movement of other particles. The problem will be to find out, under what circumstances, all the particles will cluster together at the location corresponding to a liberal democratic society. It is thus a problem of clustering of active walkers in a two-dimensional deformable landscape.[7]

Such a problem has been studied before in physics in another context, and clustering of active walkers indeed occurs [Schweitzer & Schimansky-Geier 1994]. The corresponding investigation in history as outlined above will bring Fukuyama's historical study one level up, as scientific level is concerned, and serve as an example in other cases.

Note that AW is not the only paradigm possible in modeling history. And, in rare occasions, modeling of a system can be done analytically without the use of computers.

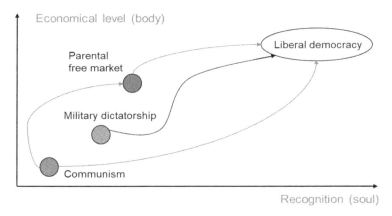

Fig. 3. Sketch of an active walk model for the evolution of political systems.

2.3 Computer simulation

Another approach is the method of computer simulations, which usually are at the *bottom-up* level—with the mechanisms incorporated, even only simplified mechanisms. Here is a very interesting example.

Simulating the growth of a historical society

A simulation of the development of a society in the Long House Valley in the Black Mesa area of northeastern Arizona, USA, was carried out by Axtell et al. [2002]. The simulation results show agreement with the quantitative historical data, which are reconstructed from paleoenvironmental research based on alluvial geomorphology, palynology, and dendroclimatology. For example, between the years *anno Domini* 400-1400, the number of households has two peaks; this is reproduced in the simulation (Fig. 4). So is the evolution of the spatial

220   *Lui Lam*

distribution of settlement. In this study, heterogeneity in both agents and the landscape, hard to model mathematically, is found to be crucial.

The modeling starts with a landscape reconstructed from paleoenvironmental variables, which is then populated with artificial agents representing individual households. Five household attributes are specified, together with household rules guessed from historical data. The model involves 14 reasonably chosen parameters, plus 8 adjustable parameters for optimization. The model is very detailed. It is interesting to see whether the model can be simplified to its bone, with fewer parameters, that can still produce essentially the same results.

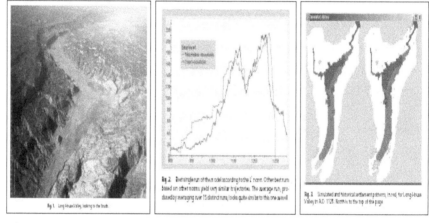

Fig. 4. A historical study of the growth and decline of a village in Arizona, USA [Axtell et al. 2002].

2.4 The Zipf plot

Sometimes, very interesting results can be obtained from some very simple techniques which are quite well known in the field of complex systems. An example is the use of *Zipf plot*, which is at the *empirical* level. To obtain this plot for a given sequence of numbers, there are four steps:

1. The sequence of numbers is rearranged in a decreasing order.

2. Redundancy is removed by keeping only one number among those of same magnitude, resulting in a sequence of monotonic decreasing numbers.[8]

3. The largest number is assigned rank 1 (R =1), the second largest rank 2, etc.

4. The Zipf plot is the curve appearing in the plot of the number vs. rank (with rank as the horizontal axis).

Here is an example concerning the Chinese dynasty lifetimes [Lam, 2006a, 2006b]. The lifetime of a Chinese dynasty τ_D is the sum of regime lifetimes τ_R, corresponding to all the emperors within the same dynasty. The Zipf plot of τ_D is given in Fig. 5, with the presence of 26 data points, which is less than 31 (the number of available dynasties), due to the adopted procedure of removing redundancies in the data sequence. The data points fall on two straight lines—a result named the *Bilinear Effect* [Lam 2006b; Lam et al. 2008]. It implies that

1. The "curse of history," as Chinese dynasties are concerned, does exist.

2. A dynasty can survive every 3.5 ± 0.1 years if it lasts 57 ± 2 years or less; beyond that, every 25.6 ± 0.1 years—dynasty lifetime is discrete, or "quantized."

3. There is a transition point separating these two different behaviors.

This is the *second quantitative law concerning Chinese history*. Whether the discreteness in τ_D results from some periodic external conditions in the Chinese history or is a self-organizing phenomenon resulting from some nonlinear dynamics remains to be investigated.

The bilinear effect is the phenomenon that an adaptive system becomes stronger after existing for a period of time. The mere fact of survival reinforces its strength, through learning, restructuring and so on. Similar behavior is known to exist in the case of restaurants, corporations, or biological species. What is surprising here is the presence of two linear lines and a sharp transition point (in the Zipf plot).

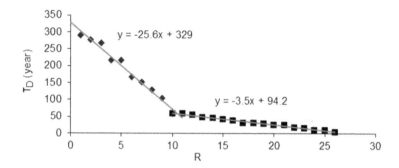

Fig. 5. The Zipf plot of Chinese dynasty lifetime τ_D, an example of the Bilinear Effect. R represents the "rank."

A *quantitative* prediction could be inferred from Fig. 5. Under the *assumption* that Chinese dynasties remain in the bilinear-effect class, any dynasty after Qing, if exists, will either

1. Last 303 ± 1 years or less, and fall more or less on the two lines in Fig. 5; or

2. End definitely and exactly in its year 329 ± 1.[9]

Note that the second law, corresponding to Fig. 5 (in contrast to the first law in Fig. 2), is *not* statistical in nature; and this prediction is not a statistical prediction. These two laws and the prediction concerning Chinese history are both quantitative and model independent. As far as we know, no other quantitative, non-statistical historical laws and predictions are known, to the historians or others.

The essence of a dynasty is not much the succession mechanism within a family, but the way of governance resulting from that mechanism. The "curse of history" spelled out here in Fig. 5 can be avoided only if one is willing to move the country away from the trajectory of the two straight lines, by abandoning the old ways of doing things.

It turns out that this regularity in Chinese history is only a particular case of the bilinear effect; more examples are subsequently found in other human affairs and complex systems (Fig. 6) [Lam et al. 2008]. Figure 6a is the Zipf plot of the number of votes for Chinese *Xiaopin* actors.[10] Figure 6b comes from the airline quality ratings in the year 2005.[11]

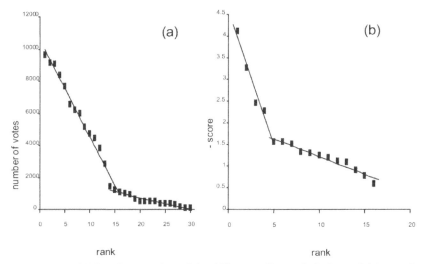

Fig. 6. Two additional examples of the bilinear effect. Zipf plots of (a) popular votes for *Xiaopin* actors, and (b) airline quality data.

In other words, the bilinear effect is a *new* class of Zipf plots, apart from the other two well-known classes—power laws [Newman 2005] and stretched-exponent distributions [Laherrère & Sornette 1998].

The generic mechanism behind the bilinear effect is not yet understood. But we do know there are more than one way to obtain the bilinear effect [Lam et al. 2008]. For example, first pick N_p number of points according to a one-peak probability distribution function $p(x)$, say, to obtain a sequence of numbers $\{x_i\}$, with $i = 1, 2, \ldots, N_p$. That is, the probability that x_i being picked is proportional to $p(x_i)$. A Zipf plot is then performed with this sequence $\{x_i\}$. With luck, the Zipf plot is bilinear.[12] The chance of obtaining bilinear effect increases as N_p is increased. Such a mechanism is applicable to the case in Fig. 6a, wherein, the votes were cast *independently* by people on the web from a *xiaopin* list. It is not applicable to the Chinese dynasties, the lifetimes of which are obviously correlated.

3 History in the Future

The importance of history can be seen, for example, through its negative impact on human lives. Powerful political leaders could mistake an unproven historical hypothesis as firm theory, apply it to a confined population and cause millions of death in a few short years [Lam 2002]. Another example is the massive protests in China few years ago, due to different interpretation of past history involving two countries (Fig. 7). Yet, in spite of its importance, the physical basis of history is unrecognized by most historians.

Fig. 7. Protests in China, April 2005. (a) "Face Up to History" is the slogan on the left placard. (b) "Protect Diaoyudao" is a historical issue also raised in the protests. Diaoyudao, or Diaoyutai, is a group of tiny islands in the East China Sea. The protect-Diaoyutai movement was started by overseas Chinese students in the United States at the end of 1970 [The Seventies Monthly 1971].

For instance, in the historiography textbook *The New Nature of History* [Marwick 2001: 248], the alleged "fundamental" differences between history and the sciences are listed:

1.   Fundamental difference in the subject of study: Natural sciences concern natural world and physical world; history concerns human beings and human society, very different in character.

2. No controlled experiments by historians.

3. Historians develop theories and theses, but not concerned with developing laws and theories like that in sciences.

4. History studies do not have prediction power.

5. Relations and interactions in history studies are not expressed mathematically.

6. Historians report their findings in prose (articles or books), not in terse research articles.

Unfortunately, all six points are wrong, for the following reasons:

1. As explained in Sec. 1, human beings and thus human society are material systems, which are part of natural science. Human society share same characteristics as other inanimate complex systems, as demonstrated in Figs. 1, 2, 5 and 6.

2. Some physical disciplines like astronomy and archeology also do not have controlled experiments.

3. It is untrue that all historians are not concerned with developing laws and theories in history. Some tried, not very successfully, partly due to their inadequate training in using scientific tools. Historical laws do exist, as shown in Secs. 2.1 and 2.4 and in Figs. 1, 2 and 5.

4. History studies, like that in Sec. 2.4, do have prediction power.

5. Relations and interactions in history studies can be expressed mathematically. An example is the landscape theory of Axelrod and Bennett [1993] to show how and why 17 European nations in World War II aligned themselves into two large groups. The pairwise propensities between nations are assigned numerical values, and the configuration energy in the (fixed) landscape is given in equations.[13]

6. Historians do report their findings in research articles, terse or not. That is why there exist quite a number of history journals, such as *History and Theory* and *American Historical Review*. It is true that

many historians still skip the peer-reviewed journals and directly report their findings in books—not a healthy thing for the history discipline, epistemologically speaking [Lam 2002]. These are actually popular history books, like the popular science books written by physicists.[14]

In the case of the history profession, there are at least three reasons behind this practice. (1) Many research results in history are still at the data gathering and empirical analysis stage, not very technical and can be presented in narratives. (2) There is enough number of readers out there who is willing to pay to find out what happened to their ancestors or their own kind in the past. In contrast, not that many will pay to learn what happened to the electrons. Bad for physics. (3) Historically, before history became a professional discipline in the universities in the second half of the 19th century, historians had to earn their living by writing books that are readable and salable to the public [Stanford 1998: 228]. In other words, writing popular history books was a survival need for historians, a tradition carried over up to now.

These errors are due to misunderstanding of the nature of science, and the neglect of the material basis of the historical system itself. The inadequate science training received by historians, past and present, explains why they failed to find historical laws. For example, the Chinese dynasty data have been lying there after 1912; the plots in Figs. 2 and 5 could be carried out by hand without computers, and even by high school students. But unless one knows about power laws and the existence of the Zipf plot, there is no motivation to do so. And these are current topics in the study of complex systems. Ironically, Zipf plots were first done by George Zipf (1902-1950), a Harvard linguist, with data from the humanities and social science.

Quite recently and surprisingly, while the importance of history is well recognized in Hong Kong, the history department of the University of Hong Kong is threatened with closure because it fails to attract enough number of students [Xie 2005: A15]. Anyway, it is time for all history departments to revamp their curriculum, by increasing the mathematical skills of their students, going beyond storytelling and making history research more technical and scientific, and creating a course on the physics of history (or *Histophysics* [Lam 2002, 2004]). This revamp will help current students to become better historians after they graduate, and may appeal to a new class of incoming students who have a technical background but feel attracted more to the humanities than the traditional sciences.

4 Conclusion

As shown above, human history can indeed be studied scientifically, using techniques borrowed from physics and complex systems. Human history is thus a true example of Science Matters (Scimat), the new multidiscipline that treats all human matters as part of science [Lam 2008a, 2008c]. This, of course, does not imply that the conventional studies by other historians or the existing history departments should be abolished. On the contrary, they are very valuable. They are doing splendid jobs at the empirical level, the first level in the scientific development of any discipline, which is to collect data, analyze and summarize data, and come up with "explanations" in understanding them. Yet, the explanations are usually educated guesses—"hypotheses" but not yet "theories," in the sense that a theory is a confirmed hypothesis. The single-event, unrepeatable nature of history makes the direct confirmation of a hypothesis very difficult, unlike the case in many "natural sciences." It is at this point that history as a science matter enters, as illustrated in Sec. 2.

As demonstrated, with a little bit of luck, the right perspective, and right tools, a hidden historical *law* (not merely a historical trend) might suddenly jump out and meet the eyes of the investigator. With a lot of luck, this historical law might even lead one to the discovery of a general principle in nature (such as the Bilinear Effect).[15]

What we try to do is to raise historical studies to a higher scientific level— to the phenomenological and bottom-up levels. To this end, collaborations between traditional historians and physicists are strongly urged. And everybody gains.

Notes

1. "Stochastic" is a technical word in physics, meaning that probability appears somewhere in the process; a random process is a special case [Paul & Baschnagel 1999].

2. For history, there is also the artificial level—artificial history [Lam 2002].

3. In a power law, two variables x and y relate to each other through $y = Ax^\alpha$, where A and α are constants. Equivalently, the plot of log x vs. log y shows up as a straight line.

4. In this chapter, "natural science" with quotation marks is defined as the science of mostly simple systems [Lam 2008a].

5. Other examples are the competition between Apple computers and the PCs, as well as Beta and VHS videotapes. In each case here, the second product is the inferior product.

6. The "sensitive zone" is a region in the parameter space, within which, for the same set of parameters, different runs of the computer model may result in different patterns due to the use of a different sequence of random numbers in each run.

7. The model can be generalized to include the possibility that two particles may combine into one, and some particles may split into two or three—some kind of chemical reactions—corresponding to the case in history that countries may get unified or fragmented in time.

8. Other people might retain all the numbers of the same magnitude in the sequence, resulting in a Zipf plot which could contain horizontal parts. Our version here is more reasonable, because one would like to fit the plot to a smooth curve.

9. The number 303 is the height at rank 1 on the straight line in Fig. 5; 329, that at "rank" 0.

10. ent.sin.com.cn/2004-09-30/1050521359.html (Oct. 7, 2004). Xiaopin is a popular form of short drama performed by a cast of usually two actors in China.

11. www.aqr.aero.

12. Experience shows that the one-peak shape of $p(x)$ is not a necessity in obtaining bilinear effect this way. Sometimes, a monotonic decreasing $p(x)$ also works; but so far, a decreasing power-law $p(x)$ does not seem to work.

13. See [Galam 1998] for a comment on this work, and the following response by the original authors.

14. The unique characteristics of popular science books and how to integrate them into science teaching are discussed in [Lam 2001, 2005b, 2006c, 2008b].

15. Starting with a particular case and ending with a general principle, is the common route of discoveries in physics and other disciplines. We ourselves have experienced this before: The modeling of filamentary patterns found in thin cells of electrodeposit experiments led us to Active Walk, a general paradigm for complex systems [Lam 2005a, 2006b]. Similarly, that was how Charles Darwin (1809-1882) found his evolutionary principle for living systems.

References

Arthur, B. [1990]. Positive feedbacks in the economy. Sci. Am., Feb. 1990: 92.

Axelrod, R. & Bennett, D. S. [1993]. A landscape theory of aggregation. British J. Political Sci. **23**: 211-233.

Axtell, R. L., Epstein, J. M, Dean J. S., Gumerman, G. J., Swedlund, A. C., Harburger, J., Chakravarty, S., Hammond, R., Parker, J. & Parker, M. [2002]. Population growth and collapse in a multiagent model of the Kayenta Anasazi in Long House Valley. Proc. Natl. Acad. Sci. USA **99** (Suppl. 3): 7275-7279.

David, P. A. [1986]. Understanding the economics of QWERTY: The necessity of history. *Economic History and the Modern Economist*, Parker, W. N. (ed.). New York: Blackwell. pp 30-49.

Feder, T. [2005]. Lab webs brain research and physics. Phys. Today, April, 26-27.

Fukuyama, F. [1989]. The end of history? The National Interest **16** (summer): 3-18.

Galam, S. [1998]. Comment on "A landscape theory of aggregation." British J. Political Sci. **28**: 411-412.

Gardiner, P. (ed.) [1959]. *Theories of History*. Glencoe, IL: The Free Press.

Gould, S. J. [1989]. *Wonderful Life: The Burgess Shale and the Nature of History*. New York: Norton.

Han, X.-P., Hu, C.-D., Liu, Z.-M. & Wang, B.-H. [2008]. Parameter-tuning networks: Experiments and active-walk model. EPL **83**: 28003.

Huang, R. [1997] *China: A Macro History*. Armonk, NY: M. E. Sharpe.

Johnston, A. C. & Nava, S. [1985]. Recurrence rates and probability estimates for the New Madrid seismic zone. J. Geophys. Res. **90**: 6737-6753.

Krakauer, D. C. [2007]. The quest for patterns in meta-history. Santa Fe Inst. Bull. (Winter, 2007): 32-39.

Laherrère, J. & Sornette, D. [1998]. Stretched exponential distributions in nature and economy: 'fat tails' with characteristic scales. Eur. Phys. J. B. **2**: 525-539.

Lam, L. [1997]. Active Walks: Pattern formation, self-organization, and complex systems. *Introduction to Nonlinear Physics*, Lam, L. (ed.). New York: Springer. pp 359-399.

Lam, L. [1998]. *Nonlinear Physics for Beginners: Fractals, Chaos, Solitons, Pattern Formation, Cellular Automata and Complex Systems*. Singapore: World Scientific.

Lam, L. [2001]. Raising the scientific literacy of the population: A simple tactic and a global strategy. *Public Understanding of Science*, Editorial Committee (ed.). Hefei, China: Science and Technology University of China Press.

Lam, L. [2002]. Histophysics: A new discipline. Mod. Phys. Lett. B **16**: 1163-1176.

Lam, L. [2004]. *This Pale Blue Dot: Science, History, God*. Tamsui: Tamkang University Press.

Lam, L. [2005a]. Active walks: The first twelve years (Part I). Int. J. Bifurcation and Chaos **15**: 2317-2348.

Lam, L. [2005b]. Integrating popular science books into college science teaching. The Pantaneto Forum, Issue 19 (2005).

Lam, L. [2006a]. How long can a Chinese dynasty last? Preprint.

Lam, L. [2006b]. Active walks: The first twelve years (Part II). Int. J. Bifurcation and Chaos **16**: 239-268.

Lam, L. [2006c]. Science communication: What every scientist can do and a physicist's experience. Science Popularization (No. 2, 2006): 36-41. See also Lam, L., in *Proceedings of Beijing PCST Working Symposium*, June 22-23, 2005, Beijing, China.

Lam, L. [2008a]. Science Matters: A unified perspective. *Science Matters: Humanities as Complex Systems*, Burguete, M. & Lam, L. (eds.). Singapore: World Scientific.

Lam, L. [2008b]. SciComm, PopSci and The Real World. *Science Matters: Humanities as Complex Systems*, Burguete, M. & Lam, L. (eds.). Singapore: World Scientific. pp 89-118.

Lam, L. [2008c] Science Matters: The newest and biggest interdicipline. *China Interdisciplinary Science*, Vol. 2, Liu Zhong-Lin (刘仲林) (ed.). Beijing: Science Press.

Lam, L., Bellavia, D. C., Han, X. P., Liu, A., Shu, C. Q., Wei, Z. J., Zhu, J. C. & Zhou, T. [2008]. Bilinear effect in complex systems. Preprint. (Expanded and published as Lam, L. et al. [2011]. Bilinear effect in complex systems. EPL **91**: 68004.)

Lam, L. & Pochy, R. D. [1993]. Active walker models: Growth and form in nonequilibrium systems. Comput. Phys. 7: 534-541.

Levy, J. S. [1983]. *War in the Modern Great Power System, 1495-1975*. Lexington: University Press of Kentucky.

Marwick, A. [2001]. *The New Nature of History*. Chicago: Lyceum.

Morby, J. E. [2002]. *Dynasties of the World*. Oxford: Oxford University Press.

Newman, M. E. J. [2005]. Power laws, Pareto distributions and Zipf's law. Contemp. Phys. **46**: 323-351.

Paul, W. & Baschnagel, J. [1999]. *Stochastic Processes: From Physics to Finance*. New York: Springer.

Richardson, L. F. [1941]. Frequency of occurrence of wars and other fatal quarrels. Nature **148**: 598.

Roberts, D. C. & Turcotte, D. L. [1998]. Fractality and self-organized criticality of wars. Fractals **6**: 351-357.

Schweitzer, F. & Schimansky-Geier, L. [1994]. Clustering of active walkers in a two-component system. Physica A **206**: 359-379.

Sornette, D. [2000]. *Critical Phenomena in Natural Sciences*. New York: Springer.

Stanford, M. [1998]. *An Introduction to the Philosophy of History*. Malden, MA: Blackwell.

The Seventies Monthly (ed.) [1971] *Truth Behind the Diaoyutai Incident* (The Seventies Monthly, Hong Kong).

Xie, Xi [2005]. Historical feel regained. World Journal (Millbrae, CA), April 16, 2005.

Zipf, G. K. [1949]. *Human Behavior and the Principle of Least Effort*. Cambridge, MA: Addison-Wesley.

Published: Lam, L. [2008]. Human history: A science matter. *Science Matters: Humanities as Complex Systems*, Burguete, M. & Lam, L. (eds.). Singapore: World Scientific. pp 234-254.

12

Bilinear Effect in Complex Systems

*Lui Lam, David C. Bellavia, Xiao-Pu Han, Chih-Hui Alston Liu,
Chang-Qing Shu, Zhengjin Wei, Tao Zhou and Jichen Zhu*

The distribution of the lifetime of Chinese dynasties (as well as that of the British Isles and Japan) in a linear Zipf plot is found to consist of two straight lines intersecting at a transition point. This two-section piecewise-linear distribution is different from the power law or the stretched exponent distribution, and is called the *Bilinear Effect* for short. With assumptions mimicking the organization of ancient Chinese regimes, a 3-layer network model is constructed. Numerical results of this model show the bilinear effect, providing a plausible explanation of the historical data. The bilinear effect in two other social systems is presented, indicating that such a piecewise-linear effect is widespread in social systems.

1 Introduction

A common way to characterize and classify complex systems is through Zipf plots [Zipf 1949]. Given a sequence of numbers, the corresponding Zipf plot is obtained in four steps: (1) The sequence is rearranged in a decreasing order. (2) For numbers of the same magnitude, retain only one of them in the sequence. (3) The largest number is assigned rank 1, the second largest rank 2, etc. (4) The Zipf plot is the curve of number vs. rank (R). Note that as a result of the decreasing order, the Zipf plot so defined is always a monotonically decreasing curve. Since the Zipf plot can give an inverse function of the cumulative distribution from original data [Newman 2005], it is widely used in the statistical analysis of small samples [Reed 2001; Han et al. 2010].

There are two well-known non-Poisson types of Zipf plots: power laws [Zipf 1949; Newman 2005] and stretched exponents [Laherrère & Sornette 1998]. The power law distribution has been widely observed in a large number of self-organizing systems. In the last several decades the power law distribution has attracted the attention of many scientists. It is at the center of complex systems research because of its special mathematical and dynamical properties [Albert 2000; Newman 2002; Pastor-Satorras & Vespignani 2001], and physical implications [Bak et al. 1987]. In the last decade, the rise of research in the network sciences makes the power law more prominent [Watts et al. 1998; Barabasi & Albert 1999].

The stretched exponent distribution generally can be viewed as an intermediate form between a scaling form (power law) and homogenous types of distributions (such as Poisson distribution). It has also been widely observed in many social and material systems [Xulvi-Brunet & Sokolov 2002; Holanda et al. 2004; Sturman et al. 2005; Han et al. 2008]. The typical form of both the power law and stretched exponent is that of a continuous curve in a Zipf plot.

Here we investigate the rank distributions of the lifetime of the dynasties in ancient China, British Isles, and Japan. The surprising result is that these distributions in the linear Zipf plots obey neither power law nor stretched exponent type, but a special two-section piecewise-linear function. The two monotonic decreasing straight lines intersect at a transition point; the slope of the curve is not continuous. In the rest of this paper, this type of two-section piecewise-linear form in a linear Zipf plot is called *Bilinear Effect* for short. Previously, multiple piecewise-linear forms have been discussed as a phenomenon or an assumption in many natural and technological systems [Sinha & Chakrabarti 1998; Batista et al. 2002; Chan et al. 2002; Carbone & Palma 2007; Thomas & Luk 2007]. For example, [Sinha & Chakrabarti 1998] and [Batista et al. 2002] have respectively studied the stochastic resonance and synchronization dynamics in piecewise-linear maps; [Chan et al. 2002] has analyzed the economic problem in the distribution of products in supply chain with piecewise-linear cost structures; [Carbone & Palma 2007] has investigated the discontinuity correction in piecewise-linear models of oscillators; and [Thomas & Luk 2007] has proposed a random number generator based on piecewise-linear approximations. In this regard, bilinear effect is a special piecewise-linear form—a two-section version; our findings indicate that it widely exists in complex systems, especially in social systems. Below, we present a dynamical model to explain its possible underlying mechanism, and introduce two more examples of such bilinear rank distribution in other social systems.

2 Bilinear Effect in the Lifetime of Dynasties

The dynasty cycle in Chinese history has been studied mainly by nonlinear dynamical models in the last two decades. One of the features noticed by these early researches is the periodic alternation of society between despotism and anarchy. Usher [1989] proposed a differential model based on the interaction between three basic classes in ancient China: farmers, bandits, and rulers. This model includes three simple three-variable differential equations, denoting, respectively, the evolution of the three classes. This model successfully generated the alternation between despotism and anarchy, which is similar to the basic feature in dynasties cycle. Based on this work, several differential game models were proposed [Feichtinger & Novak 1994; Chu & Lee 1994; Feichtinger et al. 1996], and the generalized version of this class of models was further investigated [Gross & Feudel 2006]; some of them extended the discussions to the evolution of population in ancient China [Chu & Lee 1994]. Most of these works paid much attention to the nonlinear dynamical properties of these models, but the statistical patterns of real-world dynasty cycles are rarely discussed. Here in this paper, the distribution of lifetime of Chinese dynasties is investigated for the first time, to the best of our knowledge.

The record of the history of each dynasty in ancient China from Qin dynasty (221 BC) to Qing dynasty is exhaustive and reliable. The Zipf plots of the lifetimes of these dynasties (τ) are presented in Figs. 1A and 1B. Both of the two sets of data range from Qin to Qing dynasty (221 BC to 1912 AD). The data for Fig. 1A is obtained from [Morby 2002], which includes 31 main Chinese dynasties. The data for Fig. 1B is from *Cihai* [Xia 1979], a Chinese encyclopedia, which not only includes the 31 main dynasties, but also many local powers and provisional governments, resulting in a total of 74 dynasties. The two data sets are listed on the website.[1]

These two sets of data depict similar behavior: the decreasing two-section piecewise-linear form, or, the bilinear effect. The transition point in these two Zipf plots is $\tau = 57 \pm 2$ years. This implies that if a Chinese dynasty survives longer than 57 years, it will have a greater chance of surviving longer, and the chance that it will be destroyed is sharply reduced. For example, Fig. 1A implies that a dynasty can survive 3.6 ± 0.1 years if it lasts 57 ± 2 years or less; beyond that, every 22.2 ± 0.1 years. In other words, the distribution of the lifetimes of Chinese dynasties is discrete, or "quantized." Moreover, this is the phenomenon that a human entity, a dynasty in this case, becomes stronger or more stable after existing for a period of time. The mere fact of survival reinforces its strength, through adaptive learning, restructuring or other means.

Bilinear effect can also be observed in the lifetime distribution of dynasties of some other countries. Figure 2 shows two examples: the dynasty lifetime distributions of Britain and Japan. In contrast to the Chinese data, the transition point of British and Japan, respectively, is $\tau = 68 \pm 2$ and 268 ± 10, which are larger than that of China. However, the number of data points for these countries is less than those of China, and this is why the bilinear effect is less certain in these systems. In the following, to understand the underlying mechanism, a governmental structure giving rise to the bilinear effect is introduced.

3 The 3-layer Network Model

Roughly speaking, the government structure of a Chinese regime in the last two thousand or so years since the Qin dynasty consists of three layers: the emperor court (the central government), the provinces, and the cities/villages. They are represented schematically by layer A, B and C, respectively, in Fig. 3. Every year, the cities/villages submit part of their income, in the form of "taxes" to the upper layer, the provincial governments. And the provincial governments in turn submit a certain amount of their revenues to the emperor court, the top layer. At the same

time, the emperor court maintains its control by allocating funds/resources
to the governments in the middle layer as it pleases. But there will be no
downward flow of resources from layer B to layer C. The communication
between local governments in each layer is indirect and will be ignored in
our model.

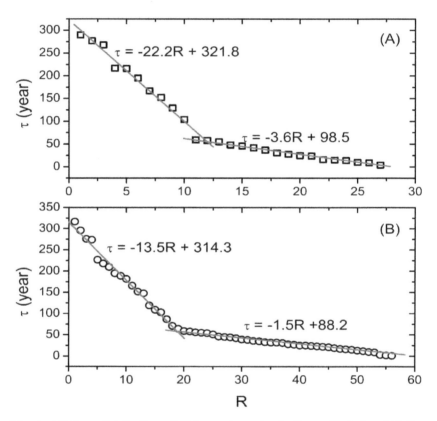

Fig. 1. Lifetime distribution of Chinese dynasties in linear Zipf plot. (A) Data
from [Morby 2002]; the corresponding dynasty of each data point is (from left to
right): Tang, Ming, Qing, Liao, Western Han, Eastern Han, Northern Song,
Northern Wei (abreast with Southern Sung and Yuan), Chin, Eastern Chin, Liu
Sung, Wu, Liang, Western Chin, Wei, Minor Han (abreast with Western Wei),
Sui, Chen, Northern Chi, Northern Zhou, Southern Chi, Hsin (abreast of Later
Liang), Qin, Later Tang, Later Chin, Later Zhou, Later Han. (B) Data from *Cihai*
[Xie 1979].

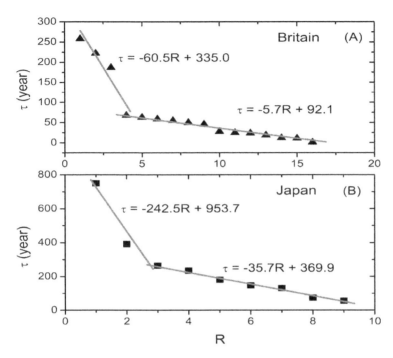

Fig. 2. Lifetime distributions of dynasties of Britain (A) and Japan (B). Data from [Morby 2002].

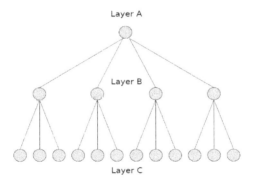

Fig. 3. The structure of the 3-layer network model ($N_B = 4$ and $N_C = 3$).

In our 3-layer network model, the upper layer A has one node. Node A is connected to all the nodes B_i ($i = 1, ..., N_B$) in the second layer (layer B). And each node B_i is connected to nodes C_{ij} ($j = 1, ..., N_C$) in the third layer

(layer C). So, the number of nodes on layer B and layer C is N_B and N_C, respectively. There is no connection between the nodes in the same layer.

Let $F(B_i, t)$ be the amount of resources (or fitness) possessed by node B_i at time t; similarly, $F(A, t)$ for node A, and $F(C_{ij}, t)$ for node C_{ij}. Each node will transfer part of its own resource to the other nodes that are connected to it, according to the following rules.

1. **From A to B_i:** At time t, a node in layer B (k, say) is randomly picked and an amount $T_A(t)$ is transferred from node A to node B_k such that

$$T_A(t) = aF(B_k, t - 1). \tag{1}$$

2. **From B_i to A:** An amount T_{BA} is transferred from node B_i to node A such that

$$T_{BA}(B_i, A, t) = bF(B_i, t - 1). \tag{2}$$

3. **From C_{ij} to B_i:** An amount T_{CB} is transferred from node C_{ij} to node B_i such that

$$T_{CB}(C_{ij}, B_i, t) = cF(C_{ij}, t - 1), \tag{3}$$

where a, b and c are constants (each one is less than 1).

To keep itself running, node A does consume its own resources; the amount is denoted by $eF(A, t)$ with e (< 1) being a constant. It follows that the time evolution of the fitness at each node is given by

1. **For node A:**

$$F(A, t) - F(A, t - 1) = -T_A(t) + \sum_i T_{BA}(B_i, A, t) - eF(A, t - 1). \tag{4}$$

2. **For node B_i:**

$$F(B_i, t) - F(B_i, t - 1) = T_A(t)\delta_{ik} + \sum_j T_{CB}(C_{ij}, B_i, t)$$

$$-T_{BA}(B_i, A, t). \tag{5}$$

3. **For node C_{ij}:**

$$F(C_{ij}, t) - F(C_{ij}, t - 1) = -T_{CB}(C_{ij}, B_i, t). \tag{6}$$

Starting with initial fitness for the nodes, the computer run is stopped when $F(A, t) \leq 0$ for the first time (at time $t = \tau$, say), which mimics the exhaustion of the resources of the central government. The lifetime of the regime is taken to be τ.

For a set of given parameters, many computational runs of this model are performed, giving rise to the normalized probability function $p(\tau)$—such that $p(\tau)$ is the probability that τ is found among all the runs. A sequence of numbers, $\{\tau_i\}$ with $i = 1, 2, ..., N_p$, are picked according to this $p(\tau)$. The Zipf plots derived from particular sequences so picked are depicted in Fig. 4. All of these obviously show bilinear effect, indicating that this property can indeed emerge from the government resource assignment process. We further investigate the impact of each parameter on the bilinear effect. The main parameters of our model include N_B, N_C, a, b, c and e. Simulation results indicate that this model can generate a bilinear τ-rank distribution with large parametric variations. What we focus on here is the value of the transition point.

The value of τ of the transition point are sensitively impacted by parameters a, b and e. As shown in Fig. 4, τ of the transition point increases along with the reduction of a and e, and the rise of b. Parameter c has no obvious impact on the transition point. Parameters a and b denote the strength of the transfer of resources between the central government and several local powers, and e denotes the consumption of the central government; and thus, the transition point is mainly impacted on by the resource flow in the upper layers. This conclusion could be a key to understand the difference of the transition point in different countries. In parallel, when $a > 0.5$, $b < 0.2$ or $e > 0.8$, the bilinear effect does not readily manifest itself in our model. In contrast, the impact of other parameters (N_B, N_C) on the transition point is insensitive.

4 Other Examples of Bilinear Effect

In addition to the distribution of lifetime of dynasties, two other interesting examples of such bilinear effect in social systems are found. One example is the number of online votes for Chinese *Xiaopin* actors.[2] In this vote, each voter can choose their favorite actor from a list of 30. The more number of votes an actor gets, the higher the popularity. The bilinear effect is obvious in Fig. 5A. This implies these actors can be divided into two groups. In other words, the social reputation of actors could be dichotomous: If an actor can pass a critical popularity, he/she will achieve greater popularity more easily. Also, the discreteness of the data points contradicts the common understanding that the social reputation of people is continuously distributed.

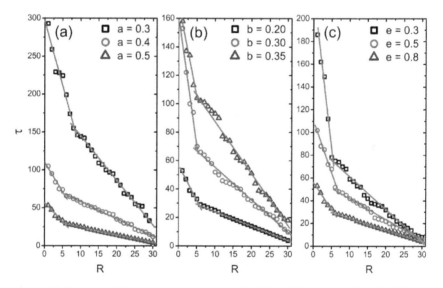

Fig. 4. Zipf plot resulting from a particular pick ($N_p = 30$) generated by the 3-layer network model ($N_B = 4$, $N_C = 3$, $c = 0.3$). (a) Results for different a; other parameters used: $b = 0.2$, $e = 0.8$. (b) Results for different b; other parameters used: $a = 0.5$, $e = 0.8$. (c) Results for different e; other parameters used: $a = 0.5$, $b = 0.2$. The initial conditions are: $F(A, 0) = 100$, $F(B_i, 0) = 50$ and $F(C_{ij}, 0) = 30$. These results show a bilinear effect similar to that in Fig. 1A.

Another example is the 2004 airline quality ratings,[3] as shown in Fig. 5B. These examples imply that that bilinear effect could be widespread in some complex systems, especially in social systems.

5 Conclusion

We present examples from three different types of social systems that show a special piecewise-linear distribution called the Bilinear Effect.

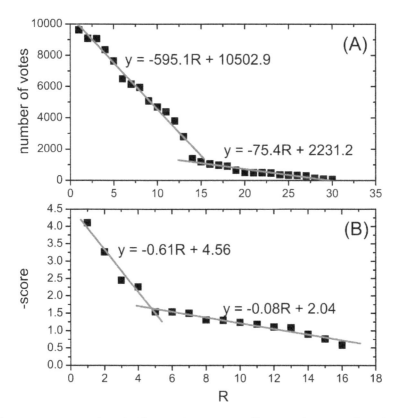

Fig. 5. Two examples of bilinear effect. (A) Online popular votes for Xiaopin actors. (B) Airline quality data. In both plots, y in the equation corresponds to the quantity on the vertical axis.

The most interesting characteristic of the bilinear effect is that the samples are divided into two distinct linearly distributed groups which are connected by a sharp transition point. It indicates that the statistical properties of some complex systems could be discrete. What is the meaning of the transition point? It could be an indication of a "phase transition." However, this supposition needs more empirical evidence and theoretical understanding.

Our research of the bilinear effect covers several different realms, including the lifetimes of dynasties of several countries, online votes of actors, and airline quality ratings. These empirical results imply that bilinear effect could have some universal significance.

We propose a 3-layer network model to investigate the underlying mechanisms of the bilinear effect in the lifetimes of dynasties. This model can generate bilinear effect in wide parameter settings, in agreement with our empirical data. This model also makes it possible to understand the difference of the transition point in different countries. While this model does provide a plausible explanation of the origin of the bilinear effect in dynasty lifetimes, it is not a microscopic model, and it does not explain the bilinear effect in other social systems. A more sophisticated model is needed in all these cases. Other models and explanations are not precluded a priorily. The fact that there could be more than one mechanism in producing the bilinear effect is not that surprising. A similar case exists in the case of power laws in Zipf plots [Newman 2005].

In summary, we report the empirical statistics and modeling for the bilinear effect in Zipf plots in several social systems. The dynasty lifetime results contribute to the advancement of *Histophysics* [Lam 2002] and *Scimat* (Science Matters) [Lam 2008]. This letter investigates only a few examples; the research is just at its beginning. There are still many open questions and more insightful research are needed.

Notes

1. www.sciencenet.cn/upload/blog/file/2010/8/2010817181447879502. pdf.

2. Xiaopin is a popular form of short drama performed by a cast of usually two actors in China; the data is obtained from http://ent.sina.com.cn/2004-09-30/1050521359.html (Oct. 7, 2004).

3. Data obtained from http://www.aqr.aero/reports/2005aqr.pdf.

References

Albert, R., Jeong, H. & Barabasi, A.-L. [2000]. Error and attack tolerance of complex networks. Nature **406**: 378-382.

Bak, P., Tang, C. & Wiesenfeld, K. [1987]. Self-organized criticality: An explanation of 1/f noise. Phys. Rev. Lett. **59**: 381-384.

Barabasi, A.-L. & Albert, R. [1999]. Emergence of scaling in random networks. Science **286**: 509-512.

Batista, A. M., de S. Pinto, S. E., Viana, R. L. & Lopes, S. R. [2002]. Lyapunov spectrum and synchronization of piecewise linear map lattices with power-law coupling. Phys. Rev. E **65**: 056209.

Carbone, A. & Palma, F. [2007]. Discontinuity correction in piecewise-linear models of oscillators for phase noise characterization. Int. J. Circuit. Theor. Appl. **35**: 93-104.

Chan, L. M. A., Muriel, A., Shen, Z.-J. M., Simchi-Levi, D. & Teo, C.-P. [2002]. Effective zero-inventory-ordering policies for the single-warehouse multiretailer problem with piecewise linear cost structures. Manag. Sci., **48**: 1446-1460.

Chu, C. Y. C. & Lee, R. D. [1994]. Famine, revolt, and the dynastic cycle: Population dynamics in historic China. J. Popul. Econ. **7**: 351-378.

Feichtinger, G., Forst, C. V. & Piccardi, C. [1996]. A nonlinear dynamical model for the dynastic cycle. Chaos, Solitons Fractals **7**: 257-271.

Feichtinger, G. & Novak, A. J. [1994]. Differential game model of the dynastic cycle: 3D-canonical system with a stable limit cycle. J. Optim. Theory Appl. **80**: 407-423.

Gross, T. & Feudel, U. [2006]. Generalized models as a universal approach to the analysis of nonlinear dynamical systems. Phys. Rev. E **73**: 016205.

Han, X.-P., Hu, C.-D., Liu, Z.-M. & Wang, B.-H. [2008]. Parameter-tuning networks: Experiments and active-walk model. EPL **83**: 28003.

Han, X.-P., Wang, B.-H., Zhou, C.-S., Zhou, T. & Zhu, J.-F. [2010]. Scaling in the global spreading patterns of pandemic Influenza A (H1N1) and the role of control: empirical statistics and modeling. arXiv:0912.1390.

Holanda, A. J., Pisa, I. T., Kinouchi, O., Martinez, A. S. & Ruiz E. E. S. [2004]. Thesaurus as a complex network. Physica A **344**: 530-536.

Laherrère, J. & Sornette, D. [1998]. Stretched exponential distributions in nature and economy: "fat tails" with characteristic scales. Eur. Phys. J. B **2**: 525-539.

Lam, L. [2002]. Histophysics: A new discipline. Mod. Phys. Lett. B **16**: 1163-1176.

Lam, L. [2008]. Science Matters: A unified perspective. *Science Matters: Humanities as Complex Systems*, Burguete, M & Lam, L. (eds.). Singapore: World Scientific. pp 1-38.

Morby, J. E. [2002]. *Dynasties of the World.* Oxford: Oxford University Press.

Newman, M. E. J. [2002]. Spread of epidemic disease on networks. Phys. Rev. E **66**: 016128.

Newman, M. E. J. [2005]. Power laws, Pareto distributions and Zipf's law. Contemp. Phys. **46**: 323-351.

Pastor-Satorras, R. & Vespignani, A. [2001]. Epidemic spreading in scale-free networks. Phys. Rev. Lett. **86**: 3200-3203.

Reed, W. J. [2001]. The Pareto, Zipf and other power laws. Econ. Lett. **74**: 15-19.

Lui Lam

Sinha, S. & Chakrabarti, B. K. [1998]. Deterministic stochastic resonance in a piecewise linear chaotic map. Phys. Rev. E **58**: 8009-8012.

Sturman, B., Podivilov, E. & Gorkunov, M. [2003]. Origin of stretched exponential relaxation for hopping-transport models. Phys. Rev. Lett. **91**: 176602.

Thomas, D. B. & Luk, W [2007]. Non-uniform random number generation through piecewise linear approximations. IET Comput. Digit. Tech. **1**: 312-316.

Usher, D. [1989]. The dynastic cycle and the stationary state. Am. Econ. Rev. **79**:1031-1044.

Watts, D. J. & Strogatz, S. H. [1998]. Collective dynamics of "small-world" networks. Nature **393**: 440-442.

Xia, Z.-N. [1979]. *Cihai*. Shanghai: Shanghai Lexicographical P. H.

Xulvi-Brunet, R. & Sokolov, I. M. [2002]. Evolving networks with disadvantaged long-range connections. Phys. Rev. E **66**: 026118.

Zipf, G. K. [1949]. *Human Behavior and the Principle of Least Effort*. Cambridge, MA: Addison-Wesley.

Published: Lui Lam, David C. Bellavia, Xiao-Pu Han, Chih-Hui Alston Liu, Chang-Qing Shu, Zhengjin Wei, Tao Zhou and Jichen Zhu [2010]. Bilinear effect in complex systems. EPL **91**: 68004.

PART III

ART

13

Arts: A Science Matter

Lui Lam

The nature and origin of arts, and its relationship to "science" have been under much debate since Plato about 2,400 years ago. Here, a new perspective on these issues is presented. Science is to understand how nature works, while nature consists of (human and nonhuman) living systems and nonliving systems. Consequently, all human matters are part of science—the premise underlying the new multidiscipline called Science Matters (*Scimat*), which covers all topics in the humanities and social science, arts in particular. (Arts here refer to visual arts, literature, film, performing arts, music, architecture, new media arts and so on.) In fact, arts are a subset of humans' creative activities that aim to excite the receiver's neurons in a certain manner, through that person's senses, with or without significant consequences.

The usual kind of "science" is to understand mostly nonliving, simple systems and how the world/universe works; it is part of science in general. Arts as a science matter is to find out everything about arts, including arts' origin and nature, and how and why arts work at both ends of the creator and the receiver. Like physics and any other discipline, arts can be classified into two types: *pure* arts and *applied* arts. Some arts, such as drawing and performing arts, could start a million years ago. All arts evolved over time and space, and the contents kept on changing as humans invented language and writing and as they migrated out of Africa and spread over the world; arts contain both global universal elements and local features. Here, all these issues as well as how arts as a science matter could be studied are elaborated, after a brief introduction to Scimat and humans' developmental history and inheritance mechanism (genes and epigenes) is given.

1 Introduction

Arts in this chapter refer to visual arts, literature, film, performing arts, music, architecture, new media arts and so on. The origin and nature of arts, and its relationship to "science"[1] are under much debate.[2] The confusion arises from many factors which will become clear later as our discussion goes along.

Here we will try to clear up these confusions by reexamining the problem and presenting a new perspective on all these issues. We will first clear up the definition of the word Science and introduce the new multidiscipline *Scimat* [Lam 2008a] within which arts belong (Sec. 2). Since arts are human activities, it is important to understand where we came from and how we developed evolutionarily (Sec. 3). The origin and nature of arts are then discussed, respectively, in Secs. 4 and 5. Arts as a science matter are presented in Sec. 6, where the relationships of arts to "science" and how arts could be studied are given. Discussion and conclusion in Sec. 7 conclude the chapter.

2 Science and Science Matters

The scope and nature of Science, and the new multidiscipline Science Matters are presented here.

2.1 What is science?

Science is about the study of nature and a means to understand it in a unified way. Nature consists of everything in the universe—all material systems: humans and nonhumans. Consequently, the only logical conclusion about the scope of science is:

Science = Natural Science

$$= \text{Physical science} + \text{Social Science} + \text{Humanities} \tag{1}$$

where the three items on the right-hand side of Eq. (1) are in decreasing level of scientific development; they are *not* classified according to the nature of the objects under study [Lam 2008a].

That "Everything in nature is part of science" was well recognized by Aristotle and Da Vinci and many others. However, it is only recently, with the advent of modern science and experiences gathered in the study of evolutionary and cognitive sciences, statistical physics [Lam 1998; Paul & Baschnagel 1999], complex systems [Lam 1997, 1998] and other disciplines, that we know how the human-related disciplines can be studied scientifically.

2.2 Three misconceptions about science

The contents of science can be divided into two parts: the human system and the nonhuman systems. The study of the human system (humanities and social science) was hindered by three misconceptions in science. In fact, the miserable part of human history (e.g., ideological massacre and religious burning at stake) is related to these three misconceptions:

Misconception 1: Natural systems include nonhuman systems only (i.e., humans are excluded). This misconception started, at least, from the early Greek time some 2,400 years ago. It is wrong because all material systems in nature, humans included, are made up of atoms (mostly) created in the stars some 300 million years ago [Turner 2009].

Misconception 2: Physics is about deterministic systems only (i.e., stochastic systems[3] are excluded). This misconception is due to the tremendous success of Newtonian physics in the past 300 years and people's ignorance of physics developments (Fig. 1).

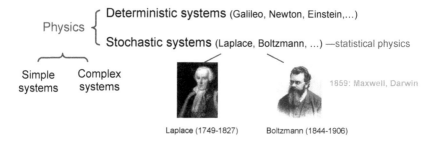

Fig. 1. Two ways to classify the contents of physics: (1) It is made up of deterministic systems and stochastic systems; (2) it consists of simple systems and complex systems.

Misconception 3: Science is about (mostly) simple systems only (i.e., complex systems are excluded). This misconception started from Galileo's time 400 years ago even though science before that actually includes the study of both simple and complex systems (Fig. 2).

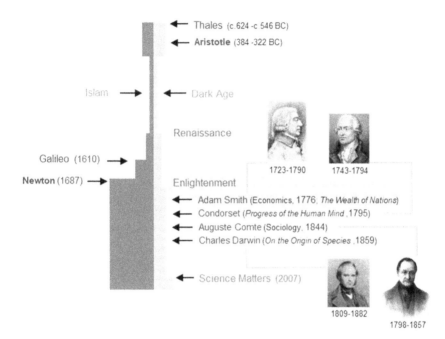

Fig. 2. A brief history of science in the last 2,600 years since Thales. The left (right) column corresponds to simple (complex) systems; the column width represents roughly how much the development activity was during different time periods.

2.3 Science Matters

Science Matters is the new multidiscipline that treats all human matters, the humanities in particular, as part of science (by avoiding the three misconceptions above) (Fig. 3). It was originated by Lam in 2007 [Lam 2008a, 2008b]. Accordingly, Eq. (1) is replaced by

Science = Natural Science

$$= \text{Nonhuman-system Science} + \text{Science Matters} \qquad (2)$$

Scimat has a number of important implications [Lam 2008a]. One of them is that the usual usage of "Science and Art" is misleading, since they imply that Science and Art are two different things while, according to the reasoning above, art is contained within science.[4]

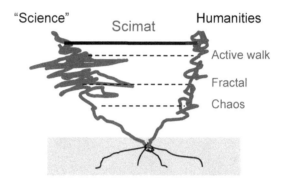

Fig. 3. Scimat links up the humanities and "science" completely while active walk, fractal, and chaos, respectively, does that partially. The humanities and "science" share the same root, growing up like two branches of the same plant.

3 Human

Arts are created by humans. To understand arts we have to understand humans. Here are the basic facts about humans [Hoagland & Dodson 1998]. A human body is composed of 5×10^{12} cells. Each cell is made up of molecules, a combination of atoms (coming mostly from the stars). One of these molecules is the DNA molecule which is the same inside each cell. It is the DNA that passes biological information from generation to generation. However, a human being's thinking and behavior are controlled by the 10^{11} neurons in her brain. And the neurons could be influenced by external media (e.g., artwork, sunset) through the bodily sensors or substances (e.g., marijuana smoke) absorbed into the body.

There are several basic facts about human development [Mithen 1999] that are relevant to the discussion of arts (Table 1).

Equally important is how humans pass on their genes and the question of "nature vs. nurture" [Ridley 2003; Moore 2003]. According to the British naturalist, Charles Darwin (1809-1882), and putting it in modern language, human inheritance is stored in the genes and passed on from generation to generation. However, *random* mutation of the genes happens from time to time, resulting in the appearance of new species. Different species compete with each other for resources and "the fittest wins." This is called "natural selection" or the evolutionary pressure; the winner keeps the (new) genes that help it win—an adaptive trait in the evolutionary sense. Moreover, the evolutionary process is very slow and continuous; no learned skills can be passed on to the next generation.

Table 1. A brief history of human development. Data source: www.newscientist. com/movie/becoming-human (June 18, 2010).

Years ago	Evolution	Migration	Lifestyle	Art related
6 million	Chimpanzee and human lineages split			
3.5-1.8 million			First hominids move from forest to savannah; meat eating begins	
2.5 million	*Homo habilis* appears			
2 million	*Homo erectus* appears; mimesis begins			
1.8 million		First wave of migration out of Africa begins		
1.6 million			First use of fire; more complex stone tools created; art could begin	
400,000			Earliest evidence of cooking	
195,000	*Homo sapiens* (early modern humans) appears			
150,000				Language begins
120,000				Pigment use gives first evidence of symbolic culture
72,000				Clothing invented and earliest evidence of jewelry
60,000		Second wave of migration out of Africa (Fig. 4)		
50,000				Cultural revolution: ritualistic burials, clothes-making, invention of complex hunting techniques
35,000				Oldest known cave art (in France, Spain …)
10,000		Agriculture begins; first villages appear		
5,500			Bronze Age begins.	
5,000			Earliest known writing	

The present understanding is that although we do inherit stable genes, we also inherit alterable epigenes [Shenk 2010]. Epigenes are molecules external to the genes that can switch on and off particular genes (Fig. 5). More importantly, an epigene's switching state could be influenced by the environment and could be passed on to the next generation, for many generations. For example, this pass-on ability has been demonstrated in fruit flies. When exposed to a drug, fruit flies show unusual outgrowths on their eyes that can last through at least 13 generations of offspring when no change in DNA occurred [Cloud 2010]. Similarly, experiments on roundworms fed with a kind of bacteria show changes that last at least 40 generations [Jablonka & Raz 2009].

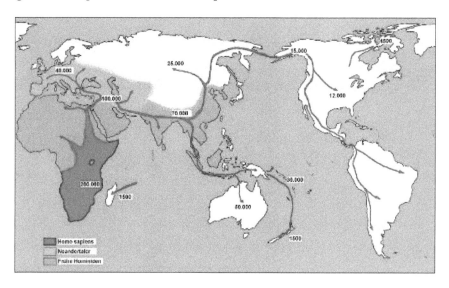

Fig. 4. Spreading of *Homo sapiens* out of Africa. (The numbers in this map differ slightly from those in text.) According to this map, the ancestors of the Chinese are Indians who (and everyone else) in turn are descendants of Africans.

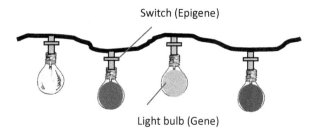

Fig. 5. Presently, a DNA chain can be visualized as a chain of light bulbs in four different colors plus switches for the light bulbs. The light bulb represents a gene; the switch, an epigene.

What all these imply is that the debate of "nature vs. nurture" is losing its importance since both nature (genes) and nurture (environment acting through epigenes) are both inheritable, with important implications for the origin-of-arts problem (see Sec. 6.2).

4 Origin of Arts

The cave art 35,000 years ago [Aczel 2009] is pretty sophisticated and mature (Fig. 6), as painting goes, not something you and I can likely do.[5] The recent discovery of Neandertal[6] jewelry and body paint in Spain suggests that modern human behavior has ancient roots [Wong 2010]. Even carving of 250,000 years ago had been found [Appenzeller 1998]. Fire use (starting 1.6 million years ago) and cooking (400,000 years ago), two sophisticated inventions, occurred long before modern human appeared. All these point to the fact that arts could start a million years ago.

Fig. 6. Painting in Lascaux cave, southwestern France (c.15,000 years ago).

No one could be sure how it happened since there is no record left from that period in time so long ago; moreover, arts like pantomime and dancing would leave no marks. Yet, since evolution happened very slowly, the basic instinct of our ancestors could not differ that much from our own. We could thus guess reasonably. And it could happen like this.

Let us say, somewhere in Africa, a community of few or ten people confined to a habitat during a rain of three days and nights, what would they do? And that happened after a good kill; food was plentiful and there was no rush to make preparations for the next hunt. Sex activities could use up just that much time;[7] and pretty soon, all females could be pregnant. So some of them might start doing something purely "useless," just to *kill time*.[8] Someone could be tracing another person's shadow on the soil surface with a piece of stick—early *drawing*, hitting a piece of wood with a stick—early *music*, telling something with his hands and body— *pantomime* performance, and balancing the body on a piece of wood put on top of a rock—an early *performing art*. Of course, all these could serve as entertainment, too. And it took time to perfect these skills—with more time killed. The ability to handle all these activities is *mimesis*, i.e., to mime, imitate, gesture, and rehearsal of skill, which was already there 2 million years ago [Donald 2006: 7]. Or, something like these might happen later, when there were tens or hundreds of persons living together in a larger group.

All these acts were not for sexual selection (which came later), because that was basically a "free sex" community. And they are never about aesthetics alone, since nature provides plenty, and better, aesthetic experiences (like the sunset).

In any case, sooner or later, someone (or a few in the group) would emerge as an expert in something and might become the first professional "artist," earning his food by entertaining others and staying at home instead of going out to hunt. In other words, *being an artist was the first safe job in history*. This would happen when the population expanded to a certain size and could afford to keep such an artist in their group. The first expert could be male or female, in principle.[9] The important points are:

1. Being a safe job, competition is keen and innovation was called for in the arts profession from the very beginning.

2. To keep the good job and reduce competition, the first artists would tend to maintain secrecy of their trades, pass their skills to their own sons only,[10] or inject mysterious elements into their practices—giving birth to the new profession of *sorcery*, perhaps.

3. The market demand and positive-feedback effect guaranteed that arts as a profession, once established, would not vanish.

As time went by and human advanced, more varieties of arts were created. For example, with the invention of pigments, we had color painting; with language, singing; with writing, literature. And only with plenty of leisure time and a large enough market that *pure* arts appeared. Before that it was all *applied* arts,[11] which, of course, coexisted with pure arts after that. Here, pure arts mean "arts for arts' sake," and applied arts are done with some practical applications in mind, such as group dancing in ritual ceremonies which is to increase group adhesion.

Only in the last two hundred years that the word "art" was associated with aesthetics and fine art [Shiner 2001]. Art in this narrow sense is part of pure arts.

5 Nature of Arts

Arts are a subset of humans' creative activities that aim to excite the receiver's neurons in a certain manner, through that person's senses, with or without significant consequences. This is pretty strange since, while good research works in all other disciplines are also creative activities, only arts as a discipline—with the exception of entertainment—aims at someone's neurons. For instance, pure science is to understand how nature works; it does not aim at anybody's neurons; it does not even need anybody out there (apart from the creator) to receive the end results.[4]

5.1 Applied arts

Applied arts, by definition, are explicitly useful. For instance, a well-decorated vase will help to sell more of those vases, apart from increasing its aesthetic value when placed in a sitting room. A skillfully written novel could change the worldview of the reader, turning her into a fighter for a noble cause or a revolutionary. Obviously, architecture is one of applied arts. It is pure arts that are puzzling. What are they good for?

5.2 Pure arts

As observed by Immanuel Kant (1724-1804), art is useless [Kant 2007]. Here, Kant is referring to pure art, and useless does not mean that it is completely void of consequences. A beautiful landscape painting, for example, could put the receiver into a serene mood. Da Vinci's *Mona Lisa* (1503-1506) could jumpstart the receiver's neurons to wonder what that lady is smiling about. Yet, apart from exercising the receiver's brain—perhaps providing pleasure to and stimulating the thinking or creativity of this person—pure arts do not seem to have any important consequences. This is not exactly true (see Criterion 2 below). Here, five criteria on *lasting* pure arts are presented.

Criterion 1: Aim at receiver's neurons

This is the basic characteristic of any art.

Criterion 2: Kill time

An important function of pure arts is to kill time,[8] the time of the *receiver*.[12] If it is indeed an important piece of art, it is always the case that the receiver has to spend a lot of time contemplating it, while experiencing it and *afterward*. That is what happens to Marcel Duchamp's *Fountain* (1917) (see Fig. 7 below), as the receiver is concerned. That is also the case for an art movie or a good play, even though the viewing time of each is about two hours only. In short, pure arts kill time on the part of the receiver; *good pure arts kill a lot of time*. And that is an important criterion on pure arts.

Criterion 3: Kill time gently and harmlessly

But "kill time" by itself is not enough for something to be called pure art. Entertainments and drugs could kill time, too. The difference is that pure arts kill time gently and harmlessly while entertainments such as a World Cup football game jerks your neurons every 10 or 15 minutes if it is good. Similarly, drug effects are usually not gentle and drug use could land you in jail. (Moreover, drugs may not satisfy Criterion 5 below.) By the same reasoning, classical music is art; heavy metal music, bordering on entertainment. My guess is that Napoleon (1769-1821) would not hang *Mona Lisa* in his bedroom if the smile in the portrait was not that gentle.

In other words, pure arts allow us to kill time in ways that make us feel good, without exciting our neurons too violently, and thus, *encourage us to revisit them frequently*.

Criterion 4: Passivity

It is not true that people want to get involved actively in everything they do. After a day's hard work, most people would like to relax themselves passively by watching TV, for example, and, for those artistically inclined, listen to classical music or doing something, again passively. On the weekend, they might read a book or go to an art museum, enjoying arts passively.

In fact, passivity on the part of the receiver is the signature of all great arts, from painting to literature and to performing arts. All pure arts (and some applied arts) have served the receivers this way in the past many, many years, building up a habit or tradition that we humans still keep. That is why interactive arts never caught on, and perhaps will never be in the future. *Too much interactivity is bad for pure arts.*

Criterion 5: Human creation or intervention

Arts, by definition, have to be something created or intervened by humans. It does not mean that the artists cannot use materials—natural or human-made—or do their work with the aid of machines or computers. Of course, they do, and have been doing it all the time.

By this criterion, a piece of rock lying on the roadside is not a piece of art, no matter how beautiful it is. However, if you take a photo of that rock, the photo could become a piece of art—*photographic art*, because the creation of the photo involves your intervention, assuming that Criterion 1 is also satisfied (which you can help by making the photo interesting, e.g., by bettering the camera angle and using artificial lighting).[13] You might also bring that rock home, put a frame around it and become an artist instantly, because the frame is your way of telling the receiver that you want that person to look at it from a certain angle, a human intervention. Of course, there is no guarantee that this *geographic art* is a piece of great art.

As a result of the tradition a million years in the making, we treasure more those arts that are created with the less external aids. Paintings and sculptures, created practically without external aids, are high on the list. For this reason, we will never consider computer graphics (such as fractals) or ape's "painting" as high art.[14] Similarly, mass reproductions of an art piece are considered commercial products but not art pieces because they are too many steps away from the human creator, the artist. That explains why Andy Warhol (1928-1987) reproduced his silk-screen art pieces in a very limited number, and did it with human hands.

In summary, *pure arts are created by humans or with human intervention, to kill time gently and harmlessly, and let the receiver to experience it (preferably) passively.* With this understanding, it is obvious that the content or form of a pure-art piece is secondary;[15] they are there to serve Criteria 2 to 4.

Since the system of neurons of human beings is an extremely complex system that is not yet well understood, there are not yet sure ways to create an artwork that would satisfy Criteria 2 and 3, i.e., arousing the interest and getting the repeated attention of the receiver. All feasible ways had been tried by artists, such as appealing to human's deep emotions about love and motherhood, and religious upbringing. However, since the brain's neuron connections are shaped not just by nature but by culture, something that worked in a previous era may not work for the present generation. What is clear is that pure arts do not always work on people's sense of aesthetics (Fig. 7, left); they are also about all kinds of emotions such as fear (Fig. 7, upper right) and other things (Fig. 7, lower right), too. Just like physics, arts are about the representation, description or interpretation of everything in nature. Sometimes arts abstract the real (Fig. 8)[16] or play on people's affinity for ambiguity (Fig. 9). Apparently, there are endless ways of doing good arts; we just do not know exactly what they are. This problem of arts is both its strength and difficulty, and obviously is an open problem in science.

As humans migrated out of Africa 60,000 years ago, the contents of arts assumed local features, in addition to global universal elements developed in Africa. The fact that we treasure artworks (more than old stamps, say) implies that they do touch human's deep emotions, needs, values, or something uniquely human. This was exemplified clearly in the French's national effort to hide the Louvre's artworks—and not something else— outside of Paris before the occupation of the Germans during World War II [Nicholas 1995].

Fig. 7. *Left*: Jean-Auguste Dominique Ingres, *The Spring* (1820-1856). *Upper right*: Edvard Munch, *The Scream* (1893-1910). *Lower right*: Marcel Duchamp, *Fountain* (1917).

6 Arts as a Science Matter

Arts, a part of science and a topic in Scimat, can be and should be studied scientifically. To study something is to understand it as thoroughly as possible, with all possible methods and using all appropriate tools. Therefore, knowledge and experiences from other disciplines could be borrowed; theoretical and experimental approaches are both allowed.

6.1 Three lessons from physics

There are three lessons that physics can offer to arts:

Fig. 8. Charlene Lam, *Petals (Longing for Light)* (2009), discarded paper strips and thread, 27 x 27 x 3 cm.

1. How to define a field

The domain of physics keeps on changing. For example, presently, the matured Newtonian dynamics drop out of physics and are picked up by engineering; new subdisciplines such as Econophysics [Mantegna & Stanley 2000], Histophysics [Lam 2002, 2008c], and Complex Systems [Lam 1998] are added. To accommodate this ever shifting scene, *Physics Today*, the monthly magazine published by the American Physical Society, cleverly defines physics like this: "Physics is what physicists do" [Lubkin 1998: 24]. Arts, with its content and form ever changing due to cultural and technological advancements, are like physics in this regard. It is thus not surprising that efforts to find the necessary and sufficient definition of arts all fail [Carroll 1999].[17]

2. The existence of subdisciplines

In physics, different aspects of the same material are studied in different subdisciplines. As an example of the latter, for solids, we have Optics of Solids, Mechanics of Solids, Thermodynamics of Solids, etc. Every such subdiscipline can describe only certain properties of solids. They complement each other; when combined, a full understanding of solids is achieved. When viewed this way, the different approaches in art studies (such as Marxism, feminism, biography and autobiography, semiotics, psychoanalysis, and aesthetics) [Adams 1996] are actually studies at the phenomenological level (see below); each approach is a subdiscipline that specializes on a narrow aspect of arts. No one should expect that any one of them to be encompassing and be able to tell the whole story.

3. There are three levels of study

In *any* scientific study, after observing and collecting data, and analyzing the data, there are three approaches or levels—empirical, phenomeno-logical, and bottom-up—that one can adopt to go further [Lam 2002]. These three approaches in the cases of physics and arts are sketched in Table 2. Empirical studies always happen first. Phenomenological studies are done without knowing the mechanism underlying a phenomenon; they are very powerful and sometimes undervalued. Fundamental understanding of a phenomenon is reached through the bottom-up studies in which the mechanism will reveal itself and become understood.

Fig. 9. Zhuang Wei-Jia, *Guru and the Little Woman* (2010), ink on rice paper, 22 x 24 cm each. Story unfolds from left to right, top to bottom.

Table 2. The three approaches in the study of physics (gas as an example) and arts.

Approach	Gas	Arts
Empirical	Gas law	Empirical rules discovered by artists; empirical studies by art critics/historians/ scholars; fractal analysis of paintings and music by physicists
Phenomenological	Navier-Stokes equations	Interpretations of nature of art by art philosophers/ historians/scholars; evolutionary arguments via Darwin
Bottom-up	Molecular picture (called microscopic method)	Studies through biology: evolution theory (genetics), cognitive science, neuroscience; through physics: statistical analysis

6.2 Arts studies in three approaches

Here examples of arts studies done with the three approaches are given.

1. Empirical

In arts studies, at the empirical level, there are various empirical rules worked out by the artists and empirical analyses due to arts scholars. An example of the latter is the work by André Leroi-Gourhan (1911-1986); he regarded *all* the signs *and* animals depicted in the prehistoric cave paintings as sex symbols and classified them as either female or male (e.g., bison for female, horse for male), reflecting the supposed worldview that the world is divided into two types or two kinds of things, two genders, akin to the Chinese's yin-yang philosophy (see [Aczel 2009]). Another example is Taylor's fractal analysis [2002] of the dripping paintings of Jackson Pollock (1912-1956), and the fractal studies of music (see [Barrow 1995]). The third example is the statistical analyses of literature by Gottschall and his coworkers [Gottschall 2008]; e.g., romantic love is shown to be a literary universal by counting the number it occurs in folktales from different cultures.

2. Phenomenological

At the phenomenological level, apart from the different interpretations of arts such as those summarized in Adams' book [1996] mentioned above, the origin of arts was studied by Dissanayake [1988] through Darwin's evolutionary theory. Here, arts are argued to be an adaptive behavior that benefits human's survival, passing from generation to the next through the genes. The problem with this approach is that, in view of the new findings in epigenetics (Sec. 3), it is not clear that the genetic route is at all necessary. Maybe the cultural effects on arts could be inherited too, through the epigenes; or both. The crucial problem is that unless there is experimental proof of the adaptive nature of arts, such a conjecture remains a conjecture. Similar considerations plague the debate of the nature vs. culture origin of arts. Further examples of the phenomenological approach are the evolutionary study of literature by Carroll [2004] and film by Anderson [1996].

3. Bottom-Up

The bottom-up approach in the study of physics of materials is to start from the molecular description and work out the macroscopic properties; that is because molecules are at the lower level of materials. The case in the humanities is more complicated and slightly different. Even though the immediate lower level of a human body is consisting of cells, no one ever tried to understand humans by considering trillions of cells together. Instead, the bottom-up approach in the humanities starts from a much lower level, from either the neurons or the genes. The former leads to *Neurohumanities*, with subdisciplines such as Neurophilosophy and Neurotheology, and, for arts, Neuroesthetics [Skov & Vartanian 2009], Neuromusicology, Neuroarthistory [Onians 2007], and Neurocinematics [Hasson et al. 2008]; this rapid development of neuro-based studies is sometimes called the Neuro Revolution [Lynch 2009]. The latter leads to the genetic study of the humanities and is less developed.

That neuroscience and cognitive science [Turner 2006] are important in arts studies is not at all surprising, since the brains of the artist and the receiver are obviously important in the creation and appreciation of artworks. And neuroscience helps us to understand what makes the human brain unique [Gazzaniga 2008]. Yet, in the neuro-studies of arts, some progresses are made but success is limited, perhaps due to the fact that the development in neuroscience and cognitive science [Kolak et al. 2006] is still in its infancy. More nontrivial results are needed to get the attention of the artists.[18]

Apart from the books mentioned above, neuroscience in connection with literary studies is discussed in [Hogan 2003a, 2003b]. Applications of neuroscience in movie (and advertisement) studies are reported in [Hamzelou 2010]. Examples of specific works connected with global art history [Onians 2011] and the paintings of Su Dong-Po and Paul Cézanne [Lam & Qiu 2011] as well as other topics can be found in the book *Arts: A Science Matter* [Burguete & Lam 2011].

7 Arts and "Science"

The issue of arts and "science"[1] is of great interest to many people. What this actually means is that whether there is any connection between the arts and the study of the (nonhuman) physical world since both are creative processes. Unfortunately, there exist a lot of confusion concerning this issue, due mainly to the misunderstanding of the domain of science, and the origin and nature of arts. Our takes on this issue are summarized in Table 3.

Furthermore,

1. There is a common misunderstanding on how artists and "scientists" use their brains. The truth is that these two kinds of professionals both use intuition and rational thinking in their trades [Lam 2004]; they use both their left and right brains. Real-time brain scans using fMRI (functional Magnetic Resonance Imaging, started in 1990) could help to clarify this issue.

2. Since artists, painters in particular, quite often depict the world as it appears to them, the same physical principles that govern the working of the physical world could show up in the artists' works. This is indeed the case; some of these principles such as symmetry and broken symmetry are discussed in [Leibowitz 2008].

3. Occasionally, artists are inspired by "science" in creating their work (see Fig. 10 and [Lam 1998: 19; Burguete 2011]). The reverse case is discussed in [Burguete 2011].

Table 3. Comparison between arts and "science."

Characteristics	Arts	"Science"
Both are part of science	Arts are part of science	"Science" (about nonhuman systems) is also part of science
Different aims	Arts aim at receiver's neurons	"Science" aims to understand how nature works
Receiver	Arts need a receiver to appreciate the artwork (but need no comparison with nature)	"Science" needs no receiver (but has to compare with nature, the ultimate judge)
Different history	Arts started at least 35,000 (and could be a million) years ago	"Science" started about 2,600 years ago since Thales (c.624-c.546 BC), after the invention of language and writing
Relationships between arts and "science"	<ul><li>Both involve creative process (for different reasons)—but same in many other human activities</li><li>Arts are human's creation, reflecting on the world of human and nonhuman systems; the principles governing this world are the same principles (e.g., symmetry, spontaneous symmetry breaking, fractal, chao, active walk) studied by "scientists"</li><li>Progress in "science" (and related technology) advances the development of arts; e.g., pigments → color painting, film/camera → photographic art, electricity → cinema, laser → laser art, computer → digital art</li></ul>	

4. For artists and scientists living in the same era, will their works influence each other? There is no direct evidence that this happened, but they, like Picasso and Einstein, certainly were aware of each other's work [Miller 1996, 2001].

5. Since artists are those with supreme bodily sensors [Lam 2008a] and using them every day, they could know something empirically on how the brain works, long before the "scientists" find it out in their labs. Examples on this are given in [Strosberg 2001; Lehrer 2008]; see also [Gardner 1993].

6. Examples are given by Edwards [2008] that some people can cross the boundary between art and "science," and achieve breakthroughs

in both. Discussions on the link between "science" and culture are given in [Slingerland 2008; Galison et al. 2001].

7. Selected artists and "scientists" like to present their effort as the search for "truth, virtue, and beauty." But this claim could not be universally true, for the following reasons: (1) Truth is a fuzzy concept [Godfrey-Smith 2003]. (2) No virtue could be found by naturalists in their study of how insect parasites feed on their hosts;[19] virtue is claimed mostly by theoretical physicists. (3) As shown in Fig. 7, arts are not always about "beauty"; some physicists found Newton's second law of motion, $F = ma$, beautiful (or elegant), but that is because a lot of messy details are hidden from this expression.[20]

8. Some arts are known to have healing effects [McCaffrey 2007]; it links arts to medical science.

Further discussions on the issue of arts and "science" are provided in [Jaroszynski & McNeil 2000; Rodgers 2002; Shahn 2002; Crease 2003; Hidetoshi & Rothman 2008; Lucibella 2010], and references therein.

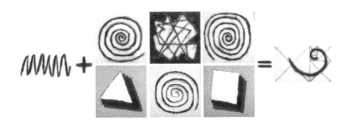

Fig. 10. George Cladis, *Line Revisited*, 10 acrylic and stitching on linen, 146" x 472" installation. The complete absence of color in this installation idealizes the heft and vastness of the conscious domain. (For more information, see "Patterns of Mind" in http://alct.com/gCladis/AbstractCD/2abs.htm).

8 Discussion and Conclusion

The nature and origin of arts is a 2,400-years-old unsolved puzzle dating to Plato's times. In this chapter, an answer to this puzzle is attempted.

In previous sections, we point out that arts consist of applied arts and pure arts (see [Lam 2024b] for more discussion). Applied arts could start a million of years ago, while pure arts emerged later when the community is large enough to support pure artists. Being an artist was the first safe job in human history. The nature of arts is then discussed; five criteria on pure arts are given. Our understanding of arts is based on a global and historical perspective; it differs from the conventional wisdom popularized by the West [Shiner 2001:6] or existing in the East. Yet, reasonable this story of arts is, it has to be substantiated by more works. Here are further discussions.

1. In a way, arts are like driving. Driving is to move a vehicle from point A to point B and deliver something. The motivation of the driver is irrelevant here; the type of vehicle used is secondary. Similarly, the essence of arts is to have the artwork created in the artist's brain (point A) and delivered to the receiver's neurons (point B). It will only confuse the issue if you search for the artist's motivation (which could be a mixture of curiosity, fame and making a living—same as for "scientists" [Tsui & Lam 2011]).[21] And the content and style of the artwork, like the vehicle in driving, is secondary as long as they sit well with the receiver's neurons.

2. Historically, the development of arts is from applied to pure (Sec. 4), similar to the case in science's early development which is from technology to applied and to pure. (The reverse order is mostly true in "science" today.)

3. The key to understand arts is to understand arts (and not just art) as a whole, both historically and globally, and not merely the small part of it existing in Europe. In fact, the latter has a short history of only two thousand years or so, including the last two hundred years of fine arts [Hoppe 2011].

4. It seems that women fall easily for male artists, but female artists do not attract men, at least not to the same degree. If this is true, it can be explained by our surmise that arts was the first safe job in prehistory. According to the evolutionary theory, women tend to choose capable men who would and could stay around to protect the mother and child. In this respect, an artist father would be favored over a hunter father, and this

preference could be carried over to the present day. More research on this issue is warranted.

5. In the West, arts (and "science") were derived from the "liberal arts" some 2,000 years ago and the ideal of liberal arts—designed for free men—was the pursuit of universal knowledge [Wu 2011]. On the other hand, artworks in Europe until a few hundred years ago was operated within the commission/patronage system; the artist was not completely free in choosing his topics, not even Michelangelo or Da Vince [Shiner 2001: 18]. Surprisingly, great arts were still produced. There were two factors that made this possible: (1) The artist was given enough artistic freedom in executing the project, and, more importantly, (2) the person who had the final say, the patron, possessed a high enough level of sophistication or artistic taste. That the artistic sophistication of the "authority" is crucial is reflected in the uniform style of the tall buildings built in Beijing, China, within a couple of years about a decade ago. They all wear a "hat"—the Chinese way of blending the Western and Chinese styles of architecture at that time; an example is shown in Fig. 11. Even in the last two hundred years when there has been a free market for artworks, in a certain sense, artists are still not completely free. The art creating process is constrained by the human nature of the artist and the receiver, and by the principles underlying nature.

6. Neuroscience in arts studies could involve the use of fMRI scan and optogenetics [Deisseroth 2010], applied to the brains of the artist and the receiver. In the optogenetic experiments, light-responsive opsin genes are inserted into the cells of the brain. Specific neurons can then be triggered to fire by a flash of light, providing precision down to the neuron level.

7. In arts studies, no matter how useful the bottom-up neuro approach (theoretically and experimentally) is, it should not be the only method one relies on in understanding arts. Empirical and phenomenological approaches as well as *common sense* should not be forgotten. The situation is like that in physics. After quantum mechanics was invented about 100 years ago, there was a rush to do quantum theory of everything interesting. However, if one wants to understand the flow of water, the advice is to start from the phenomenological Navier-Stokes equations and not

quantum mechanics, because it is unnecessary and impractical to do so. In the study of complex systems (and in any other discipline), the appropriate research tool should be selected according to the level at which the property under study emerges [Lam 2002].

Fig. 11. Pacific Digital City, a "hat"-wearing building at Peking University, Beijing, China.

8. In [Dutton 2009: 52-59], 12 features of art are listed: direct pleasure, skill and virtuosity, style, novelty and creativity, criticism, representation, special focus, expressive individuality, emotional saturation, intellectual challenge, art traditions and institutions, and imaginative experience. Unfortunately, it seems that all these features are shared by the other creative discipline called physics, and hence are not unique to art.

9. The monetary value of an art piece fluctuates in time. It has more to do with the uniqueness of the artwork, like a large piece of diamond or rare stamps, than to do with the artistic nature of the artwork itself.

10. Religion and "science" have been called the two pillars of Western civilization by the Nobel laureate in physics, Richard Feynman (1918-1988).[22] Since religion is not a prime element in societies like China, it could be said that *it is arts and "science" that are the two pillars of* any *civilization*. That is, arts are more basic than religion as civilization is concerned.

11. In principle, all modern creations by humans—like cell phones and air conditioning—could be given up and (most if not all) humans would still

survive. However, without these creations, life would become very inconvenient and many people would suffer. In contrast, if *Mona Lisa* was gone, only the Louvre museum would suffer, materially speaking. In this sense, and only in this sense, arts are inessential and "useless." Otherwise, arts are most valuable! No one should live in a world without arts, and they can't.

Notes

1. In this chapter, "science" with quotation marks means the science of nonhuman systems, identical to the conventional usage of the word.

2. There is a sizable body of literature on these topics. At the entry level, see [Appenzeller 1998; Brown & Dissanayake 2009: 43-57] for arts' origin, [Carroll 1999; Adams 1996; Dutton 2009] for arts' nature, and [Strosberg 2001; Miller 1996; Leibowitz 2008] for arts and "science."

3. Stochastic is a technical word in physics, meaning that probability appears somewhere in the process; a random process is a special case [Paul & Baschnagel 1999].

4. *Note added in 2023*. Science consists of two parts: scientific process and scientific results [Lam 2024a: 47]. Similarly, art has two parts: the art creation process and the artworks that result. Both these two parts of art aim at the receiver's neurons while that of science do not. In art, the creation process and the study of artworks are definitely part of science since science includes studies of all human matters. However, the mere act of admiring an artwork, like the admiration of Newton's gravitational law, is not part of science.

5. No cave art was found in China. For rock art in China dating to about 10,000 years ago, see [Chen 2009].

6. Neandertals, our closest relatives, ruled Europe for 200,000 years but vanished about 28,000 years ago [Wong 2009].

7. In a group of ten members, assume that there are two very old members and two childs and we are left with eight sexually active members; further assume that four are males and four are females, and they are all heterosexual. Then there are 16 possible pairing of copulation; if each copulation consumes ½ hour, it will take only two to eight hours for all copulations to finish, depending on whether the copulations take place concurrently or serially.

8. "Kill time" is a neutral word here, without value judgment; it is short for "spending one's free time"; it does not necessarily imply the person is bored. In modern society, let us say we work 5 days/week, 8 hours/day; and we sleep 8 hours/day and spend 1 hour traveling between home and work/day. In addition, we have to spend 2 hours/meal (including cooking, eating, and washing dishes),

2 meals/day, say. We are left with $24 \times 7 - [(8 + 2 \times 2) \times 7 + (8 + 1) \times 5] = 39$ hours/week, our free time after work, eat, and sleep. A million years ago and before the arts profession appeared, work consisted of hunting, gathering, and tool making, and the amount of free time could be more. How each species spends its free time differentiates one species from the other, and has important consequences in its evolutionary survival.

9. There is hint that cave arts could be made by women as well as by men. See: www.dailymail.co.uk/sciencetech/article-1197680/After-25000-years-scientists-discover-artwork-created-cave-men-AND-cave-women.html##ixzz0rJjXRF3I (June 19, 2010), and [Lane 2011].

10. This practice is maintained today in some Asian countries, in the professions such as martial arts and Chinese traditional medicine.

11. Every discipline can be divided into two parts: pure and applied, like the cases in physics and history [Lam 2002].

12. Ever since arts became a profession long time ago, perhaps a million years ago, killing time was not the motivation of any professional artist. The creative effort in arts, like that in physics, could be hard work [Lane 2011]. As in any creative profession, the time spent is partly to make a living, but mostly it is to satisfy one's personal urge to create (and ego, for most people).

13. Such a strategy is employed by Frankel in producing images from science and engineering experiments, making the two disciplines visually informational and accessible to the public and within the research community [Frankel 2002].

14. One day, if computers are smart enough to create paintings or write music and novels all by themselves (e.g., through genetic algorithm), we may have to differentiate two categories of arts: computer arts and human arts. Our bet is that the former will be valued far less than the latter due to Criterion 5. [*Note added in 2023*. With breakthroughs in AI (artificial intelligence), computer generated paintings, videos, music, novels, etc. became a reality in 2022/2023 and some earlier. They are computer arts.]

15. This does not imply that the content is immaterial. For example, in Duchamp's *Fountain*, the readymade urinal is actually a pretty complicated object with a peculiar shape that invokes all kinds of interesting thoughts, leading to the fulfillment of Criterion 2. If he replaced the urinal with a simple rice bowl, it would not work. Or, if he used a dirty urinal instead of a clean one it would not work either, because that would make the receiver uncomfortable and the resulting work would fail Criterion 3.

16. Charlene Lam's *Petals* in Fig. 8 was made during a dark February in northern Sweden, which is a visualization and representation of the contrast between the region's long, light-filled days of summer and the short days of winter. The lengths of paper used were determined by the actual and predicted lengths of daylight for the first of each month in 2009. The outer loops of each petal represent

the 24 hours in a day; the inner loops represent the hours of daylight. It should be noted that though the length of the materials were determined by scientific data, the resulting visualization is more of an artistic statement about the perception of light available in a given day. More details and additional work can be found at www.charlenelam.com.

17. Noël Carroll's practical reason [1999: 207] that we need to define art precisely so an artwork such as Brancusi's abstract sculpture *Bird in Flight* can be imported to the US duty free, is not strictly valid. This practical problem can be solved in the same way that we handle pornography, i.e., to be decided by a committee of local residents.

18. In physics, for example, the nontrivial result that maximum range of a projectile is achieved with a throwing angle of 45° is taught in high school. This knowledge is used by the teenager in the movie *Aliens in the Attic* (2009) in hitting the alien with a dart and thus saving the world!

19. For example, predatory wasps paralyze the host insect by stinging and lay eggs on it. The larvae feed on the paralyzed host and killing it in the end (http://en.wikipedia.org/ wiki/Entomophagous_parasite, Nov. 1, 2010).

20. For instance, this law is about non-existent point particles which have mass but zero size; $F = ma$ is only a special case (when m = const) of the second law which states that $F = \mathrm{d}(mv)/\mathrm{d}t$, actually a vector equation involving differential calculus; two quantities, force F and mass m, are defined with this one equation, which is completely illogical [Wilczek 2004].

21. Einstein's explanation for people's motivation in doing art and "science" is that "[to] escape from everyday life with its painful crudity and hopeless dreariness" [Einstein 1982: 225].

22. "The relation of science and religion," the transcript of a talk given by Richard Feynman at the Caltech YMCA Lunch Forum on May 2, 1956 (http:// calteches.library.caltech.edu/49/2/Religion.htm, June 18, 2010).

References

Aczel, A. D. [2009]. *The Cave and the Cathedral*. Hoboken, NJ: Wiley.

Adams, L. S. [1996]. *The Methodologies of Art: An Introduction*. Bouler, CO: Westview.

Anderson, J. D. [1996]. *The Reality of Illusion: An Ecological Approach to Cognitive Film Theory*. Carbondale and Edwardsville: Southern Illinois University Press.

Appenzeller, T. [1998]. Art: Evolution or revolution? Science **282**: 1451.

Barrow, J. D. [1995]. *The Artful Universe: The Cosmic Source of Human Creativity*. New York: Little, Brown and Co.

Brown, S. & Dissanayake, E. [2009]. The arts are more than aesthetics: Neuroaesthetics as narrow aesthetics. *Neuroaesthetics*, Skov, M. & Vartanian, O. (eds.). Amityville, NY: Baywood.

Burguete, M. [2011]. ChemArt and BioArt: Art-science interactions. *Arts: A Science Matter*, Burguete, M. & Lam, L. (eds.). Singapore: World Scientific.

Burguete, M. & Lam, L. (eds.) [2011]. *Arts: A Science Matter*. Singapore: World Scientific.

Carroll, J. [2004]. *Literary Darwinism: Evolution, Human Nature, and Literature*. New York: Routledge.

Carroll, N. [1999]. *Philosophy of Art: A Contemporary Introduction*. New York: Routledge.

Chen, Zhao-Fu [2009]. *Discovery History of Rock Art in China*. Shanghai: Shanghai People's Publishing House.

Cloud, J. [2010]. Why your DNA isn't your destiny. Time, Jan. 6, 2010.

Crease, R. P. [2003]. The Newton-Beethoven analogy. Phys. World, April 16, 2003.

Deisseroth, K. [2010]. Controlling the brain with light. Sci. Am., Nov. 2010: 49-55.

Dissanayake, E. [1988]. *What Is Art For?* Seattle: University of Washington Press.

Donald, M. [2006]. Art and cognitive evolution. *The Artful Mind: Cognitive Science and the Riddle of Human Creativity*, Turner, M. (ed.). Oxford: Oxford University Press.

Dutton, D. [2009]. *The Art Instinct: Beauty, Pleasure, and Human Evolution*. New York: Bloomsbury.

Edwards, D. [2008]. *Artscience: Creativity in the Post-Google Generation*. Cambridge, MA: Harvard University Press.

Einstein, A. [1982]. *Ideas and Opinions*. New York: Three Rivers Press.

Frankel, F. [2002]. *Envisioning Science: The Design and Craft of the Science Image*. Cambridge, MA: MIT Press.

Galison, P., Graubard, S. R. & Mendelsohn, E. [2001]. *Science in Culture*. New Brunswick, NJ: Transaction Publishers.

Gardner, H. [1993]. *Creating Minds: An Anatomy of Creativity Seen Through the Lives of Freud, Einstein, Picasso, Stravinsky, Eliot, Graham and Gandhi*. New York: Basic Books.

Gazzaniga, M. [2008]. *Human: The Science Behind What Makes us Unique*. New York: HarperCollins.

Godfrey-Smith, P. [2003]. *Theory and Reality: An Introduction to the Philosophy of Science*. Chicago: University of Chicago Press.

Gottschall, J. [2008]. *Literature, Science, and a New Humanities*. New York: Palgrave MacMillan.

Hamzelou, J. [2010]. Brain invasion at the multiplex. New Scientist, Sept. 11, 2010: 8-9.

Hasson, U., Landesman, O., Knappmeyer, B., Vallines, I., Rubin, N. & Heeger, D. J. [2008]. Neurocinetmatics: The neuroscience of film. Projections **2**(1): 1-26.

Hidetoshi, F. & Rothman, T. [2008]. *Sacred Mathematics: Japanese Temple Geometry*. Princeton: Princeton University Press.

Hoagland, M. & Dodson, B. [1998]. *The Way Life Works*. New York: Times Books.

Hogan, P. C. [2003a]. *Cognitive Science, Literature, and the Arts: A Guide for Humanists*. New York: Routledge.

Hogan, P. C. [2003b]. *The Mind and Its Stories: Narrative Universals and Human Emotion*. Cambridge, UK: Cambridge University Press.

Hoppe, B. [2011]. The Latin *Artes* and the origin of modern *Arts*. *Arts: A Science Matter*, Burguete, M. & Lam, L. (eds.). Singapore: World Scientific.

Jablonka E. & Lamb, M. J. [1995]. *Epigenetic Inheritance and Evolution: The Lamarckian Dimension*. Oxford: Oxford University Press.

Jablonka, E. & Raz, G. [2009]. Transgenerational epigenetic inheritance: Prevalence, mechanisms, and implications for the study of heredity and evolution. Quarterly Review of Biology **84**: 131-176.

Jaroszynski, D. & McNeil, B. [2000]. Science given an artistic licence. Phys. World, July 2000: 51.

Kant, I. [2007]. *Critique of Judgement*. Oxford: Oxford University Press.

Kolak, D., Hirstein, W., Mandik, P. & Waskan, J. [2006]. *Cognitive Science: An Introduction to Mind and Brain*. New York: Routledge.

Lam, L (ed.) [1997]. *Introduction to Nonlinear Physics*. New York: Springer.

Lam, L. [1998]. *Nonlinear Physics for Beginners: Fractals, Chaos, Solitons, Pattern Formation, Cellular Automata and Complex Systems*. Singapore: World Scientific.

Lam, L. [2002]. Histophysics: A new discipline. Mod. Phys. Lett. B **16**: 1163-1176.

Lam, L. [2004]. *This Pale Blue Dot: Science, History, God*. Tamsui: Tamkang University Press.

Lam, L. [2008a]. Science Matters: A unified perspective. *Science Matters: Humanities as Complex Systems*, Burguete, M. & Lam, L. (eds.). Singapore: World Scientific.

Lam, L. [2008b]. Science Matters: The newest and biggest interdicipline. *China Interdisciplinary Science*, Vol. 2, Liu Zhong-Lin (刘仲林) (ed.). Beijing: Science Press.

Lam, L. [2008c]. Human history: A Science Matter. *Science Matters: Humanities as Complex Systems*, Burguete, M. & Lam, L. (eds.). Singapore: World Scientific.

Lam, L. [2024a]. *Humantities, Science, Scimat*. Singapore: World Scientific.

Lam, L. [2024b]. On art. *Scimat Anthology: Histophysics, Art, Philosophy, Science*, Lam, L. Singapore: World Scientific.

Lam, L. & Li-Meng Qiu (邱理萌) [2011]. Su Dong-Po's bamboo and Paul Cézanne's apple. *Arts: A Science Matter*. Burguete, M. & Lam, L. (eds.). Singapore: World Scientific. pp 348-370.

Lane, H. [2011]. From curiosity to creation: The art of Holly Lane. *Arts: A Science Matter*. Burguete, M. & Lam, L. (eds.). Singapore: World Scientific.

Lehrer, J. [2008]. *Proust Was a Neuroscientist*. New York: Mariner.

Leibowitz, J. R. [2008]. *Hidden Harmony: The Connected Worlds of Physics and Art*. Baltimore: Johns Hopkins University Press.

Lubkin, G. B. [1998]. A personal look back at *Physics Today*. Phys. Today, May 1998: 24-29.

Lucibella, M. [2010]. Physics stars in theater, music and dance. APS News, Nov. 2010: 1.

Lynch, Z. [2009]. *The Neuro Revolution: How Brain Science Is Changing Our World*. New York: St. Martin.

Mantegna, R. N. & Stanley, H. G. [2000]. *An Introduction to Econophysics*. New York: Cambridge University Press.

McCaffrey, R. [2007]. The effect of healing gardens and art therapy on older adults with mild to moderate depression. Holist Nurs. Pract. **21**(2): 79-84.

Miller, A. I. [1996]. *Insights of Genius: Imagery and Creativity in Science and Art*. New York: Copernicus.

Miller, A. I. [2001]. *Einstein, Picasso: Space, Time and the Beauty That Causes Havoc*. New York: Basic Books.

Mithen, S. [1999]. *The Prehistory of the Mind: The Cognitive Origins of Art and Science*. New York: Thames and Hudson.

Moore, D. S. [2003]. *The Dependent Gene: The Fallacy of "Nature vs. Nurture."* New York: Freeman

Nicholas, L. H. [1995]. *The Rape of Europa: The Fate of Europe's Treasures in the Third Reich and the Second World War*. New York: Vintage Books.

Onians, J. [2007]. *Neuroarthistory: From Aristotle and Pliny to Baxandall and Zeki*. New Haven: Yale University Press.

Onians, J. [2011]. Neuroarthistory: Reuniting ancient traditions in a new scientific approach to the understanding of art. *Arts: A Science Matter*. Burguete, M. & Lam, L. (eds.). Singapore: World Scientific.

Paul, W. & Baschnagel, J. [1999]. *Stochastic Processes: From Physics to Finance*. New York: Springer.

Ridley, M. [2003]. *The Agile Gene: How Nature Turns on Nurture*. New York: Perennial.

Rodgers, P. [2002]. Physics meets art and literature. Phys. World, Nov. 2002: 29 (special issue on physics and arts).

Shahn, E. [2002]. Swept into the modern along with science. Science **298**: 2333-2334.

Shenk, D. [2010]. *The Genius in All of Us: Why Everything You've Been Told About Genetics, Talent, and IQ Is Wrong*. New York: Doubleday.

Shiner, L. [2001]. *The Invention of Art: A Cultural History*. University of Chicago Press, Chicago: University of Chicago Press.

Skov, M. & Vartanian, O. (eds.) [2009]. *Neuroaesthetics*. Amityville, NY: Baywood.

Slingerland, E. [2008]. *What Science Offers the Humanities: Integrating Body and Culture*. Cambridge, UK: Cambridge University Press.

Strosberg, E. [2001]. *Art and Science*. New York: Abbeville Press.

Taylor, R. P. [2002]. Order in Pollock's Chaos. Sci. Am., Dec. 2002: 116-121.

Tsui, Hark (徐克) & Lam, L. [2011]. Making movies and making physics. *Arts: A Science Matter*. Burguete, M. & Lam, L. (eds.). Singapore: World Scientific. pp 204-221.

Turner, M. (ed.) [2006]. *The Artful Mind: Cognitive Science and the Riddle of Human Creativity*. New York: Oxford University Press.

Turner, M. S. [2009]. The universe. Sci. Am. Sept. 2009: 36-43.

Wilczek, F. [2004]. Whence the force of $F = ma$? I: Culture shock. Phys. Today, Oct. 2004: 11 12. [Parts II and III appear, respectively, in Dec. 2004 and July 2005 issues of *Phys. Today*.]

Wong, K. [2009]. Twilight of the Neandertals. Sci. Am , Aug 2009: 32-37.

Wong, K. [2010]. Did Neandertals think like us? Sci. Am., June 2010: 72-75.

Wu, Guo-Sheng (吴国盛) [2011]. Science and art: A philosophical perspective. *Arts: A Science Matter*. Burguete, M. & Lam, L. (eds.). Singapore: World Scientific.

Published: Lam, L. [2011]. Arts: A science matter. *Arts: A science matter*, Burguete, M. & Lam, L. (eds.). Singapore: World Scientific. pp 1-32.

Making Movies and Making Physics

Hark Tsui and Lui Lam

The characteristics and experiences of making movies and making physics are discussed, respectively, by a movie director/producer and a physicist. Similarities and differences between the making of movies and the making of physics are presented. Discussions on the nature of movies and physics, on creativity and innovation as well as on the joy of making movies and making physics are provided.

1 Introduction

Making movies is a creative process that involves several operational stages: (1) conception of the project, (2) lining up the funding, (3) finding coworkers, (4) shooting the movie, (5) post-shooting work, and (6) marketing and distributing the movie.

Making physics means creating new physics—doing physics, or physics research, at its best. It is also a creative process and, like making movies, involves the same six stages in its operation, from the beginning to finish. The exception is that when doing pure theory or simple experiments, stages (2) and (3) may be absent.

Hark Tsui has been directing and producing movies since 1979 [Ho & Ho 2002; Morton 2001] and Lui Lam published his first physics paper in 1968 [McMillan et al. 1968]. Presently, Tsui has directed/produced over 70 movies and Lam has published over 170 research papers and 12 books. The two knew each other since 1975 when both were doing community work in the New York Chinatown, in lower Manhattan.

In the following, their background and their views on the nature of movies and physics, respectively, are given in Secs. 2 and 3. Their experience in making movies and making physics—picking a project and executing the project—are presented in Secs. 4 and 5. Some musings on creativity and innovation (Sec. 6) and the joy of working on movies and physics (Sec. 7) are also provided. Section 8 concludes the chapter with discussion.

2 Our Background

A person's background influences his personal and professional choices and working style, which shows up in the final product this person produces, whether it is his movie or physics. The backgrounds of the authors are therefore given here first.

Tsui: I think it is very difficult to trace back one's background relating to one's creativity, because most of the time ideas come from factors related to DNA instead of environmental influences. That is why some people can draw when they are not trained as a painter, and some can sing when they are not trained in music academies. It is very abstract and difficult to link up someone's creativity to his/her growing up process.

My understanding of what initiated my creativity happened in my childhood in Vietnam, starting with my sketch of a drawing. In fact, something happened before that, when I was younger. A school classmate drew a chalk drawing on a small blackboard for fun, which somehow was related to my later inclination of becoming a movie director. That drawing is about a movie that I saw before in a cinema, *Godzilla Attacking Tokyo*. In that experience I was quite frustrated because I could not draw as good as he did. In fact, I did not draw at all. But watching him draw was great fun. In fact, my classmate's drawing activity was accompanied by singing and talking. It was very much like what one experienced in a cinema and watched animation done by those classmates.

Very soon after that, I picked up a pencil and tried to draw, and it was the copy of a car show's pamphlet. It was great fun. This first trial of my creativity energized my curiosity to find out what I can draw. Then to surpass my classmates, I started to draw movie shoots. I went through the period when a lot of kids played with the flashlight shining through a plastic with drawings, with narratives—a basic kind of movie.

Later, I switched my interest to drawing buildings, which made my family believe that I was going to be an architect. It was very encouraging that my family elders asked me to draw different buildings and awarded me with petty money. Those drawings included people with activities inside buildings, a very basic video game like *Sim City*. The whole idea of using projector, drawing and narrative to tell a story was put aside when I entered the preteen period. It was later in my teen years, I, with a few friends, rented (daily when needed) a regular 8 mm camera, with my own pocket money and everybody chipped in.

At that point I moved to Hong Kong. For the first three years, school life was unlike what I had before in Vietnam; in Hong Kong, it was total book reading and many exams, only a little bit of drawing. With this trend continuing, my interest shifted to *physics* and science in general, wanting to be a scientist, being influenced by a society that respects scientists. Those years were spent in reading a lot of science books, gearing to be a scientist. Graduating from high school, I had to decide what to do in college and in the future—quite a funny idea for a 17-year-old schoolboy; he had to make such an important choice without enough knowledge. It was a big decision to make because going to college took a lot of money, and I had no money; I had to think hard.

At the end I decided to switch to the humanities and be happy. An interesting question arose: How to be happy while working as an adult. For two years I hanged around, searching for a decision. Suddenly one day on a bus, a friend asked me: "What will make you happy?" My answer: see movies, rather than working. My friend: "Then why don't you make movies?" Very inspiring conversation. After that day, I asked myself what to do if I wanted to be a movie director. With the limited knowledge I had at that age, I asked for advice. Most people went to study film or media production with ready financial support or some basic knowledge of the field. I had none. (My family did not see any promise in filmmaking and so did not really support my ambition.).

Bringing with me the little bit of money I earned in high school, I went to the United States with enough money for staying only one semester in college; I was constantly frustrated by the idea that I might be forced to return to Hong Kong soon. But the experience of being in a foreign country and living independently was very exciting. Every stage happening in life seemed to worth my effort. New perspective of what the world looked like opened up my horizon of thinking. The possibility of having to return home again did not stop me from trying my best to stay in the US. I tried very hard to do what I thought was needed to stay in college to learn what to do as a filmmaker.

The first Christmas, I went to New York looking for a job. Then, after staying in college for three years, in 1975, I went to New York again working in a film-processing lab. I was very much into New York and involved myself in community life—where people lived in Chinatown, midtown, with different cultures. There were so many things to do and to see; everything became lively and hopeful. It was very satisfactory seeing myself as a useful person.

An essential factor for me going to New York was to do a paper that was needed in my courses. The plan was to do a documentary film in New York, about the community and the society, a film about antiwar activities. I was looking for a documentary film person I had in mind but could not find him. I ended up working in a documentary company myself. Later, my involvement in the Chinatown community helped me to develop further. For example, while in New York I went to the Chinatown to see a movie from China, *Tianshan Red Flower*; the movie was in English while the audience was Chinese. Instant translation was needed, but no one could translate it well into Chinese. So I volunteered, and that happened again and again. The Chinatown turned out to be very complicated, with many groups of different interests coexisting. I looked for different channels to help people—nonprofit organization helping the locals to find jobs, Medicare clinics, and media production groups.

In 1977 I returned to Hong Kong in order to understand my race and my mother country at a close distance, since I could only get a remote feeling of these things in New York. Hong Kong was then the closest place one could get near the Mainland. I wanted to wait in Hong Kong for China to open its door so that I could enter Mainland, to explore my experience as a Chinese and my heredity. While in Hong Kong I tried to do documentary films, but was instead asked by my employer to make TV films. Fortunately for me, the experience helped. I directed my first feature film *The Butterfly Murders* in 1979 and became a movie director ever since.

In 1991, I made my first film in the Mainland, *Once Upon a Time in China*; it was a very dramatic experience. The movie systems in Hong Kong and in Mainland were not the same, not even similar to each other in mentality; it was very disturbing to me. But then those valuable experiences would ready me to work in China on a long term basis later. Seeing those experiences as the preparation for future challenges, I continued to think of how to become a filmmaker in China.

In the mid-1990s, there were a lot of invitations from Hollywood for moviemakers in Hong Kong. I became one of the chosen directors to go; I was back to the US again. Then I was offered the chance to stay in the US for good, an exciting chance indeed. But I opted for Hong Kong, more precisely, China, since Hong Kong is part of China. I wanted to see, as a filmmaker, whatever would happen in the 1997 transition when Hong Kong was reverted back to China, to open up my own horizon as a filmmaker in my own country.

My plan of making movies in China continued. However, there were a few proposed projects that did not get through the censors in China. So my dream of being a filmmaker in China was temporarily hindered. In 2002, I was involved in a company that wanted to produce a movie called *Seven Swords*. And I launched myself again, on the track of shooting a movie with permissible content. Since then, I had been in China, thinking of all possible projects for this particular market.

However, SARS (severe acute respiratory syndrome) suddenly broke out in 2003, in Hong Kong and elsewhere. My attention was turned back to Hong Kong after much of my personal relationships there had been neglected by me. I was thinking that so much more could be done for Hong Kong, by injecting creativity in that society to create more interesting culture and present them to the rest of the world. I started planning a community project called Project Hong Kong, to extend the local culture and increase the appeal of Hong Kong to the outside world. The project did not go well because of insufficient financial and social support; it was a nonprofit project, supported neither by the government nor private corporations. For two years, there was no result and I became losing interest in this project.

In 2005, *Seven Swords* was finished. I did other movies in China, like *All about Women* and *Missing* (both 2008)—the latter was shot in Taiwan, Hong Kong, and Japan—and *Detective Dee and the Mystery of the Phantom Flame* (2010).

Lam: I was born in Guangdong Province and grew up in Hong Kong, where I received my education from grade one on and graduated from the University of Hong Kong with a BS degree, spanning from 1949 to 1965. I then went abroad in 1965 to Vancouver, Canada, and then to New York City in 1966. I received my MS degree from the University of British Columbia and PhD from Columbia University (1973), both in physics.

It was during my graduate student years in the late 1960s that the anti-Vietnam-War movement erupted and the students took over the buildings at Columbia University. From the end of 1970 to 1971, I actively participated in the *Baodiao Movement* [Lam 1971];[1] and then lived in New York's Chinatown (in Manhattan) to serve the community for about two years. We started the Chinatown Food Co-op and helped the local patriotic newspaper *China Daily News* in its publication.

When at Bell Laboratories in Murray Hill, New Jersey, as a graduate student from Columbia, my mentor was Philip Platzman.[2] I did my postdoc under Melvin Lax (1922-2002)[3] at City College, City University of New York (1972-1975) before moving to Antwerpen, Belgium and then Saarbrücken, West Germany, spending about one year at each place. Starting January 1978, after the Cultural Revolution, I worked at the Institute of Physics, Chinese Academy of Sciences, in Beijing [Lam 2010]. I left China at end of 1983 due to family reasons [Li 1983]. Subsequently,

I worked at City University of New York (1984-1987) and at San Jose State University since 1987.

Essentially, my physics research ranged from nuclear physics in my Vancouver years (1965-1966) to condensed matter theory at New York and Europe (1969-1977), and then from liquid crystals research in China and New York (1978-1987) to nonlinear physics and complex systems in San Jose (since 1987). I started publishing on the humanities in 2002 [Lam 2002; Lam et al. 2010].

I do not recall that I was interested in how the physical world works when I was young. I got interested in physics research after I published a few papers on my own as a graduate student at Columbia. However, I do recall that in my high school years, probably grade 11, I wrote a short essay in class under the title, "Those who don't think ahead will have recent worries," which is a well-known Chinese idiom chosen by the Chinese literature teacher as the essay title. Instead of putting out examples that illustrate the correctness of this statement like my classmates did, I gave the mechanism that guarantees this phenomenon. The mechanism goes like this: The world is very complicated (I did not know the term "complex system" yet), and so troublesome things will keep on appearing. If one do not think about them in advance and plan on how to handle them, they would become recent worries when they do happen. I wrote it short because the mechanism is simple. And instead of getting a high mark like 90% or more that I expected, I got a pretty low grade for my perfect argument. Looking back, this is my first research work on *Science Matters*.[4]

3 What Are Movies? What Is Physics?

Tsui: Movies present a phenomenal mentality shared by the masses. I always believe that to understand movies, one has to understand the history of movies. From movies' history, for example, one can understand why some movies are made with certain social factors. Also, to study (or create) a classic movie that is enjoyed and appreciated by people for a long time, one must has a registration of the feeling or emotion of that certain period in time the movie depicts. To understand or revisit a certain era, one can go back to see old movies. Our obligation as a movie maker is to be sincere about how we feel about current events happening in that era, to give a phenomenal mentality shared by all.

Movie is a mirror from which the audience can see themselves.

Lam: "Physics is what physicists do." This is the definition of physics held to by *Physics Today*, the official monthly published by the American Physical Society [Lubkin 1998: 24]. Indeed, this is the only definition that makes sense since the domain of physics is ever changing. Mature areas like classical Newtonian mechanics shifted from physics to the engineering departments, and new areas such as *Econophysics* [Mantegna & Stanley 2000] and *Histophysics* [Lam 2002] came in.

Today, physics is not just about nonhuman or simple systems; it is also about complex systems, including all those from social science and the humanities. As advocated by the Nobel laureate Arthur Schawlow: "The task of physics is not only to understand the hydrogen atom, but to understand the world" [Wolf 1994: 53].

4 Picking a Project

Tsui: Every filmmaker always goes through history of his own growth from the childhood days, from time to time. When watching old movies, the filmmaker may come up with a new perspective and new ideas to replace the old ones—integrating it with new moralities, viewing it from a new angle, and using a new story-telling method. This is because only a member of the new generation can grasp the mentality of that new generation. New direction of movies always comes up and keeps on happening. Consequently, "remaking" a movie is not necessarily, and usually not, a simple remake of the old movie; the remake is not the same movie.

To choose a topic as a contemporary movie maker, one should make connection with the environment and the world, and look for passion and emotion that can touch you and other people. This is the basic rule in picking a project in movie making.

As a naturally born artist, the movie maker always gets nutrition from what urges him on to express himself in how he sees the world. In most of the time, the ability to express himself comes from the blood of the filmmaker. For instance, some directors may not learn story telling from academic channels; yet, they still can tell a very touching story. That is why sometimes it is DNA that makes a person what he is.

Lam: To pick a project in physics research involves two steps: (1) coming up with an idea; (2) deciding whether to go ahead with the idea. There are three ways that a good idea may come to mind, like the case that a photographer may capture a good picture.

1. She could sit still in front of her house, with the camera in hand, and capture whatever that is interesting and happening within her eyesight.

2. She could run after a rushing crowd, join them and see what is happening out there; she may get a good picture of something exciting.

3. She goes places and sometimes wanders around; she may capture good pictures from time to time.

Needless to say, method 1 is not recommended because of the slim chance of success. Method 2 corresponds to chasing a hot topic in physics research like high-T_c superconductivity when it was first discovered in 1978. The competition will be keen and one should go into it only if one thinks she has a good idea in solving the problem or enough resources that could beat out all the competitors. Method 3 is most productive and is usually preferred.

However, with method 3, there are still two problems to overcome. First, like the photographer with the camera, the research physicist should be equipped—with research skills ready or capable of picking up necessary skills quickly. Second, how does she know where to go? The solution to these two problems comes from the same source: accumulating information in your brain on a *long-term* basis—since the research topic and needed tools are not known beforehand. To do that, it involves (1) reading the monthly magazine *Physics Today*, (2) browsing all the physics journals in the library, and (3) attending weekly departmental seminars.

Reading for research is very different from reading in a physics course. Read only the abstract as well as the first and last sections of an article is enough; go back to read in detail when you need to use the material there someday with your research topic in mind. And in a seminar, paying attention to the first and last five minutes of the one-hour talk is essential, because the speaker, the expert in that topic, is summarizing for you in the first five minutes what have been done on that topic and why the research is worth doing, and in the last five minutes the new findings and open problems. Most significant works are achieved by borrowing concepts or tools from one research field to the other; completely new things are invented on very rare occasions. That is why one has to read books and journals and attend seminars which do not fall neatly into the research topic one is doing. Do not let your mentoring professor tell you otherwise.

When you have a choice in picking problems, aim high. Pick the one that will have an impact, but you still have a fair chance of solving it. To that end, you have to *guess* the level of difficulty of the solution, what tools are needed, and what your own level of expertise is. In short, you have to know yourself and your available resources pretty well; make friends and connections all the time because they are part of your resources. Of course, to play it safe before you have a tenure job, you may want to work on a familiar problem that guarantees publications while you are tackling a difficult but exciting one. The trick is always have an exciting problem under working, even if that means you have to hide it from your mentor who is using you as a cheap labor.

Wondering around from time to time, i.e., trying something new and unexplored by others, is equally important because important discoveries often show up unexpectedly. This is best described by Alexander Graham Bell (1847-1922), the inventor of the telephone: "Leave the beaten track occasionally and dive into the woods. You will be certain to find something that you have never seen before" (Fig. 1).

Fig. 1. Bust of Alexander Graham Bell at the entrance of Bell Laboratories, Murray Hill, New Jersey. Bell's quote (see text) is inscribed below the bust (July 12, 1994). [photo/Lui Lam]

5 Executing the Project

Tsui: The production of a movie involves more than one person; it is a teamwork. As such the director is the commander of the team who has to lead the way in thinking. The intention of the director has to be made clear to everybody so the group would follow the intended direction and come out with a work that has a solid vision.

When working with a group, the director's demanding control is necessary to avoid diversion of different ideas. The director has to be firm in his stand, even to confront people who do not agree with him, like a commander in an army. This is due partly to time and financial constraints, but mostly to artistic control.

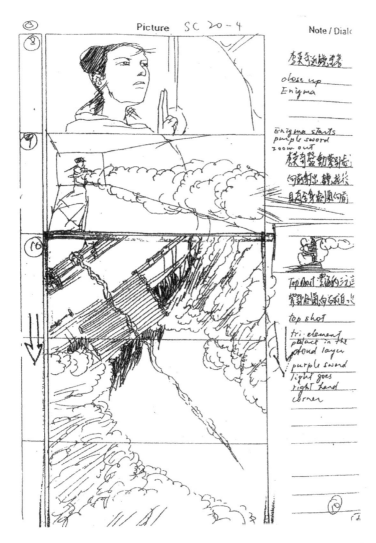

Fig. 2. Storyboards for *The Legend of Zu*, 2001 (drawn by Hark Tsui).

It is quite often that the director participates in writing the script and draws his own storyboards (Fig. 2), and involves himself heavily in post-production works. All of these, naturally, are for the director to express himself as an artist and for the artistic control of the final product, the movie. Lastly, one should not forget that the making of a movie involves not just the production component but also the financing component; a movie is a "commercial" product, in the positive sense of the word. That is why the director and actors have to participate in the promotion of a movie before or after it is finished, to help the sale of the movie.

Lam: A researcher has to be *completely* honest, trustworthy, and dependable; never work with anyone who does not meet these criteria. The reason is very simple: There are many steps in a piece of research that it is sometimes quite impossible for others to check the details, such as an extremely long and complicated calculation, the writing of a computer program, and the taking and analysis of data. Moreover, it may take a year or more to accumulate data and do the analysis, and if this person lost the records of them it will be a waste of valuable time for the whole team. In research, priority—be the first to publish—is of paramount importance; there is no second runner.

Good physics research involves two parts: The major and important part comes from guessing and imagination, or what one calls *intuition*; the other part is *logical*, coming from induction or deduction. Intuition cannot be taught as simply as logical deductions; intuition is what differentiates good physicists from less physicists. When one faces a new phenomenon, the first step is to guess what happened. Modeling (Fig. 3) or equation writing comes later; proofs of any kind come even later and are often done after the correct answer is first guessed.

Publishing a paper is the byproduct of research. The basic aim of research is to *understand* nature, not to publish papers—a point often misunderstood or ignored by many practicing physicists, which, unfortunately, might be beyond their control. Counting papers by administrators indicates the lack of qualified referees in the evaluation process. The publishing of fraudulent papers actually misleads the research direction and thus hinders the development of a research field. Anyone knowingly committing fraud and any institution tolerating fraud among its members have no business in research whatsoever. And, in fact, reporting fraud in one's own field is considered part of the duty of the researcher, irrespective of the researcher's motivation.

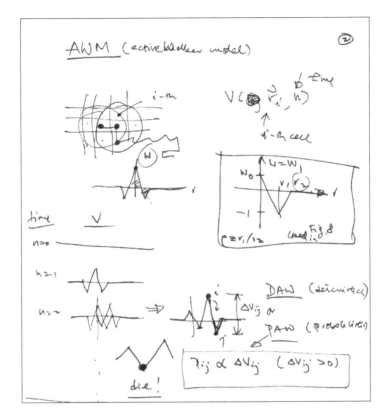

Fig. 3. Constructing an active walk (AW) model (drawn by Lui Lam, March 21, 1992). The AW model was initially designed to reproduce the filamentary patterns observed experimentally, but turned out to be a general paradigm covering many complex (human and nonhuman) systems [Lam 2005, 2006].

Reporting one's research results in seminars or conferences is part of the game, which could happen before or after the paper is published. The motivation is about advertising one's own work, but more importantly, it is to ascertain and improve one's research by facing challenges from one's peers. In physics research, while the picking and execution of the project, and even the formulation of one's results and the writing of the paper— those parts involving human action—are human dependent and could be the domain of sociology of science, it does not follow that the contents of the end scientific results, like Newton's laws of motion or Einstein's relativity theories, are also human dependent. In fact, they are not; they belong to the human-independent knowledge [Lam 2008].

6 Creativity and Innovation

Tsui: One never knows a priorily whether he has creativity or is capable of innovations. The director has to come up with the result of his work—the movie, to test with the audience by finding out the direction of their response so that the director will know what he did is indeed what he wants. For example, one might want to do a very touching story, but the resulting movie may end up with a totally different effect. The reason could be that the materials of his work clash with his method of storytelling, the two working in opposite directions. A creative person always has to explore different possibilities which might bring him to unknown realms.

The unknown realm is always a mysterious space of self-exploration, which sometimes could be very risky and uncertain. For instance, good directors do make films that end up as crap; all filmmakers do make films that the audience does not accept. Those films in the course of self-exploration may not be mature enough to establish communication between the director and the audience. Sometimes it may take more than one step to get to the destination the director wants it to be. And it is the filmmaker who has to fight all the "wars" to get to the point that he can successfully come up with, with good creativity. A director always has to focus on the audience of his choice.

Innovation is not just about techniques. It is a vision you want to pass to the audience. To achieve that, the director has to choose all the "weapons" accessible to him.

Lam: Creativity means the ability to create something novel. It involves three components: dare to think, free to speak out one's mind, and the tolerance of failures. There are material bases behind all these. Every child dares to think but such an innate ability might be suppressed by the society, at home and in school, when the child grows up. To be a good physicist, one should be able to look into Einstein's eyes and tell him he is wrong, if that is what one thinks. But this is possible only if there is a feeling of "equality" among the two persons engaged in an academic discussion, *and* any displeasure of the senior person will not jeopardize one's immediate position and future career. Furthermore, failures occur all the time during a creative process; that is why innovations are so hard to come by. The tolerance of failures is thus very important and the society must provide multiple ways for a failed person to come back, e.g., freedom to switch majors in a university, the ease of changing schools, and the existence of different types of colleges (such as the two-year community college and the four-year university) that is linked to each other. In short, we are talking about the freedom to think and speak, and the flexibility and mobility in the education system and the employment circles.

Providing pleasant and conducive working environment is equally important in encouraging creativity and innovation. Sitting the graduate students in small computer booths in a bare room is not the best arrangement. For comparison, at California's Pixar Animation Studios where award-winning movies are produced year after year, the working place is designed like a playground almost like the Disneyland.

7 The Joy of Making Movies and Making Physics

Tsui: Every time one goes for the process of self-exploration, one finds more perspectives of oneself. It is more than an excitement; it is the means to see that there is always more about yourself that you do not know about. There are always unknowns out there waiting to be discovered.

Lam: The joy of doing physics comes mostly from practical rewards: flexible working hours, reasonable income, job security, freedom to work on any topic of your choice (if you are a tenured professor), friends and colleagues around the world, and the chance to visit beautiful places (where conferences are held). Moreover, if you work on pure physics, you may discover one or two fundamental laws of nature with profound implications that might change the world—a job satisfaction hardly matched in other professions. The dream of winning a Noble Prize and the belief that you are discovering "truth" about the universe and so on help, but are not required. Lastly, if you are curious about many things in the world, including humans, and think hard enough, physics research does provide a unique training that helps you to see things with a unified perspective.

8 Conclusion

We both had received training in the United States but had worked for many years in environments with less or limited resources. Yet, we have always tried to compete and, whenever possible, outperform the best in the rest of the world. In this regard, we hope that those in similar situations will find something in this chapter useful.

Human affairs are stochastic systems that involve probability considerations. There are always exceptions to the rules. What works for us may not work for others, vice versa. However, since the objects and subjects involved in movies and physics are both part of the natural system, with constraints such as physical principles and human nature ever present, it follows that there could be more similarities than differences between movies and physics, or, at least, in movie making and physics making.

Notes

1. In 1970, after large quantities of potential oil deposits near the group of tiny islands called Diaoyudao (called Tiaoyutai in Taiwan) were announced by foreign oil companies, both Japan and China reiterated their ownership of Diaoyudao. To help keep Diaoyudao under China, overseas Chinese students in USA started the "Protect Tiaoyutai" (Baodiao) movement at the end of 1970 [The Seventies Monthly 1971]. The first large gathering was held in the basement of the Teachers College building at Columbia in Dec., 1970. Subsequently, many overseas Chinese, students or otherwise, worldwide were mobilized. See, for example: http://archives.lib.nthu.edu.tw/exhibition/diaoyun/ (April 27, 2009).

2. Phil Platzman did his PhD thesis with Murray Gell-Mann (1929-2019) in particle physics, and polaron works with Richard Feynman (1918-1988) at Caltech before he worked at Bell Labs, the place he spent his whole career life. Both Gell-Mann and Feynman are Nobel laureates. In 1997, Platzman and Peter Eisenberger shared the APS Arthur H. Compton Award (http://www.aps.anl.gov/About/Committees/APS_Users_Organization/Compton/index.htm, June 20, 2010).

3. Melvin Lax, one of the founders of quantum optics, received his PhD from MIT (1947) and worked at Bell Labs (1955-1972) before assuming the Distinguished Professorship at City College, City University of New York. He was awarded the

Willis E. Lamb Medal for Laser Physics in 1999 (photonics.usask.ca/interesting topics/files/Laser%20Invention/Melvin_Lax.pdf, June 20, 2010), and elected to the National Academy of Sciences in 1983 (www.aip.org/history/acap/ biographies/ bio.jsp?laxm).

4. Science Matters (Scimat) is the new multidiscipline that treats all human matters as part of science [Lam 2008, 2011].

References

Ho, Sam & Ho Wai-Leng [2002]. *The Swordsman and His* Jian Hu: *Tsui Hark and Hong Kong Film*. Hong Kong: Hong Kong Film Archive.

Lam, L. [1971]. Hong Kong, Taiwan, Diaoyutai. New York Hong-Kong-Students Monthly **20**. (Reprinted in: The Seventies Monthly (ed.) [1971]. *Truth Behind the Diaoyutai Incident*. Hong Kong: The Seventies Monthly. pp. 103-105; and in: Lam, L. [2022]. *China Complex*. San Jose: Yingshi Press. pp 27-29.)

Lam, L. [2002]. Histophysics: A new discipline. Mod. Phys. Lett. B **16**: 1163-1176.

Lam, L. [2005]. Active walks: The first twelve years (Part I). Int. J. Bifurcation and Chaos **15**: 2317-2348.

Lam, L. [2006]. Active walks: The first twelve years (Part II). Int. J. Bifurcation and Chaos **16**: 239-268.

Lam, L. [2008]. Science Matters: A unified perspective. *Science Matters: Humanities as Complex Systems*, Burguete, M. & Lam, L. (eds.). Singapore: World Scientific. pp 1-38.

Lam, L. [2010]. The first "non-government" visiting-scholar delegation in the United States of America from People's Republic of China, 1979-1981. Science & Culture Review 7(2): 84-94. (English translation appears in: Lam, L. [2024]. *Humanities, Science, Scimat*. Singapore: World Scientific. pp 448-464.)

Lam, L. [2011]. Arts: A Science Matter. *Arts: A science matter*, Burguete, M. & Lam, L. (eds.). Singapore: World Scientific. pp 1 32.

Lam, L., Bellavia, D. C., Han, X.-P., Liu, C.-H. A., Shu, C.-Q., Wen, Z., Zhou, T. & Zhu, J. [2010]. Bilinear Effect in complex systems. EPL **91**: 68004.

Li, Yuan-Yi (李元逸) [2003]. The unbroken China complex: The life story of a Chinese scientist. Sciencetimes, Aug. 8, 2003. (Reprinted in: Lam, L. [2022]. *China Complex*. San Jose: Yingshi Press. pp 13-20.)

Lubkin, G. B. [1998]. A personal look back at *Physics Today*. Phys. Today, May, 1998: 24-29.

Mantegna, R. N. & Stanley, H. G. [2000]. *An Introduction to Econophysics*. New York: Cambridge University Press.

McMillan, C. D., Bhargava, P., Lam, L. & Vogt, E. W. [1968]. A calculation of the production of pions by 450 MeV protons on nuclei. Can. J. Phys. **46**: 1141-1144.

Morton, L. [2001]. *The Cinema of Tsui Hark*. Jefferson, NC: McFarland & Co.

The Seventies Monthly (ed.) [1971]. *Truth Behind the Diaoyutai Incident*. Hong Kong: The Seventies Monthly.

Wolf, W. P. [1994]. Is physis education adapting to a changing world? Phys. Today, Oct. 1994: 48-55.

Published: Hark Tsui & Lui Lam [2011]. Making movies and making physics. *Arts: A science matter*, Burguete, M. & Lam, L. (eds.). Singapore: World Scientific. pp 204-221.

15

Su Dong-Po's Bamboo and Paul Cézanne's Apple

Lui Lam and Li-Meng Qiu

Su Dong-Po (1037-1101) of the Song dynasty is arguably the most well-known poet and writer in China. He is also a distinguished painter; he liked to paint bamboos and rocks. Unlike his contemporaries and painters before him, the leaves in Su's bamboo painting are not necessarily attached to the stem. Paul Cézanne (1839-1906), a French post-impressionist, is recognized by Picasso, Matisse and many others as the father of modern art. He went beyond impressionism and painted many things including apples. Both these two artists tried to go beyond the appearance and show the *essence* of the objects they painted, in their own new ways. It was not by accident that these two painters—one from the East and the other from the West, separated from each other by about 800 years—had the same idea about painting. There must be something basic behind this. As shown in this chapter, the mechanism behind their techniques is based on how we see things, the cognitive science of vision in the human brain. The fact that Su's style was not adopted as mainstream, unlike that of Cézanne, is discussed; it is related to the unique nature of China's ultra-stable feudal system in the past, in which science and technology are implicitly or explicitly discouraged. Finally, the possible origin of Dong-Po Pork, for which Su is also famous for, is presented in an appendix.

1 Introduction

Su Dong-Po (1037-1101) from China and Paul Cézanne (1839-1906) from France (Fig. 1) lived about 800 years apart from each other, in two different continents. Both are pioneering painters. While Su is primarily a poet and writer, Cézanne is shy in writing [Cézanne 1995: 9].

Fig. 1. Su Dong-Po (left) and Paul Cézanne (right, c.1861). Su designed the hat he was wearing.

The preferred painting subjects of these two artists are quite different. Su painted mainly bamboo and rock, while Cézanne did still life, landscape, portrait, and nude (which he did without live models, for unavailability or his shyness, or both).

Both of them were nature lovers. Yet, this is not the only thing the two had in common, in spite of their very different personal character and career (see Secs. 2.1 and 3.1 below). In their paintings, they were not satisfied to reproduce the likeness of the objects they painted, but instead strived to capture the very *essence* of those painting objects. And they invented new ways to do that.

Technically, Su used ink brushes and his black-and-white paintings were always done with simple strokes. In his bamboo paintings, some of the leaves are detached from the stem, creating a style quite different from those of his contemporaries. For Cézanne, he used oil brushes. His paintings were done with swift strokes with "planes" of color, reducing the objects to their basic geometric forms as he visualized them. And in so doing, he ushered in the era of modern arts.

In the following, the life and career, art view and their innovative painting style are presented, respectively, in Secs. 2 and 3. The neurological basis of their art is discussed in Sec. 4. In Sec. 5, the fate of these two artists' painting styles, with the social background, is contrasted and explained. Finally, since many Chinese learn about Su's name through Dong-Po Pork, if not his arts, our findings about the origin of this delicious Chinese dish—including the cooking recipe—is given in an appendix.

2 Su Dong-Po and His Bamboo

Su Dong-Po's career, art view, and painting innovations are presented.

2.1 Su's life and career

Su Dong-Po,[1] a poet, writer, painter, and calligrapher, lived in the Northern Song dynasty (960-1127). His life and career are summarized in Table 1.

Table 1. Chronology of Su Dong-Po's life and career.[2]

Year	Age (yr)	Event
1037	0[+]	Born Jan. 8 in Meishan, Sichuan Province
1054	17	Married first wife, Wang Fu, 14 years old
1056	19	Ranked no. 2 in *Juren* national exam. in capital, Chang'an
1057	20	Ranked no. 2 in *Jinshi* national exam. in Chang'an; mother died; returned home
1061	24	Got his first job as assistant magistrate in Fengxiang
1064	27	Met Wen Tong, a bamboo painter
1065	28	Became Secretary of Dept. of History in capital; wife Wang Fu died at age 25
1066	29	Father died; returned home
1068	31	Married second wife, Wang Run-Zhi, 19 years old; moved family to Chang'an; resumed Secretary of Dept. of History
1071	34	Opposed emperor's reform movement; sent to Hangzhou as deputy magistrate and stayed there for 3 years
1074	37	Took in Wang Chao-Yun, age 10, as servant; moved to Mizhou as magistrate
1077	40	Became Xuzhou magistrate
1079	42	Became Huzhou magistrate; in prison (about 6 months) for writing poems and essays deemed politically incorrect
1080	43	Banished to Huangzhou, with a job but no pay;[1] took in Wang Chao-Yun as concubine
1084	47	Assigned to Ruzhou; visited Lu Mountain in transit
1085	48	Assigned to Changzhou, then Dengzhou, finally back to Chang'an as Secretary to Premier's office
1086	49	Became Secretary to Emperor (*Hanlin*) for 3 years

1089	52	Asked and got transferred to Hangzhou (due to political conflicts in office)
1091	54	Back to Chang'an as Minister of civil service (for 3 months); asked and got transferred to Yingzhou
1092	55	Transferred to Yangzhou; returned to Chang'an as Minister of war and Minister of education; Empress, his political support, died
1093	56	Second wife Wang Run-Zhi died at age 44; transferred to Dingzhou as Commander
1094	57	Exiled to Huizhou, Guangdong Province (his old writings in his *Hanlin* days were judged politically incorrect), bringing concubine Wang Chao-Yun (became third wife later) and third son along;[2] started building an elaborate mansion the next year
1096	59	Third wife Wang Chao-Yun died at age 32
1097	60	Mansion completed; banished to Qiongzhou, Hainan Island (the most southern part of China) when the emperor thought Su was too happy in Huizhou after reading Su's verse[3]
1100	63	Returned to mainland, with permission to live anywhere
1101	64	Died of illness in Changzhou, Aug. 24, with entire family by his side

[1] [Yang 2006: 59].   [2] [Yang 2006: 83].   [3] [Lin 1947: 368].

Su was a very smart person. He was not a dogmatist in philosophies, unlike many of his fellow countrymen. He exercised judgment in selecting and applying whatever he found suitable from Confucianism, Daoism, and Buddhism. Confucianism helped him in handling interpersonal relationships. Daoism kept him calm, in lieu of psycho-therapists, when he was banished from the capital (to nice places like Hangzhou and not so nice places like Huizhou); and directed him to embrace nature. Buddhism smoothed out his agonies when his three wives died and when unpleasant things were encountered.

And Su was a very good scholar, innovative and flexible when needed. The excellence of his poems and essays had been well recognized since his early age. After he was released from jail at age 42, he stopped writing poetry that got him into trouble.[3] Instead, he turned to writing Song prose. Breaking with tradition, he expanded the scope of prose to cover anything and everything that he felt for—nature, daily life, festivals, expired wives and other women, philosophical thoughts, and so on. All were very touchingly and skillfully done. Su became a pioneer and the foremost person as Song prose is concerned.

Furthermore, throughout his life, Su was surrounded by very gifted friends. Su, together with the younger Mi Fei, established the impressionistic style of Chinese painting, called *Scholar Painting* (initiated by Wang Wei in Tang dynasty) [Jiang 1993: 162-163]. And bamboo is one of the major themes in their paintings.

2.2 Su's art view

Here, we examine his thoughts and attitude on art and literature around the time of his exile. As he opposed the political reform by Wang An-Shi, the prime minister, he was impeached for writing many verses which were regarded as lese-majesty. He was exiled to Huangzhou in 1080. Though he did not write many poems at this time, he devoted himself to painting. Through this experience he began to realize that both painting and composing poems were originated from the same source of creation. Su's art view can be sum up by four major points.

1. Both Dao and art are integral for painting

All paintings are unconscious reflection of a philosophy. Su was heavily influenced by his mother's love for Dao. One of Zhuangzi's well-known quotations, "Heaven and Earth coexist with me and everything is combined with me," can be considered as the basis of his view on painting bamboos. Su was totally aware that Dao itself cannot bring art to perfection. He realized there are men who can possess both Dao and art; others who possess Dao but not art—although the subject takes shape in their heart, it does not take shape in their hands that hold the brushes [Sirén 1935: 440]. In fact, it is never easy to make the hand to cooperate with the heart. The spirit of the subject can only be fully expressed when the necessary technical skills have been mastered thoroughly by the painter.

2. Painting is to bring out the "inherent reason" behind the subject

Su objected to painting bamboo with too many leaves, because that would be just a copy of the object. At that time and even now, lots of people judged a painting only by its resemblance to the real object. Su did not agree with this criterion. Su believed the purpose of a painting is to bring out the "*inherent reason*" (*li*) behind the subject [Sirén 1935: 439-441]. Su is regarded as the first one who extracted the noumenon from the conventional conception theory [Leng 2004: 500].

3. Rapid rhythmic strokes done with a unifying conception

The style of "scholar painting" emphasizes rapid rhythmic strokes done with a unifying conception of the subject [Lin 1947: 274]. To Su and other scholar-painters, painting served mainly as a symbolic means of expressing visual ideas or reflexes of the mind [Sirén 1935: 434].

Lin Yu-Tang believes art are all about rhythm, whether in painting, sculpture, or music. As long as beauty is movement, every art form has an implied rhythm [Lin 1947: 279]. In this sense, the brush movement follows Su's mind and the nature's rhythms can easily be seen from the brush movement. The lines and contours are results of a process of growth and serve a definite purpose [Lin 1947: 280]. The pictures that contain nature's rhythms may be considered the impressionistic school of Chinese art, in which the artist is much more concerned about expressing in a controlling rhythm than making copy of the objects. Su described the painting process as "Then grasp the brush, fix your attention, so that you see clearly what you wish to paint; start quickly, move the brush, follow straight what you see before you, as the buzzard swoops when the hare jumps out. If you hesitate one moment, it is gone" [Su 1986: 365].

Su sometimes painted after drinking a lot of wine. To him, wine kindles inspiration and improvement of brush speed; he is then no longer bound by any imposed rules. Wine is just the method that leads Su to the purified world, from which he can get spiritual energy. However, it cannot guarantee him to reach the perfect state for painting. Only a small number of bamboo paintings created by Su while he was a little drunk were preserved.

4. Absence of irrelevant matter

Su's painting of a few sprays of bamboo leaves with a barely visible moon shining from behind creates two effects. First, the absence of irrelevant matter stimulates the imagination of the spectator; second, it implies that these few bamboo leaves are worth looking at forever and ever in the delight of the simple rhythms they express [Lin 1947: 281].

2.3 Su's bamboo painting

Su's bamboo painting is done with three guiding principles.

1. Bamboo is the emblem of hermits

Bamboo, considered as one of the noble plants by the Chinese, expresses a peculiar combination of "suppleness and strength" [Sirén 1935: 437]. The thin bamboos are like hermits [1935: 441]. Their power to remain green even in the cold season and their habit of yielding and bending before the storm, without breaking and always coming back, appeal to Su and Chinese intellectuals like him. Yielding and bending can suggest life, and life can easily be seen from Su's unique bamboos.

2. Bamboo has no constant form but inner spirit

Some of Su's bamboos were painted in one stroke from the ground up to the top without any bamboo joints. For Su, a painting's ultimate value is justified by how well it satisfies the painter's mind, not by the painting skills or the highly-finished painting itself, although he once had a great admiration for this type of work. He fully expressed this view in two lines: "To judge a painting by its verisimilitude is to judge it at the mental level of a child" [Su 1986: 367]. Fewer details make it easier to express nature's rhythm. Correct proportions and vividness are not as important as the *"constant principle"* [1986: 367], which was interpreted as the "inner spirit" by Lin Yu-Tang [1947: 282]. The constant principle, or in other words, inner spirit, is the rule of nature.

3. Constant form and constant principle are against each other

The bamboos that Su painted have no *constant form* but inner spirit. "The lack of constant form does not spoil the whole thing, but the whole thing would lose spiritual significance without the inner spirit" [Su 1986: 367]. Without the constant form, one can break away from the material form and rule, so that he can truly grasp the reason for the nature. For Su, the lack of constant form is "the constant form," which embodies the rule of nature. Su kind of regarded the constant form and the constant principle as opposing each other, with the result that some of his bamboos (Fig. 2) are thinner than usual, the bamboo leaves are not always linked to the stem, and the bamboo joints' bulging parts are less tangible, unlike the bamboos painted by his contemporaries such as Wen Tong (Fig. 3).

3 Paul Cézanne and His Apple

Paul Cézanne's career, art view, and painting innovations are presented here.

3.1 Cézanne's life and career

Paul Cézanne was born on January 19, 1839 in Aix-en-Provence in southern France.[4] His father is a well-to-do banker who supported Cézanne financially all his life; unlike other impressionist painters of his time, Cézanne had no need to worry about selling his paintings.

Upon the insistence of his father, Cézanne studied law for three years while taking classes at the drawing academy. At age 22, he went to Paris to study painting, with his father's permission, but was unhappy there. He returned home shortly and worked in his father's bank; could not stand it and went back to Paris in 1862. He went home again, for good, in 1870; the motivation was to evade the military draft [Jennings 1986: 44]. Cézanne kept his marriage and a son secret to his father for ten years, but not to his mother, for fear of offending the father.

While in Paris, Cézanne was in the company of the other impressionists but insisted to paint differently. It was Camille Pissarro (1830-1903), a father figure to Cézanne, who introduced him to outdoor painting; and Cézanne loved it for the rest of his life.

His joined the first Impressionist exhibition in 1874, skipped the second, and rejoined the third. He kept sending his paintings to the Salon, *the* annual government-sponsored juried exhibition in Paris [Romano 1996: 27], and was rejected each time, until the year 1882, at age 43.

It was in his forties that Cézanne found his own style of painting; he flourished and received wide recognition in his fifties. He loved his art so much that he wrote, "I have sworn to myself to die painting…" [Cézanne

1995: 330]. And he got it. On Monday, October 15, 1906, Cézanne, aged 67, was out painting as usual. But he got soaked in the rain for several hours, and was brought home in a laundry cart by others. The *next* day, he went to the garden to work on a portrait, and then went home collapsed. Six days later, Cézanne died at home [Lindsay 1969: 341-342].

Fig. 2. Su Dong-Po, *Bamboos and Stones*, 106 x 28 cm (partially shown).

Fig. 3. Wen Tong, *Ink Bamboo*, 131.6 x 105.4 cm.

Cézanne, together with Seurat, Gauguin, van Gogh and others, pioneered post-impressionism [Strickland 2007: 112], but it was Cézanne who opened the door to modern painting—in fact, modern art (see Sec. 3.2). Pablo Picasso (1881-1973) praised him as "my one and only master." Both Picasso and Henri Matisse (1869-1954) called him "the father of us all."[5]

3.2 Cézanne's art view

Cézanne hardly wrote anything on his art view. But in the letters (1889-1906) to his young friends, especially the painter Emile Bernard (1868-1941), he did express some of his thoughts on this subject [Cézanne 1995]. Here are some quotes from these letters and elsewhere.[6]

1. <u>On the nature of painting</u>

   - Painting from nature is not copying the object; it is realizing one's sensations.

   - Pure drawing is an abstraction. Drawing and color are not distinct; everything in nature is colored.

   - The man of letters expresses himself in abstractions whereas a painter, by means of drawing and color, gives concrete form to his sensations and perceptions [Cézanne 1995: 303].

2. <u>On painting nature</u>

   - The truth is in nature, and I shall prove it.

   - In order to make progress, there is only nature, and the eye is trained through contact with her. (p 306)

3. <u>On painting method</u>

   - There are two things in the painter, the eye and the mind; each of them should aid the other.

   - A work of art which did not begin in emotion is not art.

   - Let us go forth to study beautiful nature, let us try to free our minds from them, let us strive to express ourselves according to our personal temperament. (p 315)

   - I believe in the logical development of everything we see and feel through the study of nature and turn my attention to technical questions later. (p 330)

   - Treat nature by means of the cylinder, the sphere, the cone. (p 301)

4. <u>On painting assessment</u>

   - Taste is the best judge. It is rare. Art addresses itself only to an excessively limited number of individuals. (p 302)

   - When I judge art, I take my painting and put it next to a God object like a tree or flower. If it clashes, it is not art.

- Michelangelo is a constructor, and Raphael an artist who, great as he may be, is always tied to the model—when he tries to become a thinker he sinks below his great rival. (p 309)

5. <u>On being a painter</u>

- Talking about art is almost useless. (p 303)

- The Louvre is a good book to consult but it must be only an intermediary. The real and immense study to be undertaken is the manifold picture of nature. (pp 302-303)

- Genius is the ability to renew one's emotions in daily experience.

From these quotes, one can see that Cézanne considered that

- Nature is the most worthy as painting subjects are concerned.

- The painter should start with his emotion and depict what he feels (not the impression) about the subject; copying the subject is out of the question.

- Those who can appreciate good art are rare.

- Some great artists may not be so great in his eye.[7]

3.3 Cézanne's apple painting

Cézanne once said, "With an apple I will astonish Paris."[8] Well, he did not really paint an apple significantly different from others before him. But he did find a new way to paint a set of apples, by arranging them (usually with other fruits and other objects on a table) in such a way that different subsets of the arrangement are shown in different perspectives. In other words, *multiple perspectives* coexist in the same painting, breaking the single-perspective tradition.[9]

Figure 4 is an example of this multi-perspective design. For instance, the basket handle is shown from the perspective of looking from the right side of the picture, while the jar and the pear on the right front is seen from the front [Loran 1943]. This fusing of different viewpoints on a single canvas was used by Picasso [Miller 1996: 414].

The second innovation by Cézanne is equally important. To render what he saw as the *essence* behind the subject matter, neither the subject's shape nor the painter's impression, he structurally ordered whatever he perceived into simple forms and color planes.[10] This is what he meant when he wrote, "Treat nature by means of the cylinder, the sphere, the cone." This technique was adopted and pushed to the limit by Picasso in his Cubism paintings [Miller 1996].

In fact, near the end of his life, Cézanne's paintings got simpler and simpler and became more abstract, like the one shown in Fig. 5, which is at the brink of abstract painting.

Fig. 4. Paul Cézanne, *The Kitchen Table* (1888-1890).

4 Neurological Basis of Their Artworks

That Su Dong-Po's detached bamboo leaves and Paul Cézanne's color-plane brushes (as well as his multi-perspective composition) actually work is due to the complex mechanism of seeing in the viewer's brain. The old understanding of vision is that our brain simply tells us how the pixels of light received by the eye are spatially arranged. The new understanding based on neurological studies tells a different story [Kolak et al. 2006: 82-86; Livingstone 2002].

Fig. 5. Paul Cézanne, *Mount Sainte-Victoire Seen from Lauves* (1904-1905).

Essentially, light enters the cornea, focused by the lens and projected to the retina, where photoreceptors convert light signals into electric signals. The signals are processed first in the occipital lobe (at the back of the brain) and then the parietal lobe and the temporal lobe (Fig. 6). In so doing, the brain analyses the signals, recalls information from memory for comparison, presents a final interpretation of the image that the eye receives, and excites other neurons as a response.

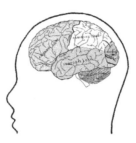

Fig. 6. The brain lobes: Frontal (left), parietal (middle up), temporal (middle down), and occipital (right).

Since we do know from observing nature that all tree leaves are connected to the stems, our brain was able to interpret Su's detached leaves correctly as part of the bamboo and not something that flow in air. The same goes for Cézanne's paintings. That is, our brain connects the dots for us. And since the brain of each person is different, it follows that no two persons see the same thing identically; minuscule difference in their interpretation and appreciation of the same artwork is always present.

5 Discussion and Conclusion

It is interesting to compare and contrast the life and career paths as well as the subsequent influence of these two giants in the humanities, Su Dong-Po from the East and Paul Cézanne from the West.

1. Su made his living as a public servant[11] while Cézanne was supported mainly by his father. This shows that it is important to have financial security as an artist. Moreover, Su's stormy and dramatic career allowed him to move around the country with plenty of spare time, which benefited greatly his arts.

2. That two painters living almost a thousand years apart from each other with very different cultural backgrounds came to the same conclusion that the aim of painting is to depict the essence behind the object and not the appearance or impression of it, illustrates the same human nature that the two shared, in consistent with the slow variation of human nature according to Darwin's evolutionary theory and the out-of-Africa understanding that all human beings shared the same ancestors who came out of Africa millions of years ago [Lam 2011].

3. As for the bamboo painting style that Su initiated, unfortunately, with few exceptions, it was not adopted by other Chinese painters. This is in sharp contrast to the fate of Cézanne painting style, which opened up the era of modern arts in the West. The reason behind this difference in fate is due to the driving force of science and technology that existed in Cézanne's Europe but not in Su's China. In Europe, color photography appeared in 1873, thirty six years after photography was invented by Daguerre [Strickland 2007]. These science/technology advances made the birth of Impressionism almost inevitable. (In fact, the first official

Impressionism exhibition was held in 1874.) On the other hand, China was an ultra-stable society [Jin & Liu 1992, 1993], in which innovation was discouraged by Confucianism. Furthermore, mathematics was not included in the national examination in old China in spite of the country's few ingenious mathematical achievements. This resulted in the lack of interest and advance in science in the society. Arts in China suffered, unfortunately for Su's bamboo.

Appendix: Dong-Po Pork

There is no absolute proof that Dong-Po Pork (Fig. 7) was indeed invented by Su.[12] However, the following poem, embodied in *Complete Works of Su Shi* [Su 2000: 1047], could be considered as a sort of "proof."

Fig. 7. A piece of Dong-Po Pork at Grandma's Home, Beijing (May 24, 2009). [Photo/Lui Lam]

Ode to Pork

> Ensure continued low burning with a little water after washing the pot clean.
>
> The flavor of the pork is amazing if the lid is lifted at the right time.
>
> The price of pork in Huangzhou is dirt cheap.
>
> The rich disdain eating it, and the poor don't really know how amazing it could be with the right cooking methods.
>
> I eat two bowls of pork after getting up every morning.
>
> Leave me alone; I am stuffed.

It is obvious that "Ode to pork" was written in Huangzhou, not Hangzhou. Su stayed in Huangzhou for four years from 1080 and went again to his post in Hangzhou in 1089. The recipe could be brought to Hangzhou by Su and ended up becoming one of the best-known Hangzhou dishes.

Both *Records of the West Lake* [Shi 1995] and *The Integration of West Lake Documents* [Wang 2004] embody the specialty products of Hangzhou. However, Dong-Po Pork is recorded in neither of these two books. He Yin-Jie, the guide in Su Dong-Po Memorial Hall, stated positively that no existing official history book on Hangzhou records Dong-Po Pork.[13] As a matter of fact, Dong-Po Pork can be found at many places where Su ever stayed, such as the provinces of Sichuan, Jiangxi, Jiangsu, and Hubei. These dishes share the same name, but the recipes are different. The foregoing poem named "Ode to pork" is the only one of all Su's works that explain definitely how to cook pork. This clue leads to the days that Su was in Huangzhou, Hubei Province. Rao Xue-Gang, who studied Su's life in Huangzhou for over 23 years and published his research results as *Su Dong-Po in Huangzhou* in 1999, confirmed[14] that "Ode to pork" was written in Huangzhou in April, 1082.[15]

Although "Ode to pork" was written in Huangzhou, the pork was not named Dong-Po Pork yet. Su loved "stewed pork with brown sauce,"[16] but Dong-Po Pork was not named by Su himself. In "Ode to Dong-Po thick soup," Su brought forth "Dong-Po" definitely in the title and the poem: "Dong-Po thick soup, is the vegetable thick soup cooked by Dong-Po Hermit" [Su 2000: 1046].

In the folklore of Dong-Po Pork's Hangzhou version, Su received a lot of rice wine and pork from graceful local people after he directed the building of the Su Causeway. He told his family to cook "stewed pork with brown sauce" and distribute the pork to the workers who took part in the building. But his family heard "cook pork with rice wine" by mistake. The local people were deeply touched by his generous action and named the pork after Dong-Po [Lin & Shen 1993: 49]. This folklore can be considered as a circumstantial evidence that Su himself did not name "Dong-Po Pork." And it is quite notable that this recipe is different from the Huangzhou recipe, because "rice wine" takes the place of "water."

The cooking time is not explicit in the folklore. "Ode to pork" mentioned "the flavor of the pork is amazing if the lid is lifted at the right time." "The right time" is present in *Scattered Records of the Fairy in Cloud*, edited by Feng Zhi. A man named Huangsheng cooked 1 kg venison from morning every day. He would say it is the right time by nightfall [Feng 2008: 75]. The original story was recorded in *Records of Cheng'an*, which is no longer extant. Feng lived in Later Tang dynasty (923-936); Su might have read Feng's book.

The two best restaurants in Hangzhou (Lou Wai Lou and Zhi Wei Guan) are unwilling to reveal the recipe of Dong-Po Pork. The following is the recipe of Dong-Po Pork displayed at the Su Dong-Po Memorial Hall (May 18, 2010).

Ingredients

　　Streaky pork[17] 1500 g
　　Green onion 50 g
　　Knotted green onion 50 g
　　Refined white sugar 100 g
　　Shaoxing rice wine 250 g
　　Ginger 50 g
　　Soy sauce 150 g

The cooking

　　Dice the pork after washing.[18] Boil pork in boiling water for five minutes. Put the pork skin downwards into a big earthen pot with green onion and ginger on bamboo grid. Get the lid on the pot after adding refined white sugar, soy sauce, Shaoxing rice wine and knotted green onion in turn. Tuck "peach blossom paper"[19] into the aperture between the lid and the earthen pot. Put the earthen pot over "high" heat. Adjust stove heat to "low" after the cooking liquid starts boiling. Turn off stove when pork is tender.[20] Skim fat and put the small clay pot with pork skin upwards in a bamboo steamer. Adjust stove heat to "high" and steam for 30 minutes.

As set forth, the recipe of Dong-Po Pork is different at different places. The way of cooking Hangzhou Dong-Po Pork is a typical localism method. Yuan Mei (1716-1797), a well-known poet, scholar, and artist in Qing dynasty, was born in Hangzhou. He is an epicure and an expert on Hangzhou Style of cooking. What is interesting is that unlike Su, Yuan himself did not cook. However, his book *Sui Yuan Cooking Methods* does describe the way to cook pork. According to Yuan, the cook should be patient and cook pork with the skin downwards first. Oil and fat will soak through the skin. It is the way to keep the flavor and make the skin tastes crisp. Otherwise the skin will get hard and the flavor will disappear [Yuan 2000: 26]. Yet, Yuan did not mention Dong-Po Pork in his book. All he recorded are the "universal" methods of cooking.

Notes

1. Su is the family name of Su Shi; his literary name is Su Dong-Po, which is spelled differently in old literature as Su Tungpo [Lin 1947], Su Tung-p'o [Sirén 1935; Watson 1994] or Soo Doong-Bô (The Columbia Encyclopedia, 2004, Columbia University Press). A short account of his life is given in [Watson 1994: 3-12]; a long account, [Lin 1947]. (Note that in Lin's book, he records lunar calendar dates in solar calendar terms.) A thin Chinese book on his life and arts, with many of his calligraphy reproduced and a very useful chronology included, is [Yang 2006]; a thick Chinese book is [Li 1996].

2. Adapted from [Yang 2006: 110-121]. For the same year, the number listed in our Age column differs from that in Yang's book, since he follows the Chinese custom in age counting (e.g., he lists Su as one year old in 1036). Similarly, in our Event column, we reduce the age number by two from that in Yang's book.

3. After all, poetry reached its zenith in the Tang dynasty (618-907) or even before. And the very form of poetry allowed his enemies to read his mind too easily.

4. A short and very readable account of Paul Cézanne's life and art is presented in [Jennings 1986: 41-52], with color pictures; a long account is [Lindsay 1969]. A useful reference is: http://en.wikipedia.org/wiki/Paul_C%C3%A9zanne. The one-hour long video "Cézanne in Provence" (pbs.org) is enjoyable.

5. http://en.wikipedia.org/wiki/Paul_C%C3%A9zanne (May 12, 2010).

6. http://mypaulcezannepainting.com/Blog/category/quotations-from-paul-cezanne (May 12, 2010).

7. Belittling fellow colleagues, past or present, is a common trait in all creative professions. For example, if you worship Einstein, you probably could not

become a great physicist; the reason is pretty complicated (see [Tsui & Lam 2011]).

8. http://mypaulcezannepainting.com/Blog/category/quotations-from-paul-cezan ne (May 12, 2010).

9. *Mona Lisa* by Leonardo da Vinci (1452-1519) is a famous example of the single-perspective design, in which all lines converge to a point behind Mona Lisa's head [Strickland 2007: 34].

10. http://en.wikipedia.org/wiki/Paul_C%C3%A9zanne (May 12, 2010).

11. Su was not a rebel and definitely not a maverick in politics. He went through the national examinations and worked in the government for two reasons: First, being an officer was the only way for an intellectual to get a decent living in a feudal society. Second, according to Confucius teaching, the foremost duty of an intellectual is to serve his country. To that end, the most efficient way was getting oneself near the emperor and let him listen to all your smart advices; less than that, became the governor of someplace oneself and served the people directly. Unfortunately, the ever changing political climate of the Song dynasty was beyond what Su's limited political skills could handle. The good news for Su is that no other political figures survived the roller-coaster Song court much better than he did. The reason is that Northern Song emperors came and gone frequently (9 emperors in 168 years), like the rapid shuffling of CEOs in a big, modern company—and the stock price plunges, with one important difference; i.e., arts flourished in the former as a result but not in the latter.

12. The folklore has two versions: (1) One day, Su was cooking stewed pork himself (cooking is not an unusual habit for some Chinese male artists, not to mention the Italians), a good friend came in for a visit. The two played Chinese chess for several hours, and Su forgot his cooking. When he remembered it, the pork had been over cooked, but it actually tasted better, and Dong-Po Pork was born. [Source: http://en.wikipedia.org/wiki/Su_Shi (March 3, 2010).] (2) Su directed the building of the Su Causeway in 1090, Hangzhou, which favored the citizens. Su received so much pork in return that he cooked Dong-Po Pork and distributed it to the workers; see below for more details. (Source: Display in Su Dong-Po Memorial Hall, Hangzhou.) These two tales could both be right but are never verified.

13. Interview of He Yin-Jie by Li-Meng Qiu, Su Dong-Po Memorial Hall, May 18, 2010.

14. Telephone interview of Rao Xue-Gang by Li-Meng Qiu, May 26, 2010.

15. The name of the poem listed in the two chronologies of Su in Rao's book is "Ode to boiling pig's head" [Rao 1999: 17 and 157]. Whereas the name of the same poem embodied in one of the articles of the book is "Ode to pork" [Rao 1999: 331]. Rao confirmed "Ode to pork" was called "Ode to boiling pig's head" as well. If it is true, the main ingredient of cooking Huangzhou Dong-Po Pork was

quite different from that of Hangzhou Dong-Po Pork while the poem was written, because the latter uses streaky pork. *Su Shi Corpus* [Su 1986] and *Poetry of Su Shi* [Su 1982] embody this poem as "Ode to pork." According to the footnote to "Ode to pork" in *Su Shi Corpus*, this poem has another name, "Ode to boiling pork thick soup" [Su 1986: 597]. The same poem is embodied as "Eating pork in bamboo slope random notes on classical poets and poetry," written by Zhou Zi-Zhi [He 1981: 351]. Zhou was born in 1082 and also lived in Song dynasty. "Eating pork" is a little bit different in order and characters, but the meaning is the same. The only thing notable besides the name is, "I eat one bowl of pork after getting up every morning", not "two bowls of pork." Although this version might be more reliable, it is not the version widely adopted, probably because the approximate number used in Chinese language is usually expressed by "two."

16. "Fu Yin cooks pork to treat Dong-Po" appears in "Answering Fu Yin playfully," written by Su in Huangzhou [Su 1982: 2654]; it clearly shows that how much Su loved stewed pork.

17. The best ingredient for cooking Dong-Po Pork is the Jinhua pig (alias Liangtouwu pig), which has white trunk and black head and black rump. The pigskin is thin, so are the bones. Jinhua pig has been farmed since the West Jin dynasty (266-316).

18. According to the recipe in *West Lake Cyclopedia* [Shen & Zhang 2005: 177], the diced pork should be 75 g each.

19. Peach Blossom Paper, thin and tough, is a type of semitransparent paper used for making window paper and kites.

20. According to the recipe in *West Lake Cyclopedia* [Shen & Zhang 2005: 177], it takes 2 hours to get the pork tender on simmer.

References

Cézanne, P. [1995]. *Paul Cézanne Letters*, Rewald, J. (ed.). New York: Da Capo.

Feng, Zhi (ed.) [2008]. *Scattered Records of the Fairy in Cloud*, Beijing: Zhonghua Book Company.

Jennings, G. [1996]. *Impressionist Painters*. London: Chancellor Press.

Jiang, Xun [1993]. *History of Chinese Art*. Taibei: East China Books.

Jin, Guan-Tao & Liu Qing-Feng [1992]. *The Cycle of Growth and Decline: On the Ultrastable Structure of Chinese Society*. Hong Kong: Chinese University of Hong Kong.

Jin, Guan-Tao & Liu Qing-Feng [1993]. *The Transformation of Chinese Society (1840-1956): The Fate of Its Ultrastable Structure in Modern Times*. Hong Kong: Chinese University of Hong Kong.

Kolak, D., Hirstein, W., Mandik, P. & Waskan, J. [2006]. *Cognitive Science: An Introduction to Mind and Brain*. New York: Routledge.

Lam, L. [2011]. Arts: A science matter. *Arts: A Science Matter*, Burguete, M. & Lam, L. (eds.). Singapore: World Scientific. pp 1-32.

Leng, Cheng-Jin [2004]. *Su Shi's View on Philosophy and Art*. Beijing: Xueyuan Press.

Li, Yi-Bing [1996]. *A New Biography of Su Dong-Po*. Taibei: Linking.

Lin, Yu-Tang (林语堂) [1947]. *The Gay Genius: The Life and Times of Su Tungbo*. New York: John Day.

Lin, Zheng-Qiu & Shen Guan-Zhong (eds.) [1993]. *Hangzhou, China, Lou Wai Lou*. Hangzhou: Zhejiang Photography Press.

Lindsay, J. [1969]. *Cézanne: His Life and Art*. New York: New York Graphic Society.

Livingstone, M. [2002]. *Vision and Art: The Biology of Seeing*. New York: Abrams.

Loran, E. [1943]. *Cézanne Composition*. Berkeley: University of California Press.

Miller, A. I. [1996]. *Insights of Genius: Imagery and Creativity in Science and Art*. New York: Copernicus.

Rao, Xue-Gang [1999]. *Su Dong-Po in Huangzhou*. Beijing: Jinghua Press.

Romano, E. [1996]. *The Impressionists*. New York: Penguin Studio.

Shen, Guan-Zhong & Zhang Wei-Lin (eds.) [2005]. *West Lake Cyclopedia: Lou Wai Lou Restaurant*. Hangzhou: Hangzhou Press.

Shi, Ji-Dong (ed.) [1995]. *Records of the West Lake*. Shanghai: Shanghai Ancient Book Press.

Sirén, O. [1935]. Su Tung-p'o as an art critic. Geografiska Annaler, Vol. 17, Supplement: Hyllningsskrift Tillagnad Sven Hedin, pp 434-445.

Strickland, C. [2007]. *The Annotated Mona Lisa: A Crash Course in Art History from Prehistoric to Post-Modern*. Kansas City: Andrews McMeel.

Su, Shi [1982]. *Poetry of Su Shi*. Wang Wen-Gao (ed.). Beijing: Zhonghua Book Company.

Su, Shi [1986]. *Su Shi Corpus*. Beijing: Zhonghua Book Company.

Su, Shi [2000]. Ode to pork. *Complete Work of Su Shi*, Vol. 3, punctuated by Fu Cheng & Mu Chou. Shanghai Ancient Book Press, Shanghai: Shanghai Ancient Book Press.

Tsui, Hark (徐克) & Lam, L. [2011]. Making movies and making physics. *Arts: A Science Matter*, Burguete, M. & Lam, L. (eds.). Singapore: World Scientific. pp 204-221.

Wang, Guo-Ping (ed.) [2004]. *The Integration of West Lake Documents*. Hangzhou: Hangzhou Press.

Watson, B. [1994]. *Selected Poems of Su Tung-p'o*. Port Townsend, WA: (Copper Canyon.

Yang, Pin [2006]. *Su Dong-Po*. Taiyuan: Shanxi Educational Press.

Yuan, Mei (袁枚) [2000]. Boiling pork. *Sui Yuan Cooking Methods*, Yuan Mei. Nanjing: Jiangsu Ancient Book Press.

Zhou, Zi-Zhi, [1981]. Eating pork. *Bamboo Slope Random Notes on Classical Poets and Poetry*, *All Ages Random Notes on Classical Poets and Poetry*, He Wen-Huan (ed.). Beijing: Zhonghua Book Company.

Published: Lui Lam & Li-Meng Qiu [2011]. Su Dong-Po's Bamboo and Paul Cézanne's Apple. *Arts: A science matter*, Burguete, M. & Lam, L. (eds.). Singapore: World Scientific. pp 348-370.

On Art

Lui Lam

Art refers to visual art, literature, film, performing arts, music, architecture, multimedia art, etc. We discuss here the origin and nature of art, relationships between art and (natural) science, the importance of art and science, interactions between artists and scientists, the foresight of Su Dong-Po, cultural confidence and other problems related to art. In fact, art can be divided into pure art and applied art. Art originated a million years ago while science appeared only 2,600 years ago, after language and writing were invented. The appearance of art helped importantly humans separating themselves from other animals. Art and science are two of the three major pillars supporting modern civilization (the other pillar is ethics).

1 Introduction

Modern civilization has three pillars: ethics, art, and science (see Fig. 8), and the importance of art is obvious. Art includes visual arts (painting, sculpture, photography, video, etc.), literature, film, performing arts (dance, drama, performance art, language arts, etc.), music, architecture, new-media art (optical art, digital art, artificial intelligence art, etc.).

In China, the earliest discussion of "art and science" (*Artsci*) came from Cai Yuan-Pei (蔡元培, 1868-1940). In the 1950s and 1960s, art emerged as a topic in philosophy of science or history of science in the discussion of the dialectics of nature. Beginning in 1987, the China Center for Advanced Science and Technology, founded by Tsung-Dao Lee (李政道) and the Chinese Academy of Sciences, held a series of international conferences on "science and art" (*Sciart*), participated by both scientists and painters, and started a new tradition.

In fact, the School of Humanities of the Graduate School of the Chinese Academy of Sciences held an international science-and-art conference on June 10-11, 2010 at the Yuquan Road Campus in Beijing. (The Graduate School of the Chinese Academy of Sciences was renamed the University of Chinese Academy of Sciences in 2012.) Tsung-Dao Lee wrote a congratulatory letter to the conference. More than 60 papers were presented by participants from more than 10 countries, covering the relationships between science and art, design and architecture, and the integration of science and art in the field of new media, which were published by Tsinghua University Press in 2011 (Fig. 1). The speakers of the 10 invited talks (30 minutes each) came from Canada (John Blouin, Martin Legault), China (Burnberg), France (Ghislaine Azémard, Jean-Louis Boissier, Charles Camberoque), Germany (Hans Dehlinger), Japan (Yoichiro Kawaguchi), Switzerland (Daniel Pinkas), and United States (Lin Lei, 林磊). For a summary report on this conference and the development of art in China in the last 40 years, see [Liu 2011].

This article discusses art and aesthetics (Sec. 3), the origin and nature of art (Secs. 5 and 6), the relationships between art and (natural) science (Sec. 7.1) , the importance of art and science (Sec. 7.2), the interaction between artists and scientists (Sec. 8.2), Su Dong-Po's foresight in art and cultural confidence (Sec. 9), and other issues related to art (Sec. 10). Note that the origin and nature of art are two of the most basic and important topics in art research, which were discussed in my articles 10 years ago [Lin 2010a, 2010b; Lam 2011].

2 Ask the world, who knows art?

Art has a long history. Cave paintings appeared 35,000 years ago, and even earlier signs of art have been found, such as carvings from 250,000 years ago [Lin 2010a; Lam 2011].

The word Art is related to but differs from the Latin word Artes, and the two's evolutionary relationship is described in [Hoppe 2011]. For more than 2,000 years, art has been mostly associated with beauty or aesthetics, and in the 17th century the term "fine art" appeared. However, the emergence of modern art a hundred years ago has subverted people's old notion of art, resulting in different opinions on what art really is.

Fig. 1. The international conference on *Science and Art, Intersection and Integration*, organized by the Graduate School of the Chinese Academy of Sciences on June 10-11, 2010 at the Yuquan Road Campus in Beijing. *Upper left*: Book cover of conference proceedings, *Science and Art, Intersection and Integration* (2010). *Lower left*: Copyright page of proceedings. *Upper right*: Tsung-Dao Lee's congratulatory letter. *Lower right*: First page of proceedings' Contents. The first article is Lin Lei's "Arts: A science matter" [Lin 2010a].

In Fig. 2, the two books, *Philosophy of Art* (1999) and *The Methodologies of Art* (1986), collect failed theories on the philosophy and methodology of art, respectively, and *The Art Instinct* (2009), written by the art philosopher Denis Dutton (1944-2010), argues that art cannot be defined in one or two sentences and lists instead 12 characteristics about art: (1) direct pleasure, (2) technique and affection for art, (3) style, (4) novelty and creativity, (5) critical, (6) expressiveness, (7) special focus, (8) personality expression, (9) emotional saturation, (10) challenge of knowledge, (11) artistic tradition and professional recognition, and (12) experience with imagination. Unfortunately, all of these are found in another subject that also requires creativity—Physics, and are not unique to art. (For further discussion of item 11, see Sec. 6.)

Fig. 2. *Left*. A collection of failed attempts to define art (1999). *Middle*: A collection of failed attempts to interpret art (1986). *Right*: A newer book that fails to define art (2009).

Figure 3 is a commonly used university textbook on the nature of art, which collects the discourses of many scholars from different fields (including philosophers Plato, Kant, and Derrida; writer Tolstoy; psychologist Freud; historian Collingwood) on art, showing that each person only describes a certain aspect of art, and no one knows the whole picture of art, just like a blind man touching an elephant, no one knows the whole picture of an elephant. In other words, none of these people know what art is.

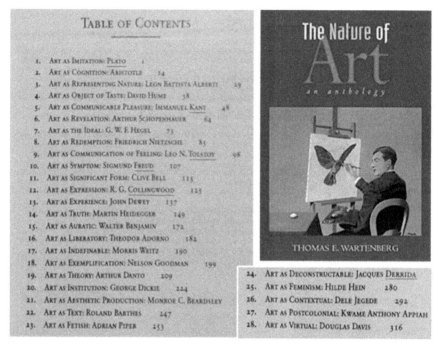

Fig. 3. Book cover and Contents of Thomas E. Wartenberg's *The Nature of Art: An Anthology* (2001).

3 Art and Aesthetics

For more than 2,000 years, art has been linked to aesthetics, so much so that some people equate art with fine arts, as can be seen from the fact that many art academies and art museums in China are still called fine-art academies and fine-art museums. In fact, the situation already changed a hundred years ago, as can be seen from the paintings of Wu Da-Yu (吴大羽, 1903-1988). Wu paints mainly abstract paintings, and the subject he wants to express is clearly not beauty, but something more (Fig. 4).

Art is not always about beauty; it is also about emotions (like in Edvard Munch's *The Scream*; see Fig. 6). It can also be about the expression, description, and elucidation of everything in nature, just like Physics. In fact, it can even be *not* about anything specific.

Fig. 4. Wu Da-Yu, his album (2015), and one of his abstract paintings.

A work of art reflects the artist's perception or worldview of the real world. As a reflection or articulation of nature, both nonliving and living systems (such as human feelings and relationships), art is conditioned by the human nature of the artist and the receiver (the viewer/listener/participant of the artwork). In other words, art is not completely free, and neither is the artist; otherwise, a work of art can be completed in a matter of minutes or hours. Of course, some important works of art are made in a matter of minutes, but that doesn't count into the amount of time the artist spends conceiving them.

The turning point of the change in the perception of the nature of art occurred in 1917 (incidentally, the year of the establishment of the Russian Soviet Federative Socialist Republic, USSR). In that year, the *Fountain* of Marcel Duchamp (1887-1968; see Fig. 6) was accepted as a work of art for exhibition in New York. The work is a ready-made male urinal bought from a store, and the author simply signed the (fake) name and year on it, and decided how to place it for display. The appearance of *Fountain* abolished all previous art theories, especially the theory that art is necessarily related to aesthetics.

As mentioned in Sec. 2, the confusion of thought about the nature of art lasted for 2,400 years since Plato, and entered a state of crisis after 1917, until the emergence of a new theory in 2010, which is the interpretation of art from the perspective of Scimat—a multidiscipline proposed by the author in 2007 [Lam 2008].

4 Classification of Art: Pure Art and Applied Art

To understand what art is, one can put everything that is recognized as art on the table and see if one can find something in common. For illustrative purposes, Fig. 5 lists some representative, easy-to-see examples, including painting, sculpture, dance, and architecture. (The results of the discussion of art below cover also the non-visual arts, such as music, literature, etc., that are not in Fig. 5.) At first glance, it is hard to see that everything in Fig. 5 has something in common, but just as physics (and other subjects) can be divided into pure physics and applied physics, we can also divide the things in Fig. 5 into pure art and applied art (Fig. 6).

Fig. 5. Some representative examples of art.

There are some practical considerations in the creation process of applied arts. For example, a vase can be beautifully decorated to increase its aesthetic value in the living room and as a commodity to increase sales, and a group dance in a tribal ritual is meant to strengthen group cohesion. Obviously, architecture belongs to the applied arts.

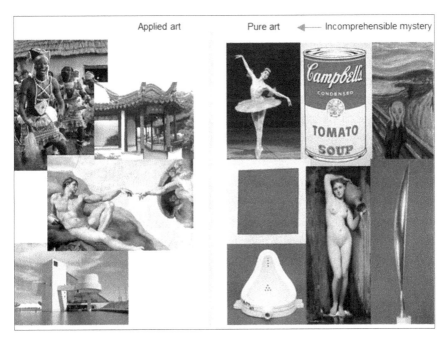

Fig. 6. The various arts in Fig. 5 are divided into two categories: pure art and applied art. In the pure-art category on the right, Edvard Munch's *The Scream* (1893-1910) appears in the upper-right corner and Marcel Duchamp's *Fountain* (1917), the lower-left corner.

Pure art is all art other than applied art. There are two classes of pure art: A. Purposeful; B. Purpose unknown. Ballet and Da Vinci's *Mona Lisa* are examples of Class A; Duchamp's *Fountain* and abstract painting are examples of Class B. Both classes appear in the pure-art category in Fig. 6. Class A is relatively easy to understand; Class B of pure art is puzzling. Why does it exist and what is its use? To answer these two questions, we have to go back to the origin of art.

5 Origin of Art

There used to be two schools of thought about the origin of art: evolutionary (innate) theory and cultural (acquired) theory. Innate theory emphasizes that in the process of evolution, art is beneficial to the continuation of human beings, so it is passed on from generation to generation through genes. The acquired theory holds that after the emergence of art, it can be maintained as a part of culture, and the role of genes is not necessary. There is not necessarily a direct contradiction between these two hypotheses, they can coexist and work at the same time. The problem is that the two theories only says why art can continue after it appears, but do not say how art began, do not talk about the raison d'être of pure art, and cannot prove that it is correct. In this context, a new perspective emerged a decade ago, which was to discuss the origin of art from a Scimat perspective [Lin 2010a, 2010b; Lam 2011].

The Scimat view is that human is a creature constrained by human nature and influenced by its evolutionary history, so it is often necessary to understand the deep questions related to human by going back in time, sometimes millions of years ago [Lam 2011]. The origin of art falls into this category. To understand the origin of art, we must first understand the origin of human.

5.1 A brief history of human

Everything in the universe began with the big bang 13.7 billion years ago. Subsequently, atoms were formed, some of which coalesced into the solar system 4.7 billion years ago, and one of the planets is the Earth. A billion years later, life appeared on Earth. After a long evolutionary process, human and chimpanzee (both descendants of fish) separated 6 million years ago, until modern humans (*Homo sapiens*) appeared 195,000 years ago.

From the perspective of painting, the cave art of 35,000 years ago is not simple; it is quite mature. The not-so-simple inventions of fire (1.6 million years ago) and cooking (400,000 years ago) predated the appearance of *Homo sapiens* (195,000 years ago). These all point to the fact that art may have originated more than a million years ago.

5.2 Emergence of art

No one can be sure how art originated because it is too old to be recorded. But because of the slow pace of human evolution, the instincts of our ancestors cannot differ much from our own. Therefore, it is reasonable to guess that art may have started like this:

Let's imagine that one million years ago, somewhere in Africa, there was a group of a dozen people, say, living together in caves, and what would they do when a heavy rain lasted three days and three nights? Out of boring, some of them may start doing "useless" things just to *kill time* [Lam 2024a: 359-361].

One might use a tree branch to draw the shadow of another person on the ground along the outline on the mud floor—an early *drawing*, or a tree branch striking a tree trunk to mimic the sound of raindrops outside the cave—early *music*, or dancing to tell others about his hunting experience—*pantomime*, or balancing on a tree trunk resting on a rock—an early *performing art*. Performing these activities requires *mimesis* (acting, imitation, skill drilling, etc.) which was already available two million years ago [Donald 2006].

When the rain stopped, everyone had to go out hunting again, and because women who were pregnant or had to take care of children had to stay behind, they asked someone who had performed well before to accompany them, killing time. This man or woman was the first professional *artist* in the history of humankind, entertaining others in exchange for food. He or she stays at the residence and did not have to go out hunting. In other words, artists are the first *safe* profession in human history. (There was only one other profession before—hunting, and that was a dangerous work.)

This first artist, if not appearing in the cave earlier, would have appeared a million years ago, because there was so much spare time in human beings at that time, and all kinds of modern methods of killing time (newspapers, television, etc.) had not yet appeared. Finally, this would happen even more so when the population reached a certain size, because a large enough group can support such an artist. Equally important, because being an artist is a safe and good job, there has been fierce competition in this profession from the beginning. And the effective way to compete is through innovation, so innovation has existed in the art industry from the beginning. Market needs and positive feedback effects ensure that art, as a profession, does not disappear once it is established.

With the passage of time and the development of human beings, more types of art have emerged. For example, the invention of pigments gave rise to color painting, language to singing, and writing to literature. It is only when humankind has a large amount of time to spare and sufficient market demand that pure art finally emerges, and until then it has only been applied art (see Sec. 6.1).

The emergence of art millions of years ago was based on three conditions: (1) Humans already have the mimesis ability. (2) Tons of leisure time. (3) Art doesn't necessarily need language and writing (that's after the advent of *Homo sapiens*).

What separates human from other animals in the process of evolution is the appearance of upright and enlarged brains in terms of hardware, and the emergence of art in terms of software. The emergence and development of art has accelerated the breadth and depth of human thinking and imagination. Software is just as important as hardware.

6 Nature of Art

Art is a type of human creative activity whose purpose is to stimulate the recipient's neurons through the recipient's senses. In fact, good-quality work in other fields also requires creativity, so creativity is not unique to art, but art is aimed at the recipient's neurons, and this is what makes art special.

By definition, applied art is useful art, while Class B pure art does not seem to be useful other than exercising the recipient's brain—providing pleasure or stimulating the person's mind. But this is not the case. To understand this question, we must first look back at art history to see how pure art emerged.

6.1 A brief history of art

Although art first appeared millions of years ago as a simple attempt to kill time, over time, the content and form of art have become more and more complex due to the innate competitiveness of the artist's profession (see Sec. 5.2) and the complexity of human life (such as the belief in ghosts and gods, the emergence of language and writing). For example, there are landscape paintings, figure paintings, and religious paintings, and sculptures include bronze sculptures, stone sculptures, and jade sculptures [Strickland 2007].

In Europe, however, artists are free to choose their subject matter only after the advent of the free art market in the last 200 years. Before that, the subject matter of the artwork (e.g., figure painting or religious content on the walls) was determined by the employer. It is the free selection of topics that allows artists to make art for art's sake, which leads to the emergence of pure art's Class B.

Specifically, advent of photographic technology more than 100 years ago forced the Impressionist style of 1874; modern art pioneered by Paul Cézanne (1839-1906) led to abstract art; visual arts (painting, sculpture) moved away from realism towards abstraction, and Ducham's *Fountain* of 1917 made art almost free from the artist (see below). It is this kind of non-realistic or "nonsensical" pure art of Class B that is particularly puzzling. What is their essence?

6.2 Class B of pure art

To answer this question, let's compare the time-killing functions of various artworks. While all art can kill time, for applied arts, its time-killing function has taken a back seat. For example, adding a nice pattern to a coffee cup is just to make the cup look better and sell for a higher price, not to prolong your coffee time. For pure art's Class A works such as Leonardo da Vinci's *Mona Lisa*, the author not only satisfies the basic requirements of the employer (depicting the specified person), but also uses mixed techniques to make the character's smiles have multiple interpretations, which increases the viewer's return rate.

Pure art's Class B goes one step further: Although the painting may be simpler or ultra-simple on the surface (see, e.g., Fig. 4, right; middle-left of pure-art category in Fig. 6; Fig. 7, left), it can arouse various associations in the viewer and thus kill more time. In other words, the essence of pure art's Class B is to return to the function of art at its very beginning millions of years ago: to kill time.

Fig. 7. *Left*: *Crossing the Sky* (橫空, 2008) by Wu Guan-Zhong (吳冠中, 1919-2010) [photo/Lui Lam]. *Right*: *Stone Behind the Frame* (2020) by Lui Lam (林磊). The piece of stone is picked up from the roadside and put behind a frame—an example of human intervention in art creation.

6.3 Criteria of a good pure art

Here we give five criteria for a good (and long-lasting) pure art: (1) Aim at the receiver's neurons. (2) Kill a lot of the receiver's time. (3) Kill time gently and harmlessly. (4) Passivity. (5) Human creation or intervention.

Explanations of the criteria are given in [Lam 2011]. Here, we just note that Criterion 3 is to distinguish art from entertainment or drugs, say. And an example of Criterion 5 is given in Fig. 7, right.

To sum up, art is created by humans or created with human *intervention*; pure art is to kill time *gently* and *harmlessly*, and let the receiver to experience it *passively*. According to this understanding, it is clear that the content or form of pure art (which changed with time and place anyway) is secondary; they are there to serve Criteria 2 to 4.

7 Art and Science

Science is human's effort and outcome to understand all things in the universe (including the human system and all nonhuman systems) without introducing God/supernatural considerations. The study of nonhuman systems, generally referred to as "natural science," is part of science. The rest of science is the study of human, called Scimat, which is scattered in the humanities, social science, and medical science. In other words, Science = Scimat + "Natural Science" [Lam 2008, 2014, 2024b].

The meaning of the word Art includes the conceiving and process of art creation, artwork, and art research. The study of art definitely is part of science, and the thinking and process of art creation can also be a part of science if sorted out, but artwork is not.

7.1 Relationships between art and science

When people talk about "art and science," they are generally referring to the relationships between art and natural science, and this section is no exception. There are many books on this subject, such as *Art and Science* [Strosberg 2001], *Art and Science* [Ede 2005], *Artscience* [Edwards 2008], and more in [Lam 2011]. Unfortunately, all of these writers have misunderstood this issue, mainly because they have a wrong understanding of the nature and scope of science, and the nature and origin of art.

Art and science are concerned with the same things, everything in the universe, but with different emphasis. So far, art has been primarily concerned with the human system—a complex system, while science has been primarily concerned with nonhuman systems—especially simple systems. Therefore, it is more difficult to do art well than to do science well, which is why there are fewer good artists than good scientists in the world.

The relationships between art and science are expressed at four levels: (1) They are both human's innovative activities (for different reasons). (2) Art reflects the artist's recognition of the world and her worldview which are both constrained by nature's principles, the same principles studied by scientists. (3) Scientific (and related technological) advances pushed the development of art, but not vice versa. (4) There are examples of mutual inspiration between artists and scientists.

For more discussion on these issues from the Scimat perspective, see [Lam 2011].

7.2 Relative importance of art and science

Many people have talked about the importance of art and science. In China, Cai Yuan-Pei pointed out a hundred years ago, "Those who value morality in the world all depend on art and science, just as a cart has two wheels, and a bird has two wing" [Gong 1998]. Later, Nobel laureate

Tsung-Dao Lee put it another way, comparing art and science to two sides of the same coin [Lee 2000: 138].

Unfortunately, these metaphors are all problematic for the following reasons:

1.   Art and science are not like the wings of a bird, the two wheels of a cart, or the two sides of a coin, because art and science did not appear at the same time.

2.   Art and science are not equally important in the history of human development.

In fact, a few millions of years ago, primitive humans living together must have had ethics such as "it is forbidden to eat people while they are in sleep," otherwise no one will feel safe enough to join the group. In other words, ethics was the *only* pillar that sustained society at that time, until the advent of art one million years ago (see Sec. 5.2). After that, ethics and art became the *two* pillars that supported society until 600 BC. 2,600 years ago, ancient Greeks invented science, and in the past 400 years, modern science has advanced by leaps and bounds, greatly affecting the society's economy and people's daily life. Since then, all modern civilizations have *three* ingredients: ethics, art, and science (Fig. 8).

Relatively speaking, ethics is the *most* important of the three categories. The Chinese say it well: Etiquette collapses, nation no more (礼乐崩, 国安在). That is, ethical destruction leads to social instability. The *second* most important is art, which, in addition to playing an important role in the history of human evolution (Sec. 5.2), has in modern times ensured the quality of man's spiritual life and made life worth living. Science is the *least* important, because its main function is to improve people's material living standards and prolong their lives; think about life before we have mobile phones: inconvenient but more leisurely.

Therefore, about the importance of art and science, the best way to describe it is this: Art, science, and ethics are the three pillars that underpin modern civilization!

Fig. 8. The three pillars that underpin modern civilization: ethics, art, and science. Ethics began millions of years ago, art one million years ago, and science 2,600 years ago. Art was born long before science.

8 Artists and Scientists

There are scientists who play the violin after working (Einstein), and there are artists who paint and dissect corpses at the same time (Leonardo da Vinci), but no one works in art and science professionally at the same time because each requires concentration to do well. However, good artists and scientists are creative people, and it is not uncommon for them to pay attention to and inspire each other in their work [Miller 2001], and even become friends.

8.1 Scientist's art and artist's science

Some scientists have done amateur drawings or paintings, usually after they have become famous, such as Nobel laureates Richard Feynman (1918-1988) [Feynman 1995; Popoya xxx] and Tsung-Dao Lee [Ji 2010]. Without exception, their artistic level is far below that of professional artists. Feynman explained his motivation like this [Popoya xxx]:

> I wanted very much to learn to draw, for a reason that I kept to myself: I wanted to convey an emotion I have about the beauty of the world. It's difficult to describe because it's an emotion. It's analogous to the feeling one has in religion that has to do with a god that controls everything in the universe: there's a generality aspect that you feel when you think about how things that appear so different and behave so differently are all run "behind the scenes" by the same organization, the same physical laws. It's an

appreciation of the mathematical beauty of nature, of how she works inside; a realization that the phenomena we see result from the complexity of the inner workings between atoms; a feeling of how dramatic and wonderful it is. It's a feeling of awe—of scientific awe—which I felt could be communicated through a drawing to someone who had also had that emotion. I could remind him, for a moment, of this feeling about the glories of the universe.

It's not sure that Feynman really meant it. Or, he's trying to convey his emotion about "beauty of the world" through drawing beautiful women.

Similarly, there were also artists who dabbles in science, notably Leonardo da Vinci (1452-1519). He was trained professionally in art but not so in science, so he was a folk scientist. And that is why he failed to sublimate his scientific observations into scientific theories.

In general, the artist's science is better than the scientist's art, much better.

8.2 Artist-scientist interaction

In 1987, Tsung-Dao Lee and the Chinese Academy of Sciences held an international conference on science and art with the participation of scientists and painters. In May 1988, the China Advanced Science and Technology Center held the International Conference on Two-Dimensional Strongly-Correlated Electronic Systems. After reading Lee's popular introduction, the painter Wu Zuo-Ren (吴作人) deformed the *Liangyi* (两仪) pattern to express the two-dimensional strongly-correlated electronic system, and created the *Endless Infinity* (无尽无极) (Fig. 9, left).

In 1989, Li Ke-Ran (李可染) created the theme painting for the International Symposium on Relativistic Heavy Ion Collisions held by the China Advanced Science and Technology Center: *Nuclear Weights Like a Cow, Collision Creates a New State* (核子重如牛，对撞生新态; Fig. 9, right).

These interactions are at the most superficial level, which is one-way; i.e., the artist, induced by a scientist or science, expresses some scientific principles in painting which has a bit of a popular-science flavor. The resulted paintings ignore the modern trends in art, and do not contribute much to the development or mutual promotion of art and science.

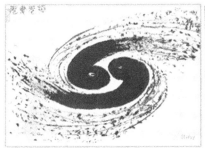

Fig. 9. *Left*: Wu Zuo-Ren's *Endless Infinity* (1988). *Right*: Li Ke-Ran's *Nuclear Weights Like a Cow, Collision Creates a New State* (1989).

8.3 Recommendations

Here are two workable suggestions:

1. Artists and scientists make friends and go watching the sunset together, and something good will come out.

2. Artists and scientists buy two of my books and peruse the first chapter, which concisely and clearly clarifies the most fundamental questions about art and science (Fig. 10).

Fig. 10. Lui Lam's two books on art and science: *Arts* (2011) and *All About Science* (2014); publisher: World Scientific (Singapore).

9 Cultural Confidence: Su Shi's Proposition

Su Shi (苏轼, 1037-1104), also known as Su Dong-Po (苏东坡), was a writer, calligrapher, and painter of the Northern Song dynasty, who was famous for his poetry and painting and the Dong-Po Pork [Lam & Qiu 2011]. Su left behind only two paintings: *Dead Wood and Strange Stone Drawing* (枯木怪石图, entered Japan during the Sino-Japanese War) and *Xiaoxiang Bamboo Stone Scroll* (潇湘竹石图卷, now in the National Art Museum of China, Beijing) [Li 2012].

Su mainly paints ink bamboo and strange stones. He advocated that painting bamboo should not be realistic, but should draw the *essence* of bamboo, the so-called "gain outside the image" (得之象外) [Wang 2006]. He believes that the essence of bamboo is *ethereal*, so the bamboo he painted has no bamboo knots, and the leaves can be separated from the bamboo branches and suspended in the air, just to show the ethereal (Fig. 11, left). On the contrary, his contemporary Wen Tong (文同) painted bamboos with each leaf connected to the branch, which was dense and quite tacky.

Su Shi's claim that "to paint is to show the essence of things" is exactly the same as Paul Cézanne's claim (Fig. 11, right). But Su was 800 years before Cézanne and was really ahead of his time. The fact that Su failed to become the pioneer of modernism is because there was a lack of successors to carry on his torch. In contrast, Cézanne had a lot of fans and converts such as the two masters, Picasso and Matisse. *Lesson*: Be culturally confident and tell well your own story.

10 More on Art

Here are more thoughts on art.

10.1 The importance of art

1. 【Life Path】 A person's life path is best decided by each person for herself. Opinions worth referring to are from, first of all, poets, literati, and artists, not scientists, and not philosophers.

Fig. 11. *Left*: Su Shi's *Xiaoxiang Bamboo Stone Scroll* (潇湘竹石图卷), 105.6 cm x 28 cm, partial. *Right*: Paul Cézanne's *Mount Sainte-Victoire Seen from Lauves* (1904-1905), a painting leading to modernism in art.

2. 【Importance of Art】 The three pillars of modern society are: ethics, art, and science. Science affects not just the spirit; its proper application is related to economics, medicine, and peace. Ethics is not only spiritual, but also a tool for maintaining social order and peaceful coexistence (Confucius was committed to this, in order to maintain stability). Art helps us willing to live in this (philosophically) meaningless world.

3. 【Two Sorts of Priority】 As far as civilization is concerned and marked by happiness, society is divided into two categories according to two different worldviews (called three-views in China). Depending on its assessment of science's importance, some societies are: science (economic promotion) > art > ethics; others are: ethics > art > science. In both cases, art is in the middle.

4. 【Four Major Leads】 In addition to the four major inventions (compass, papermaking, gunpowder, and printing), China also has four major leads: (1) Guanzi precedes Thales: all things originate from water. (2) Confucius precedes Plato: civilian education. (3) Su Shi precedes Cézanne: modernism in art. (4) Cai Yuan-Pei precedes Lowell: general education. The four leading figures: Guanzi, Confucius, Su Shi, and Cai Yuan-Pei; the four leads involve three areas: science, education, and art. Unfortunately, the lack of successors made it impossible to pass down in China, and so was surpassed by the West.

5. 【 Humanities 】 The humanities are more important than "natural science" (the study of nonhuman systems) because human happiness is heavily affected by the success of the humanities, which include the study of art, interpersonal relationships, and human decision-making.

10.2 Modern art

6. 【Modern vs. Classical】 Modern art, compared with classical art (e.g., abstract painting to realistic painting, modern dance to ballet), can better express the author's thoughts and personality, or arouse the thoughts and associations of the recipient. Chinese artists or decision-makers may not understand this, or they may understand but dare not and cannot have the courage to emancipate themselves and their minds, so they have not been able to step into the door of modern art. Do you have to enter this door? No. It would be better to skip the stage of modern art (after all, it has been 114 years since Cézanne's death in 1906), but it is more difficult to invent a new style, just as it is to invent the special theory of relativity without going through Newtonian mechanics. Can China do it? No one can say that it is impossible, in art or science. The question is: If this innovation needs to wait a hundred or a thousand years, are you willing to wait? Can you wait?

7. 【Modern Art】 Modern art is not for the viewer to understand; it is different from the classical art approach. Modern art is to stimulate your thinking, and the viewer can think in whatever way he wants. Few hundreds of years ago, Leonardo da Vinci's *Mona Lisa* had already used this approach—making sure the woman's smile in the portrait hard to decipher so that the viewer never get tired of watching it.

8. 【Death of the Author】 Unlike classical art, abstract art does not need to be understood or requires the viewer's understanding; it only wants the viewer to feel, think, and appreciate. Here, thinking is arbitrary and free, which does not have to include or be limited to conjecting what the painter thinks or wants to say. As someone puts it: When a work appears, the author is dead; i.e., the viewer does not have to care about the author. Freedom and emancipation are the themes of modernism. Personal choice is entirely possible and respectful, and is part of freedom.

10.3 Art governance

9. 【Art Censorship】Outside of China, art censorship is decided by local residents. Once in history, New York has set up a temporary committee of experts to determine whether an imported product, *The Bird* (Fig. 6, bottom-right corner), is a work of art or just a metal product (the latter is subject to taxes). In the past, there were committees in some cities to judge whether a film was artistic or pornographic, and the committee was composed of ordinary people and "experts" (such as clergies) who represented public opinion and the local moral standard at the time, which was basically a jury system.

10. 【Art Film】Mainland China's films (and television dramas) are not rated; each drama is for anyone (one to one hundred years old, say) to watch together, and the audience is supposed to be able to understand it. Consequently, only very mild films such as fairy tales and historical dramas are produced and shown in the cinemas, similar to the case that no nude sculptures or paintings are created in the art world. Film and television, like other media (painting, photography, etc.), can either be an entertainment work or an artwork. Yet, this understanding has not yet been popularized in the mainland. Films are rated in Korea and thus art films are produced and screened, which, however, was a historical process; it takes times. Future development of the art-film industry depends on the decision-maker's understanding of the nature of film—a mass medium and an art medium.

11. 【Public Art】The artistic level of public art is determined by the artistic taste/sophistication of the authority approving it; it does not necessarily reflect the artistic level of the artist or the local public (Fig. 12) [Lam 2011].

10.4 Art history

12. 【Reason and Romance】The Italian Renaissance provided good art for humankind (Leonardo da Vinci, Michelangelo), followed by modern science (Galileo). Britain provided reason for humanity (Magna Carta, Newton, Darwin). The French Enlightenment initiated the science of

human, and then provided humanity with universal values (freedom, equality, fraternity) and romance (ballet, cinema, modernism in art).

Fig. 12. Two buildings in China: The artistic taste of the authority determines the artistic level of public art.

13. 【Dong-Po's Life】 Su Dong-Po (aka Su Shi), the father of Dong-Po Pork, is China's top literati. Su's life seems simple and free. The first reason is that he has no shortage of beautiful women, and his life has never been without food. Simple and free should be his choice (but I don't really know because he has never been rich). He was able to survive, in addition to his father guiding him to study early and be sensible in life early, thanks to close woman friends (in the family, and the queen). But more importantly, he lived and applied Confucianism, Buddhism, and Daoism all together, whatever suited the occasion—his way of doing self-heal and self-preservation, in the era when there were no psychotherapists [Lam & Qiu 2011].

14. 【Simple → Complex → Simple】 Western cave painting is simple/ elegant, and the spirit of modern art is also elegancy. In the history of Western art, the art *form* is a history of "simple → complex → simple," but artistic thinking is "simple → complex," becoming more and more complex. In China, the jade carvings unearthed in the Shang dynasty were elegant; the later art became more complex. In the history of Chinese art, the artistic *thinking* is "simple → complex → simple," and the art form is "simple → complex," which has not yet returned to simplicity/elegancy (Fig. 13). Will it?

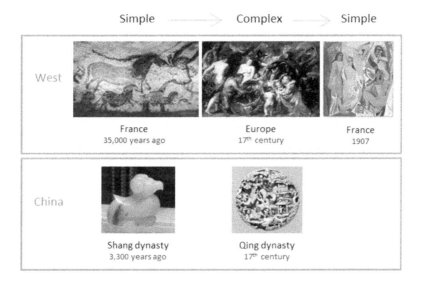

Fig. 13. The road of "simple → complex → simple" in art history.

10.5 The dao of art

15. 【Art Seeks Truth】 Science and art are not necessarily about beauty and goodness: The big fish in the ocean swallowing small fish is not beautiful or good, neither is Duchamp's *Fountain*. But both seek truth in feelings and things, not necessarily the "truth" philosophers talk about, which no one knows what it is.

16. 【Sunset Sunrise】 Some people say that art has seven functions: memory, hope, sorrow, balance regain, knowing oneself, growth, and appreciation. The problem is, it's not art-specific. Watching good scenery (like sunset and sunrise) can accomplish the same.

17. 【Analogy and Logic】 There are two ways of scientific thinking. (1) Analogy, imagination, intuition: The first step in explaining a new phenomenon or experimental data is to guess the reason or effect involved, and these three will be used. (2) Logical thinking: This is the second step. So, to say that ancient Greek or Western science relies on (Euclidean) logical thinking (without paying attention to analogy and imagination) is knowing only one of these two; it is incomplete. Analogy is as important

as logic, and many scientists are only good at one of the two. For example, Edward Teller, father of the hydrogen bomb and PhD mentor of Nobel laureate Chen-Ning Yang, is good at the former; Chinese theorists, including Tsung-Dao Lee and Yang, are good at the latter. Only those good at both can be great masters, such as Archimedes, Einstein, and Feynman. Analogy should be combined with logic to form good theory, and then be confirmed or disproved by experiments. Because of the reality check provided by experiments, no matter how imaginative or crazy the theories are, science will not go astray. Artists use analogies a lot (especially in poetry), but logic is not necessarily absent (e.g., Leonardo da Vinci used geometric perspective in his *Mona Lisa*). The difference between scientific research and artistic creation: Long-term deviation won't happen in science; there's no such thing called deviation in art.

18. 【Imperfect Beauty】 All roads lead to Rome, and Rome—the purpose of pure art—is to kill time (see Sec. 6). For art to reach Rome, route #1: Seek beauty. The pursuit of beauty in art is a shortcut; it is easy to get a high return rate from the viewer by taking advantage of human's evolutionary preference for beauty. However, the quest for beauty was only one of the routes to Rome, although it did work out fine for more than 2,000 years (see Sec. 3). Route #2: Pursue the perfect combination of art and science. Route #3: Imperfect beauty, such as porcelain burned out cracks (called crackle glaze), breaking the cylindrical symmetry of the vase. Some Japanese artists have carried it forward and deliberately designed it by controlling the growth of cracks.

19. 【Nudity in Art】 (1) To make a human-body sculpture (whether dressed or not), one should start with a naked model when learning and practicing, to know the position, size, and direction of each muscle and blood vessel so that the clothes, if that is the case, will not be out of shape when engraved on the surface of the statue. Leonardo da Vinci went one step further by dissecting corpses to gain more accurate understanding (as others of his generation also did). (2) If you mean to show a human being in a work of art (in photography, painting, sculpture, or film, say), rather than a person of a certain era or social status, you must use a naked human body, otherwise the theme and meaning conveyed will be deviated

(because clothing will give information other than a human being). (3) Most Chinese do not understand these principles of art and disdain nudity in any art form. If a country decides to outlaw nude art in any public place, then of course it has the right to do so, just like some Middle Eastern countries do not allow women to drive, according to their customs and laws. (4) Yet, any country (including China) has pornographic/erotic drawings/paintings in their heritage, kept in museums or privately owned if not available in bookstores. In fact, sex and nudity have never been absolutely taboo in history; the Chinese philosophy Mencius says it well: The desire for food and sex is part of human nature. But how much nude art can be displayed in public is a matter of debate, and the answer varies with time and place. For example, brothels were legal in China before 1949, but not later; in some European countries, the opposite happened. (5) All taboos are human-made, which, through discussion, argument, and struggle, could be removed, just like the forward-moving process in physics—called innovation. National innovation, apart from scientific innovation, should include artistic innovation.

20. 【iPad Art】 For artists, the iPad has two uses: (1) *Make a draft*. Oil painting materials are very expensive; it is recommended to buy a large iPad, download a (paid) software, click on the optional oil painting or ink, etc., and use an Apple pencil or finger to work. Use the iPad to do sketch, draw/paint a few versions, store them (so the iPad storage capacity should be the largest, not that expensive), and choose one or two after comparison, before really painting on the canvas. Save money, save time and be environmentally friendly. (2) *iPad painting*. There are professional painters who have used their fingers to paint on the iPad instead of the canvas, and the effect is super good. I don't know why it has not become a trend or a branch of art later. Holding more iPad painting competitions and exhibitions, in addition to the consideration of artistic diversity, is to popularize painting. Currently, most computers have touch screens, and so many people including students without an iPad can participate.

21. 【Good and Right】 Not all art is good; not all science is right. There are good and bad arts, and there is right and wrong science. But what can stay in the end, without exception, is good art and right science.

22. 【Art Ethics】 Don't do arts that harm or invade the body (including the body of your own, others, butterflies, and other animals), like in so-called biological art. Art has an ethical bottom line, just like any human activity.

23. 【Sadness without a Trace】 During *The Gengzi Epidemic* (庚子大疫, 2020), no artwork reflecting the Covid disease appeared, no trace of sadness?! What is important is something that everyone, including artists, has to decide. In my opinion, a good work of art is produced when the artist is deeply touched and moved, with an unstoppable urge of creating it and without considering the non-artistic issues related to publication or exhibition. It's like someone madly in love...without consideration of any consequences (like in *Romeo and Juliet*).

10.6 Art education

24. 【Read Philosophy】 A painting that can make the receiver (viewer) to view and reminisce repeatedly is because there is a "thought" behind it, although the receiver does not necessarily need to know or guess what the painter wants to say. It is recommended that all artists read easy-to-read popular books on philosophy, starting with this one: *The Questions of Life: An Invitation to Philosophy* (2002) by Fernando Savater, a Spaniard.

25. 【Thought Matters】 A good artwork often offers more than just being good to look at. In addition to the technique (which is not difficult to learn), all works of art will convey the artists' inner feelings and thoughts in addition to the superficial appearance, and it is this thought, not the technique, that determines the artistic level of the artwork. How can an artist improve his depth of thinking and literacy? I advocate reading novels written by philosophers, such as *The Stranger* (1942) by Albert Camus (1913-1960). If you want to read philosophy, you can read Laozi and Zhuangzi (vague enough), and Russell's *A History of Western Philosophy* (clear enough), and that would be enough. Example: The songs of literary Nobel laureate Bob Dylan are not particularly good singing and beautiful tunes, but deep and complex lyrics. Those with both good singing and lyrics are Joan Baez's anti-war songs.

26. 【Art Academy】 The shortcomings of art academies in China and elsewhere are: They teach a lot of techniques but not enough other subjects to help the students to think beyond art, because they are not comprehensive universities. Therefore, after graduation the students should enhance their own thought-training by experiencing life, like working different kinds of jobs, doing portrait for money on the side of streets, delivering takeaways, and falling in love a number of times. However, falling out of love is deeper than falling in love, and it is especially useful for artists. Don't you see that many good pop songs are about after falling out of love?

27. 【Can't Understand】 The charm of modern art is that different people see different things, and each time they see something different. It's a pity that the public and many art decision-makers don't understand this, and always say: I can't understand. Or ask questions that shouldn't be asked: What is this work about? People in art schools urgently need to learn from the popular-science community and do some art popularization. For example, every art center or college should have an art popularization platform on the web.

28. 【Artpop and Scipop】 Artists create works of art, and generally will not engage in art popularization (*artpop*, 艺普). Artpop is related to art creation but is a separate entity/profession, just like science popularization (*scipop*, 科普) has its own identity and is separated from science. Art museums are in the artpop business, in addition to preserving artworks, but science museums are mostly in scipop; preserving science is the business of the science community, through science textbooks, teaching, and grad-student mentoring. The role of artpop workers to artists is equivalent to scipop workers to scientists.

The focus of artpop is on art appreciation, not on the art production mechanism/process (although it is a part of artpop). Popular science can learn from this and open up a new direction of "science *appreciation*." At present, one of the misconceptions of popular-science workers is that everyone will enjoy thinking and understanding the details. No, thinking is a very tiring thing—this is one of the reasons that religions thrive, with a large number of believers and increasing every year.

The job of explaining scientific phenomena and theories accurately is better left to science students/professors—not all of them though, since not everyone is good at communicating with people. Popular science, whether it is to explain the scientific mechanism or promote scientific appreciation, involves communication and exchange between people, which inevitably will involve both the humanities and science—the domain of humanities-science synthesis (文理交融). Scipop is a humanities-science endeavor while artpop involves only two branches of the humanities: mass communication and art studies.

29. 【Art Appreciation】 Newton couldn't draw; not everyone can paint (and it is not necessary). But I hope that everyone could appreciate art, and that could be done. France does it very well: Starting with primary school students, the schoolteachers take them to art museums at least once a month.

30. 【Art Research】 Nobel laureate Richard Feynman (doctoral mentor of my doctoral mentor) believes that philosophy is useless to science. He is right because philosophy (or philosophy of science) has not advanced enough to guide scientific research, just as art studies have failed to guide the creation of art. Maybe that will change in the future; no one knows.

10.7 Artistic innovation

31. 【Inheritance and Innovation】 To move forward, art needs innovation and new ideas, no matter what new media or means are used in art. For example, the instruments of Chinese music seem to have not changed for hundreds or thousands of years. Why can't you change it? Don't dare to change it? Attitude towards existing art, no matter which kind, should be a two-legged walk: (A) Preserving what is old and good is called inheritance. (B) Developing the good ones and improving the bad ones is called innovation. The two do not conflict. If there is only A and no B (in storytelling, 说书; Peking Opera, 京剧; Kunqu Opera, 昆曲; etc.), it will decline and eventually disappear, and we will end up losing the inheritance. Art schools should train two kinds of talents at the same time: (1) Only do B's. (2) Do both A and B. There should not be only A's, which in fact ruins talents and has the side effect of not being able to recruit first-

class students because doing A's alone is a bit boring. Unfortunately, this is exactly what's going on in China's practice of cultural-heritage preservation.

32. 【Weed in the Crack】 Like in many other things, the level of artistic creation depends on the space or crack in the societal/educational system that allows for the emergence of different or weird species. Crack-free cement slabs don't grow weeds, but innovation is the kind of weeds that turn out to be good and useful. Of course, even if there exists space/crack but people keep pulling weeds out when they see them, there is no way that good arts will appear. As far as art is concerned, the level/sophistication of the artist and the authority are equally important.

33. 【Eat Art】 Artworks using food as the medium is a new art variety. I name it *Eat Art*. It is artistic and delicious, viewable and eatable (Fig. 14).

Fig. 14. Two examples of Eat Art. *Left*: Fruit carving. *Right*: Raspberry glazed doughnut.

11 Conclusion

Here are the major conclusions.

1. Art appeared at least one million years ago (before cave painting 35,000 years ago).

2. The emergence and development of art has accelerated the breadth and depth of human thinking and imagination, which has distanced humans from other animals in the process of evolution.

3. Art can be divided into pure art and applied art.

4. Art is aim at the receiver's neurons. Art is created by human, or by human's intervention; pure art's purpose is to kill the receiver's spare time gently and harmlessly, and allow the receiver to experience it passively.

5. The content and form of pure art change with time and place, and are subjectively determined by the artist.

6. How to kill time is an important marker that distinguishes human from other animals.

7. Scientific and related technological innovations often promote artistic innovation, but not vice versa.

8. Friendship between artists and scientists is beneficial to both parties, stimulating thinking and creativity.

9. About 800 years before Paul Cézanne, Su Shi advocated that painting should depict the essence of objects, and almost became the founder of modern art. Be culturally confident.

10. Art, science, and ethics are the three pillars that underpin modern civilization!

11. Art makes life worth living, and science makes life more comfortable. Art is more important than science!

Acknowledgments

On September 7, 2019, I gave a talk on "The nature of art and its relationship to science" at the *Question of Art and Science Forum* at Yanqi Lake in Beijing, which was organized by the University of Chinese Academy of Sciences. This article was written on the basis of my talk, which was finished on May 12, 2020 when California's Silicon Valley was locked down in the year of *The Gengzi Epidemic* (庚子大疫). Thanks to the organizers of the forum for arranging my talk, and members of the WeChat groups (Yanqi Lake/Art group in particular) for their helpful discussions.

References

Donald, M. [2006]. Art and cognitive evolution. *The Artful Mind: Cognitive Science and the Riddle of Human Creativity*, Turner, M. (ed.). Oxford: Oxford University Press.

Ede, S. [2005]. *Art and Science*. New York: I. B. Tauris.

Edwards, D. [2008]. *Artscience: Creativity in the Post-Google Generation*. Cambridge, MA: Harvard University Press.

Feynman, M. [1995]. *The Art of Richard P. Feynman: Images by a Curious Character*. Guernsey, British Virgin Islands: G & B Arts International.

Gong, Zhen-Xiong (龚镇雄) [1998]. Cai Yuan-Pei: Science and Art. Democracy and Science No.3, 1998: 36-37.

Hoppe, B. [2011]. The Latin "artes" and the origin of modern "arts." Arts: A Science Matter, Burguete, M. & Lam, L. (eds.). Singapore: World Scientific.

Ji, Cheng (季承) [2010]. *Biography of Tsung-Dao Lee*. Beijing: International Culture Publishing Company.

Lam, L. [2008]. Science Matters: A unified perspective. *Science Matters: Humanities as Complex Systems*, Burguete, M. & Lam, L. (eds.). Singapore: World Scientific. pp 1-38.

Lam, L. [2011]. Arts: A science matter. *Arts: A Science Matter*, Burguete, M. & Lam, L. (eds.). Singapore: World Scientific. pp 1-32.

Lam, L. [2014]. About science 1: Basics—knowledge, nature, science and Scimat. *All About Science: Philosophy, History, Sociology & Communication*, Burguete, M. & Lam, L. (eds.). Singapore: World Scientific. pp 1-49.

Lam, L. [2024a]. Art and killing time. *Humantities, Science, Scimat*, Lam, L. Singapore: World Scientific. pp 359-361.

Lam, L. [2024b]. *Humanities, Science, Scimat*. Singapore: World Scientific.

Lam, L. & Li-Meng Qiu (邱理萌) [2011]. Su Dong-Po's bamboo and Paul Cézanne's apple. *Arts: A Science Matter*. Burguete, M. & Lam, L. (eds.). Singapore: World Scientific. pp 348-370.

Li, Jian-Ya (李健亚) [2012]. Su Shi's *Xiaoxiang Bamboo and Stone Painting Scroll* was unveiled. Beijing News, January 17, 2012.

Lee, T. D. (ed.) [2000]. *Science and Art*. Shanghai: Shanghai Science and Technology Press.

Lin, Lei (林磊) [2010a]. Arts: A science matter. *Science and Art, Intersection and Integration*, School of Humanities of the Graduate School of the Chinese Academy of Sciences & Beijing Digital Science Popularization Association (eds.). Beijing: Tsinghua University Press.

Lin, Lei [2010b]. Arts' origin and nature: The Scimat perspective. *A New Perspective on Modern Technology and Modern Art*, Jin Lin-Lang (金琳琅) (ed.). Chongqing: Sichuan Science and Technology Press.

Liu, Bing (刘兵) [2011]. Science and Art in China. *Arts: A Science Matter*, Burguete, M. & Lam, L. (eds.). Singapore: World Scientific.

Miller, A. I. [2001]. *Einstein, Picasso: Space, Time and the Beauty That Causes Havoc.* New York: Basic Books.

Popoya, M. [xxx]. The art of Richard Feynman: The great physicist's little-known sketches and drawings, collected by his daughter. https://www.themarginalian.org/2013/01/17/richard-feynman-ofey-sketches-drawings/ (Feb. 18, 2024).

Strickland, C. [2007]. The Annotated Mona Lisa: A Crash Course in Art History from Prehistoric to Post-Modern. Kansas City: Andrews McMeel.

Strosberg, E. [2001]. *Art and Science.* New York: Abbeville.

Wang, Jin-Shan (王金山) [2006]. Wen Tong & Su Shi. Shijiazhuang: Hebei Education Press.

Unpublished: Lam, L. [2020]. On art. Based on my talk given at the *Question of Art and Science Forum*, Yanqi Lake, Beijing, September 7, 2019, with Secs. 1 and 10 added.

Chinese Painting: Innovation and Educational Training

Lui Lam

1 Introduction

In addition to the four major inventions in technology (compass, papermaking, gunpowder, and printing), China also has four major leads: (1) Guanzi "precedes" Thales: all things originate from water.[1] (2) Confucius precedes Plato: civilian education. (3) Su Shi precedes Paul Cézanne: modernism in art. (4) Cai Yuan-Pei precedes James Conant: general education (Fig. 1) [Lam 2022a, 2024].

Additionally, Chinese traditional medicine fully conforms to the modern definition of science (1867) [Lam 2014, 2024], so it is a science, and ancient China has science [Lam 2008, 2022a; Lam & Jian 2021]. The four leads involve science, education, and art, but due to the lack of successors at that time, they could not be passed down in China, allowing the West to catch up.

All this shows that Chinese traditional culture does have a few bright spots, and the cultural confidence of the Chinese people has a solid foundation. In recent years, the content of China's cultural export consists of Chinese language, Chinese writing, and Confucianism; we call it the old model. The new model could come from cultural innovation, and Chinese-painting innovation is an important part of cultural innovation.

Here, we first briefly review the innovation of Chinese painting in ancient, modern, and contemporary China. We then address the issue of the lack of art innovation in recent years, and discuss the cultivation methods of innovative talents. Recommendations in reforming China's education system in art are suggested. The hope is to strengthen the old culture by innovating a new one.

Fig. 1. China's four major leads in science, education, and art. Guanzi's case is a bit complicated; see footnote 1 for discussion.

2 Past Innovations in Chinese Painting

Three periods of innovation, from ancient to contemporary, in Chinese painting are presented here. (The source materials on painters and paintings are from Baidu Baike.)

2.1 Ancient period

Chinese painting is also called "national painting" (国 画), a term originated in the Han dynasty (206 BC-220). It mainly refers to paintings painted on silk or rice paper, which is "framed" and rolled into a scroll, making them easy to carry around. They are thus called "scroll paintings."

The painting method is to use a brush dipped in water, ink, or color pigment to paint on silk or paper. The subject matter could be human characters, landscapes, flowers and birds, etc. The techniques used can be figurative or freehand. The content reflects the ancients' cognition of nature, society and related politics, philosophy, religion, morality, and literary/artistic matters [Wang 2018; Xue & Shao 2000].

An important innovation in national painting appears in the style and content of a class of paintings, the so-called "literati painting" (文人画). These are "scholar freehand paintings" (士 大 夫 写 意 画), which are different from the folk paintings and palace paintings. It began with Wang Wei (王维) in the Tang dynasty (618-907) and flourished in the Song dynasty (960-1279). Generally, literati paintings deliberately avoid social reality. Most of the content are taken from landscapes, flowers and trees to express personal spirituality, and sometimes there is also a sense of resentment against national oppression or decadent politics.

Literati paintings pay attention to the "lightness" of pen and ink, delineate the resemblance of form, and emphasize the charm and essence of the subject matter (Fig. 2, left).

Fig. 2. *Left*: Su Shi's *Xiaoxiang Bamboo Stone Scroll* (潇湘竹石图卷). *Right*: A bird painting by Bada Shanren.

Another important innovation is due to one person: Zhu Da (朱耷, 1626-1705). Zhu, formerly known as Zhu Tong-Gui (朱统鍫), was known as Bada Shanren (八大山人). He lived to the age of 79. Zhu was the tenth-generation descendant of the founder of the Ming dynasty (1368-1644). After Ming fell and replaced by the Qing dynasty, he shaved his hair and became a monk, and later converted to Daoism.

Zhu is skilled in calligraphy and painting, and able to write poetry and prose. His flowers and birds are mainly freehand brushwork, with exaggerated and unique images, and with very few strokes (Fig. 2, right). His landscape strokes are concise and have a sense of tranquility.

2.2 Modern period

There are four important innovative painters in the modern period: Qi Bai-Shi, Xu Bei-Hong, Lin Feng-Mian, and Zao Wou-Ki.

1. Qi Bai-Shi

Qi Bai-Shi (齐白石, 1864-1957), born in the Qing dynasty, grew up in the Republic of China, and died in new China, at the age of 93. He used to be doing wood decorative carving, but soon turned to painting and poetry. In 1919, at the age of 57, he settled in Beijing as a professional painter. He is good at painting flowers and birds, insects and fish, landscapes, and people (Fig. 3, left and middle top). His style is influenced by Bada Shanren.

Fig. 3. *Left and middle top*: Two paintings by Qi Bai-Shi. *Right and middle bottom*: Two paintings by Xu Bei-Hong.

2. Xu Bei-Hong

Xu Bei-Hong (徐悲鸿, 1895-1953), born in the Qing dynasty, grew up in the Republic of China, and passed away in new China at the age of 58. In

1917, he served as the mentor of the Peking University Painting Research Association; 1919-1921, studied in Paris at the age of 24; 1921-1925, 27 years old, studied abroad in Berlin; 1926-1927, visited France, Switzerland, and Italy. After 1949, he served as the director of the Central Academy of Fine Arts in Beijing.

Despite his six-year stay in Paris and Berlin (1919-1925) when abstract art was already common, upon his return, Xu advocated *realism*. Yet he did advocate the integration of Western painting techniques into the reform of traditional Chinese painting, light and form in painting, the accurate grasp of the anatomical structure and bones of objects, and the ideological connotation of works (Fig. 3, right and middle bottom). Overall, his artworks are less than people expected.

3. Lin Feng-Mian

Five years younger than Xu Bei-Hong, Lin Feng-Mian (林风眠, 1900-1991) was born in the Qing dynasty, grew up in the Republic of China, and died in new China. But he lived much longer, up to age 91.

Unlike Xu who spent only two years in Paris (1919-1921), Lin spent six years there (1919-1925). In 1928, Lin became the first director of the National Academy of Art (now China Academy of Art) in Hangzhou, which is China's top art institute, founded by Cai Yuan-Pei (蔡元培). Historically, the academy trained quite a number of Chinese top painters, including Li Ke-Ran (李可染), Wu Guan-Zhong (吴冠中), Ai Qing (艾青), Zao Wou-Ki (赵无极), and Chu Teh-Chun (朱德群).

Lin is praised as the "father of Chinese modern painting," and the advocate, pioneer, and the most important representative of "Chinese and Western art integration" (Fig. 4). In particular, he advocated the educational idea of "inclusiveness and academic freedom."

4. Zao Wou-Ki

Zao Wou-Ki (赵无极, 1921-2013), born in the Republic of China, died in Switzerland at age 92. In 1935, at the age of 14, he entered National Academy of Art in Hangzhou and studied under Lin Feng-Mian. In 1948, one year before the new China was established, at the age of 27, he went to France to study and ended up settle down there.

From 1954 onwards, Zao's style of painting shifted from Cézanne and Picasso to Klee, from constructing a complete world to creating randomly: abstract, floating in the void of space and changing colors (Fig. 5).

After 1958, Zao's work was no longer titled, and the date of creation was on the back of the canvas.

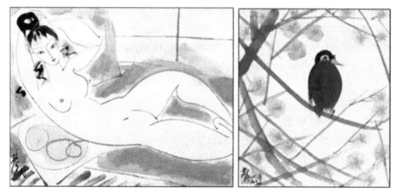

Fig. 4. Two paintings by Lin Feng-Mian.

Fig. 5. Two paintings by Zao Wou-Ki. The one on the right is kept at Fragrant Hill Hotel, Beijing.

In Table 1, the source of innovation of these four painters is surmised and summarized. The source can come entirely from national painting, like in the case of Qi Bai-Shi, because national painting has enough nutrients and occasionally the simplicity style, consistent with modern art's spirit of essence/simplicity/abstraction. Examples are "paint the essence of things" advocated by Su Shi (aka Su Dong-Po), the freehand style of literati painting, and the simplicity of Bada Shanren.

Whether or not to absorb the nutrients of Western painting and whether to settle abroad definitely has an impact on the style of painting, but has no absolute impact on innovation, as shown in the three cases of Xu, Lin, and Zao.

Table 1. Source of innovation of four Chinese painters in modern period. In Xu and Zao's case, borrowing from national painting is less important, unlike the case of Lin.

Painter	Study abroad/Settle abroad	Source of innovation
Qi Bai-Shi	No/No	National painting
Xu Bei-Hong	Yes/No	Western painting/National painting
Lin Feng-Mian	Yes/No	Western painting/National painting
Zao Wou-Ki	Yes/Yes	Western painting/National painting

2.3 Contemporary period

Two contemporary painters with two different ways of innovation, are presented here.

1. Wu Guan-Zhong

Wu Guan-Zhong (吴冠中, 1919-2010) was born in the Republic of China and passed away in new China, at the age of 91. In 1936, at the age of 17, he entered National Academy of Art in Hangzhou to study Western painting, Chinese painting, and watercolor painting; 1946-1950, studied in Paris at the age of 27. In the 1950s and 1970s, Wu dedicated to landscape oil painting; 1964, taught at the Central Academy of Arts and Crafts in Beijing; 1970-73, delegating rural labor in Hebei Province during the Cultural Revolution (1966-1976); 1979, elected as executive director of the Chinese Artists Association. He created a new art style that is distinct from that of all others (Fig. 6).

2. Gu Wen-Da

Gu Wen-Da (谷文达) was born in 1955, in new China, making him different from all the painters mentioned above. In 1981, he obtain a MA degree from China Academy of Art, Hangzhou, specializing in national painting. From 1981 to 1987, he taught at the Department of National Painting of the same school.

Also, differing from the older generation of Chinese painters who lived and worked in new China, Gu has worked abroad, as an associate professor in the Department of Fine Arts at the University of Minnesota. He was one of the leaders of the "'85 Art New Wave" movement. In the early 1980s, he made ink paintings of dislocated and dismembered calligraphic characters that "breaks through the shackles of beauty" (Fig. 7). Unfortunately, it seems that they aren't great art—not yet.

Fig. 6. Three paintings by Wu Guan-Zhong.

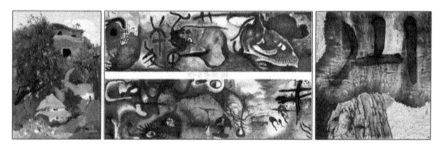

Fig. 7. Three paintings by Gu Wen-Da.

3 The Dao of Innovation in Chinese Painting

A new scientific theory needs experiments to guide it and verify it, which could involve complex and expensive equipment. In contrast, a new art style needs only an idea from the artist and nothing else, and so can be done much easier in any country. But somehow, in recent years, despite numerous art innovations in the past (see Sec. 2), no breakthrough in art innovation appears in China. Why?

It seems that there are two major reasons behind this:

1. The degree of freedom to explore condoned by society/authority and enjoyed in the past by the literati painters (Tang/Song dynasty); Bada Shanren (Ming/Qing); Qi Bai-Shi, Xu Bei-Hong, Lin Feng-Mian, and Wu Guan-Zhong (Republic of China); and Zao Wou-Ki (Republic of China/France) has shrunk tremendously or almost vanished in new China.

2. The concern and debate about whether an artwork counts as Chinese art is distractive and counterproductive in art innovation.

While not much can be done about point 1 at the individual level, point 2 is different. Good art is good art, no matter what it is called. If an artist produces world-class arts which are not recognized as Chinese art, so what? Expressionism is never called French art even though it is invented in France and produced mostly by French painters in the beginning. (On the other hand, Japanese artists do produce arts that can be easily recognized from Japan; i.e., they succeeded in blending new art with old culture. This example is worth exploring.)

Well, the first step to produce new Chinese painting is produce good painting. And like in any kind of innovation, the way to do it is to forget the two words "innovation" and "national" while doing it; those are afterthoughts. Just follow the well-known recipe in inventing anything: Don't paint anything that someone else has painted; don't paint what you've painted. Of course, you can improve on them a few times, but not for long. Learn from Picasso; he reinvented himself every few years. Also, don't worry whether your arts will be exhibited or sold; learn from Van Gogh. (This is similar to the case of good physicists; they won't worry whether their peers will understand or appreciate their new theory; history will tell.)

Technically, use iPad to do drafts. Don't start painting until you really feel the urge to create. Read some books on philosophy and literature to make your inner thinking interesting and complex; that will show up in your arts.

4 Educational Training of Innovative Talents in Chinese painting

It is quite impossible to change the thinking or habit of grown-up artists (or anybody, in fact). Thus, it is important to train young talents properly before they venture out on their own. For any creative subject, like art or physics, learning from the masters are important [Lam 2022b]. But there are not that many art masters in China today and what can be done is to improve the education system of the art academies and art department of the universities.

At present, the art academies are designed to train art scholars or teachers rather than artists. They only have bachelor's degree (4 years), master's degree (3 years), and doctoral degrees (3 years). They are all too long to train artists. Painting techniques can be taught but innovation can't be taught, and the teaching of techniques (where to buy the tools, how to mix paints, etc.) needs only one year.

1. In order to cultivate artistic *creative* talents, it is proposed to establish *two* new degrees, 2-year long each: Associate degree in Chinese painting, and Master degree in Chinese painting. The two degrees are essentially the same; the former is for undergrads; the latter, grad students. The master's degree students could be from any undergraduate majors.

2. Additionally, for associate/master's-degree students: Teach them art techniques in the first year, philosophy/literature and English language in the first semester of the second year, and send them to Paris or New York (where art galleries are concentrated) in the second semester of the second year, so that students can live freely, see galleries, and make (painter) friends there. If the financial conditions do not allow, the art schools can send their students to Beijing, second to the best. For both degrees, do graduation projects but *not* graduation thesis, which is a waste of time.

3. One more thing: Should encourage humanities (not science) students to pursue a master's degree in Chinese painting after they received their bachelor's degree. Reason: The difficulty of innovation in Chinese painting (or any art) is not in the technique, but in the painter's lack of humanistic thinking and courage to break away from tradition (refer to Bada Shanren). See Table 2 for a comparative example of two artists.

4. Post-graduation facilities: Imitate Paris and provide free/cheap studios, residences, and living allowances for painters, so that they can concentrate on experiencing life and painting.

Table 2. Comparing two good and famous film directors: Ang Lee and Hark Tsui. Tsui lacks the 4-year undergrad education in the humanities that Lee has. Apart from personal characters and upbringings, that could be an important factor leading to the differences in content and quality of their films. For artistic creation, humanistic cultivation is more important than technology.

Item	Ang Lee	Hark Tsui
Year of birth	1954	1950
Living place before university	Taiwan	Vietnam, Hong Kong
(USA) BA/MA	Drama (4 yr)/Film (2 yr)	Film (4 yr)/
Representative films	*Pushing Hands* (1991), *Brokeback Mountain* (2005), *Life of Pi* (2012)	*Butterfly Transformation* (1979), *Once Upon a Time* (1991), *Changjin Lake* (2021)
Film distinctives	Skill + Humanity	Technology + Hilarious

5 Conclusion

1. Innovation of Chinese painting is an important part of cultural innovation.

2. Unlike scientific innovation, there are no cultural barriers to innovation in Chinese painting.

3. The way to innovate Chinese painting is to forget the words "Chinese painting" and "innovation"—methodology is useless in any field.

4. The definition of "national painting" should be expanded and broad (refer to the definition of "physics": Physics is what physicists do).

5. The difficulty of innovation in Chinese painting lies in the restrictions in subject matter and the lack of humanistic cultivation of painters.

6. It is proposed to establish an associate degree and a master's degree in Chinese painting, each two-years, to cultivate innovative painters in Chinese painting.

7. It is recommended to provide free/cheap studios, residences, and living allowances for Chinese-painting innovators.

Notes

1. The book *Guanzi* is presumably done (or finished) by people in the Warring States period (475-221 BC) [Hu (1919) 2018: 12], not precluding *Guanzi*'s claim that "All things originate from water" appearing earlier. Thales left no writings and his proposal of "Everything is water" was first written down by Aristotle (384-322 BC) [Editors of Encyclopedia Britannica 2024]. In both cases, we don't know whether Guanzi and Thales themselves actually say those things and, if so, when did they say it. Since the lifespan of Aristotle is embedded in the Warring States period, and the beginning of the latter (475 BC) precedes the birth of the former (384 BC) by 91 years, one can still reasonably claim that the Chinese (via *Guanzi*) invented science, ahead or independently of the Greeks (via Aristotle).

References

Editors of Encyclopedia Britannica [2024]. Thales of Miletus: Greek philosopher. https://www.britannica.com/biography/Thales-of-Miletus (April 10, 2024).

Hu, Shih (胡适) [(1919) 2018]. *Outline of the History of Chinese Philosophy*. Beijing: Zhonghua Book Company.

Lam, L. [2008]. Science Matters: A unified perspective. *Science Matters: Humanities as Complex Systems*, Burguete, M. & Lam, L. (eds.). Singapore: World Scientific. pp 1-38.

Lam, L. [2014]. About science 1: Basics—knowledge, nature, science and Scimat. *All About Science: Philosophy, History, Sociology & Communication*, Burguete, M. & Lam, L. (eds.). Singapore: World Scientific. pp 1-49.

Lam, L. [2022a]. *Science and Scientist*. San Jose: Yingshi Workshop.

Lam, L. [2022b]. *Research and Innovation*. San Jose: Yingshi Workshop.

Lam, L. [2024]. *Humanities, Science, Scimat*. Singapore: World Scientific.

Lam, L. & Jian Xiao-Qing (简小庆) [2021]. *Lam Lectures: New Humanities, Science, Hawking*. San Jose: Yingshi Workshop.

Wang, Bo-Min (王伯敏) [2018]. *General History of Chinese Painting*. Beijing: SDX Joint Publishing Company.

Xue, Yong-Nian (薛永年) & Shao, Yan (邵彦) (eds.) [2000]. *History and Aesthetic Appreciation of Chinese Painting*. Beijing: China Renmin University Press.

Unpublished: Lam, L. [2021]. Starting from cultural confidence: The innovation of Chinese painting and its education and training. Based on my undelivered talk prepared for a conference on art and science.

PART IV

PHILOSOPHY

Philosophy: A Science Matter

Lui Lam

Any philosophical thought system is built upon the scientific knowledge established at the time when the system is proposed. To remain relevant at the present time, the philosophical system has to be consistent with the confirmed scientific knowledge presently known. Consequently, it is extremely important for philosophers to know correctly what science really is and what the current knowledge of science is. Science is to understand the universe which includes all material systems such as the human system, without bringing in any supernatural. Scimat (Science Matters) is the new multidiscipline that treats all studies about humans as part of science, from the perspective of complex systems. Here, the nature of science, the concept of Scimat, philosophy as a science matter, the proper way to integrate the humanities and "science," the fallacy of scientism, and the misconceptions that lead to Zeno's Paradox are discussed.

1 Introduction

Any philosophical system is built on the scientific knowledge known at that time. Similarly, if a philosophical system is to remain valid presently it has to be consistent with the scientific knowledge currently known. Thus, philosophers need to understand correctly what science is and what the current scientific knowledge is.

This chapter[1] explains that "Everything in nature is part of science"; in particular, everything related to humans are part of science. And this is the basic premise of *Scimat* (Science Matters), the new multidiscipline proposed by Lui Lam in 2007 [Lam 2008a, 2008b]. Thus, Philosophy,[2] a discipline in the humanities, is also part of science.

This chapter tries to clarify, from the Scimat perspective, the common misunderstandings about the humanities and science, and discuss problems related to philosophy, the synthesis of the humanities and "science,"[3] and scientism. Finally, the mistakes about Zeno's Paradox are pointed out.

2 Science

Science is humans' serious effort to understand nature *without* bringing in God or any supernatural [Lam 2014], a process starting with the ancient Greeks about 2,600 years ago. And nature includes all material systems in the universe. Presently, our understanding of the universe [Turner 2009] is that it began with the big bang which occurred 13.7 billion years ago. Protons and neutron appeared 10^{-5} second later, which bind together to form nuclei 0.01-300 second after the Bang. Stars were formed 13.4 billion years ago (from which almost all the atoms on earth came [Berkowitz 2012]). The solar system which includes Earth appeared 4.7 billion years ago; life on earth, 3.7 billion years ago. The important point is that all living and nonliving systems on earth are formed of atoms.

The origin of life on earth remains an unsolved problem. But we know from Darwin's evolutionary theory (1859) that humans did evolve from much simpler living systems. Additionally, we know that six million years ago, the lineages of human and chimpanzee split from each other, and our kind of *Homo sapiens* appeared only 195,000 years ago [Lam 2011: 6].

Consequently, all studies about humans (the humanities in particular) are part of science because humans are a material system and science includes the study of all material systems. In other words, "Everything in nature is part of science" [Lam 2008a].

Sometimes, control experiments are considered one of the characteristics of any science. Yet, for historical disciplines such as astronomy and archaeology this is not true. Similarly, if we insist that science must have the ability to predict then we will be forced to give up the early stages of any science. For example, when Thales (c.624-c.546 BC) proposed that everything is made of water he was not able to predict anything but we still call him the "father of science."

3 Scimat

Knowledge in the world can be divided into two types: one type is human-dependent (about humans); the other type, human-independent (about nonhuman systems). Literature and the applications of science are examples of human-dependent knowledge. Newton's three laws of motion are human-independent, meaning that they can be discovered by smart aliens, too, if they exist. Human-independent knowledge is those commonly called "natural science"; human-dependent knowledge, the humanities and social science (and medical science). However, this is not entirely correct. As described above, humans are a material system called *Homo sapiens* that is made up of atoms, the same atoms that are studied in "natural science."[4] Thus, all human-dependent knowledge should be part of natural science since the latter covers all material systems. Consequently, a logical conclusion is that

Science = Natural Science

   = Nonhuman-system Science + Human-system Science   (1)

Human-system science is called *Scimat* [Lam 2008a, 2014],[5] which is about probabilistic, complex systems.[6] Scimat includes all the topics in the humanities and social science as well as medical science. Thus, Philosophy, part of the humanities—derived mostly from human's thinking—is part of science.[7]

Nonhuman-system science since Galileo (1564-1642) is known as modern science, which is mostly about simple systems. In last two hundred years or so, modern science has progressed drastically, resulting in the underdevelopment of Scimat (except medical science) which are about complex systems.

The reason is that the behavior of complex systems is quite different from that of simple systems and is more difficult to study, even though there do exist three "universal" principles (chaos, fractals, and active walks [Lam

1998]) that are equally applicable to simple and complex systems [Lam 2008a]. Moreover, the human system is most complex and probabilistic.[8] In other words, deterministic predictions about human matters are impossible (except that every human being is predicted to die someday, but we cannot tell which day). Mistakenly, many scholars and laypeople alike identify science with the science of simple systems, and thus identify probabilistic, complex systems with deterministic, simple systems. This common misunderstanding is unfortunate and has grave consequences. Most human tragedies can be traced to this mistake.

Scimat aims to understand everything related to humans; every appropriate research methods and tools could be used, including evolutionary theory, cognitive science, neuroscience, statistical physics, complex-system science, etc. Scimat effectively connects the humanities and "science." From the Scimat perspective, we have studied human history [Lam 2008c] and established the new discipline called Histophysics [Lam 2002] (see an example on Chinese history in Sec. 6). Additionally, we have proposed a new explanation of the origin and nature of arts—an unsolved problem in art studies for 2,400 years since Plato [Lam 2011].[9]

4 Philosophy: The Scimat Perspective

The word philosophy means "love wisdom," the wisdom about all kinds of things. Thus, at its very beginning in ancient Greece, Philosophy is to understand not just humans but all things, small or big, such as whether the universe is finite or infinite in size, and the nature of time.[10] Later, when many topics originally studied in philosophy were hijacked by physics [Morris 2002], philosophy retreated[11] and shrunk to the study of highly abstract, unsolved problems of great difficulty (truth, meaning of life, etc.) as we know it today.[12]

The correctness of scientific knowledge underlying the philosophical system is extremely important; when it is wrong, the whole system crumbles. As an example let us consider the philosophical arguments of Immanuel Kant (1724-1804). In Kant's days, *the* science is Newton's deterministic mechanics. Thus (1) he first assumed, incorrectly, that science belongs to the deterministic domain; (2) he also realized, correctly, that morality is nondeterministic and belongs to the "freedom" domain; i.e., humans have freedom of choice in morality matters. (3) Then he asked: If everyone is free to choose what he or she wants to do, how can we guarantee that world is rational and meaningful? To solve this problem, (4) he claimed: We need "religion" (or a higher principle), which will govern morality matters and bring meaning to our lives [Ye 2007: 6-9].

This series of arguments is no longer valid, judging from what we know today. Kant's philosophy, at least this part, is outdated. The reason is that we now know that science is not just about deterministic systems but is also about probabilistic systems. Moreover, even though it is not yet final, morality can be linked to the evolutionary advantages of the humans [Harris 2010; Shermer 2015] and could even be hard-wired in the brain [Waal 2015]; thus, religion is not absolutely needed here. Finally, to say that life must be meaningful is an assumption slipped in by Kant.[13] Like many other philosophers, Kant's arguments are full of metaphysical presumptions.[14]

For a long period of time, in philosophy, the mind is considered to be completely separated from the body—the so-called mind/body dualism thesis (or conjecture). In other words, thinking could not be reduced to the brain's physiochemical functions. However, this duality view is no longer in fashion, for two reasons. (1) As pointed out by the philosopher Jaegwon Kim, "[I]mmaterial nonspatial minds would be totally impotent, and this renders them explanatorily irrelevant and useless" [Kim 2005: 151]. (2) The development of cognitive science [Kolak et al. 2006] makes it possible to connect human's many sensations and thinking with the operation of the brain's neurological system. That is, the mind has a physical (or biological) basis. All these point to the fact that human-related matters are physically based and thus can be studied scientifically—the premise of Scimat.

Like any other discipline, Philosophy can be studied with three different approaches (or at three different levels): empirical, phenomenological, and bottom-up [Lam 2002, 2011]. These three approaches complement and mutually support each other. What approach to use should depend on the specific problem under study. Most of the philosophy studies are at the empirical or phenomenological level, as is the case in the early stages of any scientific study. The recent Neurophilosophy [Churchland 1989], a branch of *Neurohumanities* [Lam 2011: 22], is an example of the bottom-up approach in philosophy.

5 Synthesizing Humanities and "Science"

The synthesis of humanities and "science" is an important topic in both the East and West. In the West, this topic appears in the form of the "two-culture problem."[15] In China, this problem was recognized early on by Cai Yuan-Pei (蔡元培, 1868-1940),[16] and later on by Gong Yu-Zhi (龚育之, 1929-2007) [2005] (Fig. 1) and others.

Fig. 1. *Left*: Cai Yuan-Pei. *Right*: Gong Yu-Zhi.

In fact, Cai's understanding on this issue is very deep, more in line with that of Scimat. In 1918, he already recognized that there is "science" in the humanities, vice versa; he opposed separating "science" from the humanities. And he was a man of action. As the president, he reorganized Peking University into departments according to the disciplines, and reformed the curriculum [Sun 2004]—the beginning of a truly general education in Peking University but, unfortunately, not in the rest of China.

In the West, with rapid advance of "natural science" in the last 200 years, the humanities were underdeveloped and two cultures (humanities and "science") were formed. In 1959, C. P. Snow pointed out the two-culture problem (in United Kingdom). To bridge the gap he suggested that each side to learn something about the other side—a strategy adopted in general education today in the West and in China. However, this method is ineffective and insufficient because it does not go to the root of the problem, which is the misunderstanding of the definition of science [Lam 2008a]. The effective way out is to start from the fundamentals: (1) Recognize that the humanities are part of science; (2) let everybody know about the three common principles governing both sides, viz., chaos, fractals, and active walks [Lam 1998, 2005]; (3) raise the *scientificity* (scientific level) of the humanities by encouraging collaboration between humanists and "scientists" (as advocated by Scimat). Better, teach everybody the unified perspective of the humanities and "science" through a new general-education course [Lam 2017].

6 Scientism

There are two definitions of scientism: (1) Science is almighty; (2) the so-called Scientific Method can be used in all research disciplines [Gong 2005: 384]. But both these two assertions violate the scientific spirit, which is to keep an open mind. No conclusion or method is valid before it is done and proven to be "true" or applicable. Scientism could be coming from people's experience in dealing with simple systems but the world consists of not just simple but complex systems, too.

As an example, let us consider the distribution of lifetime of Chinese dynasties, from Qin to Qing. As shown in Fig. 2, historical laws do exist (at least in this case). The lesson is that the existence of historical laws usually cannot be predicted; their existence is ascertained only when they are found. What is needed is that someone has to go out and look for them, by assuming (even just temporarily) that they do exist.

In Scimat we say only that "Everything in nature is part of science." We insist that the researcher in any discipline has to keep an open mind, be serious, be honest, and is willing to admit mistakes [Lam 2008a]; i.e., in the words of Gong Yu-Zhi [2005: 372], "be realistic." And this is the well-tested scientific approach. We have not insisted that a particular research method would lead to success. Thus, Scimat is *not* scientism. We welcome anybody to use her own method of choice to do her research. The scientific spirit is tolerant and pluralistic, with the "reality check" as the final judge

[Lam 2014]. Only the funding agencies have to decide who they want to support.

Yet, it should be pointed out that a good research method should be able to solve a large number of problems. The scientific approach or process is the one we know of that was capable of doing this, and the *only* one that enabled us to make cell phones, say.

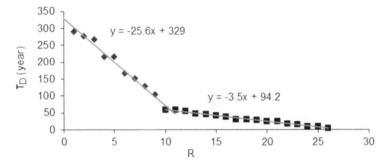

Fig. 2. Distribution of lifetime of Chinese dynasties, from Qin to Qing (221 BC-1949) [Lam 2006]. The dynasty lifetimes τ_D are arranged from large to small, from left to right—the Zipf plot, where R = rank. Only one of the overlapping data is kept. The result shown here is a particular example of the "universal" *Bilinear Effect* [Lam et al. 2010]. The historical law exemplified here is that (1) the dynasties lifetimes are discrete; (2) a dynasty can last another 3.5 years once it survives 3.5 years; but once it passes 57 years, it can survive every 25.6 years. It also predicts that the dynasty after Qing, if exists, will end exactly in its 329 year if it lives longer than the Tang dynasty. The "curse of history" indeed exists in the case of Chinese dynasties.

7 The Zeno's Paradox

It is beneficial for philosophers to know a little bit more science (especially that of complex systems) and mathematics. For instance, this could prevent them from making the mistakes committed by Zeno of Elea (c.490-430 BC), an early Greek philosopher. The birth of the Zeno's Paradox was due to his deficiency in mathematics. He mistakenly thought that an infinite series (i.e., the sum of an infinite number of numerical terms) must be infinite—not his fault though, since infinite series was not well studied at his time.

The Zeno's Paradox, one of his many paradoxes, can be expressed in more than one way. The simplest version is related to the problem raised in *Zhuangzi: The Society Chapter*. In this Chapter, Zhuangzi (庄子, c.369-286 BC) remarks: A one-foot long stick, take half each day, and you will never exhaust it. Similarly, Zeno remarked that a man could not move from point A to point B because, starting from point A, he had to move to the half point first, then the half point of the rest distance, ad infinitum; and because there is an infinite number of steps, he could never reach point B. These two descriptions are equivalent to each other. What is wrong with their arguments, mathematically speaking?

Even if we accept Zeno's presumption that this person is a point mass with zero size and the distance from A to B is a continuum, his arguments are still wrong for the following reasons. Let the distance from A to B be L, the person moves with constant speed v, and the total time traveling from A to B be t. Then

$$t = t_1 + t_2 + t_3 + \ldots \tag{2}$$

Here t_1 is the time moving from A to the half point between A and B, t_2 the time to cross half of the remaining distance, etc. It is easy to see that $t_1 = \frac{1}{2} L/v$, $t_2 = \frac{1}{4} L/v$, etc. Equation (2) becomes

$$t = L/v \,(1/2 + 1/4 + 1/8 + 1/16 + \ldots) = L/v \tag{3}$$

That is, the total time of travel is L/v, a finite number, even though there is an infinite number of terms in Eq. (3).[17] The fact that the sum of an infinite number of terms is finite is because each term decreases fast enough. Zeno's mistake is that he thought that the sum of an infinite number of terms must be infinite (i.e., $t = \infty$).

Zeno was already wrong at the mathematical level. But when he and other philosophers (and laypeople) applied his argument to the real world, they were wrong by ignoring the fact that everything, the walker or the stick, is made up of atoms (of finite size) and not point particles (of zero size). Any real thing has a finite size and the Zeno procedure cannot be carried out by real objects in practice [Gong 2005: 138-140; O'Carroll 2014].

8 Conclusion

Ultimately, all serious and honest pursuit of knowledge is to understand nature (i.e., the universe). But nature includes two kinds of material systems: the human system and nonhuman systems. They are all legitimate objects/subjects in research. Thus all of these researches are within the domain of science even though the method used may be different. For example, literary scholars use their bodily sensors and their brain as information processor while "natural scientists" use, additionally, measuring instruments and computers. Yes, all these activities can be view from a unified perspective: They are scientific developments at various stage of maturity. There is a lot that they can share with and learn from each other.

Many scholars, from Aristotle (384-322 BC) to Da Vince (1452-1519), are fully aware that "Everything in nature is part of science." Yet, not until recently with the successful experience gained in modern science—from evolutionary, cognitive and neuro- sciences, statistical physics, complex-system science, and other disciplines—that we know how to study humans scientifically. Scimat follows the Aristotelian tradition of treating the human-system and all nonhuman-system matters, philosophy included, as part of science.

We concur with Bertrand Russell when he writes:

> Most philosophers...profess to be able to prove, by a priori metaphysical reasoning, such things as the fundamental dogmas of religion, the essential rationality of the universe, the illusoriness of matter, the unreality of all evil, and so on....This hope, I believe, is vain. [Russell (1912) 2010: 99]

But we beg to differ when he claims:

> Philosophy is to be studied, not for the sake of any definite answers to its questions, since no definite answers can, as a rule, be known to be true, but rather for the sake of the questions themselves. [Russell (1912) 2010: 112]

This was definitely neither the attitude of the ancient Greek philosophers—the pioneers, nor the late comers like Kant, even though it could be the attitude shared by many contemporary philosophers when they are left to deal with the most difficult problems not yet hijacked by the "natural scientists." The prospect of never reaching the "truth"—the ultimate answer, if true, never deters the "natural scientists" from looking for an answer, even only a provisional answer. And there is no reason to expect less from the philosophers. In this regard, raising the scientific education of philosophy students and combining theoretical approach with empirical approach in philosophical research (see, e.g., [Churchland 1989; Churchland 2007; Knobe & Nichols 2008]) are two things that should be encouraged.

Notes

1. The content of this chapter was presented in Chinese at the High Forum on Natural Dialectics and Advancement of Science (自然辩证法与科学发展高阶研讨会), Renmin University of China, Beijing, May 27, 2011. In this English version, a few places and references are updated and additional footnotes are added.

2. Philosophy about 2,600 years ago was the single discipline created by the ancient Greeks. It split into 'Philosophy' (single-quotation marks; about humans), Natural Philosophy (about nonhuman systems) and Theology around the 14th century. In the 18th century 'Philosophy' was further split into the Humanities and Social Science. "Philosophy" (with double-quotation marks), the Philosophy commonly understood today and the one referred to in this chapter, is a discipline within the humanities; it is about "metaphysical" problems and is a small subset of the ancient Greeks' Philosophy. See Figs. 9.2 and 9.4 in [Lam 2014].

3. In this chapter, "science" with double-quotation marks means science in the conventional sense, which is the sum of "natural science" and social science but excludes the humanities (see [Lam 2014]).

4. Knowing that all materials are made up of atoms does not imply that for every problem under study we have to start from the level of atoms. This is not even so in physical problems. In fact, for any discipline there always exist three different research levels (see Sec. 4).

5. Scimat is *Renke* (人科) in Chinese, meaning Human Science [Burguete & Lam 2013].

6. For complex systems, see [Lam 1998]; for philosophy of complexity, see [Liu 2008].

7. Karl Marx (1818-1883) was not always right. But he was right when he said in his *Economic and Philosophic Manuscripts of* 1844 that human science and natural science would merge into a *whole* science [Marx & Engel 1979: 128]. See also [Liu 2003].

8. It is probabilistic because a human being is an open system, which continuously interact with its environment.

9. The Scimat Program to advance Scimat was started by Maria Burguete and Lui Lam in 2007. Presently, we have (1) set up a biennial international conference series (held 2007, 2009, 2011, and 2013 in Portugal); (2) set up a Scimat book series (publisher: World Scientific; founder and editor: Lui Lam); (3) set up an international Scimat committee. Scimat website: www.sjsu.edu/people/lui.lam/scimat. See also [Burguete & Lam 2016].

10. The Sept/Oct 2014 issue of the magazine *Philosophy Now* focuses on the early Greek philosophers: Thales, Diogenes, Heraclitus, and the "crazy Aegean gang" including Democritus. Fun to read. See [Steinbauer 2014] for an introduction.

11. Similarly, the Christian religion retreated when "science" advanced [Lam 2004].

12. For beginners, a gentle entry to "Philosophy" could be Fernando Savater's *The Questions of Life* [2002]. It contains, in an Appendix provided by the translator Carolina Ospina Arrowsmith, a brief introduction to the lives and ideas of important philosophers since Socrates. A quick summary of philosophers and their ideas, arranged chronologically starting with Thales and Laozi, is given in *The Philosophy Book* [Buckingham et al. 2011]. A clear and authoritative survey is *The History of Western Philosophy* by the mathematician and philosopher Bertrand Russell (1872-1970) [1945], a Nobel laureate in Literature (1950). An academic resource is Stanford Encyclopedia of Philosophy (plato.standford.edu).

13. If life does not have any a priori meaning, it does not follow that life has no meaning. An individual can find meaning to her life by helping others or observing sunsets, say. The world is full of wondrous things.

14. Here is how the historian and philosopher Robin Collingwood (1889-1943) rated Kant's works: "So long as he confines himself to drawing the distinction between philosophical method and mathematical, his touch is that of a master; every point is firm, every line conclusive. But when he turns to give a positive account of what philosophy is, his own distinction between a critical propaedeutic and a substantive metaphysics, hardened into a separation between two bodies of thought, becomes a rock on which his arguments splits" [Collingwood 2008: 25]. As for Bertrand Russell, he says, "Immanuel Kant (1724-1804) is generally considered the greatest of modern philosophers. I cannot agree with this estimate, but it would be foolish not to recognize his great importance" [Russell (1945) 1972: 704]. Thus, Collingwood considered Kant's work on philosophy mostly wrong and Russell thought Kant was overrated.

15. See [Lam 2008a] for a detailed discussion of the origin, nature, and solution of the two-culture problem.

16. In 1907, Cai Yuan-Pei went abroad to study in Germany (for 4 years) at age 40. He went again in 1913 to France at age 46 and stayed in Europe for 3 years. As president of Peking University (1916-1927), he reformed the university and started the tradition of putting "academic" and "freedom" first in a university education. In 1918 he pointed out that "university is a place purely for the pursuit of knowledge, not the place to produce professional certificates nor the place for selling knowledge. A scholar should be interested in researching knowledge and should train oneself to be a knowledgeable person with high character" (http://baike.baidu.com/view/2008.htm?fr=ala0_1_1#3, Aug. 28, 2010). In 1928 he established the National Academy of Art (now called China Academy of Art) in Hangzhou and the National Academia Sinica (the forerunner of the Chinese Academy of Sciences) in Nanjing. Cai was president of the Academia Sinica from 1928 to 1940; went to Hong Kong from Shanghai in 1937 due to Japanese invasion; died of illness in Hong Kong in 1940 (http://zh.wikipedia.org/zh-cn/蔡元培, Aug. 28, 2010).

17. Here an algebraic proof is given. Let $x = 1/2 + 1/4 + 1/8 + 1/16 + ...$ Then $x = \frac{1}{2} + \frac{1}{2} (1/2 + 1/4 + 1/8 + ...) = \frac{1}{2} + \frac{1}{2} x$. Thus $x (1 - \frac{1}{2}) = \frac{1}{2}$, resulting in $x = 1$. QED.

References

Berkowitz, J. [2012]. *The Stardust Revolution: The New Story of Our Origin in the Starts*. New York: Prometheus Books.

Buckingham, W., Burnham, D., Hill, C., King, P. J., Marenbon, J. & Weeks, M. [2011]. *The Philosophy Book: Big Ideas Simply Explained*. New York: DK Publishing.

Burguete, M. & Lam, L. (eds.) [2013]. *Renke: Humanities as Complex Systems*. Beijing: Renmin University of China Press.

Burguete, M. & Lam, L. [2016]. The Scimat Program: Towards a better humanity. *Humanities as Science Matters: History, Philosophy and Arts*, Burguete, M. & Riesch, H. (eds.). Luton, UK: Pantaneto Press.

Churchland, P. [2007]. *Neurophilosophy at Work*. Cambridge: Cambridge University Press.

Churchland, P. S. [1989]. *Neurophilosophy: Toward a Unified Science of the Mind-Brain*. Cambridge, MA: MIT Press.

Collingwood, R. G. [2008]. *An Essay on Philosophical Method*. Oxford: Clarendon Press.

Gong, Yu-Zhi [2005]. *Natural Dialectics in China*. Beijing: Peking University Press.

Harris, S. [2010]. *The Moral Landscape: How Science Can Determine Human Values*. New York: Free Press.

Kim, J. [2005]. *Physicalism, or Something Near Enough.* Princeton: Princeton University Press.

Knobe, J. & Nichols, S. (eds.). [2008] *Experimental Philosophy.* Oxford: Oxford University Press.

Kolak, D., Hirstein, W., Mandik, P. & Waskan, J. [2006]. *Cognitive Science: An Introduction to Mind and Brain.* New York: Routledge.

Lam, L. [1998]. *Nonlinear Physics for Beginners: Fractals, Chaos, Solitons, Pattern Formation, Cellular Automata and Complex Systems.* Singapore: World Scientific.

Lam, L. [2002]. Histophysics: A new discipline. Modern Physics Letters B 16: 1163-1176.

Lam, L. [2004]. *This Pale Blue Dot: Science, History, God.* Tamsui: Tamkang University Press.

Lam, L. [2005]. Active walks: The first twelve years (Part I). Int. J. Bifurcation and Chaos 15: 2317-2348.

Lam, L. [2006]. Active walks: The first twelve years (Part II). Int. J. Bifurcation and Chaos 16: 239-268.

Lam, L. [2008a]. Science Matters: A unified perspective. *Science Matters: Humanities as Complex Systems*, Burguete, M. & Lam, L. (eds.). Singapore: World Scientific. pp 1-38.

Lam, L. [2008b]. Science Matters: The newest and biggest transdicipline. *China Interdisciplinary Science*, Vol. 2, Liu Zhong-Lin (刘仲林) (ed.). Beijing: Science Press. pp 1-7.

Lam, L. [2008c]. Human history: A science matter. *Science Matters: Humanities as Complex Systems*, Burguete, M. & Lam, L. (eds.). Singapore: World Scientific. pp 234-254.

Lam, L. [2011]. Arts: A science matter. *Arts: A science matter*, Burguete, M. & Lam, L. (eds.). Singapore: World Scientific. pp 1-32.

Lam, L. [2014]. About science 1: Basics—knowledge, nature, science and Scimat. *All About Science: Philosophy, History, Sociology & Communication*, Burguete, M. & Lam, L. (eds.). Singapore: World Scientific. pp 1-49.

Lam, L. [2017]. Humanities, Science, Scimat: A new general-education course. *Interdisciplinarity and General Education in the 21st Century*, Burguete, M. & Connerade, J.-P. (eds.). Cascais, Portugal: Science Matters Press.

Lam, L., Bellavia, David C., Han Xiao-Pu, Liu Chih-Hui A., Shu Chang-Qing, Wei Zhengjin, Zhou Tao & Zhu Jichen [2010]. Bilinear effect in complex systems. EPL 91: 68004.

Liu, Da-Chun (刘大椿) (ed.) [2003]. *Development of Humanities and Social Science at Renmin University of China, 2002.* Beijing: Renmin University of China Press.

Liu, Jin-Yang (刘劲杨) [2008]. *Complexity: A Philosophical View*. Changsha: Hunan Science and Technology Press.

Marx, K. & Engels, F. [1979]. *The Complete Works of Marx and Engels*, Vol. 42. Beijing: People's Press.

Morris, R. [2002]. *The Big Questions: Probing the Promise and Limits of Science*. New York: Henry Holt.

O'Carroll, F. [2014]. Gerry's paradox. Philosophy Now, Issue 104, Sept/Oct 2002: 53-54.

Russell, B. [(1912) 2010]. *The Problems of Philosophy*. Simon & Brown (www.simonandbrown.com).

Russell, B. [(1945) 1972]. *The History of Western Philosophy*. New York: Simon & Schuster.

Savater, F. [2002]. *The Questions of Life: An Invitation to Philosophy*. Cambridge, UK: Polity.

Shermer, M. [2015]. *The Moral Arc: How Science and Reason Lead Humanity toward Truth, Justice, and Freedom*. New York: Henry Holt and Company.

Steinbauer, A. [2014]. Step out of my sunlight! Philosophy Now, Issue 104, Sept/Oct, 2014: 4.

Sun, Xiao-Li [2004]. *The Synthesis of Humanities and Science: A Scientific Trend in the 21ˢᵗ Century*. Beijing: Peking University Press.

Turner, M. S. [2009]. The universe. Sci. Am., Sept. 2009: 36-43.

Waal, F. B. M. de [2015]. Hard-wired for good? Science **347**(6220): 379.

Ye, Xiu-Shan (叶秀山) [2007]. Philosophy as the spiritual home. *Humanities General-Education Lectures: Philosophy (I)*, Lu Ting (陆挺) & Xu Hong (徐宏) (eds.). Beijing: Culture and Art Publishing House. pp 1-22.

Published: Lam, L. [2016]. Philosophy, science, Scimat. *Humanities as Science Matters: History, Philosophy and Arts*, Burguete, M. & Riesch, H. (eds.). Luton, UK: Pantaneto Press. Based on a talk given at the *High Forum on Natural Dialectics and Advancement of Science*, Renmin University of China, Beijing, May 27, 2011.

19

Philosophy: East and West

Lui Lam

Chinese philosophy differs from Western philosophy in many ways. In particular, Western philosophical propositions are clearly written, an essential element enabling meaningful debates, while fuzziness is the norm in Chinese philosophy. Here, the two are compared and the reasons behind them are explained. Some important philosophers in ancient Greece and China are introduced. The Scimat perspective on philosophy in general is presented. Finally, three topics are discussed: the Zeno's paradox, the science of Fuzzyism, and how some philosophers consider the role of "philosophy" today.

1 Nomenclature

The word Philosophy today is not the Philosophy in the ancient Greek times. The former is a subset of the latter and so should not be written identically; we denote it by "Philosophy" (double-quotation marks) (Fig. 1).

In more detail, the ancient Greek Philosophy, invented by Thales (c.624-c.546 BC) and named by Pythagoras (c.570-c.495 BC), is about everything in the universe and was the only discipline then. Later, in the 14th century, Philosophy was divided into three parts: 'Philosophy' (single-quotation marks), Natural Philosophy, and Theology. 'Philosophy' is the study of the human system; Natural Philosophy, nonhuman systems; Theology, everything related to God (then supernatural) under the premise of God's existence. The content of 'Philosophy' and Natural Philosophy, respectively, contains two parts: "no God" and "invoke God."

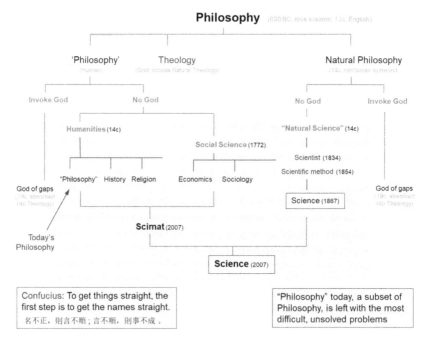

Fig. 1. A brief history of the disciplines. Medical science is included in Scimat.

Subsequently, the no-God part of 'philosophy' was divided into the Humanities and Social Science, and the no-God part of natural philosophy (also known as "natural science") was formally defined as Science in 1867 [Lam 2014, 2024]. The invoke-God parts in 'philosophy' and natural philosophy are incorporated into theology. The humanities contain "Philosophy" (double-quotation marks)—the kind of philosophy currently found in universities' philosophy department.

In short, Philosophy > 'Philosophy' > "Philosophy."

2 "Philosophy" Today

The differences of "Philosophy" and its relevance to ordinary people in the West and China today are summarized in Table 1. West = Europe + USA + Canada; each country is like a province in China, with same cultural system. Presently, the main cultures in the world are exemplified, respectively, in the West, China, India, Middle East, Japan, etc.

Table 1. "Philosophy" and its relevance in the West and China.

West	China
"Philosophy" is mostly about intractable, very difficult questions (truth, reality, etc.), but also some pressing, relevant questions such as "justice"—related to maintaining social stability.	Confucianism (because there is a vacuum in ethics/morality to fill; in West, filled mostly by Christianity)
USA invented Pragmatism.	China invented *Fuzzyism*.
Ordinary people in the West don't care much about "Philosophy" (except in France where high school students still have to pass a philosophy exam before graduation, a requirement laid down by Napoleon).	Care.

Fuzzyism is the style of mainstream philosophy in China, starting with Confucius. There are two possible reasons for the ambiguity: (1) It cannot be explained clearly because the author has not thought it through. (2) It is inconvenient, for political reasons or otherwise, to explain clearly. Reason 1 is more likely than reason 2.

Naturally, Fuzzyism won't solve any problem; it just delays the problem, allowing the problem to reappear, again and again, and so history just repeats itself. And this is why the German philosopher Georg Hegel (1770–1831) says "Chinese history is essentially historyless, it is just a repetition of the overthrow of the monarch, and no progress can be produced from it," if history, according to his teleological account, must involve progress in the social system or humanistic values.

In the West, starting with the ancient Greek times, philosophy strives to understand the universe, human included. The method to do philosophy then is to express it clearly and through face-to-face debates, culminating in the so-called Socratic method [Lam 2024]. After hundreds of years of empty talks in Europe's modern philosophies, Pragmatism—meaning reality check of any theory—was invented in the United States, resulting in its dominant position in science and technology, among other things, as we see it today.

After science and other useful disciplines split away from Philosophy, "Philosophy" was left with the hard-to-crack topics (like truth and reality) to explore, without much success. That's why it is not in the minds of ordinary people in the West. In contrast, because Confucianism was the orthodox philosophy mandated by the Chinese emperors for almost 2,000 years, "Philosophy" in the form of Confucianism (and Daoism) is still cared by the Chinese people. More comparisons between philosophies in the West and China are given in Sec. 4 below.

3 Ancient Philosopher

Some important philosophers in the classical period of Greek philosophy and Chinese philosophy are introduced here. The lifespan of each one of them is given in Fig. 2.

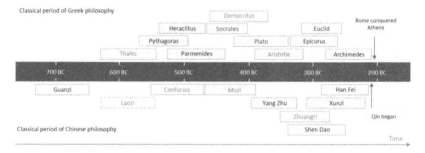

Fig. 2. Lifespan of philosophers in ancient Greece and China. The existence of Laozi as a real person is uncertain.

3.1 Greek philosopher

In ancient Greece, three successive generations of mentor-student relationships with each student overthrowing the philosophical thesis of the mentor, occurred *two* times: (1) The Milesian school: Thales → Anaximander → Anaximenes. (2) Socrates → Plato → Aristotle (Fig. 3). This sets in the tradition in the West that a good student is not the one who inherits and propagates the mentor's teaching but to overthrow it; i.e., innovation is the signature of good scholarship. (This tradition is absent in China, starting with Confucius and his students.)

An interesting philosopher who lived between Thales and Aristotle, though less famous than them, is Democritus (Fig. 3). His argument, called "Democritus' cone," for the existence of atoms, confirmed more than 2,000 years later in 1905 by Einstein, is ingenious and still valid today.

| Thales
(c.624-c.546 BC) | Democritus
(c.460-c.370 BC) | Aristotle
(383-322 BC) |

Fig. 3. Three ancient Greek philosophers. Thales is the father of philosophy (and science) [Lam 2024]. The red dot in the map is Miletus in Greek Ionia (now in Turkey), where Thales was born.

Democritus' cone

Democritus' argument: If a cone is divided by a plane parallel to its base, are the surfaces of the segments equal or unequal? If they are equal, then the cone becomes a cylinder; if they are unequal, then the surface of the cone must be stepped. The fact that the cone is not a cylinder implies all matter is composed of small indivisible particles, called atoms (Fig. 4).

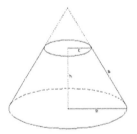

Fig. 4. The Democritus' cone.

3.2 Chinese philosopher

Four important Chinese philosophers during the Spring and Autumn period (770-476 BC) and the Warring States period (475–221 BC) are presented here.

1. Confucius

Confucius (551-479 BC) (Fig. 5) is a political scientist (at the empirical level; see Sec. 5), concentrating on ethics/morality with the aim of maintaining the social order [Lam 2024]. His saying "What you do not wish for yourself, do not do to others" sounds good and innocent, something like from the Bible. But his other saying "Officials obey the emperor (absolutely); sons obey fathers (absolutely)" is lethal, which underpins the feudal system in China for 2,000 years since the Han dynasty (AD 25-220) and is still practiced by people today.

The "tian" (天, Heaven) in Confucianism and other Chinese philosophies is a supernatural, not the physical sky, since it somehow knows how the emperor rules and punishes him (the country, in fact) by administering, for example, flood/famine. With supernatural and rituals, Confucianism is a religion.

Fig. 5. Four Chinese philosophers. *Top*: Confucius (right); *Bottom*: Mozi (left), Laozi (middle), and Zhuangzi (right). The red dot in the map is Lu state (now in Shandong Province) during the Spring and Autumn period, where Confucius was born.

2. Laozi

Laozi (6th century BC) (Fig. 5) is a mystery, whose existence is an open question [Hu (1919) 2018]. Under debate by academic scholars are the questions: Did he really exist? If so, did he live before or after Confucius?

In any case, judging from the writings in the book *Tao Te Ching* (*Daodejing*, 道德经) attributed to him, Laozi is a master of Fuzzyism. For example,

> The Way that can be told of is not an unvarying way;
> The names that can be named are not unvarying names.
> It was from the Nameless that Heaven and Earth sprang;
> The named is but the mother that rears the ten thousand creatures,
>   each after its kind.

This famous saying is so ambiguous, resulting in multiple interpretations in the literature with no way to settle on a definite answer. Why didn't the author write it clearer?

His other writings are clearer but still fuzzy, such as "Try to change it and you will ruin it; try to hold it and you will lose it" and "Those who know do not say; those who say do not know."

3. Mozi

Mozi (c.476 – c.390 BC) (Fig. 5), a peasant and carpenter skilled in creating devices, apart being a philosopher, was a pioneer in science and technology. He invented mechanical birds and mobile "cloud ladders" used to besiege city walls. He developed principles of logic and his works on optics, recorded in *Mozi*, are still good today. He kept asking "how" and "why" on things, as a good scientist would do.

Differing from Confucianism that encourages love among relatives, Mozi advocated universal love (兼愛): "We begin with what is near." However, he is not a rebel. Like Confucius, he aimed to improve and preserve the kingdom. He proposed that every level of officials collect people's thinking and pass it up one level, eventually to the King so he can make wise decisions.

Mohism was a strong contender against Confucianism then. Unfortunately, it was suppressed in the Qin dynasty and died out in the Han dynasty. While politically conservative, Mozi is incisive and creative—a successful innovator in science and technology. His approach to the world is rational-empirical—very modern. Mozi, not Confucius, is most relevant to China today!

4. Zhuangzi

Zhuangzi (c.369 – c.286 BC) (Fig. 5) advocated skepticism, relativism (in systems of value), and anarchism. He said, "Good order results spontaneously when things are let alone"—the mechanism behind this is self-organization. He also said, "Life is limited and knowledge to be gained is unlimited." His thinking of interdependence of things fore-shadows modern ecological thinking.

The two stories attributed to him are hearsays: Butterfly in dream, and drumming after wife died.

His thesis that the authority for ethical judgments comes from dao (道) and not from tian (天), liberating people from superstitions, resonates with the idea of Renaissance that came about 1,800 years later in Europe.

Furthermore, the writing in *Zhuangzi*, "Heaven, earth and I are born of one, and I am at one with all that exists" (天地与我并生，万物与我为一), could lead to the most important concept in Chinese philosophy: Heaven-Man Oneness (天人合一), proposed by Dong Zhong-Shu (董仲舒), a Confucian in the West Han dynasty.

In my interpretation, the Oneness means Heaven and man share the same organizing principles and can influence each other, which obviously is correct. The bearing in answering the so-called Needham Question—Why modern science didn't arise from China despite its past success in science/technology?—is discussed elsewhere [Lam 2014, 2022, 2024].

4 Ancient Philosophy: East and West

The basic differences between the ancient philosophies in Greece and China are in the degree of freedom in choosing topic, the topics chosen, and the freedom to argue/debate among the philosophers as well as between the mentor and disciple. And there are economic reasons behind these, as explained in Table 2.

Note that Philosophy in the Greek times was the only discipline of learning in the world, which actually was very successful as can be seen from the fact that all disciplines today (including science) did branch out from it.

In contrast, it is hard for the ancient Chinese to make progress since they neither wrote clearly nor debated rigorously. What the philosophers uttered are very general and vague, amount to "chicken soup" sometimes (called Chinese wisdom by others). When pressed, the philosopher will appeal to tradition or the will of Heaven. Yet, it is "useful" to a certain extent; it seems to benefit the emperors since the longest dynasty—the Tang dynasty (AD 618–907)—did last 289 years.

Table 2. Differences in ancient Philosophy in Europe and China.

Item	Europe	China
Contents	About anything in the universe (human included)	Mostly about morality/ethics
Aim	Care about everything in daily life (and the rest of universe); want to understand thoroughly and solve problems; find out the "truth" and advance knowledge	Maintain social harmony/stability, except Zhuangzi (who worships freedom)
Degree of freedom	Freedom of speech (living comfort and leisure supported by slavery) in a democracy	Lack of freedom of speech in feudal kingdoms
Topic picked	Ancient Greek philosophers didn't need a regular job, and so were free in picking topics in pursuing acknowledge.	The only job available to ancient Chinese intellectuals is to be officials in the government, and so they concentrated in designing ways in helping the King in political matters, except Zhuangzi (who was not poor).
Approach	Analytical (philosophers express themselves clearly)	Fuzzy/circular arguments (philosophers never wrote clearly or argued convincingly)
Research method	Debate (Socratic method)	No (or not much) debate; read literature, think alone, and teach/write

5 The Scimat Perspective

"Philosophy" is built upon or be consistent with the available scientific knowledge. "Philosophy," like any other discipline, has three research levels: empirical, phenomenological, and bottom-up [Lam 2002, 2011]. Here are some examples.

The philosophy of Immanuel Kant (1724-1804) is at the empirical level. In Kant's times, science was Newton's science of deterministic systems: Results are certain, given initial conditions. Thus, after noticing that human's free will is not deterministic and assuming that the world must be rational and meaningful, he finds it necessary to introduce the faith system (i.e., religion) into his philosophy.

Kant's argument is full of metaphysical assumptions [Lam 2016]. Like many other philosophies, Kant's philosophy is outdated. Note that Kant's philosophy was rated unfavorably by the two philosophers, Robin Collingwood (1889-1943) and Bertrand Russell (1872-1970) [Lam 2024].

Another empirical-level example is Experimental Philosophy (Fig. 6, left two). They conduct systematic experiments to understand people's intuitions about philosophically significant questions. It succeeds in challenging a number of cherished assumptions in both philosophy and cognitive science.

Fig. 6. Four books on contemporary "philosophy." *Left to right*: *Experimental Philosophy* (2008), *Experimental Philosophy* (2012), *Neurophilosophy* (1989), and *Braintrust* (2012).

Neurophilosophy is at the bottom-up level. It tries to link the classical mind-body problem to functional neurobiology and in the connectionist models within artificial intelligence research. It seems to be an ongoing project.

Finally, we are not aware of any philosophical works done at the phenomenological level. Experimental Philosophy and Neurophilosophy are two of the few examples of doing philosophy through humanities-science synthesis. And this is the approach recommended by Scimat [Lam 2008, 2024].

6 More

Three more topics in philosophy are presented here.

6.1 The Zeno's paradox

The Chinese philosopher Zhuangzi (c.369-286 BC) once said: A one-foot long stick, take half each day, and you will never exhaust it. Before him and independently, the Greek philosopher Zeno (c.490-430 BC) said the same thing: A man could not move from point A to point B because, starting from point A, he had to move to the half point first, then the half point of the rest distance, ad infinitum; and because there is an infinite number of steps, he could never reach point B. This is called the "Zeno's paradox" (Fig. 7).

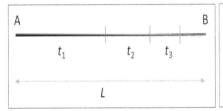

$$t = t_1 + t_2 + t_3 + ...$$

$$t_1 = \tfrac{1}{2} L/v, \ t_2 = \tfrac{1}{4} L/v, \text{ etc. } (v = \text{velocity})$$

$$t = L/v \left(1/2 + 1/4 + 1/8 + 1/16 + ...\right) = L/v$$

Fig. 7. The Zeno's paradox. The total time of travel t is a finite number, even though there is an infinite number of terms in the sum. Zeno's mathematical mistake is that he thought the sum of an infinite number of terms in a series must be infinite ($t = \infty$). Zeno's procedure for a zero-size particle cannot be carried out in practice because real objects have finite size.

Both of them are talking a person (or particle) of zero size, and that is the crux of the fallacy: A human body has finite size (since it is made up of atoms and atoms have finite size no matter how small), and so the observation by Zeno/Zhuangzi, applicable to a zero-size particle, cannot be applied to the real world.

This is a common mistake committed by many philosophers that the ideal world presupposed in their proposition will never match the real world. That is why so many philosophical theories (called metaphysics) are wrong and why Pragmatism, as a philosophy, was invented.

6.2 The science of Fuzzyism

Fuzzyism, as a philosophy and management style of more than 2,000 years in China, should be taken seriously and studied scientifically. Possible questions to explore: How, when and why it works? How long to wait before seeing it work? When will it fail? The way to study it, in our opinion, is through the use of active walk modeling. *Active walk* is a paradigm for complex systems invented by Lam in 1992 [Lam 2005, 2006].

An active walk model consists of a particle coupled to a (real or abstract) deformable potential, called the landscape. In an active walk, the walker (the particle) changes the landscape at every step; the changed landscape affects the next step of the walker (Fig. 8). However, the landscape can also be modified by external forces—the changing environment.

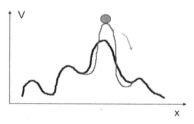

Fig. 8. An active walk model.

As management style is concerned, the Western analytic approach tends to rely on the walker's action. The Chinese Fuzzyism approach is to minimize immediate action and wait for the landscape to change favorably later, giving the impression of inaction or burying the contradictions. Obviously, this runs the danger of prolonging the bad situation and suffering if the favorable conditions take a long time to come or may never come.

6.3 Philosophers on "Philosophy"

We concur with Bertrand Russell (1872-1970) when he writes in *The Problems of Philosophy* (1912):

> Most philosophers...profess to be able to prove, by a priori metaphysical reasoning, such things as the fundamental dogmas of religion, the essential rationality of the universe, the illusoriness of matter, the unreality of all evil, and so on....This hope, I believe, is vain. [Russell (1912) 2010: 99]

But we disagree with his claim:

> Philosophy is to be studied, not for the sake of any definite answers to its questions, since no definite answers can, as a rule, be known to be true, but rather for the sake of the questions themselves. [Russell (1912) 2010: 112]

This was definitely not the attitude of the ancient Greek philosophers [Steinbauer 2014]—the pioneers, nor the late comers like Kant, even though it could be the attitude shared by many contemporary "philosophers" when they are left to deal with the most difficult problems not yet hijacked by "natural science."

7 Conclusion

1. "Philosophy" today is a small subset of ancient Greek's Philosophy; this distinction must be made.

2. Ancient Western and Chinese philosophies differ in their choice of topics and research methods.

3. Philosophical concept/construction is built upon the best available scientific knowledge at the time they are proposed.

4. Thus, contemporary philosophers have to be aware of the current scientific results.

5. Students in philosophy should have some scientific training.

6. Philosophers are strongly urged to collaborate with others in natural science and medical science.

References

Hu, Shih (胡适) [(1919) 2018]. *Outline of the History of Chinese Philosophy*. Beijing: Zhonghua Book Company.

Lam, L. [2002]. Histophysics: A new discipline. Modern Physics Letters B **16**: 1163-1176.

Lam, L. [2005]. Active walks: The first twelve years (Part I). Int. J. Bifurcation and Chaos **15**: 2317-2348.

Lam, L. [2006]. Active walks: The first twelve years (Part II). Int. J. Bifurcation and Chaos **16** ; 239-268.

Lam, L. [2008]. Science Matters: A unified perspective. *Science Matters: Humanities as Complex Systems*, Burguete, M. & Lam, L. (eds.). Singapore: World Scientific. pp 1-38.

Lam, L. [2011]. Arts: A science matter. *Arts: A science matter*, Burguete, M. & Lam, L. (eds.). Singapore: World Scientific. pp 1-32.

Lam, L. [2014]. About science 1: Basics—knowledge, nature, science and Scimat. *All About Science: Philosophy, history, sociology & communication*, Burguete, M. & Lam, L. (eds.). Singapore: World Scientific. pp 1-49.

Lam, L. [2016]. Philosophy, science, Scimat. *Humanities as Science Matters: History, Philosophy and Arts*, Burguete, M. & Riesch, H. (eds.). Luton, UK: Pantaneto Press.

Lam, L. [2022]. *Science and Scientist*. San Jose: Yingshi Workshop.

Lam, L. [2024]. *Humanities, Science, Scimat*. Singapore: World Scientific.

Pfaff, D. W. [2015]. *The Altruistic Brain: How We Are Naturally Good*. Oxford: Oxford University Press.

Russell, B. [(1912) 2010]. *The Problems of Philosophy*. Simon & Brown (www.simonandbrown.com).

Steinbauer, A. [2014]. Step out of my sunlight! Philosophy Now, Issue 104, Sept/Oct, 2014: 4.

Waal, F. B. M. de [2015]. Hard-wired for good? Science **347**(6220): 379.

Unpublished: Lam, L. [2015]. Philosophy. Lecture presented on July 23, 2015, in my general-education course "Humanities, Science, Scimat," International Summer School 2015, Renmin University of China, Beijing.

PART V

SCIENCE

About Science 1: Basics
—Knowledge, Nature, Science and Scimat

Lui Lam

There is a lot of confusion and misconception concerning Science. The nature and contents of science is an unsettled problem. For example, Thales of 2,600 years ago is recognized as the "father of science" but the word science was introduced only in the 14th century, and so it is obviously wrong if science is understood as modern science only, which started with Galileo about 400 years ago. If science is mainly about nonliving systems, then social science cannot be part of science. And if social science is part of science, then why the humanities, which are also about humans, are not part of science? All these confusions and dilemmas concerning science could be traced to the historical evolution of the word and concept of Science and the many misconceptions perpetuated by various philosophers and historians of science, due to the lack of an agreed-upon definition of science.

This chapter aims to clear up all these confusions by retracing the historical development of Science—the word, concept, and practice. The nature of knowledge, nature, religion, and philosophy are covered. A simple definition of Science according to Scimat, the new multidiscipline that treats all human-system studies as part of science, is provided. Three important lessons learned about science, including the required Reality Check (which differentiates science from other forms of knowledge) are given. Important ramifications from this definition concerning antiscience and pseudoscience in particular are discussed.

1 Introduction

Science is one of the three pillars that support an advanced civilization, East and West. While the other two pillars, ethics/religion and arts, have an extremely long history of at least one million years [Lam 2011], science, counting from the days of Thales (c.624-c.546 BC)—the "father of science"—has a "short" history of only about 2,600 years. Short as it is, it is long compared to the span of modern science, a mere 400 years or so since Galileo (1564-1642).

While the tremendous success of modern science did lead to positive results (and important applications like the cell phone), unfortunately, it also led to all sorts of confusion among the philosophers, historians, sociologists, and communicators whose works are related to science; e.g., the definition of science is often avoided in philosophy of science books [Oldroyd 1986; Godfrey-Smith 2003]. The confusions concern three aspects of science: (1) the contents of science, (2) the existence and nature of the different stages in the development of science, and (3) the scientific research process.

1. The crux of the problem is that the concept and practice of science are not constants but have evolved over time, with many twists and turns, contributing to the *lack of an agreed-upon definition of science*. Thus, most statements made on science more than 56 years ago (the year 1957, see Sec. 7.1) by scholars and even by scientists turn out to be no longer valid. To clear up the confusion, the historical development of science over the last 2,600 years since Thales is retraced. And since the essence of science is to gather knowledge about nature, the idea of nature and the nature of knowledge are reexamined, too. It turns out that, based on our current scientific knowledge and the historical record, two simple conclusions are reached: (1) Humans are part of nature, and (2) the aim of science was never to challenge the existence of God but merely to see how far humans can go in understanding what is around them by reasoning without appealing to God. These two recognitions, missed by many others, form the premise of *Scimat* (Science Matters), the new multidiscipline initiated by Lam in 2007 [Lam 2008a, 2008b].

2. In Sec. 2 below, two kinds of human knowledge and the *Knowscape*, a metaphor for the landscape of knowledge, are introduced. The rationale behind Scimat, which treats all human-system studies (covered in the humanities and social science) as part of science, and the importance of the humanities are presented in Sec. 3. And since science first appeared (without the name) as part of philosophy in early Greek times, which was human's first attempt to break away from superstitions (the Greek gods/goddesses in this case), the connection between religion and philosophy is examined in Sec. 4. The historical developments of the idea of nature and of the idea of science (a very recent concept) follow (Sec. 5).

3. In Sec. 6, the very definition of science and of scientist according to Scimat are presented, which are simple, clear and historically correct. *Science Room*, the metaphor for science, is also included. Section 7 outlines how science was actually done, for both simple (mostly inanimate) systems and complex systems (which include human and nonhuman animate systems). The contents of this Section heavily influence our discussion on the philosophy, history, sociology, and communication of science, presented in [Lam 2014]. Three important lessons learned about science are given in Sec. 8. A summary of the spirit and essence of Scimat (in the format of Q & A), and the ramifications from the basics about science are given in Sec. 9. Finally, Section 10 concludes with discussion and a take-home message.

2 Human Knowledge and the Knowscape

In spite of the many different opinions, we could probably agree that "Science is to understand nature."[1] Then what is nature? The present understanding is: Nature consists of all material systems including humans and (living and nonliving) nonhumans. That humans are part of nature is a relatively new recognition. It follows from Darwin's evolutionary theory [1859] that humans are a living system like the chimpanzees and evolved from the fish, say [Shubin 2009]. Humans are thus one of the many kinds of animals[2] and a material system, and, like all material systems, are made up of atoms. Note that the existence of atoms was established only about 100 years ago due to Einstein's work on Brownian motion [1905].

Consequently, most discussions on the contents of science published 150 years ago or before are simply wrong or misleading, resulting in many misconceptions.

All knowledge about nature accumulated through science by humans could be divided into two parts: A human-independent part[3] and a human-dependent part [Lam 2008a]. An example of the former is the law of gravity, which could be discovered by aliens, too, if they exist. The latter includes knowledge of human-made materials such as semiconductors and all the topics in social science and the humanities.[4]

To facilitate discussion in this chapter, we introduce the *Knowscape*, the landscape of knowledge. It is meant as a metaphor, and, as a metaphor it is far from perfect and has its limitations; e.g., not everything in the picture should be taken literally. In the Knowscape (Fig. 1) there are hills/mountains, valleys, and plateaus, all linked to each other in a vast terrain, and some human-made lakes. Each hilltop represents a highlight in human knowledge; the height of the hill corresponds to the difficulty of reaching it. And, as in real mountain climbing, the explorer usually has to pass a lower hill before reaching a higher one. A researcher (whether you call her/him a scientist or not) is the explorer.

However, there are two types of mountains. One type is human independent which is out there whether humans exist or not (and could be found by aliens, say), represented by the upper curve in Fig. 1. The other type is human dependent, the knowledge found in the humanities and social sciences (lower curve). The isolated lakes represent artificial, nonhuman systems (such as semiconductor, computer, and artificial life) which are human dependent, too (ellipses).

3 Scimat 1: The Humanities

And so we have these three recognitions:

1. Science is to understand nature.
2. Nature includes all material systems.
3. Humans are a material system.

The logical conclusion derived from these three statements is that "all things related to humans are part of science." What are "all things"?

Let us consider bees which, like humans, are a kind of animal. When we say "all things related to bees" we mean the biological property of a single bee, the behavior of a single bee or a group of bees living together (how they communicate with each other; how they divide jobs among themselves; how their society is organized; etc.), the competition or cooperation between different groups of bees, and so on. Everything! The same goes for chimpanzees. And so, the same goes for humans.

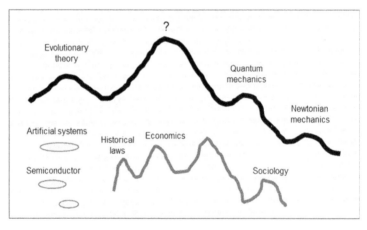

Fig. 1. The *Knowscape*: landscape of knowledge. The upper curve represents the human-independent part (nonhuman systems; so-called "natural science"); lower curve, human-dependent part related to human matters (human system; the humanities, social science, and medical science); ellipses, human-dependent part related to artificial systems.

Historically, the study of different aspects of humans was classified into medical science (including human biology), social science, and the humanities. Medical science is about the biology of a single human or a group of humans; an example of the latter is epidemiology. Social science (e.g., Economics and Sociology) is the study of some, but not all, aspects of a group of humans. Cultural study, a branch of the humanities, is about a group of humans, too. It is thus clear that the division of social science and the humanities is not according to the number of humans under study.

Rather, as pointed out before [Lam 2008a: 13], it is due to the scientific level achieved in these two large group of studies. Yet, for some people, it is due to the belief that the humanities can *never* be a part of science, either (1) because humans are so complex that it cannot be handled by science or (2) because humans are fundamentally different from the bees and chimpanzees.

These issues form the core of this chapter and will be clarified later. At this point, let us note that this classification of human studies was established well before the times of Charles Darwin (1809-1882) and Albert Einstein (1879-1955), and the separation of the humanities from science has been maintained, for many people, even today. As shown below, it is due mostly to the misconceptions about science.

Why is it crucial to recognize the humanities, as it should, as part of science? The answer lies in the importance of the humanities. This point could never be overstated even though it is usually overlooked in every country, perhaps with the exception of France where Philosophy is a required examination for graduating high school students since the Napoleon days. The humanities' importance could be seen from these two considerations:

1. If all the present research projects in "science"[5] were frozen or eliminated, the world would still be the same—chaos and tragedies would continue—because it is the humanities which include decision making, underdeveloped in the last 2,400 years since Plato (427-347 BC), that matter in human affairs.[6]

2. Apple company is successful because they put a humanist, Steven Jobs (1955-2011), in charge of the engineers—good for the economy.

Further discussion on the humanities is given in Sec. 10.

Scimat is the multidiscipline initiated by Lam [2008a, 2008b] that treats all human-dependent matters, the humanities in particular, as part of science. (See Secs. 6 and 9 for more discussion of Scimat.) The Scimat Program[7] is the latest concerted (international) effort in reviving the Aristotelian tradition of treating human and nonhuman systems alike in the pursuit of knowledge. (See Fig. 3 and Sec. 10 for a discussion of the past efforts that failed.)

That such a simple and almost trivial conclusion that the humanities are part of science is not immediately and universally accepted by every learned person is at first puzzling. It turns out that the reasons lie in people's understanding of what science is and what the word Science represents to them. A little bit of research shows that (1) not just laypersons but even some good "scientists" hold the wrong ideas about science, (2) the root of the problem is mostly historical, and (3) misconceptions about science could be traced to the inadequacies in the four disciplines concerning science, viz., the philosophy, history, sociology, and communication of science. Below, the historical root of the problem is discussed; each of these four disciplines is examined in the next chapter [Lam 2014].

4 Religion and Philosophy

The human and chimpanzee lineages split from each other six million years ago. Four million years later *Homo erectus* appeared and already possessed the ability of *mimesis* [Donald 2006]. Then, 1.6 million year ago, fire and complex stone tools were invented. It is thus not hard to imagine that about a million years ago, our ancestors though primitive were sophisticated enough to think and wonder about things they saw and the happenings in their lives. They might ask: Why does the sun rises and disappears every day? Why am I sick and recovered but not the one next to me? They might even ask: Who am I? This was likely to happen because, apart from hunting and mating, there was plenty of leisure time; there were no televisions or football games to watch [Lam 2011]. Besides, curiosity helped in survival [Lam 2004: 36].

We do not know what answers they came up with since there was no record to show; writing had not been invented yet. But we do know what answers their descendants, *Homo sapiens* who appeared in the scene 195,000 years ago, came up with. More precisely, we mean the later generations; their answer: Everything is due to "something out there"—the supernatural.

This is understandable. In the absence of scientific knowledge at that time, it is natural to explain everything by using analogies and lessons learned in their daily lives. Consequently, the ascent and descent of the sun or the moon is governed by a human-like god, like the way a piece of rock could be moved by a human being. When one is sick, without the benefit of any medical knowledge, an easy explanation is that a certain god was offended. To get well again, the god's anger had to be removed, and that could be done by bribery—in the form of animal or human sacrifices—in the same way that it works with humans themselves.

What we call superstition was refined by the ancient Greeks in the form of mythology, with numerous gods with specific names. For example, Apollo takes care of the Sun while multitasking in light, knowledge, music, healing, and the arts; Boreas, the north wind; Notus, the south wind; Zephyrus, the west wind; Apheliotes, the east wind. Similarly, human matters are governed by gods; e.g., Eros is the god for love and sexual intercourse; Athena, the goddess for intelligence, skill, and wisdom [Buxton 2004]. Two points about this "theory" of gods: (1) It is consistent with everything they know at that time.[8] (2) It treats human and nonhuman systems alike, by the same mechanism.

This romantic theory of gods is clumsy and suffers from the lack of evidence and predictive power, even though predictive power is not the necessary quality of a new theory. In fact, the gods hypothesis was already being criticized in the 6th century BC by what is later called philosophers[9] who preferred to explain natural phenomena in terms of natural causes. Thales is the most important and famous one who claims that "all things are water" and is said to have predicted the eclipse of the sun happening in 585 BC [Cornford 2004: 1]. But he also maintained that the universe is alive with soul in it, full of daemons or gods [Cornford 2004: 127; Lloyd 1970: 9]. Aristotle (384-322 BC) suggested that the inquiry into the causes of things began with Thales [Lloyd 1970: 1] who is the first to define general principles and set forth hypotheses, and has thus been dubbed the "father of Science."[10] (The word Science first appears in the 14th century; see Fig. 2.)

Meanwhile, the number of gods was reduced from many to one [Armstrong 1993], reflecting humans' desire for simplicity. This principle of simplicity remains in the core of modern science and is called Occam's Razor: What can be done with fewer is done in vain with more; i.e., the simpler the better (as long as the simpler works). With one God, organized religion as we know it today emerged.

Philosophy, starting with Thales, is the effort to understand and explain things in the universe—the nonhuman systems and even the human system[11]—through reasoning without bringing in the Greek gods. It has two parts: one part that God is not brought in explicitly and another part that God is purposively and explicitly put there [Buckingham et al. 2011]. Thales' water hypothesis is an example of the former; Aristotle's "unmoved movers" in *Metaphysics*, the latter [Collingwood 1960: 87]. It is the former (but not the latter) that is identified as science in modern times.

Philosophy is a transition from the "narrative" or "story telling" to "reasoning"—a big step forward. Note that what the ancient Greeks abandoned are the primitive supernatural and the relatively simple gods; they did keep the more sophisticated supernatural ("soul") which actually is a central part of their knowledge system. In fact, philosophy *never* challenges the existence of God; it just assumes *a priorily* that part of the universe could be understood through reasoning (which turns out to be correct as shown by later developments). That this is at all possible is quite trivial since no one knows how God runs the universe after he creates it. For example, God could lay down the rules or laws and go fishing and comes back occasionally to burn down a city if he finds a sizable portion of the residents' attitudes are God-incorrect, or he could be a CEO who manages minutely every happening, big or small. In either case, philosophers, if they so wish, could easily claim that what they do is just reading the mind of God. And this explains why, e.g., Plato, Aristotle, and Immanuel Kant (1724-1804) could do their philosophy while still believing in God.

Since philosophy appeared after religion which includes mystic considerations, there exist in philosophy two lines of abstract speculations among pre-Socratic thinkers, leading to two traditions: scientific and mystical. The scientific trend puts the gods completely away in reasoning, reduce the "soul" to material particles, and concentrate in inanimate systems, more in line with modern science. It starts with the Milesian school (Thales, Anaximander, and Anaximenes) and leads to Democritus (c.460-c.370 BC) who proposes that everything is made of atoms. The mystical trend is "rooted in certain beliefs about the nature of the divine and the destiny of the human soul" and tries "to justify faith to reason." It is exemplified by Pythagoras and Plato [Cornford 2004]. The two traditions coexist and influence the development of philosophy for a long, long time, and are still among us today.

5 Nature and Science

Both the concept of nature and the concept of science evolved with time which, when ignored, resulted in many misconceptions and a lot of confusion among the scholars on science.

5.1 The idea of nature

The word "nature" has two meanings, same in ancient Greek times and modern times: (1) the sum total of natural things; (2) the principle (or source) governing natural things. It is mainly the first meaning, the contents of nature, which concerns us here. In particular, (1) Do the natural things include humans? (2) If so, do humans differ from other animals? (3) And if so, could the distinctions be explained completely on a material basis?

First issue first. In the 6th and 5th centuries BC, philosophers of the Ionian school (of which the Milesian school is a subset) believe that (1) there is such a thing as "nature"; (2) nature is "one"; (3) the thing which in its relation to behavior is called nature, is itself a substance or matter [Collingwood (1945) 1960: 46].

For Aristotle of the 4th century BC, nature is the essence of things which has a source of movement in them [Collingwood (1945) 1960: 81], which, therefore, should include humans. It is in later years that humans are excluded from the domain of natural things. This issue is settled with

Darwin's discovery [1859] that humans, like other animals, evolved from other more primitive creatures and organisms; *humans are thus part of nature.*[12]

The second issue of whether humans are distinct from other animals has two levels. At the first level, we all know that through adaptation and heredity humans have evolved to be quite distinct from other animals; e.g., our brains are larger; we have invented written language; etc.[13] [Suddendorf 2013; Pollard 2009]. The second level is more sophisticated: The existence of "soul," consciousness, and free will that are thought to be uniquely human (which could be wrong). This is fine. We all have a vague idea of what these are since we could experience it ourselves.

The third issue of where do soul and free will come from has been investigated intensively by philosophers, past and present, and more recently, by neurobiologists. The majority opinion currently is that they are emergent phenomena/features derived from the neurons and their connections [Gazzaniga 2012; Tse 2013]. In other words, the mind-body problem will be solved by neuroscience; it also means that it is not yet a settled issue.

5.2 The idea of science

The word Science first came into English in the 14th century and its present usage appeared even later, in 1867, soon after the Age of Revolution (1775-1848, such as the Industrial Revolution and the French Revolution). Science comes from Latin *Scientia*, meaning "knowledge" or "the pursuit of knowledge," and especially "established theories" when it first appears [Williams 1983; Ferris 2010: 3].

Since Thales is called the "father of science," naturally, Science (no matter what it means) should already exist in 600 BC. What was it called then? It was called philosophy; more precisely, it is that part of philosophy that supernatural is not mentioned explicitly (see Sec. 4).[14] For example, Aristotle's works on biology are science; his works on the human system such as ethics and arts, in our opinion, are also science (at the empirical level; see Sec. 7.1). Another example: The works of Archimedes (c.287-

c.212 BC) on buoyancy, the Archimedes' Principle, is obviously science, even by the present definition of the word.

However, in the 14th century the term "natural philosophy" appears and Philosophy is split into three parts: 'Philosophy', Theology (including natural theology), and Natural Philosophy (Fig. 2). Here, *'Philosophy'* with single-quotation marks means philosophy in a restricted sense, the study of deep questions about humans (e.g., ethics and metaphysics). *Natural theology* is a branch of theology based on reason and ordinary experience, in contrast to revealed theology based on scripture and religious experience.[15] *Natural philosophy* is the study of (living and nonliving) nonhuman systems and non-religious issues, which is divided into two parts: one part without invoking God and another part with God invoked. The former is later absorbed into "science"; the latter falls into the "God of the gaps"[16] category and later becomes part of theology. A noted example of the latter is provided by Isaac Newton (1642-1727) who, knowing well that the stars always attract each other due to gravity, has invoked God's active intervention to prevent them from sticking to each other—a notion that was ridiculed by his competitor, Gottfried Leibniz (1646-1716) [Alexander 1956: xvii].

Modern science, beginning with Galileo Galilei (1564-1642) who died the year Newton was born, has blossomed rapidly in the last 400 years, especially in physics (see Sec. 7). After Galileo, a major player in the so-called Scientific Revolution, Newton's (deterministic) mechanics was so successful that people in the *Enlightenment* (1688-1789) wanted to make human matters a science, too [Porter 2001]. They succeeded partially. Before the Enlightenment ended prematurely by the French Revolution (1789-1799), Adam Smith (1723-1790) published *The Wealth of Nations* [(1776) 1977], ushering in Economics, the first discipline in Social Science (Fig. 3).

By 1840, with enough number of people working full time in "science" the English scientist and theologian William Whewell (1794-1866) felt the need to coin the word "scientist" to describe these professionals; he was inspired by the word artist. After all, arts have been in existence long before science does, for at least 35,000 or perhaps a million years [Lam

2011]. And the word scientist was not fully accepted until the early 20th century [Ross 1962].

Somehow, the word "science" in its present usage (written with double-quotation marks in this chapter) was not in place until April, 1867 [Harrison et al. 2011: 2] even though the word science, as remarked above, appeared already in the 14th century. "Science" initially means the study of nonhuman systems (in 1867) but later expanded to mean the sum of "Natural Science"[17] and Social Science, after 'philosophy' split into two parts in 1772: the humanities and social science.

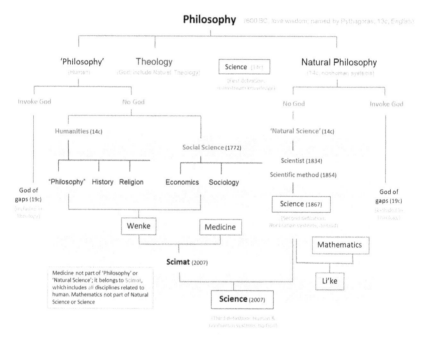

Fig. 2. A brief history of academic terms and disciplines, from Philosophy to "Science" and to Scimat. "Philosophy" and "Science", with double-quotation marks, correspond to philosophy and science, respectively, in their present, restrictive use (see text). The invoke-god part in Natural Philosophy is huge and frequently motivates the basic assumptions (such as Newton's absolute space) adopted by individuals working on the no-god part [Henry 1997: 73-85]. The year of 1834 for Scientist comes from the Merriam-Webster dictionary while 1840, attributed to William Whewell, appears in [Williams 1983: 279]. *Wenke* (文科, the humanities plus social science) and *Li'ke* (理科, 'Natural Science' plus Mathematics) are two terms uniquely used in the Chinese language.

1687	Newton's *Mathematical Principles of Natural Philosophy* published
1688	Enlightenment begins
1762	Probability used in gambling and insurance business
1776	Adam Smith (Economics; *The Wealth of Nations*)—birth of Social Science
1789	Enlightenment ends
1812	Pierre-Simon Laplace (*Analytical Theory of Probabilities*)
1844	Auguste Comte (Sociology—the 2nd discipline in Social Science)
1859	Probability used in Darwin's evolutionary theory and Maxwell's kinetic gas theory—first time in science

Fig. 3. The birth of Enlightenment, Social Science, and probability theory. The Enlightenment (1688-1789) aims to create a Science of Man, which succeeds in creating Social Science but fails with the humanities. The reasons: (1) Human matters are not deterministic like in Newtonian mechanics but are probabilistic; (2) the necessary tools of a probabilistic science were not yet there [Lestienne 1998].

"Natural Science" differs from natural philosophy in that all theological and metaphysical considerations are excluded in the former but not necessarily in the latter. By these definitions, Newton was doing neither "science" nor "natural science" but natural philosophy.[18] However, we are so fond of Newton that we want to make him one of our own, a "scientist." The way to do that is to retain only parts of his book *Mathematical Principles of Natural Philosophy* (1687), later editions in fact, in which God is not mentioned, notably, the mathematical description of the three laws of motion and the law of gravity, as we are doing it today in every physics textbook.

Presently, "natural science" consists of the study of nonhuman systems: physics (including astronomy), chemistry, biology, Earth sciences (e.g., geology and atmospheric science), etc. Social science consists of economics, sociology, psychology, linguistics, law, anthropology (which could include archaeology), etc., while the humanities are made up of "philosophy", religion, history,[19] arts (e.g., literature, visual arts, and performing arts), languages, etc.

However, as shown elsewhere, this historical classification is unreasonable and unscientific [Lam 2008a]. It is harmful to the healthy development of not just the humanities but also of social science and physical science. Since the humanities, social science, and medical science are all about humans, a logical and systematic approach would be to group them together in one (umbrella) multidiscipline—*Scimat*.

6 Scimat 2: Science, Scientist and the Science Room

With the historical developments in mind, here are the new definitions of science and scientist proposed by Scimat, followed by an introduction to the Science Room.

6.1 Science defined

According to Scimat: *Science is humans' pursuit of knowledge about all things in nature, which includes all (human and nonhuman) material systems, without bringing in God or any supernatural.* Some explanations are in order.

1. The pursuit of knowledge is the common denominator among philosophy, "philosophy", "science" and science (Figs. 2 and 4). "Pursuit" here means earnest and honest research aiming to find out what and why.

2. Nature here, after Darwin and Einstein as discussed in Secs. 3 and 5.1, includes all the material systems—the human system and all (living and nonliving) nonhuman systems. It is identical to the universe.

3. By excluding God in the scientific process we are in disagreement with Philosophy and 'Philosophy' but in agreement with modern science.

4. That God is excluded is based on two considerations: (1) God, if exists, is beyond nature since, e.g., according to the Bible, he creates everything. (2) God is a "game stopper." For example, if you got a funding from the National Science Foundation (NSF) in USA, to find the mechanism of high-T_c superconductors and you reported that "God did it," NSF would ask you to stop the project and return the money immediately.

5. Scimat holds no position on whether God exists or not. The aim of science, from its beginning with Thales through Galileo and Newton, was

never to challenge the existence of God. Religion and science is in conflict with each other only when religion does not retreat fast enough as science advances, and when either side overclaims [Lam 2004].

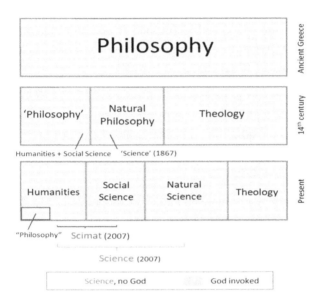

Fig. 4. Splitting of the discipline Philosophy in the last 2,600 years and the retreat of God/supernatural in the disciplines over times (based on Fig. 2). The domain of Science, defined as humans' research in understanding nature (human and nonhuman systems) without bringing in God/superstition, exists within the single discipline of Philosophy at Thales' times, the early Greek times. Science expands to about half of the discipline, Natural Philosophy, in the 14th century while Philosophy split to cover 'Philosophy', which further shrinks to "Philosophy" as we know it today (see text). Religion in the humanities differs from Theology in that God's existence is not assumed in the former but in the latter.

6.2 Scientist defined

Following the definition of Science in Sec. 6.1, here is Scimat's definition of scientists: *A scientist is a person who honestly seeks knowledge about nature without bringing in God or any supernatural.* In other words, a scientist is simply a *researcher* (res for short). Here are some remarks.

1. By this definition, apart from the natural scientists and social scientists, the humanists are also scientists, because nature includes human and nonhuman systems.

2. If the humanists do not look like scientists to some people, it is because most humanists are still carrying out their research at the empirical level, similar to what Aristotle did long time ago. (See Sec. 7 for the three levels/approaches of research, any research.)

3. Honesty is a must in real research. It refers to the researcher's honesty in collecting and handling data, and in reporting results. It also refers to the res' readiness in admitting mistakes when the res knows a mistake has been made, even though we understand that researchers, being humans, could find it difficult to do so.[20]

4. If one wants to, scientists could be separated into two classes: professionals and amateurs. Both could contribute to the progress of science, like in astronomy. The distinction should be of concern to the administrators in universities and funding agencies since it is their job to bet on the success chance of a researcher, and to the general public who has to decide whom to trust more (which is not a simple matter [Lam 2014]). For scientists, the works of professionals and amateurs are judged critically alike anyway.

5. When one is enjoying the beauty of a rainbow in the sky, one is not a scientist. But when the same person starts to wonder where those rainbow colors come from she is taking the first step in doing science. If she goes further and records the shape of the rainbow and the distribution of the rainbow colors, she is doing science at the empirical level and becomes a scientist. When she tries to figure out, by theory or experiment, the mechanism of rainbow formation she is doing science at a higher level. If she succeeds she is a good scientist. If she is the first one in history who discovers the mechanism and gets it published, she is a "successful" scientist. Publish or not, the real joy in doing science is to have fun in discovering or understanding something in nature.

6. Similarly, an artist is not a scientist when she is going through the motion of creating an artwork. But before or during the process, if she tries

to figure out seriously by herself how to make things work; e.g., what techniques applied will achieve the effect she wants or how the receiver's brain (through the senses) can be stimulated the way she intends it to be, she is doing science and could be called a scientist [Lam 2011: 24]. That is why Leonardo da Vince (1452-1519) is both an artist and a scientist even though he did not have formal training in science [Capra 2007].

6.3 The Science Room

With scientists so defined in Sec. 6.2, we could also come around and say "science is what scientists do," borrowing the dictum from physics that "physics is what physicists do" [Lubkin 1998: 24]. Science thus has two parts: (1) the results obtained by scientists, and (2) the process of doing science. (See [Warren 2014] for an example.) Both parts evolve over time and are full of surprises (see Sec. 7).

To capture visually the essence of the above, we introduce the *Science Room* as a metaphor for science, which is represented by the right box in Fig. 5. For comparison, the conventional "natural science" room, which excludes the humanists and social science, is shown on the left. In the science room, over time, we see new scientists entering and old scientists leaving when they die or quit prematurely. Some, like Newton and Einstein, are bigger than others. Inside the room, results obtained by the scientists are kept, too. There are two kinds of results: those that are still in use (like the law of gravity, quantum mechanics, and the special and general relativity theories) and those that are outdated (like Aristotle's mechanics). The latter are kept in a storage room (not shown). We do throw out junks (like alchemy[21]) from time to time.

The number of researchers inside the science room increases with time and also the size of the room. The former could be seen by counting the number of physicists: In ancient Greece the number could be in the tenths; 100 years ago, in the hundreds; today, 50,000 from the American Physical Society alone. The latter is evidenced by the rapid increase of research journals. Overall, the density of scientists in the room (i.e., number of scientists divided by room size) seems to become pretty high in the last 30 years, signaled by the difficulty of finding a worthwhile project and the "publish or perish" pressure felt by the professionals.

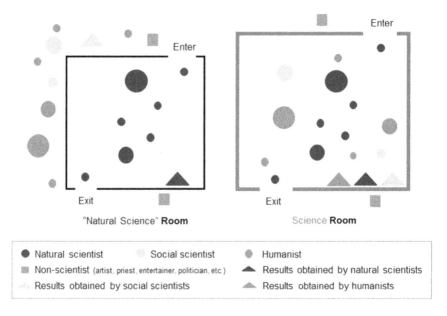

Fig. 5. The Science Room (right box) and the "Natural Science" Room (left box). The difference between the two is that humanists and social scientists (and their fruitful results) are included in the former but not in the latter. Note that "natural science" is mistakenly identified as science by many people. The walls of the room do not imply boundary of knowledge.

Observing the science room is like viewing a movie that never ends or not yet ends (we do not know which case it is), like the *Star War* series, with a new cast on the screen every 60 years or so (in historical time) if you happen to fall sleep intermittently. Like a good movie, science never repeats itself and is as intriguing as *War and Peace* while sometimes looks like the *Life of Pi*.

7 How Science Is Done

We here focus on the second part of science, i.e., how science was or is done.[22] A good way to explore this subject is to divide our discussion into two parts according to the subjects being studied, viz., simple systems and complex systems, since the methods of study though related, are sometimes quite different.

7.1 Simple systems

Only a subset of nonliving systems and a small number of living systems could be considered simple systems (SS). They are the subjects studied in physics (including astronomy), Earth science (excluding meteorology and climatology), and chemistry. Other living things such as vertebrate animals, humans included, belong to complex systems.[23]

The historical development of research in SS, particularly in physics and in later years, is rather well documented [Mason 1962; Lindberg 1992; Chen 2009]. Essentially, it goes something like this.

1. Early period: from Thales to Galileo

With plenty of time and patience, ancient people *observed* and *recorded* what they saw in the sky about the movement of heavenly bodies and astronomy was born—science at the *empirical* level. Then with the aid of arithmetic and geometry, miraculously, they succeeded in predicting eclipses, for example, taking astronomy a big step forward.

After observation and data analysis, *theorizing* emerged, through guessing and logical thinking. Aristotle did it without experimentation about moving bodies, in the sky and on earth, and got it all wrong. Yet, the important point is that, in the absence of quantitative data, he was "right" at his time—an important criterion we use in judging scientists. Furthermore, through his classification works in biology, Aristotle did show us the need of empirical inquiry. His pioneering thinking and perceptions, the Aristotelian tradition, was maintained for about 2,000 years after his death, helped by the ruling class. But this has nothing to do with Aristotle and is not part of the scientific process.

Archimedes (c.287-212/211 BC) was different. As the story goes, after the "eureka" moment triggered by water-level rising when he got into the bathtub, he run naked in the street and became the first streaker in history, which, luckily, did not become part of the scientific tradition. Importantly, after streaking, he did a few experiments in the lab and established the

Archimedes' Principle about buoying bodies, which is still correct today [Hirshfeld 2010]. Archimedes' Principle is an example of science at the *phenomenological* level, i.e., obtaining rules/laws about a phenomenon without knowing the mechanism.[24]

The Aristotelian tradition was finally broken by Galileo.[25] It happened not because Galileo was much smarter than those before him but because he (1) picked simple systems (such as a small ball rolling down an inclined plane) to study, (2) did drastic and daring approximations in constructing theories (e.g., simplifying the body to a size-zero point particle), and (3) used detectors other than his bodily sensors to observe and record (e.g., using water fall as a timer in inclined plane experiments[26] and improving the telescope to look at the moon and beyond) [Lam 2008a]. *Experimentation* became the hallmark of modern science; *Mathematization*,[27] modern physics.

2. Modern period: four lessons

Everyone was happy, except for those living in the Vatican. And modern science flourished in every discipline (including biology and medical science in the complex-system domain) except the humanities. Subsequently, for SS, two great theories, thermodynamics (1824) and Maxwell's electromagnetism (1873), were discovered. Near the end of the 19th century, it seemed to some that the "end of physics" has arrived. The sudden emergence of quantum physics in 1900, the black-body radiation experiments by Max Planck (1858-1947) [1900], was the party crasher. Newtonian physics no longer works and every able physicist was scrambling to find a new theory. Here is the *first* important lesson every modern scientist learns: *Nature is full of surprises and every dear theory may not be the final theory.*[28]

As physicists are concerned, the next heart breaker showed up 57 years later: Madame Wu and her collaborators discovered that parity is violated in weak interactions [Wu et al. 1957]. Parity is the mirror symmetry taken for granted by every physicist in the past, which says that if you set up an apparatus and put a mirror near it, construct a second apparatus according to what you see in the mirror, then the two apparatuses will give you identical results. Parity symmetry was regarded as basic as the time- and space-translational symmetries (which are still good, so far). Thus the *second* important lesson: *Never take* any *basic assumption for granted.*

Nature is kind to us. As if for compensation, in the same year of 1957, superconductivity discovered by Heike Kamerlingh Onnes (1853-1926) 46 years ago was satisfactorily explained by the BCS theory proposed by John Bardeen (1908-1991),[29] Leon Cooper, and Robert Schrieffer. Here is the *third* lesson: *If you wait long enough the answer* may *come*. However, it also shows that answer to your "prayer" does not come by keep on praying, but by more experiments and harder thinking. Also, Kamerlingh Onnes did not live to see his discovery explained; the answer may not come within a lifespan. We kind of know this already since it took about 2,300 years for Democritus' atoms to be confirmed, which is the case of theory preceding experiments while superconductivity is a case in reverse.

When the high-T_c superconductors were discovered in 1986 by Karl Müller and Johannes Bednorz, and the Nobel Prize was awarded the next year—a lightning speed, quite a number of theoretical papers from famous authors were rushed to print in the prestigious physics journal *Physical Review Letters*. They turned out to be all wrong or unbelievable as more experimental results came in. In fact, instances like this happened before that a beautiful theory is not sustained by experiments and has to be rejected. The *fourth* lesson: Reliable *experimental results have the ultimate say in deciding the fate of theories in science.*

3. Three remarks

1. Disciplines (or subdisciplines) in SS could be divided into two types: non-historical disciplines (e.g., condensed matter physics and particle physics) and *historical* disciplines (e.g., astronomy and geology). Control experiments are possible in the former but not in the latter. Then how do we study the historical disciplines? It is done by comparing data collected in the historical disciplines with knowledge gathered through controllable, repeatable experiments, and confirmed theories in the non-historical disciplines. For example, by comparing the color spectrum of light coming

from a star with those from known elements obtained in the lab we could tell what kinds of elements existing in the star.

2. We never check *all* consequences/predictions from a hypothesis before we accept it as a confirmed theory. For practical reasons, we just do enough number of checking to convince ourselves that it is correct. And that is why an established theory *could* be broken later when new contradictive findings (usually experiments), if any, show up. *This strategy is what makes the rapid progress of science possible* in the last few hundred years.

It is like how countries are being conquered: The invading army occupies some strategic cities/places and then the capital, never the whole country, and declares the job done; then they move on to the next country— Napoleon did that; Hitler did that. In science, we do the same in the Knowscape (see Sec. 2) instead of the landscape. Consequently and occasionally, disturbance might suddenly burst out from a not-yet-occupied place in a conquered country. The conqueror would be forced to look back and suppress it or tolerate it. And that was what happened to the conqueror, the physicists, in the Knowscape, in the case of parity nonconservation, except that they could not suppress it because parity nonconservation sits in the human-independent part of Knowscape (see Fig. 1). Instead, they update their map of the Knowscape.

3. The existence of atoms and the discovery of quantum mechanics made it possible to study many-body systems with a new approach, the *bottom-up* approach, called the "microscopic picture" in physics. This is the third approach apart from the empirical and phenomenological approaches. For closed systems (such as gas in a jar) the theory to accomplish this is Statistical Mechanics, which links the microscopic world to the macroscopic world. Computer simulations could also do the job, for both open and closed systems, but are less enlightening [Tuckerman 2010]. Each approach or level of study complements and reinforces the level above it. Thus, results from the phenomenological level, if correct, will have to reproduce that from the empirical level and give more; results from the bottom-up level will do the same for the phenomenological level above it [Lam 2002]. The availability of the three approaches is like that of the

army, navy, and air force in a war situation; you want to use all of them, if necessary, to do a quick and thorough job.

7.2 Complex systems

Complex systems (CS) consist of nonliving (e.g., the weather or climate system) and living systems. The latter consists of humans or human-related systems (e.g., the economy) and other nonhuman, biological systems (e.g., plant, insect, fish, bird, and coyote). A central, time-honored method of tackling CS (and SS) is the *Socratic Method* due to Socrates (470/469-399 BC); i.e., "to solve a problem, it would be broken down into a series of questions, the answers to which gradually distill the answer a person would seek." [30] Later methods are described below.

1. Difficulties and successes

Complications and difficulties in studying CS arise from five sources.

1. *The potential of chaos.* Chaos is the phenomenon that the future of a system depends sensitively on the initial conditions [Lam 1998]. It could happen in a system of three bodies (like the Earth, Moon and a rocket) or many bodies (like the weather, even though, in this case, only a simplified model of three variables has been proved). Given a CS, the potential of chaos is always there but is hard to prove. That is why chaos is so frustrating.

2. *Heterogeneous complexity.* Complex systems are mostly heterogeneous, in terms of its components and interactions among them. For instance, the human body consists of a large number of different organs, interacting directly or indirectly with each other; in a society, no two persons are the same. (In contrast, in the SS of electrons, all electrons are identical to each other and every two electrons interact the same with each other.)

3. *Ethical limitations.* There are ethical problems in experimenting with living systems [Rollin 2006]. That has been always the case with humans as the experimental object (with well-known exceptions like dealing with war prisoners). It extends to experiments with nonhuman animals in recent years. In other

words, not every informative experiment (like human cloning) that could be done can be done, a problem not associated with simple systems. But it does not mean that we cannot and do not experiment with humans. Psychologists do that all time, harmlessly, by passing out questionnaires; hospitals try new drugs or new treatments on volunteers every day. (See [Venter 2013] for an interesting discussion on this issue.)

4. *Historical irreproducibility.* Human-related happenings are all historical, like in astronomy, which are irreproducible, with some exceptions in medical research. Therefore, on-site and real-time collection of data is rare except in mass demonstrations these days; even so, the data are always incomplete. And being humans, recounting of events by the participants, the so-called first-hand data, is subject to memory deterioration and intentional personal considerations.[31]

5. *Temporal-spatial localization.* Since humans are influenced by culture and environment (in addition to human nature [Wilson 1978; Machery 2008]), research on humans based on observation/data from a local place for a particular historical period may not be applicable to other periods or humans in other countries or continents. Ignoring or ignorance of this localization feature results in overclaim by humanists and social scientists, West and East. The problem is lessened but still present in the globalization era.

How and to what extent these complications could be overcome will be discussed below.

While the Holy Grail in CS research is to find the universal law(s), similar to the Second Law of Thermodynamics, say [Waldrop 1992], it is very difficult to do so and fails so far. Instead, much progress has been made in the study of individual systems or classes of systems. Two "universal" organizing principles applicable to a large number, but not all, of CS are found: fractals and active walks [Lam 2008a]. In lieu of general laws, computer simulations are heavily used [Mitchell 2009] and techniques developed in simple-system studies are borrowed [Castellano et al. 2009].

The former overcomes the heterogeneous complexity since heterogeneity can be programmed easily in computers.

2. Medical science and biology

Among CS studies, the transdisciplinary medical science (more a basic science than applied) is the most developed despite and because it is about humans. Medical science could easily be the most important among all the disciplines, for the obvious reason. Its development benefits from early start,[32] continuous attention, and heavy funding.[33] It also benefits from the rapid advances in physics, chemistry, and biology since the Scientific Revolution. But more importantly, medical research in the West keeps a very open mind and employs all the three research approaches (i.e., empirical, phenomenological, and bottom-up) as soon as it is feasible to do so. (Drug designs at the molecular level, and genetic and stem-cell treatments under test are examples of the bottom-up approach.) Unfortunately, the same cannot be said about Chinese traditional medicine [Lam 2008a: 32-33].

In biology, the discovery of evolutionary theory (1859) by Darwin and the double helix (1953) by James Watson and Francis Crick (1916-2004), leading to the prospect of synthetic life [Venter 2013], demonstrates the workings of the empirical, phenomenological, and bottom-up research approaches in successive actions. For many people, what we are witnessing in biology is comparable to what happened in physics in the early 20th century. Moreover, the fact that Crick is a physicist-turned-biologist exemplifies the early trend of an s-res morphing into a c-res. (Here, s-res means a "researcher in simple systems"; c-res, in complex systems.)

3. Social science

As humans are concerned, it is in fact easier to study a large number of humans than a single human being because many approximations can be made and justified in the former but not in the latter.[34] And that is why the scientific level achieved is much higher in social science than in the humanities.

Economics is the most developed discipline in social science, not merely because it was the first discipline invented in this field (Fig. 3) but because (1) a lot of data are generated and kept (think stock index), and (2) the financial reward is huge. The economy being a CS, it is not surprising that no universal theory about macroeconomics has yet been found, not to mention that no one is able to predict the rise or fall of a stock market. This is in contrast to the success of the meteorologists who are able to predict pretty accurately the *local* weather, also a CS, of the next day or a few days. The difference is that the equations involved are known in the latter (solved by supercomputers) but not in the former. We therefore see our top economists revising their "prediction" of the national economy from time to time, if not every day, which in fact is a good sign showing that they are honest scientists [Lam 2008a: 28]. But then the question: If the economists fail so miserably, why is there a Nobel Prize given out every year in economics? Well, the Nobel Prize rewards the solution of individual, significant problems, which is possible even in CS.

There is a long string of successful stories of physicists making contributions in economics and finance [Weatherall 2013]. But Wall Street has been blamed for hiring physicists who helped to bring down the global economy in 2008 [Patterson 2010]. This is an issue of s-res morphing into c-res, unsuccessfully in this case. But that is because Wall Street has hired the wrong kind of physicists.[35] There are different kinds of physicists, like there are different kinds of engineers. If one wanted to design a new bridge, one would not go out and hire electric engineers to do the job; right?

Fortunately, there are also many successful examples of s-res doing fine in Sociology.[36] And we see the emergence of a new field called Computational Social Science [Cioffi-Revilla 2014; Epstein 2014].

4. The humanities

Development in the humanities is lacking behind social science, not to mention natural science, due partly to the intrinsic complexity of humans as individuals but also to the inadequate scientific training of the researchers involved. Thus, occasionally, we hear people saying that the humanities cannot be part of science. They are wrong, for two reasons: (1) Many humanists mistakenly identify science with Newtonian mechanics; (2) the humanists ask the wrong questions.

The mis-identification of science with Newtonian mechanics started early (since the Enlightenment) and is still with us today,[37] due to the failure of science education and science communication [Lam 2014]. Since each human being is an open system (that exchanges energy and materials with the environment) and the factors that could affect a human or a group of humans cannot be completely account for, the human system is a *stochastic* system (i.e., probability/chance is involved). If one identifies science with the deterministic Newtonian mechanics then, of course, science is inapplicable in handling the human system. But scientific theory (physics in particular) and techniques dealing with both deterministic and stochastic systems are available (see Fig. 3 and [Lam 2011]).

Given a stochastic system, the question one should ask is not what will surely happen in the future but with what probability something may happen in the future [Lam 2002]. For example, in History, the question one can ask about the longevity of a dictatorship is not in which year it will end definitely, but what is the chance it will still be there five years later, say. Surprisingly, amid all the contingencies and historical irreproducibility a *law* about the lifetimes of Chinese dynasties has been found [Lam 2006; Lam et al. 2010]. And an e-journal dealing with the theoretical and mathematical aspects of history from the Scimat perspective [Lam 2008c], *Cliodynamics*, has been published by the University of California since 2010. What this demonstrates is that human matters can indeed be studied scientifically (by going beyond the narratives), irrespective of all the complex thinking and so-called "free will" going on in human's brain.

Recently, a DNA study of 1,000 descendants of Cao Cao, an important Chinese general in the Thee Kingdoms period (220-280), eliminates the possibility that Cao was the descendant of a famous aristocrat [Wang et al. 2013; Jiang 2013]. This work showcases the bottom-up approach in studying history.

Studies at the empirical and phenomenological levels in the humanities have been going on for more than 2,400 years since Plato and Aristotle. What is new is that in the last decade or so we see the bottom-up approach being advocated by the humanists themselves (including some in English Literature, e.g., [Hogan 2003]). This includes the emergence of *Neurohumanities* and efforts to understand human matters from the cognitive and evolutionary perspectives (see [Lam 2011] for details). As an example, the neurobiologist and Nobelist Eric Kandel's *The Age of Insight* [2012] shows how the Vienna portraiture from 1900 to present could be understood at the three research levels.

Even in "philosophy", the toughest discipline where metaphysics is studied, one finds serious attempts by its practitioners to do their trade with non-traditional methods. For instance, in addition to Neurophilosophy that tries to solve the mind/brain problem using cognitive science [Churchland 1986; Churchland 2007], Experimental Philosophy tries to solve philosophical questions by using empirical data (usually from surveys of ordinary people) [Knobe 2011; Knobe & Nichols 2008]. All these are very encouraging, from the Scimat perspective.

5. Living with uncertainty

For inanimate CS, the existence or absence of chaos can be ascertained in rare cases while oversimplified models are used (e.g., the three-variable climate model of Edward Lorentz (1917-2008) [1993]). But the human system is different; we have no choice but to live with the potential of chaos. Similarly, the human system's intrinsic probabilistic nature cannot be circumvented. Worse, a small but finite probability for something to happen does not mean that it would not happen; on the contrary, it means it *could* happen. Some probabilistic events (like not winning a lotto) if happen, would be harmless; but others (like global warming or the danger of genetically modified foods) could have dire consequences.

It is true that consilience (i.e., the convergence of evidence) implies a conclusion is most likely to be correct [Wilson 1998]. But since nature is subtle and full of surprises, "most likely" means it is not safe to assume it is 100% valid. Uncertainties, big or small, are with us every day.[38]

The research history of simple and complex systems shows that there exists no such thing called the *Scientific Method* [Bauer 1994] even though this term is invoked frequently by scholars [Thurs 2011; Gower 1997] and laypeople alike, if the method means a recipe of steps that guarantee success when followed, like in the case of a cooking recipe. Instead, what we have are equally valuable, viz., "scientific experience" and "scientific tradition." The reason behind all this is that research is a very complex process that does not easily yield to a simple summary. Besides, there are emergent properties manifested on many levels in a given system which could be attacked through one or all of the three research approaches; there is no place for an oversimplified, universal scientific method. The absence or de-emphasis of a scientific method is exemplified by the criterion used in the Nobel Prizes in the sciences, which are awarded for confirmed, grand discoveries, irrespective of the method and reasoning used. In fact, if there was really such a powerful scientific method, research would be pretty boring and we would see many creative scientists walking away— doing art or go fishing.

8 The Essence of Science

The three features about science listed here, though incomplete, are the most important.

1. Science advances and lives with approximations

It is a myth that "exact science" ever exists, for the following three reasons:

1. Every theory constructed turns out to be an approximated theory. First, a theory is the model of the real thing; e.g., in Newton's laws of motion a body is approximated by a zero-size point mass. Second, the theory is the approximation of a bigger, later theory; e.g., Newtonian mechanics is an approximation of Einstein's theory of special relativity. The Standard Model in particle physics is the low-energy approximation of something not yet known [Weinberg 2011].

2. Even if a theory is exact, the solutions obtained usually involve approximations because exact solutions are rare.

3. Even if the theory and the solution are both exact, it still has to be checked by experiments. And experiments involve instruments and measurements which always have finite resolutions. For example, to check the equation $A = B$ we measure each quantity and try to show that the left-hand side equals the right-hand side. But let us say our measurements give $A = 2.5 \pm 0.1$ and $B = 2.5 \pm 0.1$, we can only conclude that $A = B$ within experimental uncertainty (due to that ± 0.1). Measurement's unavoidable finite resolution prevents a rigorous proof of any theory.

In other words, *science never proves anything, rigorously speaking* (in the mathematical sense of proof). There is nothing absolute and final in science, with profound implication for "philosophy" [Lam 2014]. Thus, there *could* be room for improvement in any theory, which is known for sure when the new improvement appears. Approximation works because, like when lost in a forest, a rough map is all one needs to get out of the forest. More importantly, as mentioned above, it is precisely due to the use of approximations that science progressed so fast in the last few hundred years.

Science lives and thrives with approximations. Coupled with what we say in Sec. 7.2, we humans (i.e., everybody, scientists and non-scientists alike) simply have to go on living with uncertainty, more *wisely* and *humbly*.

2. Scientists are humans

Scientists are humans, like you and me.[39] Most if not all scientists, including Einstein [Kennefick 2005], do make mistakes [Youngson 1998; Livio 2013]. But some, like Newton, are not like you and me. They are not just brighter, but are tormented and darker—not entirely their fault.[40] Some genius not just contributed more but suffered more, too. And humanity owes them a lot.

3. There is always the Reality Check

If science is not and could not be exact, and scientists are just humans, why should anyone take science seriously? Everyone should, for two reasons: (1) *Science delivers.* We owe our air conditioning and cell phone plus many other goodies to science. Science is valued because it works.

(2) *Science is the best game in town.* As Newton [(1730) 1952: 404] puts it, "And although the arguing from Experiments and Observations by Induction be no demonstration of general Conclusions; yet it is the best way of arguing which the Nature of Things admits of."

Why is science able to deliver? It is because it has to pass the *Reality Check* (RC), like every legitimate driver has to prove herself by passing the road test. Reality check means "confirmed" by experiments or practices, or, at the minimum, is consistent with established data. Social theories, due to the imprecise and incomplete data collected, are hard to be 100% confirmed [Lam 2008a]. Reality Check is the *necessary*, crucial step for a theory to be recognized as part of the Knowscape. It is the RC that makes scientific knowledge unique among all forms of "knowledge."

Among the RCs, the *Cell Phone Test* (CPT) stands out because the working of a cell phone depends on the validity of a large number of theories (Fig. 6). Any new theory, if it conflicts with what are behind the working of a cell phone, has to explain why and why it is better. A good answer would be the new theory contains the old ones as special cases. We strongly recommend the CPT to check the advocations of anything new.

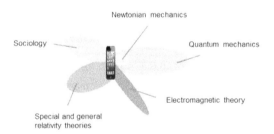

Fig. 6. Cell Phone Test: established, interrelated theories behind a cell phone. The working of a cell phone depends on Maxwell's electromagnetic equations, quantum mechanics, semiconductor theory, general relativity, and the sociological fact that there is enough number of humans who want to interact with others.

9 Scimat 3: Q & A and Ramifications

Scimat's spirit and essence as well as the action plan in the immediate future are summarized in Sec. 9.1. Some of the ramifications, including a new perspective on antiscience and pseudoscience, are discussed in Sec. 9.2.

9.1 Q & A

Q: What is Scimat?
A: It is a new multidiscipline that recognizes "Everything in nature is part of science."

Q: What do you mean by nature?
A: Nature includes all living and nonliving material systems, humans in particular.

Q: What do you mean by Science?
A: Science is humans' pursuit of knowledge about all things in nature, which includes all human and nonhuman systems, without bringing in God or any supernatural.

Q: What is Scimat's position on God and religion?
A: Scimat holds no position on whether God exists or not. Personal choice of religion is respected.

Q: What are the topics that Scimat covers?
A: All topics related to humans. That is, all the topics in the humanities and social science (and related topics in medical science).

Q: Why does Scimat put its emphasis on the humanities?
A: Because social science has been recognized as science but not yet the humanities.

Q: Why are the humanities so important?
A: All the world tragedies (poverty, war, race cleansing, injustice, corruption, etc.) are human-dependent matters (due to poor decision making) and could be traced to the underdevelopment of the humanities in the last 2,400 years since Plato and Confucius.

Q: What new method or new tool is used by Scimat in its research?
A: None. Scimat advocates the use of any method or tool that is available and applicable as long as honesty and ethics are respected. Reason:

Scimat is about the search for knowledge which knows no boundaries and no pre-determined routes.

Q: Anything more?

A: Well, we do point out and want to emphasize that in any discipline there are three approaches—empirical, phenomenological (i.e., without knowing the mechanism), and bottom-up; they supplement and reinforce each other, like army, navy, and air force in a war situation. This is well known in nonhuman studies but less so in the humanities.

Q: Then what is new about Scimat?

A: The concept that "all human-dependent matters are part of science" is new.

Q: So what?

A: Humanity advances through new concepts. Examples: "All men are born free" brings down slavery; "all men are born equal," royalties and totalitarian regimes; "all women are born equal, too," restrictions on women's rights in education, employment, and voting.

Q: Will all the world problems be solved if enough number of people become Scimatists?

A: We don't know. But, in our judgment learned from history, Scimat is the best, practical way to make the world better and more peaceful since people would be more enlightened and, hopefully, act more rationally, *humbly* and kindly toward others and the environment. The next step is to educate the decision makers.

Q: What is Scimat's action plan for the future?

A: To set up 100 Scimat Centers worldwide—to ensure the development of and sustain Scimat's ideals. The first step is to set up one such center. What the center can do is spelled out in "The *Science Matters* Program and a Proposal" (see: www.sjsu.edu/people/lui.lam/scimat).

Q: How can I help?

A: Buy this book. And tell others to do so.

Q: How can I help more?

A: Buy the other two books in the Scimat series: *Science Matters* (2008) and *Arts* (2011). Or, ask your library to do that; better, do both.

Q: How can I help much more?
A: Be a sponsor or co-sponsor of our next Scimat conference. It takes 20,000 USD to run a good conference, 10,000 USD to run a conference.

Q: What if I can't wait?
A: Gather, from your friends if necessary, 50,000 USD and contact the author: lui2002lam@icloud.com. We will help to set up a Scimat Center in the university or city of your choice (tax exempt in the US). The Center could be named after the person you prefer if more money is donated.

Q: How can I help without money involved?
A: Visit the Scimat website and help spread the word.

Q: What is Scimat's take home message?
A: The humanities are part of science.

9.2 Ramifications

Important implications of Scimat, including a new answer to the Needham Question, have been given before [Lam 2008a]. Here are more.

1. Since science includes the humanities, "philosophy" in particular, the early Greek philosopher Socrates was a scientist, too. Socrates, as a social and moral critic while practicing his philosophy, became a "gadfly" of the state, in Plato's words. He was sentenced in court for "guilty of both corrupting the minds of the youth of Athens and of impiety (not believing in the gods of the state)" and persecuted.[41] Thus, the c-res *Socrates was the first martyr of science* while the s-res (in astronomy) Giordano Bruno (1548-1600) was the second one.[42] The former was ordered to swallow poison; the latter, burned at the stake.

2. Does antiscience really exist? It depends on how science is defined. Let us consider marriage and antimarriage first. The conventional definition of marriage is that (1) it is a legal piece of paper, (2) signed by a woman

and a man (who promise to take care of each other). Antimarriage usually refers to the objection of item 2 but not item 1; i.e., an antimarriagist advocates that the legal paper could also be signed by a woman and a woman, or a man and a man. Similarly, it is hard to imagine anyone in her right mind would object to doing science per se unless science is defined in the narrow sense; i.e., science is about nonhuman systems only. *Antiscience* usually refers to the objection of certain things such as doing certain kinds of scientific research and the abuse of science; instead, putting the money in education, fighting poverty, and human development are recommended [Holton 1993]. All these debates are about overclaims by some scientists and human's choices, e.g., the priority and allocation of resources. These are legitimate, human-dependent matters. Thus, if science is to include all human matters as it should, according to Scimat, there is no such thing as antiscience. On the other hand, there does exist antiscientists: Those who commit science frauds (in the humanities and other sciences) since they violate the first rule in doing science, i.e., being honest in their research.

3. The *science-pseudoscience demarcation* is a complex issue [Shermer 2011] which is less about science per se [Pigliucci & Boudry 2013] but more about the competition for attention, prestige, and resources [Gordin 2012]. In particular, the debate on "intelligent design" (ID, a form of creationism) being science or pseudoscience is due to the fuzzy definition of science used, either in the media or in court [Shermer 2006]. If God is explicitly excluded from the definition of science, as in Scimat, this debate would never happen—case closed. In fact, ID could be discussed in "philosophy" or theology classes. The problem for the ID advocates, of course, is that in America's public high schools, there is neither "philosophy" (unlike in France) nor theology classes. If religion-government separation works (in the US at least), why not religion-science separation?

4. The objectivity in and the reconstruction of past "reality" through historical narratives are considered impossible by some historians [Hayden 1973] and deconstructionists [Derrida 1976] because, they argue, the exact meaning of any writing is undecidable. These and other

postmodern attacks led to a crisis in the history profession in the 1990s [Gilderhus 2000; Evans 1997]. That language, spoken or written, is an imprecise mode of communication is a well-known fact. The good news is that since science advances by approximations, history, a branch of science, also does not need to be exact in its narratives; i.e., research at the empirical level. (See also [Lam 2008c].)

10 Discussion and Conclusion

Here are some discussions and elaborations concerning what are written above.

1. The most negative effect derived from misconceptions about science is that the topics studied in the humanities are excluded from the domain of science. If that was the case, Aristotle would have two hats to wear. He would don on the "science hat" while he was studying the cosmos, plants, and animals; and the "non-science hat" while he was talking about human affairs. Won't that drive Aristotle crazy? You bet!

2. A positive effect of recognizing the humanities as part of science, apart from helping to reduce or eliminate human tragedies (such as ideological massacre or hunger), is to reverse the rapid decline of enrollment in the humanities in the universities. According to a newspaper report (in *San Jose Mercury News*), from 1966 to 2009, the percentage of students receiving bachelor degrees in the humanities at Stanford University drops from 37% to 11%; for all institutes in the US, from 17% to 7%. By treating their disciplines as part of science, the humanities will attract science-inclined students who normally would opt for "natural science" and engineering for the wrong reasons; these students will also help to raise the scientific level of the humanities disciplines and create more jobs.

3. We have come a long way to be able to declare freely that humans are animals made up of atoms. Only about 400 years ago, scientists in Naples were arrested and tried for maintaining "that there had been men before Adam composed of atoms equal to those of other animals" [Jacob 1997: 28].

4. The existence and scale of the god-invoked part in natural philosophy and its influence on the no-god part is often overlooked by scholars in science. Consequently, some experts simply identify, wrongly, natural science with natural philosophy, while only the no-god part of natural philosophy could be identified with natural science. (For more see [Cunningham & Williams 1993: 421].)

5. Even some practicing scientists' characterization of today's science is wrong. When they say that controllable and reproducible experiments are a must in science they forget all those historical (scientific) disciplines like astronomy and archaeology (see Sec. 7.1). An example is provided by the chemist Igor Novak [2011 p. 31] when he says, "Science uses reproducible, controlled experiments (as part of the 'scientific method') and draws conclusions on the basis of experimental results *and* logical reasoning *alone*" (emphasis in original). And, of course, he excludes social science, not to mention the humanities, in his use of the word science.

6. Some experts identify science with modern science. This choice causes unnecessary confusion and even some inner contradictions. If only modern science is science, what is the Archimedes' Principle which is still being used today? This problem could not be overcome by dividing the development of science into three levels, viz., prescience (e.g., Aristotle), parascience (Newton), and euscience (Galileo), as done in [Jaffe 2010]. If so, Thales would be the "father of prescience" and not the "father of science." And it is odd to label Newtonian mechanics as parascience; it is good science, a special case of Einstein's special relativity theory which in turn could be the special case of something else in the future. A better way is to view Aristotle's physics (which was correct at his time but is obsolete today) as one of the numerous outdated theories in the history of science. (For more see [Lam 2014].)

7. Scientism has various definitions.[43] But essentially, *Scientism* implies the universal applicability of the so-called "scientific method" and the supremacy of scientific knowledge. Scimat is not scientism because we do *not* believe in the existence of a scientific method (see Sec. 7) and we do respect people's right to seek knowledge by any avenue they happen to choose as long as it is safe and ethical. We just say that judging from past

history, science is the *most likely* way to succeed, from making a cell phone to curing cancer and to lessening humans' sufferings. In fact, science guarantees nothing and promises nothing except in those well-understood simple systems (like water boiling at 100 °C at atmospheric pressure). For unsettled problems in simple systems (like high-T_c superconductors) science thinks it can solve the problem given enough time; for complex problems related to humans (like love, ethics, and soul) the promise of science remains a promise, until it delivers. Thus, Scimat does not belong to what Harry Collins calls the "first wave of science studies" [Collins 2014].

8. After this chapter is essentially completed, an interview of the Nobel laureate David Gross by Peter Byrne [2013] came to our attention. The non-finality of any scientific theory and scientists' willingness to abandon what they cheered and toiled with for a lifetime when convinced otherwise (described in Sec. 7.1) are concurred by Gross when he says, "We were all looking for the next overthrow, and we were willing to sacrifice existing theories at the drop of a hat." Other common or not-so-common beliefs presented above in this chapter are consistent with Gross' sayings, too:

> New discoveries tend to be intuitive, just on the borderline of believability. Later, they become obvious. … A scientific "frontier" is defined as a state of confusion. … Philosophers who contribute to making physics are, thereby, physicists! … The public generally equates uncertainty with a wild guess. Whereas, for a scientist, a theory like the Standard Model is incredibly precise and probabilistic. *In science, it is essential never to be totally certain.* And that lesson is hammered into every scientist and reader of history. Scientists measure uncertainty using probability theory and statistics. And we have comfort zones when making predictions, error bars. *Living with uncertainty is an essential part of science,* and it is easily misunderstood… [T]he human mind is a physical object. It's put together by real molecules and quarks. (My italics.)

Nevertheless, as stated in item 2 of the *Scimat Standard*, "We will not quote anyone's writing to support our own argument" [Lam 2008a: 27]. The quotes by Gross should not be taken as evidence that the arguments in this chapter are correct, which can only be reached through the reader's own judgment.

9. Scimat provides the key in solving the two-culture problem [Lam 2008a]. It is also the theoretical foundation of scientific-culture studies [Liu & Zhang 2014] because it provides a unified perspective and basis in describing human matters, and similarly for general-education courses which try to educate students on both the humanities and "science".

10. The Enlightenment's aim of making a science of human was partially successful; it created social science but fails with the humanities as a whole (Sec. 5.2 and Fig. 3). Scimat could be viewed as *Enlightenment 2*, continuing with what the Enlightenment left unfinished but smarter.

11. In America, the way to raise the scientific level of the humanities is to ask the National Endowment for the Humanities (called the National Humanities Foundation, after NSF, in the planning stage) *and* NSF to fund more interdisciplinary research that merges the humanities and the "sciences."

Take home message: Science's characteristics are its secularity and the reality check. The necessary Reality Check is what makes science useful and distinctively different from humans' other types of inquiry.

Note added (March 10, 2024): The lack of a consensus on the definition of science has adverse effects for the two science-related disciplines, viz., philosophy of science and history of science: (1) There is no way for the practitioners to decide what exactly to be included in their studies. (2) It is impossible for people to exchange ideas and debate with each other in seminars/meetings unless each person states out their own definition of science first, which will be time consuming and inefficient. (3) It is impossible to answer questions that crucially depend on the definition of science such as these: Is Chinese traditional medicine a science? Is there science in ancient China? What is the proper answer to the Needham Question? We believe that the Scimat's *rigorous* definition of science is the only one that is consistent with the historical developments of science and the practice of scientists. For more, see my book *Humanities, Science, Scimat* (World Scientific, 2024).

Notes

1. Some people prefer "Science is to understand the universe." But since nature includes everything in the universe (see Secs. 3 and 5.1) the two statements are equivalent to each other. Here, "understand" means to understand without appealing to supernatural or God. See Secs. 5.2 and 6.1 for a historical discussion of this position, which is adopted in this chapter and in Scimat.

2. Plato seems to recognize this when he defines humans as featherless, bipedal animals with broad nails [Läertius 2011].

3. The existence of human-independent knowledge is denied by some relativists. Relativism is a branch of philosophy started by the ancient Greeks in 5th century BC. In those times, the profession of lawyer did not exist and everyone has to argue for himself in court. The relativists played the role of legal advisors. Like in today, winning the case is the only aim of arguing, not reaching for the "truth" or "reality" [Buckingham et al. 2011: 42].

4. See [Doren 1991] for a concise review of human knowledge.

5. In this chapter, "science" with double-quotation marks means science in the conventional sense, which is the sum of "natural science" and social science but excludes the humanities (see Fig. 2).

6. For example, according to Jean Ziegler [2013], the hunger and malnutrition suffered by nearly one billion people in the world results *not* from failure of agriculture, science or technology but from inhumane and shortsighted politics.

7. The Scimat Program was started by Maria Burguete and Lui Lam in 2007 with the first international Scimat conference in Portugal and the forming of the International Science Matters Committee [Burguete & Lam 2016]. For more see: www.sjsu.edu/people/ lui.lam/scimat.

8. In fact, even today's science has not proved that Apollo (or any of the Greek Gods) does not exist. We just do not need them or believe in them anymore because we have a simpler and better answer in explaining why the sun rises in the east and sets in the west.

9. The two words philosophy (meaning "love of wisdom") and philosopher are coined by Pythagoras (c.570-c.495 BC) [Bertman 2010: 244]. The word 'philosophy' with the meaning of "all learning exclusive of technological concepts and practical arts," enters the English language in the 13th century [see: Webster Ninth New Collegiate Dictionary (Merriam-Webster, Springfield, MA, 1984)]. Unless otherwise specified, all the dates of words appearing in the rest of this chapter are from this Merriam-Webster dictionary.

10. en.wikipedia.org/wiki/Thales (July 27, 2013).

11. The assertion of the historian Robin Collingwood (1889-1943) in *The Idea of Nature* [(1945) 1960: 3] that philosophy is exclusively concerned with the physical universe is wrong.

12. Of course, it took many years before Darwin's evolutionary theory was accepted by the mainstream, which, in fact, is still rejected by many laypersons on religious grounds. It is thus worthwhile to point out that in 1996, Pope John Paul II has declared that Darwin's evolutionary theory is indeed correct and covers humans, too, except that a human being, unlike other animals, is infused a soul by God when it becomes into being [Pope 1996]. And this was four years after the same Pope admitted that Galileo was mistreated by the Vatican and apologized (*New York Times*, Nov. 1, 1992; www.vaticanobservatory.org/index.php/en/history-of-astronomy/197-the-galileo-affair, Nov. 29, 2013).

13. On the other hand, there are more similarities between humans and other animals than we are ready to admit [Natterson-Horowitz & Bowers 2013].

14. Collingwood has remarked: What we call science, Aristotle called it philosophy [Collingwood 1922].

15. http://en.wikipedia.org/wiki/Natural_ theology (August 4, 2013).

16. The concept of "God of the gaps," due to the evangelist Henry Drummond (1851-1897), suggests that gaps in scientific knowledge should be taken as evidence or proof of God's existence (en.wikipedia.org/wiki/God_of_the_gaps, Nov. 30, 2013). Since new gaps keep on appearing while science closes the old gaps, this argument will never fail [Lam 2004]. For instance, when Newton's argument that it is God's hand that keeps the stars from sticking to each other is no longer needed after the expansion of the universe is established and explained by the big bang theory, Vatican embraces the big bang theory in 1951 before many scientists are willing to do so [Linder 2004]. And that is 14 years before big bang's final confirmation by Arno Penzias and Robert Wilson's experimental discovery of the cosmic background radiation comes in. The reason: Origin of the big bang is the new gap in science but Vatican has the answer: God did it.

17. In this chapter "natural science" with quotation marks means the science of nonhuman systems, the conventional definition. We prefer to call natural science the "science of all things in nature"; i.e., Science = Natural Science, by Scimat's definition [Lam 2008a].

18. As confirmed by private documents and personal papers which became known since 1936, for Newton, religion and science were two inseparable parts of the same life-long quest to understand the universe. See PBS's NOVA program "Newton's dark secrets" (www.pbs.org/wgbh/nova/physics/newton-dark-secrets.html, Sept. 1, 2013), and also [Buchwald & Feingold 2013].

19. Note that History is sometimes classified as social science. It depends on the scientific level achieved in the discipline as conceived by the classifier [Lam 2008a].

20. Honesty appears in two of the five items in the Scimat Standard [Lam 2008a: 27].

21. Alchemy was a legitimate research topic in chemistry before the emergence of nuclear physics, only about 100 years ago. From nuclear physics we know that the identity of an element, like gold, is dictated by the number of neutrons and protons inside the nucleus of the atom. Thus, one can never obtain a gold atom from another metal by chemical method only since chemistry only changes the electrons but not the nucleus.

22. Only basic science will be considered below. Applied science and technology are quite different and are excluded.

23. Note that this demarcation is not sharp and is sometimes problematic. For example, a complex system, once understood, could become a simple system [Lam 2008a].

24. The mechanism behind the Archimedes' Principle is that water molecules keep on bombarding the body immersed in it as if the volume of water displaced by the body is still there. Since the existence of molecules was not confirmed until 1905 there was no way that Archimedes could figure out the mechanism behind his Principle.

25. Attempts to move away from Aristotelian physics since the 6th century and before Galileo are discussed by Peter Berg [2012].

26. See: "Galileo's inclined plane experiment" (galileo.rice.edu/lib/student_work/experiment95/inclined_plane.html, Dec. 1, 2013).

27. Mathematization though desirable is sometimes over emphasized. Even within physics, Newton's third law of motion (action and reaction) and Thermodynamics' third law (it is impossible to reach absolute zero temperature in finite number of steps) are both written in words. Darwin's evolutionary theory, the most important result in biology, is expressed in words only, too.

28. A similar case appeared at about the same time. Einstein in 1905 showed that Newtonian mechanics is a special case of his theory of special relativity when the body moves slowly in comparison with light. That is, what we see as a high mountain in the Knowscape turns out to be part of a higher mountain.

29. Bardeen is the only person in history who won two Nobel Prizes in physics, for co-inventing semiconductor and the BCS theory of superconductivity [Hoddeson & Daitch 2002].

30. Source: en.wikipedia.org/wiki/Socrates (Jan. 1, 2014). The Socratic tradition of "questioning and debate," so central and fruitful in advancing (scientific)

knowledge, has been continued in the West but less so in the East. In China, a similar tradition starting in the Autumn and Spring period (770-476 BC) was broken early on after Confucianism became the official state ideology of the Han dynasty (206 BC-220) and has not yet been completely restored today.

31. This is demonstrated vividly in the so-called "Lee-Yang dispute." The two Nobel laureates Tsung-Dao Lee and Chen-Ning Yang, in their personal recounts of how the parity nonconservation work was done, cannot agree on which Chinese restaurant that the crucial idea was raised, not to mention who is the one who raised it [Yang 1983; Lee 1986; Chiang 2002; Zi et al. 2004].

32. In the West, it started with Shamans and apothecaries' "niche occupation" of healing and ancient Egyptians' system of medicine before Hippocrates (c.460-c.370 BC) became the "father of Western medicine" in early Greek times. In the East, medical knowledge dates equally early (en.wikipedia.org/wiki/History_of_medicine, Dec. 30, 2013).

33. In the US, the Fiscal Year 2012 funding for the National Institutes of Health is $30.860 billion, which is 4.3 times that of the National Science Foundation's $7.105 billion [Sargent Jr. 2013].

34. For example, when a windowless room with one door containing a large group of people inside is on fire, everyone will rush to the door to escape. As an approximation, the different thoughts going on in their brains could be ignored. And a strategy to avoid jamming the door could be designed from computer simulations by treating the humans as point particles with simple interactions. This kind of research that concentrates on common human nature (the desire to escape from fire) avoids the localization problem. Using active walks [Lam 2006], pedestrian modeling has been developed successfully into a science, with applications ranging from pedestrian trail formation in a German campus [Helbing et al. 1997] to crowd control in Mecca [Helbing & Johansson 2010].

35. The 2008 economic meltdown could be attributed to three causes: (1) the financiers themselves; (2) central bankers and regulators who failed to see it coming; and (3) the macroeconomic backdrop of low inflation, stable growth and plenty of cheap Asian money ("The origin of the financial crisis," *The Economist*, Sept. 7, 2013, pp 74-75). The financiers part involves their hiring of mostly high-energy physicists, the "quants" [Derman 2004], who helped to design financial "products" that were sold worldwide, making a lot of money for everyone before it collapsed. No one seemed to remember that for a new "product" to be put on the market (like a toy for children), safety tests (called stability in physics) should be performed first. In the expert's words, "the quants who devised the highly leveraged financial derivatives ignored systemic risk" [Stein 20110]. And unfortunately, stability analysis is not part of a high-energy physicist's training. These days, the stability analysis is called "stress test" mandated by the government to the banks. Recommending Wall Street to hire more physicists [Weatherall 2013] is not a bad idea, if only proper training/briefing is prescribed to these s-res before turning them into c-res and letting them play with real money.

The case of physicists doing Econophysics [Ball 2006] is a different matter since they are just creating theory, not products.

36. For example, Duncan Watts has a BS in physics, PhD in engineering and is now a professor of sociology at Columbia University; Dirk Helbing, PhD in physics, is Chair of Sociology at Swiss Federal Institute of Technology Zürich.

37. See, e.g., *The Counter-Revolution of Science* by the economics Nobel laureate Friedrich Hayek (1899-1992) [(1952) 1979] and the philosopher Peter Winch's *The Idea of a Social Science and Its Relation to Philosophy* [(1958) 2008].

38. For example, whenever we set foot on the street (or even the sidewalk) we run the risk of being hit by a car. It does happen. In the first three months of 2014 in San Francisco, six pedestrians lost their lives this way (*World Journal*, Mar. 21, 2014, p B1). The same goes for driving. Being careful yourself is not enough; it also depends on how careful the other drivers are.

39. See [Kevles 1987] and the many biographies of scientists.

40. See PBS's "Newton's dark secrets" (www.pbs.org/wgbh/nova/physics/newton-dark-secrets.html, Sept. 1, 2013).

41. en.wikipedia.org/wiki/Socrates (Jan. 1, 2014).

42. en.wikipedia.org/wiki/Giordano_Bruno (Jan. 1, 2014).

43. en.wikipedia.org/wiki/Scientism (Jan. 2, 2014).

References

Alexander, H. G. (ed.) [1956]. *The Leibniz-Clarke Correspondence*. Manchester: University of Manchester Press.

Armstrong, K. [1993]. *A History of God: The 4,000-Year Quest of Judaism, Christianity and Islam*. New York: Ballantine.

Ball, P. [2006]. Econophysics: Culture crash. Nature **441**: 686-688.

Bauer, H. H. [1994]. *Scientific Literacy and the Myth of the Scientific Method*. Urbana: University of Illinois Press.

Berg, R. [2012]. Beyond the laws of nature. Philosophy Now, Issue 88, January/February 2012: 24-26.

Bertman, S. [2010]. *The Genesis of Science: The Story of Greek Imagination*. New York: Prometheus Books.

Buchwald, J. Z. & Feingold, M. [2013]. *Newton and the Origin of Civilization*. Princeton: Princeton University Press.

Buckingham, W., Burnham, D., Hill, C., King, P. J., Marenbon, J. & Weeks, M. [2011]. *The Philosophy Book*. New York: DK Publishing.

Burguete, M. & Lam, L. (eds.) [2014]. *All About science: Philosophy, History, Sociology & Communication*. Singapore: World Scientific.

Burguete, M. & Lam, L. [2016]. The Scimat Program: Towards a better humanity. *Humanities as Science Matters: History, Philosophy and Arts*, Burguete, M. & Riesch, H. (eds.). Luton, UK: Pantaneto Press.

Buxton, R. [2004]. *The Complete World of Greek Mythology*. New York: Thames & Hudson.

Byrne, P. [2013]. Waiting for the Revolution: An interview with the Nobel Prize-winning physicist David J. Gross. Quanta Magazine, May 24 2013 (www.simonsfoundation.org/quanta/20130524-waiting-for-the-revolution/, Nov. 27, 2013).

Capra, F. [2007]. *The Science of Leonardo*. New York: Anchor.

Castellano, C., Fortunato, S. & Loreto, V. [2009]. Statistical physics of social dynamics. Review of Modern Physics **81**: 591-646.

Chen, Fong-Ching (陈方正) [2009]. *Heritage and Betrayal: A Treatise on the Emergence of Modern Science in Western Civilization*. Beijing: SDX Joint Publishing.

Chiang, Tsai-Chien (江才健) [2002]. *Biography of Yang Chen-Ning: The Beauty of Gauge and Symmetry*. Taibei: Bookzone.

Churchland, P. S. [1986]. *Neurophilosophy: Toward a Unified Science of the Mind/Brain*. Cambridge, MA: MIT Press.

Churchland, P. [2007]. *Neurophilosophy at Work*. Cambridge: Cambridge University Press.

Cioffi-Revilla, C. [2014]. *Introduction to Computational Social Science: Principles and Applications*. New York: Springer.

Collins, H. [2014]. Three waves of science studies. *All About science: Philosophy, History, Sociology & Communication*, Burguete, M. & Lam, L. (eds.). Singapore: World Scientific.

Collingwood, R. G. [1922]. Are history and science different kinds of knowledge? Mind XXXI: 443-451.

Collingwood, R. G. [(1945) 1960]. *The Idea of Nature*. London: Oxford University Press.

Cornford, F. M. [(1912) 2004]. *From Religion to Philosophy: A Study in the Origins of Western Speculation*. New York: Dover.

Cunningham, A. & Williams, P. [1993]. De-centring the "big picture": The *Origins of Modern Science* and the modern origins of science. The British Journal for the History of Science **26**: 407-432.

Darwin, C. R. [1859]. *On the Origin of Species by Means of Natural Selection, or The Preservation of Flavored Races in the Struggle for Life*. London: John Murray.

Derrida, J. [1976]. *Of Grammatology*. Baltimore: Johns Hopkins University Press.

Derman, E. [2004]. *My Life as a Quant: Reflections on Physics and Finance*. Hoboken, NJ: Wiley.

Donald, M. [2006]. Art and cognitive evolution. *The Artful Mind: Cognitive Science and the Riddle of Human Creativity*, Turner, M. (ed.). Oxford: Oxford University Press.

Doren, C. van [1991]. *A History of Knowledge: Past, Present, and Future*. New York: Ballantine.

Einstein, A. [1905]. On the movement of small particles suspended in stationary liquids required by the molecular-kinetic theory of heat. Annalen der Physik **17**: 549-560.

Epstein, J. M. [2014]. *Agent_Zero: Toward Neurocognitive Generative Social Science*. Princeton: Princeton University Press.

Evans, R. J. [1997]. *In Defence of History*. London: Granta Books.

Ferris, T. [2010]. *The Science of Liberty: Democracy, Reason, and the Laws of Nature*. New York: Harper Perennial.

Gazzaniga, M. [2012]. *Who's in Charge?: Free Will and the Science of the Brain*. New York: HarperCollins.

Gilderhus, M. T. [2000]. *History and Historians: A Historiographical Introduction*. Upper Saddle River, NJ: Prentice Hall.

Godfrey-Smith, P. [2003]. *Theory and Reality: An Introduction to the Philosophy of Science*. Chicago: University of Chicago Press.

Gordin, M. D. [2012]. *The Pseudoscience Wars*. Chicago: University of Chicago Press.

Gower, B. [1997]. *Scientific Method: An Historical and Philosophical Introduction*. New York: Routledge.

Harrison, P., Numbers, R. L. & Shank, M. H. (eds) [2011]. *Wrestling with Nature: From Omens to Science*. Chicago: University of Chicago Press.

Hayek, F. A. [(1952) 1979]. *The Counter-Revolution of Science: Studies on the Abuse of Reason*. Indianapolis: Liberty Fund.

Helbing, D. & Johansson, A. [2010]. Pedestrian, crowd and evacuation dynamics. Encyclopedia of Complexity and Systems Science **16**: 6476-6495.

Helbing, D., Keltsch, J. & Molnár, P. [1997]. Modelling the evolution of human trail systems. Nature **388**: 47-50.

Henry, J. [1997]. *The Scientific Revolution and the Origins of Modern Science*. New York: St. Martin's Press.

Hirshfeld, A. [2010]. *Eureka Man: The Life and Legacy of Archimedes*. New York: Walker & Company.

Hoddeson, L. & Daitch, V. [2002]. *True Genius: The Life and Science of John Bardeen*. Washington, DC: Joseph Henry.

Hogan, P. C. [2003]. *Cognitive Science, Literature, and the Arts: A Guide for Humanists*. New York: Routledge.

Holton, G. [1993]. *Science and Anti-Science*. Cambridge, MA: Havard University Press.

Jacob, M. [1997]. *Scientific Culture and the Making of the Industrial West*. Oxford: Oxford University Press.

Jaffe, K. [2010]. *What Is Science? An Evolutionary View*. www.dic.coord.usb.ve/WhatisScience.pdf (Jan. 2, 2014).

Jiang, Jie [2013]. Research takes NDA path to historic figures. Global Times, Nov. 13, 2023 (www.globaltimes.cn/content/824454.shtml#.UqPOWdJDtlE, Dec. 7, 2013).

Kandel, E. R. [2012]. *The Age of Insight: The Quest to Understand the Unconscious in Art, Mind, and Brain*. New York: Random House.

Kennefick, D. [2005]. Einstein versus the *Physical Review*. Physics Today, Sept. 2005: 43-48.

Kevles, D. J. [1987]. *The Physicists: The History of a Scientific Community in Modern America*. Cambridge, MA: Harvard University Press.

Knobe, J. [2011]. Thought experiments. Sci. Am., Nov. 2011: 57-59.

Knobe, J. & Nichols, S. (eds.) [2008]. *Experimental Philosophy*. Oxford: Oxford University Press.

Läertius, D. [2011]. *Lives of the Eminent Philosophers*, Hicks, R. D. (trans.). Witch Books.

Lam, L. [1998]. *Nonlinear Physics for Beginners: Fractals, Chaos, Solitons, Pattern Formation, Cellular Automata and Complex Systems*. Singapore: World Scientific.

Lam, L. [2002]. Histophysics: A new discipline. Modern Physics Letters B **16**: 1163-1176.

Lam, L. [2004]. *This Pale Blue Dot: Science, History, God*. Tamsui: Tamkang University Press.

Lam, L. [2006]. Active walks: The first twelve years (Part II). Int. J. Bifurcation and Chaos **16**: 239-268.

Lam, L. [2008a]. Science Matters: A unified perspective. *Science Matters: Humanities as Complex Systems*, Burguete, M & Lam, L. (eds.). Singapore: World Scientific. pp 1-38.

Lam, L. [2008b]. Science Matters: The newest and largest multidiscipline. *China Interdisciplinary Science*, Vol. 2, Liu Zhong-Lin (刘仲林) (ed.). Beijing: Science Press. pp 1-7.

Lam, L. [2008c]. Human history: A science matter. *Science Matters: Humanities as Complex Systems*, Burguete, M. & Lam, L. (eds.). Singapore: World Scientific. pp 234-254.

Lam, L. [2011]. Arts: A science matter. *Arts: A Science Matter*, Burguete, M. & Lam, L. (eds.). Singapore: World Scientific. pp 1-32.

Lam, L. [2014]. About science 2: Philosophy, history, sociology and communication. *All About Science: Philosophy, history, sociology & communication*, Burguete, M. & Lam, L. (eds.). Singapore: World Scientific. pp 50-100.

Lam, L., Bellavia, D. C., Han, X.-P., Liu, C.-H. A., Shu, C.-Q., Wen, Z., Zhou, T. & Zhu, J. [2010]. Bilinear Effect in complex systems. EPL **91**: 68004.

Lee, T. D. [1986]. *Selected Papers*, Vol. 3. Boston: Birhauser.

Lestienne, R. [1998]. *The Creative Power of Chance*, Neher, E. C. (trans.). Urbana: University of Illinois Press.

Lindberg, D. C. [1992]. *The Beginnings of Western Science: The European Scientific Tradition in Philosophical, Religious, and Institutional Context, 600 B.C. to A.D. 1450*. Chicago: University of Chicago Press.

Linder, D. [2004]. The Vatican's view on Evolution: The story of two popes. law2.umkc.edu/faculty/projects/ftrials/conlaw/vaticanview.html (Nov. 30, 2013).

Liu, Bing (刘兵) & Zhang Mei-Fang (章梅芳) [2014]. Scientific culture in contemporary China. *All About Science: Philosophy, history, sociology & communication*, Burguete, M. & Lam, L. (eds.). Singapore: World Scientific.

Livio, M. [2013]. *Brilliant Blunders: From Darwin to Einstein—Colossal Mistakes by Great Scientists that Changed our Understanding of Life and the Universe*. New York: Simon and Schuster.

Lloyd, G. E. R. [1970]. *Early Greek Science: Thales to Aristotle*. New York: Norton.

Lorentz, E. [1993]. *The Essence of Chaos*. Seattle: University of Washington Press.

Lubkin, G. B. [1998]. A personal look back at *Physics Today*. Phys. Today, May 1998: 24-29.

Machery, E. [2008]. A plea for human nature. Philosophical Psychology **21**(3): 321-329.

Mason, S. F. [1962]. *A History of the Sciences*. New York: Macmillan.

Mitchell, M. [2009]. *Complexity: A Guided Tour*. Oxford: Oxford University Press.

Natterson-Horowitz, B. & Bowers, K. [2013]. *Zoobiquity: The Astonishing Connection between Humans and Animal Health*. New York: Vintage Books.

Newton, I. [(1730) 1952]. *Opticks*, Cohen, O. B., Roller, D. H. D. & Whittaker E. (eds.). New York: Dover.

Novak, I. [2011]. *Science: A Many-Splendored Thing*. Singapore: World Scientific.

Oldroyd, D. [1986]. *The Arch of Knowledge: An Introductory Study of the History of the Philosophy and Methodology of Science*. New York: Methuen.

Patterson, S. [2010]. *The Quants: How a New Breed of Math Whizzes Conquered Wall Street and Nearly Destroyed It*. New York: Crown Business.

Pigliucci, M. & Boudry, M. [2013]. *Philosophy of Pseudoscience: Reconsidering the Demarcation Problem*. Chicago: University of Chicago Press.

Planck, M. [1900]. Entropy and temperature of radiant heat. Annalen der Physik **1**(4): 719-737.

Pollard, K. S. [2009]. What makes us human? Sci. Am., May 2009: 44-49.

Pope John Paul II [1996]. Message to the Pontifical Academy of Sciences: On evolution. www.ewtn.com/library/papaldoc/jp961022.htm (Nov. 29, 2013).

Porter, R. [2001]. *The Enlightenment*. New York: Palgrave.

Rollin, B. E. [2006]. *Science and Ethics*. New York: Cambridge University Press.

Ross, S. [1962]. *Scientist*: The story of a word. Annals of Science **18**, 65-86.

Sargent Jr., J. F. [2013]. Federal research and development funding: FY2013. www.fas.org/sgp/crs/misc/R42410.pdf (Dec. 30, 2013).

Shermer, M. [2006]. *Why Darwin Matters: The Case Against Intelligent Design*. New York: Owl Books.

Shermer, M. [2011]. What is pseudoscience? Sci. Am., Sept. 2111: 92.

Shubin, N. [2009]. *Your Inner Fish: A Journey into the 3.5-Billion-Year History of the Human Body*. London: Pantheon/Allen Lane.

Smith, A. [(1776) 1977]. *An Inquiry into the Nature and Causes of the Wealth of Nations*. Chicago: University of Chicago Press.

Stein, J. L. [2011]. The crisis, Fed, quants and stochastic optimal control. Economic Modelling **28**(1-2): 272-280.

Suddendorf, T. [2013]. *The Gap: The Science of What Separates Us from Other Animals*. New York: Basic Books.

Thurs, D. P. [2011]. Scientific methods. *Wrestling with Nature: From Omens to Science*, Harrison, P., Numbers, R. L. & Shank, M. H. (eds). Chicago: University of Chicago Press

Tse, P. U. [2013]. *The Neural Basis of Free Will: Criterial Causation*. Cambridge, MA: MIT Press.

Tuckerman, M. E. [2010]. *Statistical Mechanics: Theory and Molecular Simulation*. Oxford: Oxford University Press.

Venter, C. [2013]. *Life at the Speed of Light: From the Double Helix to the Dawn of Digital Life*. New York: Viking.

Waldrop, M. M. [1992]. *Complexity: The Emerging Science at the Edge of Order and Chaos*. New York: Simon & Schuster.

Wang, C.-C., Yan, S., Yao, C., Huang, X.-Y., Ao, X., Wang, Z.-F., Han, S., Jin, L. & Li, H. [2013]. Ancient DNA of Emperor CAO Cao's granduncle matches those of his present descendants: a commentary on present Y chromosomes reveal the ancestry of Emperor CAO Cao of 1800 years ago. Journal of Human Genetics **58**: 238-239.

Warren, R. [2014]. *Helicobacter*: The ease and difficulty of a new discovery. *All About Science: Philosophy, history, sociology & communication*, Burguete, M. & Lam, L. (eds.). Singapore: World Scientific.

Weatherall, J. O. [2013]. *The Physics of Wall Street: A Brief History of Predicting the Unpredictable*. New York: Houghton Mifflin Harcourt.

Weinberg, S. [2011]. Particle physics, from Rutherford to the LHC. Phys. Today, Aug. 2011: 29-33.

White, H. [1973]. *Metaphysics: The Historical Imagination in Nineteenth-Century Europe*. Baltimore: Johns Hopkins University Press.

Williams, R. [1983]. *Keywords: A Vocabulary of Culture and Society*. New York: Oxford University Press.

Wilson, E. O. [1978]. *On Human Nature*. Cambridge, MA: Harvard University Press.

Wilson, E. O. [1998]. *Consilience: The Unity of Knowledge*. New York: Alfred A. Knopf.

Winch, P. [(1958) 2008]. *The Idea of a Social Science and Its Relation to Philosophy*. New York: Routledge.

Wu, C. S. , Ambler, E., Hayward, R. W., Hoppes, D. D. & Hudson, R. P. [1957]. Experimental test of parity conservation in beta decay. Phys. Rev. **105**: 1413-1415.

Yang, C. N. [1983]. *Selected Papers 1945-1980 with Commentary*. New York: Freeman.

Youngson, R. [1998]. *Scientific Blunders: A Brief History of How Wrong Scientists Can Sometimes Be*. New York: Carroll & Graf.

Zi, Cheng (季承), Liu, Huai Zu. (柳怀祖) & Teng. Li (腾丽) (eds.) [2004]. *Solving the Puzzle of Competing Claims Surrounding the Discovery of Parity Nonconservation: T. D. Lee Answering Questions from Sciencetimes Reporter Yang Xu-Jie and Related Materials*. Lanzhou: Gansu Science and Technology Press.

Ziegler, J. [2013]. *Betting on Famine: Why the World Still Goes Hungry*. New York: New Press.

Published: Lam, L. [2014]. About science 1: Basics —knowledge, nature, science and Scimat. *All About Science: Philosophy, history, sociology & communication*, Burguete, M. & Lam, L. (eds.). Singapore: World Scientific. pp 1-49.

21

About Science 2: Philosophy, History, Sociology and Communication

Lui Lam

Within the last century, four new (sub)disciplines related to science were added to the humanities. They are Philosophy of Science, History of Science, Sociology of Science, and Science Communication. While these disciplines did contribute positively, they had also caused all sorts of problems towards people's understanding of science. What happened and why it happened? This chapter tries to answer this question with new insights gleaned from our historical and cultural heritage of thousands of years. The aim here is not to give a full review of the four disciplines but to analyze them from the perspective of Scimat, coming from a humanist and physicist with experience in simple and complex systems. In particular, the mistakes of Ernest Mach, Karl Popper, Thomas Kuhn, Paul Feyera-bend, and David Bloor and why they occurred are analyzed from a new angle. It has to do with the time-evolving nature of the scientific process, obliviousness of the differences between simple and complex systems, failure of the educational system, and the underdevelopment of the humanities. Suggestions for the near future are provided.

1 Introduction

Within the last century, three new subdisciplines and a new discipline related to science were added to the humanities. The three interrelated subdisciplines that analyze the nature and development of science are Philosophy of Science (PS), History of Science (HS), and Sociology of Science (SS). The new discipline is Science Communication (Scicomm) which depends heavily on the other three. In China these four disciplines are loosely lumped under the umbrella "scientific culture" [Liu & Zhang 2014].

While these disciplines did contribute positively and valuably, they had also caused all sorts of problems towards people's understanding of science. What happened and why it happened? This chapter tries to answer this question with new insights gleaned from our historical and cultural heritage of thousands of years. The early part of this heritage of 195,000 years is recorded in our genes (and fossils) and the last 5,000 years or so in written words [Lam 2011]. It turns out that the problem arose from intertwined issues involving both simple and complex systems. The distinction between these two kinds of systems is not always appreciated by the practitioners, partly due to the shortcomings in our educational system. The aim here is not to give a full review of the four disciplines but to analyze them from the perspective of Scimat (see Sec. 2), coming from a humanist and physicist with experience in both simple and complex systems.

To ease the discussion below we classify the practitioners into three groups according to the amount of training and experience the person has in "natural science": insider, outsider and marginer.[1] An *insider* is a "scientist"-turned-philosopher; examples are Ernst Mach (1838-1916), Arthur Eddington (1882-1944), and John Ziman (1925-2005) [2000]. An *outsider* is someone who does not have an academic degree in "natural science," such as Karl Popper (1902-1994), Paul Feyerabend (1924-1994) and most of the philosophers in science. A *marginer* is a person with professional training in "natural science" but does not become a practitioner in this field for a long period of time. The training could include a PhD in physics—like in the case of Thomas Kuhn (1922-1996), but, more often, a BS or MS in "natural science." It does not matter whether this person has published a few professional papers in "natural science" or not.

One may think that an insider should have the best chance of getting things right, but that was not the case in history. However, the nature of the mistakes they made did differ considerably according to which group they belong. And it was the marginers that got the most following even when they were wrong because people thought, mistakenly, that the marginers should know how science actually works. The reasons will become clear later after representing cases are analyzed.

In the following, the issues in these four disciplines are elaborated in turn (Secs. 3-6) after a brief summary of science is given in Sec. 2. An outline of what happened and suggested actions for the near future are provide in Sec. 7. Section 8 concludes the chapter. To our knowledge, this chapter is the first time that PS, HS, SS and Scicomm are examined together and from a unified perspective.

2 Science in a Nutshell

Science is humans' pursuit of knowledge about all things in nature, which includes all (human and nonhuman) material systems, without bringing in God or any supernatural.[2] Science thus has two parts: (1) the results obtained by scientists, and (2) the process of doing science. Both parts evolve over time and are full of surprises. The scientific process could involve induction, deduction, or intuition, but not necessarily all three of them. Scientific results that remain on the scene are those that passed the *Reality Check* (RC). Passing the RC means "confirmed" by experiments or practices, or, at the minimum, consistent with established data [Lam 2014]. *Science thus has* two *anchors (or signatures): its secularity and the reality check.*

Nature, same as the universe, includes all material systems (plus the invisibles such as the spacetime fabric and energy fluctuations). Humans, a kind of animal called *Homo sapiens*, are material systems made up of atoms. Science thus is the study of the human system and all nonhuman systems. Science of human, or human science, is divided into three parts: the humanities, social science, and medical science. Human science is also called *Scimat* (Science Matters) [Lam 2008a]. Science of nonhuman systems is commonly called "natural science"; "science," the sum of "natural science" and social science. Accordingly, an appropriate metaphor for Science is two linked animals—a *Homo sapiens* and his dog (Fig. 1).[3]

Fig. 1. A metaphor for Science: Two linked animals—a *Homo sapiens* and his dog. The human being controls the dog and can direct it to do good things (pick up a newspaper, say) or bad things (bite the neighbor). The dog represents "Natural Science" while the human being, Scimat (humanities, social science, and medical science). [A dog leash is added to a rescaled picture of Abujoy's *Size Comparison between a Beagle and a Man* (2008), Wikimedia Commons, May 11, 2014.]

Note that the aim of science is to understand nature as a whole. The boundary between different disciplines is not fundamental and should not be taken seriously. Different disciplines share common techniques and tools and can learn from each other. For any discipline there are three research levels/approaches: empirical, phenomenological, and bottom-up [Lam 2002, 2011: 20]. The three levels are routinely used in "natural science" studies while the bottom-up level in the form of computer simulations is getting more common in social science research. Most of the humanities research is at the empirical and phenomenological levels but we start seeing works at the bottom-up level in the last decade or so [Lam 2014]. In principle, the "level of scientific development" of a discipline can be measured by its *Scientificity*.[4] Generally speaking, scientificity is high for disciplines in "natural science," medium in social science and low in the humanities.

3 Philosophy of Science

Philosophy was the one and only one academic discipline in ancient Greece, starting with Thales about 2,600 years ago. It represents human's attempt to understand the world around them, from the stars in sky to their fellow citizens in the same state, by reasoning (with or without bringing in supernatural) and by abandoning the Greek gods [Lam 2014]. It was the most successful discipline in history; every discipline we know of in the universities today branched out from it, directly or indirectly [Morris 2002].

Philosophy of Science established itself as a branch of "philosophy" from 1925 to 1965 [Nickles 2013]. The latter is a discipline within the humanities after the early Greeks' philosophy was split into the humanities, social science, "natural science," and theology in the 18th century (see Fig. 2 in [Lam 2014]). The aim of PS is to understand everything about science, sometimes called "metascience" [Oldroyd 1986]; e.g., what scientists actually do in their investigations, meaning of the scientific concepts and theories, nature of scientific laws and principles as well as explanations, etc.[5] Books with the name PS are usually about philosophy of physics (and of simple systems) while biology (complex systems) [Johnstone 1914] and chemistry [van Brakel 2000] have their own extra philosophical questions since they are about emergent phenomena.

We do have prominent philosophers of science (physics actually) who are insiders, marginers, or outsiders. To understand how their ideas came about and why they are (almost) all wrong, it is imperative to know about their background, personally and historically. Here are some representative cases.

3.1 Ernst Mach (1838-1916)

Ernst Mach (Fig. 2) studied physics and for one semester medical physiology at University of Vienna and received his doctorate in physics at age 22.[6] He was a well-established physicist who also contributed to physiology and psychology. He was looking for a unified perspective on the science of inanimate systems (physics) and animate systems (psychology), and proposed a phenomenalist/positivist philosophy. A central idea is that in science, anything that cannot be observed should be excluded from physical laws which he considered as descriptions of sensations [Oldroyd 1986].

Mach wielded tremendous influence, counting the young Albert Einstein (1879-1955) among his many admirers. He is credited with inspiring the formation of the Vienna Circle, a science movement based on logical positivism-empiricism that eventually failed [Holton 1993].

Fig. 2. Four philosophers of science: Ernst Mach, Karl Popper, Thomas Kuhn and Paul Feyerabend (left to right).

What brought Mach down as a philosopher is the "trivial" mistake he made about the existence of atoms. Forgetting that the invention of a new instrument like the optical microscope can turn something un-seeable before (such as a single-cell amoeba) into seeable, he *betted* that atoms can never be observed and thus not be allowed into any theory. And he lost. These days we can directly show the image of a single atom through a scanning tunneling microscope which magnifies the atom 100 million times [Greenemeier 2013]. But in 1905 and the few years afterward, the existence of atoms was demonstrated indirectly but convincingly through Einstein's theory of Brownian motion and Jean Perrin's confirming experiments [Kennedy 2012]. The mistake committed by Mach was a little bit curious. He did not exclude atoms from theoretical grounds but on technical grounds (that they can never be observed technically) [Holton 1993: 58]. The lesson here is that *betting on the future of anything is precarious and should proceed with caution.*

3.2 Karl Popper (1902-1994)

Karl Popper (Fig. 2) was born in Vienna, in 1902. At age 17, he was attracted by Marxism and joined the workers' party but abandoned the ideology the same year; he remained a supporter of social liberalism throughout his life. At age 26, he earned a doctorate in psychology. In 1937, he immigrated to New Zealand upon Nazism's rise in Europe. Nine years later after World War II (WWII), he moved to England and became professor at London School of Economics. He died at age 92 in UK, in 1994.

Popper is famous for his idea of what constitutes a scientific theory: The criterion of the scientific status of a theory is its *falsifiability*, not its confirmability. The idea was formulated in "the winter of 1919-1920," the years he messed around with Marxism as a teenager and well before he had a professional training in psychology [Popper 1963]. He was motivated to find, in a rush, the difference between the established scientific theories and Marxism which he just abandoned and viewed as pseudoscience.[7] His "theory" looked nice on paper (ironically, like Marxism did) but it is easy to see that it could not be right and is impossible to be upheld in practice (see below). Instead of abandoning it when he grew older, Popper spent all his life in defending it.

Rigorously speaking, a scientific hypothesis cannot be falsified empirically in the manner Popper wants it. As already pointed out by Pierre Duhem (1861-1916) [1906 (1954)] when Popper was four years old, it is because auxiliary assumptions (like energy conservation) always accompany a hypothesis' explicit assumptions. Therefore, when a (reliable) experimental result is found to disagree with the hypothesis' prediction, it is impossible to decide by deduction alone whether it is the auxiliary assumptions or the hypothesis' explicit assumptions that are at fault. Unfortunately, Duhem's analysis was not "fully absorbed" by Popper even though he was aware of it [Worrall 2003: 72]. Here are more remarks on falsificationism:

1. Demarcation between science and pseudoscience is a messy business which could be a non-issue if science is defined properly and broadly [Lam 2014]. Instead, how to identify the "best available theory" should be a topic in PS [Berg 2012].

2. Falsification can only be applied to mature theories; it excludes many early theories if taken seriously. That is like defining a human being to be a person who can walk; that would make all crawling babies to be nonhumans. In fact, it may take several or hundreds generation of scientists to modify and develop a theory from its inception to its mature form of being predictable. It is a time

consuming process and more often than not, a collective process. We have to be very patient, very patient.

3. Popper had mainly physics in mind when he formulated his falsification hypothesis. It is more difficult for non-physics "theories" to make predictions, even when they are good. For example, Darwin's evolutionary theory hardly predicted anything precisely when it was first proposed [Zimmer 2001]. It did predict that the Earth has to be very old (so biological systems have enough time to evolve) but could not tell how old. For the same matter, Marxism did predict that given enough time, every society on earth will become communism.

4. The problem is that Popper ignored or was unaware of the basic differences between the theories of deterministic, simple systems (like Newtonian mechanics) and that of a stochastic, complex system (humans in this case, to which Marxism and psychology belong). For example, for a deterministic system like a piece of stone falling from sky, you can ask how long it will take to hit the ground. But for the simplest stochastic system, a random walk, you cannot ask where it will be after its next step, not to mention after 10 or 100 steps; you have to ask different questions [Lam 1982].

5. Falsification is not the way theories are developed in practice. (An example is given in [Ben-Ari 2005: 68].) It is routine to get a theoretical paper published which explains a new experimental finding without showing any predictions.

6. There are many legitimate laws of nature that cannot be falsified. Here is an example. Equation (1) is a *new* law with the H term added to Newton's law of gravity, where F is the gravitational force between two point masses m_1 and m_2 separated by a distance r. The point is that the added H term is perfectly consistent with everything known. Experimental tests, with high-enough resolutions in the future, may show the additional $1/r^4$ behavior (confirming but not falsifying this new term); or, failing to find any evidence of it, can only give an upper limit to the constant H but can never conclude that $H = 0$ (failing to falsify it). Why do scientists not keep this H

term? Because there is no need to do so, *so far*, as experimental evidence is concerned—the Occam's Razor at work.

$$F = G \frac{m_1 m_2}{r^2} + H \frac{m_1 m_2}{r^4} \tag{1}$$

3.3 Thomas Kuhn (1922-1996)

Thomas Kuhn (Fig. 2), born in 1922, earned his BS (1943), MS (1946), and PhD (1949) in (solid-state theoretical) physics, all from Harvard University. He has published a few physics papers around 1949. But he changed field to history of science before that. In 1948-1956, he taught HS at Harvard, first as a general-education instructor and later as Assistant Professor of General Education and the History of Science at Harvard.[8] In 1956, after promotion failed, Kuhn moved to University of California, Berkeley, as an assistant professor of HS in two departments: history and philosophy departments. In 1961 he was discontinued by the philosophy department but promoted to full professor in the history department [Hufbauer 2012].

Kuhn's break came in 1962 when he published *The Structure of Scientific Revolutions* (*SSR* for short; 2nd ed., 1970; 3rd ed., 1990). Two years later, he joined Princeton University as chair professor of Philosophy and History of Science. In 1979 he joined the Massachusetts Institute of Technology (MIT), not Harvard, as chair professor of Philosophy, remaining there until 1991. Kuhn was diagnosed with lung cancer in 1994 and died in 1996. Throughout his life, he published three monographs in HS and one in PS. What made him famous are not his HS writings but his PS book, the *SSR*. And he spent his life since its publication to revise it and confuse his critics and friends alike, ending, apparently, as an unhappy professor.[9] To understand why Kuhn, someone with a rigorous training in physics, failed to clarify himself in a span of 34 years (1962-1996) will be a major topic here and in Sec. 7.4. But first, let us look at the simple mistakes that Kuhn committed in *SSR*.

For one, Kuhn soon found out that the "paradigm shift" he proposed in 1962 was so problematic that, instead of abandoning the concept as a good physicist would do, he abandoned the term and replaced it by "disciplinary matrix" and "exemplar" he coined in the second edition of the book in 1970 (see [Oldroyd 1986: 324]). In other words, he started playing with words.

Another, the "incommensurability" he also proposed in 1962 was equally in trouble; the mistake of this one is easier to see. For example, contrary to what Kuhn claimed, the mass in Newtonian mechanics is recognized as the "rest mass" in Einstein's special-relativity theory, which is Einstein's mass when the velocity of the body approaches zero (Fig. 3).

$$m = \frac{m_0}{\sqrt{1 - (v/c)^2}} \approx m_0 \left[1 + \tfrac{1}{2} (v/c)^2 + \ldots \right]$$

Rest mass (Einstein)
= Mass (Newton)

Mass (Einstein)

Fig. 3. Relationship between Newton's mass and Einstein's mass. (See text.)

What happened? According to Kuhn as told to Steven Weinberg [2001: 203-204], a physics Nobel laureate and Kuhn's colleague at MIT, in 1947 he was studying Aristotle's work in physics as a young physics instructor at Harvard, and it suddenly dawned on him that Aristotle's mechanics was not "bad physics" when judged in Aristotle's times, even though it is wrong by today's knowledge. And, according to Weinberg [p 204], the (obvious) paradigm shift from Aristotle's mechanics to Newton's mechanics could be what led Kuhn to his later ideas. If so, here lies the key in understanding Kuhn's fallacy in his incommensurability. The concepts, terms, and results of Aristotle's mechanics are nowhere to be found in Newton's mechanics because the former does not pass the RC (reality check)—i.e., agreeing with the real world, while that of Newton's mechanics are retained or recognizable in Einstein's special relativity because they do. In other words, anything not passing the RC will disappear eventually and incommensurability happens. In the opposite case, anything passing the check will somehow remain and be recognized, and no incommensurability exists.[10] In fact, RC is what distinguishes science from other forms of human's inquiry [Lam 2014], but strangely, it is neither mentioned in Weinberg's discussion of Kuhn nor by Kuhn and others in PS, HS, or SS.

Two questions remain: (1) If Kuhn's premises in *SSR* are so obviously wrong why the book "has sold over a million copies in two dozen languages" [Nickles 2003: 1]. (2) If Kuhn is so wrong why he "was perhaps the most influential philosopher writing in English since 1950, even the most influential academic" [Sharrock & Read 2002: 1]? The answer to question 2 is easy: The status of a philosopher is judged not by the correctness of his ideas (which mostly are wrong by later judgment; think Plato) but by how much a stir he created during and after his lifetime. And Kuhn created quite a stir, as evidenced by the fact mentioned in question 1. The answer to question 1 is less trivial, but nothing subtle. This will be discussed in Sec. 7.4.

3.4 Paul Feyerabend (1924-1994)

Paul Feyerabend (Fig. 2) was born in Vienna, in 1924.[11] He developed an interest in theater and started singing lessons in his teens. In 1943, he joined the army; his spine was hit by a bullet. This made him impotent; severe pain accompanied him daily and he had to walk with a stick for the rest of his life [Feyerabend 1995]. At age 24, he met Karl Popper. The next year he became a founding member of the Kraft Circle, a post-WWII extension of the Vienna Circle. He studied with Popper at the London School of Economics in 1952, but he was critical of Popper's falsificationism. He started working at University of Bristol in 1955.

Since 1958, he was professor at University of California, Berkeley; he and Kuhn overlapped with each other there. He liked to travel. He has been visiting professor at University College London, in Berlin, and at Yale University. He taught at University of Auckland, New Zealand (1972 and 1974). In the 1980s, he alternated between ETH, Zürich, and UC Berkeley. He left Berkeley for good in 1989, first to Italy, then Zürich, and retired in 1991. He died of brain cancer in 1994. He has written two major books, *Against Method: Outline of an Anarchistic Theory of Knowledge* (1975) and *Killing Time* (1995), an autobiography just before he passed away.

The essence of *Against Method*, written like a long manifesto, is that there are innumerable different methods in scientific inquiry and each is worth trying. Thus, science cannot be regarded as a strictly rational enterprise; it may depend upon people thinking counter-intuitively. So far so good, even though this is common knowledge among the good scientists. But he went on to say, "[T]he time is overdue for adding separation of state and science to the...separation of state and church" [Feyerabend 2010: 164]. This is bad and wrong because the high-school science he talked about, like mathematics and English, is simply a skill the citizens need to know to survive in today's technological society (to understand the electric bills and read the nutrition labels on food products, say).

Since Feyerabend has written "I hope...the reader will remember me as a flippant Dadaist..." [Feyerabend 2010: xiv] and begged us not to take him seriously, we should honor his wish and forget his sayings except for the central statement he made against method in doing science.

4 History of Science

History of science is a branch of history. The latter is about all things happened to humans in the past [Lam 2002] and the former is about what and when scientific results are obtained and how and when scientists go about in obtaining them. Obviously, HS would and did suffer from the lack of a clear definition of science, a failing of the PS.

The beginning of History is credited to the ancient Greek, Herodotus (c.484-425 BC). Called the "father of history," he was "the first historian known to collect his materials systematically, test their accuracy to a certain extent, and arrange them in a well-constructed and vivid narrative".[12] Similarly, HS has an equally long history associated with the names of Eudemus (c.370-316 BC), Plato (427-347 BC), and Aristotle (384-322 BC).[13] But George Sarton, a Belgian and the "father of History of Science," was the one who made HS into an autonomous academic discipline in the early part of the 20th century [Liu 2008; Kragh 1987] while professionalization of HS speeded up in the 1950s [Kuhn 1984].

Sarton studied chemistry, crystallography, and mathematics; he obtained his PhD at University of Ghent (1911) with the thesis "The principles of

Newton's mechanics" [Garfield 1985]. The next year, he founded *Isis*, a review devoted to HS *and* PS aimed at philosophers, historians, sociologists, and "scientists"; he was its editor for 40 years. He immigrated to the United States in 1915 and taught at Harvard University. He founded the History of Science Society (1924) and its two official journals, *Osiris* (1936) apart from *Isis* (1912). Today, HS is a well-established discipline, with the International Union of History and Philosophy of Science's Division of History of Science and Technology (IUHPS/DHST) founded in 1947. As we see it, there are three central issues facing HS, described next below.

4.1 Scope of History of Science

The history of human-related complex systems (in social science and the humanities) is not yet included in HS. The fact that Sarton failed to recognize the humanities and social science as part of science is unfortunate since he was born long after Darwin's evolutionary theory first appeared (see Sec. 2). Consequently, he ended up advocating a New Humanism which tried to connect the "sciences" to the humanities using HS [Sarton 1962]—Sarton's way of solving the (not-yet-named) "two-cultures" problem, a viable but inefficient and insufficient approach—a mission not endorsed by Kuhn [1984].

A better approach is to increase the humanities' scientificity through collaborations between humanists and "scientists" [Lam 2008a]. Presently, partly due to Sarton's misconception of science, HS institutes cover only "natural science," medical science, and mathematics,[14] nothing on social science and the humanities.

4.2 How much detail?

How much detail the historian of science wants or needs in her historical description depends on the circumstances and personal taste. Generally speaking, more details are provided as time progresses and HS has experienced four levels of details so far. This issue is not always appreciated by HS researchers.

Since science covers everything in nature, HS is about the historical exploration of the Knowscape—the "landscape" of knowledge [Lam 2014], which includes two major mountain ranges corresponding, respectively, to human-independent knowledge ("natural science") and human-dependent knowledge (Scimat). The description of the discovery of an extended part of the landscape depends on how much details one wants. As a *first-order approximation*, Sarton's kind of HS gives a very rough description—like who had done what, in a chronological manner. It is similar to reporting who are the firsts to reach the high mountain peaks on earth, from the Carstensz Pyramid, Papua to Kilimanjaro, Africa and to Mount Everest, say. For a more local description, take Mount Everest for example. A rough history would say the British "discovered" it in 1856, and Tenzing Norgay and Edmund Hillary did the first official ascent in 1953.[15] A more detailed description would include all the previous attempts that failed. A much more detailed version would have to include the names of all the members of that expedition, and what equipment they brought with them, etc., etc.

Kuhn's HS amounts to a *second-order* approximation within which personal factors such as psychology are considered. The *third-order* approximation is provided by sociologists who emphasize interpersonal interactions and the social forces at play (see Sec. 5). The feminist approach in HS, by noticing that there are two sexes, male and female, among Knowscape's explorers, takes us to the *fourth-order* approximation [Schiebinger 1999].[16]

A higher-order approximation always brings in new insights, apart from more details. Yet *there is no need to discredit or disown the previous approximation(s) while taking it one order higher in the approximation ladder, and drop the* RC *along the way*. History of science, like any other discipline, is accumulative.

Here is the *fifth-order* approximation, not yet done by anybody: the role of supporting actors/actresses in HS. The characters who provide a scientist's personal needs (food, sex, etc.) have no effect on the scientific results obtained but do heavily affect the scientist's choice of place to live, selection of topics, and his career. For example, a detailed history of

Einstein's discovery of general relativity could include who cooked for him during those ten years (1905-1915) when the work was done, since without a good cook and good food Einstein might never made it. But, of course, the importance of a historical figure is judged by whether this person is replaceable. Einstein was not replaceable; the cook was (or could be—only Einstein would know).

4.3 Beyond narrative

Historical laws do exist and historical research has gone beyond the narrative with studies carried out at three levels: empirical, phenomenological, and bottom-up [Lam 2008c]. At the *empirical* level, statistical analysis and Zipf plot (borrowed from complex-system studies) have been used; at the *phenomenological* level, we have computer modeling; at the *bottom-up* level: computer simulations, differential equations solving, and DNA tracing (see [Lam 2014]). There is no reason that these three approaches cannot be applied in HS studies.

4.4 An open problem

The most interesting open problem in HS, in our opinion, is this one. Aristotle has proposed that heavenly bodies move in circles while territorial bodies move in straight lines [Henry 1997: 16] (Fig. 4). We know that Aristotle, unlike his teacher Plato, did get his hands dirty in empirical studies; he pioneered biology by classifying 540 and dissecting (at least) 50 animal species [Mason 1962: 44].

Why then Aristotle never hurled a piece of stone in air and observed that the path is curved, and not straight lines? More strangely, why there is no record to show that people after him in the next 2,000 years have done this "test" and put down in writing that Aristotle is wrong? It is such an easy test that can be done by a child, unlike the case of disproving the circular motion of the heavenly bodies which has to wait for Copernicus (1473-1543).

Fig. 4. Projectile path according to Aristotle (left) and in reality (right), respectively.

5 Sociology of Science

Sociology of science is the study of the scientific process and the content of scientific knowledge from the sociological viewpoint [Sismondo 2004]; it has a short history of only a few decades. It is part of the "sociology of knowledge" trend associated with the Germans: Karl Marx (1818-1883), Friedrich Nietzsche (1844-1900), Karl Mannheim (1893-1947), etc. [Oldroyd 1986].

It is perfectly alright if the sociologists decide to give a fair and "symmetrical" treatment to competing theories and competing scientists within the historical period in time they choose to study, as advocated in the "strong program in the sociology of knowledge" [Bloor (1976) 1991]—an important part of the so-called Sociology of Scientific Knowledge (SSK) [Collins 2014]. But there is an *intrinsic* limitation to this kind of approach which is ignored by the practitioners: *From the sociological study of a particular period* alone, *one can* never *generalize the finding to cover a long period of time beyond the period under study.* Here are the explanation and related issues.

5.1 The scientific process

Science is done by humans. The scientific *process* involves these steps: (1) Find a problem; (2) ask questions. Then decide (3) on a project for investigation; (4) to get funding or not; (5) who to collaborate with, if any; (6) who to discuss with, if any; (7) where to do it; (8) how to do it; (9) who's previous results to take seriously; (10) to call a news conference or not, if the result obtained looks important; (11) to publish or not; (12) if yes, the manuscript will be reviewed by referee(s) and decided by the journal's editor; (13) if not, apply for a patent or not. All these 13 steps are *human-dependent*, and thus involve sociology. In this regard, SS is a legitimate branch of science studies.

5.2 The scientific results and the Book-Drop Test

For simplicity, let us assume that the scientific *results* obtained are about nonliving systems, such as the "free" fall of a piece of rock. (Free fall means the only force acting on the rock is gravitational; it is the very opposite of the word free—physicists' humor.) The scientific result is that it takes 0.45 s (s for second) for the rock to hit the ground if it is let go from a height of 1 meter. Now, because this result was obtained after the 13 steps above which happened to involve repeated fierce competition, bitter bickering, and ugly rivalry, as observed or learned by a sociologist, this sociologist concludes that the 0.45-s result is not a "true belief" but is socially constructed. In other words, this sociologist does not believe in the existence of *human-independent* knowledge [Lam 2008a, 2014].

For such sociologists—called Class-F sociologists (F or fanciful or fantastic, your pick),[17] here is the *Book-drop Test*: Pick a heavy book around you; remove your right shoe; stand up and drop the book from your waist above your right foot; do not move for 1 s. A *true* Class-F sociologist will not try to move his right foot before the 1 s ends. (What happens to the right foot in that late 1 s involves Newton's second law of motion and medical science and is too complicated to be explained here.) Does a true Class-F sociologist really exist? You bet.

5.3 The generalization trap

The scientific process spans a period of time and is time evolving; it is thus a historical process [Lam 2014]. It is like a growing human being, *a history-dependent and time-evolving system*, unlike the case of an electron which is the same all the time. Let us say that the life of a child at age three is observed closely (24 hours per day, every day) for one year. Whatever one learns from this one-year study, it is obvious that one cannot make generalization from it alone to say much useful about the child's other years, in the past or in the future. Unfortunately, that is what some of the SS researchers did about scientific studies. What usually happens is that if one examines their writings, they look alright and conventional in what they observed but it becomes troublesome when they draw conclusions by generalizing the observations beyond their domain of validity, and fall into a trap of their own making—the *generalization trap*. This is the *intrinsic* limitation that forbids one to generalize a slice of history to the whole history of a long-time period.

Take the acceptance of Einstein's general relativity as an example. In *The Golem*, Collins and Pinch [1998] recount the happenings surrounding the 1919 eclipse experiments[18] that "prove" the validity of the general theory of relativity. They show correctly that the scientific process does involve human-dependent, professional judgments[19] and is not as neat as the so-called Scientific Method leads one to believe. But they go on to conclude that the "culture of science" is formed by the establishment's consensus, and is comparable to what happened in the Soviet Union. Here are several problems: (1) The simple-minded, recipe-like "scientific method" [Thurs 2011] does not really exist [Lam 2014]. (2) It is true that *sometimes* the mainstream consensus on a scientific topic is heavily influenced by those in powerful positions, either in the West (such as Aristotle's mechanics, for 2,000 years) or elsewhere. But (3) the difference is that, eventually, only those results that pass the RC (reality check) will remain in the scene. More importantly, (4) what *The Golem* fails to point out is that the (conclusive) acceptance of a scientific theory is not based on one experiment, even though it may seem like that to outsiders like the reporters.[20]

5.4 The outsider problem

It seems that all sociologists of science are "outsiders" (defined in Sec. 1), reflected in the kind of mistakes they committed. David Bloor, one of the pioneers of SSK, is such an example. Bloor, born 1942 in UK, was director of Science Studies at University of Edinburg. His book *Knowledge and Social Imagery* (1976) introduces the strong program of SSK, with the conclusion that knowledge is not "true belief" but whatever men take to be "knowledge" [Bloor (1976) 1991]. This may be true for some "knowledge" about human affairs that the RC is difficult or impossible to perform due to insufficient or unavailable data—a fact that sociologists were familiar with in their own discipline before they ventured into science studies. As explained above, it is absurd to call those scientific results (about inanimate systems) that pass the RC to be socially constructed, but *the RC is not mentioned by the sociologists.*

In his book, Bloor discusses Pythagoras' abandoning of the study of irrational numbers (after he found them philosophically objectionable) and concludes that mathematics is an empirical enterprise. Here, it seems that Bloor does not understand that mathematics, like science, has two parts: the process and the results. An example of the process: The abandonment belongs to the choosing/terminating of a research topic and, of course, is human-dependent. But the Pythagoras' theorem (also discovered by the Chinese) and the irrational numbers, examples of the results, are mathematical results that exist independently of Pythagoras (and could be discovered by smart aliens). In fact, the particular research topics picked by the scientists are indeed dictated by personal considerations (e.g., she wants to work with this beautiful person) and professional judgments that the time is ripe to tackle them—both are human decisions and subject to sociological and cultural influences.

As outsiders, some sociologists, like many others, have misconceptions about science. For example, they mixed up scientific proof with mathematical proof, without realizing that science is built on approximations and thrives on approximations [Lam 2014].[21] They criticized scientists for fault or overclaim when a rigorous, mathematiccal-like proof is absent.[22]

5.5 The laboratory visits

Scientific works are classified into two types: theory and experiment. Theory is further divided into three types according to how much computer is used: (1) pure theory (no computer calculation), (2) theory with computer calculation, and (3) computer simulation.[23] Experiment has three types, too, according to the size of the team: (4) big-lab experiment (hundreds or more collaborators, like those in Fermi Lab or CERN), (5) medium-size experiment (with one to a few collaborators in the same room, like those in condensed matter research in universities or the old

Bell Labs), and (6) table-top experiment (one or two persons are usually enough). From my own experience,[24] I know that these six types of research are very different. And that is why graduate students in science have to decide early on which type they want to do and are trained accordingly. But it seems this is not known to the outsider sociologists.

At a certain point, some researchers in SS decided to visit the labs, type 5 above, and observed what was going on firsthand [Sismondo 2004: 86-96]. And that is good even though what they learned is common knowledge among graduate students in experiments at good universities.[25] For example, tinkering, skills, and tacit knowledge are involved in experiments. That is exactly why the graduate training is in the form of apprenticeship, and why it is extremely difficult for science in developing countries to play catch up without sending their students abroad. For example, the students learn this when they cannot get the apparatus working and the professor comes in and makes it work by touching here and there.

Tacit knowledge and required tinkering in a lab could arise in two circumstances: (1) The measurements, such as the separation between two components or the size of a component in an apparatus, can only be specified with finite resolution (usually up to two decimal points in centimeters) but the working of the real setup requires higher precision. (2) When three control parameters, from one or a few components, are involved in an apparatus there is the possibility of chaos; i.e., the outcome depends sensitively on the initial conditions [Lam 1998]. Both cases come from the problem of finite resolution which is unavoidable in the real world. And that is a distinctive feature of delicate experiments (like replicating a laser [Collins 1992])—unlike those in undergraduate labs—which require the presence of an experienced experimentalist; not even a good theorist can handle that.[26]

This problem of course will present limitations on the development of expert system and artificial intelligence [Collins 1992; Sismondo 2004] but need not be so if the computer is capable of learning (using genetic algorithms, say). Moreover, the real world can tolerate and work with approximations. For example, you make an appointment to meet a friend at 2:30 pm; you are right on time if you arrive between 2:30:00 pm and 2:30:59 pm, say. Or, when we say this electronic appliance works with a current of 1 Ampere, it actually works with small deviations around this number. Contrary to some people's belief, the real world does not rely on mathematically exact precisions; another way of saying this is that the basin of attraction of the attractor of a real event is not fractal [Lam 1998].

Collins investigated several early gravitational-wave[27] experiments of the Weber type and came up with what he called "experimenter's regress" [Collins 1992, 2014]. It argues that "one knows an experimental system is working when it gives the right answer, and one knows what the answer is only after becoming confident in the experimental system" [Sismondo 2004: 92] and so there is a circle. That may be the case with the few experiments that Collins studied (and is short-term in time). But a good experimentalist like Chien-Shiung Wu would point out why and where the previous experimenters are wrong when presenting her own results that are different [Chiang 2014]. This practice is especially essential if similar experimental setups are used by the challengers. Collins falls into the generalization trap here. Yet, history is full of twists and turns. The latest news (March 17, 2014) is that gravitational waves have been detected via a new route, completely different from the Weber type. They were detected indirectly in the wrinkles of spacetime from the earliest moment of the universe. This time, while champagne was indeed opened at a house in Stanford, the rest of the world is waiting for independent verifications [Chodos 2014].[28]

In fact, the generalization trap applies to most lab-visit studies which are unavoidably limited in scope and time. Moreover, as in physics experiments, taking data is only half the story; *the more difficult part is in interpreting the data.* Sometimes, even a two-year field trip [Latour & Woolgar 1986] is insufficient [Oldroyd 1986: 353-356]. In the case of Aristotle's mechanics, it takes 2,000 years for the issue to be settled. And that is how science works.

5.6 The role of theory

The role of theory is commonly misunderstood by outsiders. As described above, the scientific process is like the exploration of an unknown landscape, except that it is done on the Knowscape. In the very beginning, people try to understand things visible to them (like the heavenly bodies). But after the obvious objects or phenomena are exhausted or do not yield to analyses (e.g., why the sky is blue became known only after quantum mechanics was invented), where in the Knowscape do scientists set their foot on? They cannot wait idly for the rare accidental discoveries to show up. Instead, they pick their research topics guided by the theories. Theories' predicting and retrodicting ability (e.g., Einstein's cosmology) is like a telescope that allows the explorer to look around at long distances.

If science is compared to a vertebrate (or two vertebrates; see Fig. 1) then (interrelated) theories form the backbone and control experiments the other bones of this animal.[29] Without the bones, the animal will not be full of vitality and strength. Theories enable scientists to explore swiftly the Knowscape without visiting every spot on it [Lam 2014].

Theory could appear after experiments (to explain them) or ahead of experiments (to predict them). Theory and experiments are like two brothers conversing, helping and competing with each other while running together. They correct each other's mistakes so that both will be better runners. But experiments (a form of RC) are the big brother who has the final say.

6 Science Communication

Science communication (Scicomm), called "science popularization" in China [Lam 2008b], is a discipline in its infancy. As international English journals are concerned, it has only two printed journals, *Science Communication* (quarterly, since 1979, USA) and *Public Understanding of Science* (now eight issues per year, since 1992, UK), and an e-journal, *JCOM* (quarterly, since 2002, Italy). Only a few books exist (e.g., [Gregory & Miller 1998; Sanitt 2005; Broks 2006]).[30]

Scicomm does not aim to analyze science but to communicate science to the public. It depends heavily on the success of PS, HS, and SS. Despite its obvious importance in a modern society that the citizens have to be informed and consulted in governmental decision makings, Scicomm is a discipline gravely underdeveloped. Scicomm, like any other discipline, has two aspects: pure and applied. Pure Scicomm is Scicomm studies; applied Scicomm is the public understanding (or engagement) of science. The difficulty of Scicomm, the most complex discipline on earth, derives from its very nature and comes from several sources:

1. Scicomm involves every other discipline, from physics to PS and HS, and from cognitive and learning sciences to mass communication. It is hard for anyone to master such a wide range of knowledge.

2. Consequently, we end up with many research papers of substandard quality. They are written by people with knowledge in any other field but little training in Scicomm.

3. The practical (or application) side of Scicomm is most difficult. Unlike a researcher who only has to convince one or two referees of a journal to call her work done, a Scicomm worker has to convince a large number of people, the public, and may never know when her work is done.

But the advantage of point 1 is that everyone can join in and, hopefully, most of them would improve given enough time. As for point 3, there are a lot of practical experiences but, it seems, there are no working theories; after all, we are dealing with the complex system of communication here.

The reason of doing Scicomm studies is the same as that for any other discipline, i.e., for the sake of knowledge. But the reasons of communicating science to the public are less obvious; there are several:

1. Promote the scientific spirit. (This is easier said than done, especially in societies where the scientific spirit is not generally observed by those in power.)

2. Promote the scientific way of thinking. (This is better done by telling scientific-discovery stories. The term "scientific method" is misleading and should be avoided.)

3. Popularize scientific knowledge. (Something practical like why people should floss their teeth daily and something more abstract like no one could mend the Earth after it exploded—as claimed by the founder of some new "religion." That is, the aim is to promote citizens' ability to deal with practical problems and avoid being preyed upon by bad people.)

4. Awaken citizens' inert curiosity about the world/universe. (This should be done in school, but Scicomm can supplement that since not every student has a good science teacher.)

5. Stimulate children's interest to be future humanists/"scientists." (For example, many eminent scientists decide to have a science career after reading a good popular-science book as a youth.)

6. Promote citizens' ability to participate in public-affair discussions and decisions. (This is most tricky; see Sec. 6.3.)

Not listed here are: to convince the public that the government should support scientific research; to share with the public the splendid of doing science (suggested by Carl Safina [2012]). The former is a non-issue since the public is already convinced; e.g., 5 in 10 Americans say the government's spending on science and technology research is about right [Lucibella 2014]. The latter is not restricted to science; artists and religious believers do that, too.

6.1 Two brief histories: United Kingdom and China

The importance of Scicomm was recognized only in the last few decades (as attested by the short history of its journals), even though Scicomm activities can be traced to earlier times [Broks 2014; Yin 2014]. Scicomm now exists around the world [Schiele et al. 2012]. However, the Scicomm histories in UK and in China are quite different from each other.

According to Peter Broks [2014], in UK, in the early 19th century, there was no separation between science and the public; all could join—the

Republic of Science period. In the late 19th century, rise of the experts resulted in the separation of scientific experts and the lay public; Scicomm was needed to bridge the gap. Then, in the 1950s, Soviet Union's Sputnik satellite triggered fear about the low level of scientific literary in the West. In the 1980s, the program of Public Understanding of Science (PUS) appeared, and Scicomm was viewed as the new duty of scientists. In 2000, the emphasis shifted from PUS to PEST (Public Engagement with Science and Technology); a whole new set of questions emerged:

1. What counts as being a scientist?

2. Where do we draw the boundaries between science and nonscience, between scientists and non-scientists?

3. What counts as expertise? Who are the experts? What about lay experts?

4. Why should scientists listen to what the public has to say?

5. Engagement presupposes particular social and political relationships which in turn raises questions about authority and democracy.

But the situation is more complex than these questions indicate; see below.

China's story is different. Science was first piecemeally introduced to China by Western missionaries in the late Ming dynasty (1368-1644) while Scicomm in modern China started in the 1840s [Yin 2014]. The turning point happened in the late Qing dynasty (1644-1912). The importance of technology dawned suddenly on the Chinese after the country suffered humiliating defeats in the hands of the Western powers and Japan, culminating in the invasion by the Eight-Power Allied Forces in 1900. After Qing was replaced by the Republic of China, the May Fourth Movement (1919, the same year general relativity's eclipse experiments were performed) intellectuals concluded that technology was insufficient; China needed science (and democracy). And so Scicomm was pursued earnestly amid the Republic's rough years that followed. After the establishment of the People's Republic of China in 1949, Scicomm in mainland China was transformed from grassroots efforts to government-controlled activities, resulting in a restrictive mode of operation that differs from that in the West [Li 2008].

6.2 Two modes of operation: United States and China

In the United States, Scicomm is a grassroots effort, coming from individuals (including students [Sanders 2011] and private citizens) and learned societies (e.g., the American Physical Society has an Outreach Program and *Focus* on its website—the latest physics research written by science writers for the public); it is a *bottom-up* approach. Scicomm degrees, with one exception, are not offered by the universities but a few Scicomm programs exist (e.g., at Cornell University). The exception is the Department of Life Sciences Communication at the University of Wisconsin-Madison which offers BS to PhD degrees. The grassroots approach seems to work in such a vast country.

On the other hand, the *top-down* approach is adopted in China [Li 2008; Shi & Zhang 2012]. This is due partly to the fact that most people, professionals and citizens alike, are still too busy with their daily jobs since the "big bang"—the country's "reform-and-opening up" policy change of 1978—and have not much time and energy to spare, and partly because no large non-government organizations are able to flourish.[31] No Scicomm degrees are offered in China. The major duty of the one and only one research institute, China Research Institute on Science Popularization (CRISP) under the China Association for Science and Technology, is to carry out government Scicomm projects. CRISP publishes *Study on Science Popularization* (formerly *Science Popularization*), the only Scicomm journal in China. Apart from other things, a major problem with China's Scicomm is that there is not enough number of scientists got involved, even though China is the only country with a Scicomm *law* [Zhang & Ren 2012].

6.3 Why communicating science is difficult: a few examples

There are several reasons, some technical and some fundamental, that communicating science to anybody is difficult, which are illustrated here by a few examples.

1. Newton's three laws of motion: even the physics textbooks are wrong

In all physics textbooks, Newton's three laws of motion are described as about physical "bodies," a word adopted from Newton's *Principia*. However, Newton's "body" actually means "point mass" (a mass of zero size which does not exist physically)—a point ignored by textbook writers and even some physicists, causing confusion among the students. The reason behind this is that many physicists are not aware that physics is built on approximations [Lam 2014], a fact that they should be proud of instead of sweeping it under the carpet. If even scientists cannot communicate physical laws accurately and clearly to their own students you can imagine how difficult it would be for Scicomm people to deliver them to the public.

2. The essence of Copernicus' heliocentric theory

Our universe is more like the surface of a balloon instead of the balloon itself. In other words, every point in the universe can be considered the center. Thus, we can call our Earth the center of the solar system, meaning that the Earth, like any other point in space, can be used as the vantage point in observing the sky. What Copernicus' heliocentric theory actually says is that the paths of the planets are pretty complicated if the observer is located on Earth, but the planets' orbits will be simpler to be described if the observer sits at the Sun: The planets move in concentric circles [actually ellipses as shown later by Johannes Kepler (1571-1630)].[32] That is, the essence of Copernicus' heliocentric theory is about the *convenient* choice of an observational point; it is not about the "correct" choice, a point missed in Scicomm.

3. Try to convince someone astrology is wrong

Let us say you try to explain to someone that when two particular stars line up in the sky, they will not have significant effect on a person located on Earth. For that, you have to say the two stars' only force on that person is gravitational which, according to Isaac Newton (1642-1727), is very weak and decreases inversely proportional with the square of the distance; thus, the forces can be ignored. This someone may reply: It is not zero, right? You have to admit that he is right. A more sophisticated someone may add: Don't you know the "butterfly effect" that says an extremely small force can cause big effect? You have to admit that no one has ever proved the butterfly effect, from chaos theory, does not apply to this three-body (two stars plus the person on earth) problem. We in science have great faith in Newton's laws after we spent many years learning them and using them, but not this someone. Ultimately, a non-expert—"scientist" and layperson alike—will decide to accept the conclusion on a certain topic not because he understands thoroughly the reasoning behind it (which is always technical and only the experts can ascertain it; that is why they are called experts) but because he trusts this person who is explaining it to him. And trust is not built on one day or through one conversation.

4. Lessons of the cold fusion saga

The year 1989 was eventful. China's Tiananmen "incident" occurred on June 4; California's Loma Prieta earthquake of magnitude 6.9 happened on October 17; Berlin Wall fell on November 9. Before that, on March 23, two chemists gave a press conference in Salt Lake City (SLC), Utah, and announced that they had discovered cold fusion, without mentioning their competitor in Provo of the same state. The media and the scientific community around the world went crazy. After a serious of experiments that failed to confirm the claim, the saga ended within one year, resulting in two of the professors resigning from their universities (the third one was already in retirement when the saga began) [Dewdney 1997]. Different lessons from this saga were drawn by people from various quarters, including those from Scicomm. Some say the story illustrates the sociological factors in the doing of science, which is obviously right since the scientific process is a human endeavor. Some say the press conference should not be called. This is wrong because press conferences on scientific breakthroughs have been held before, e.g., parity nonconservation in 1957 announced by Columbia University featuring Chien-Shiung Wu's experiment [Chiang 2014]. And some say the story confirms that the scientific community can seek out the bad science from good science; this is right, too, as usual.

But two fundamental lessons are missed by these commentators: (1) When the SLC professor(s) applied for funding from Department of Energy to further his research, he was opting for academic glory such as a Nobel Prize which will require him to reveal enough details of his experiments for others to confirm the discovery down the road. Submitting a proposal with referees to review it was his first act of revealing information, and he knew that. But at the press conference, his aim has changed since he refused to give out any information; he was now opting for wealth (apart from his 15 minutes of fame). In modern time, getting a Nobel Prize and getting rich could not be had by the same person at the same time because the former requires openness while the latter, secrecy (since commercial applications will be involved). Changing his aim in mid-course was fatal as the SLC professor was concerned. (2) At the press conference, he knowingly overclaimed (and misled the press by not mentioning the Provo competitor). Without detecting neutrons, what he could claim most is that there is a 99.9 % chance he had discovered cold fusion. He could remain a viable scientist in the face of a 0.1% error if he turned out to be wrong, which was indeed the case. Instead, he claimed 100% and there was no way out. In short, the cold fusion saga demonstrates bad human choices apart from bad science (sloppy science in this case). Now how could a Scicomm worker "educate" the public properly if the lessons of the story are completely misrepresented?

5. Interpretation of the Sokal hoax

The fact that the Sokal hoax [Sokal & Bricmont, 1998], amid the "science wars" of the 1990s, stirred up such a strong reaction is due to people's misinterpretation of the event itself. There is no guarantee that the editor would reject Sokal's manuscript if he added in a physics referee, because in the face of two opposing referee reports, assuming that was the case, the editor could exercise his judgment and power and choose to publish the paper. In fact, even prestigious science journals such as *Science* and *Nature* failed to filter out papers with fraudulent claims, as demonstrated in the Bell Labs incident [Agin 2006]. What Sokal did is to show that he can punch a hole in this postmodern journal, but that is the case with every other journal. No big deal! Once again, in Scicomm, as in any other matter, getting the facts right is not enough; getting the interpretation right is equally or more important.

6. Climate change: decision making

Climate change is about a complex system, the long-term weather, which comes with inevitable uncertainties [Ken, 2001] and decision making. It is perhaps the most complex and interesting issue facing the public and policymakers worldwide [Somerville & Hassol, 2011]. The debate centers on how reliable computer modeling of the climate is and, if the doom prediction of global warming is correct, what decisions we have to make to prevent it from happening or getting worse.[33] *This is the first time in human history that we are asked to make big decisions based on the computer prediction of a* messy *system.* The system addressed in climate change is mostly physical—the circulation of Earth's atmosphere and ocean—but includes uncertain chemical and biological factors. The computer simulations involve solving a huge number of coupled differential equations with probabilistic considerations. It belongs to the class of what we call "very messy physics."[34]

Decision making, also a complex system, is a branch of the humanities [Bird & Ladyman 2013] which is part of (human) science. Here, the public has to choose who to trust and the policymakers have to make risky decisions knowing the uncertainties, with potentially huge consequences. The good news is that if we act now to reduce carbon dioxide emission but the computer prediction turns out to be wrong,[35] we still have healthier air to breathe. If we do nothing and the prediction turns out to be true, there will be a lot of human hardships. Either way, there will be hardship but the human race will survive. Of course, there is the third scenario that we do nothing and the computer's prediction is wrong; we will save ourselves a lot of trouble. In climate change, we are asked to *bet* on a messy complex system with probabilistic predictions. It is unlike the making of an atomic bomb, a simple system where $E = mc^2$ is an exact result from well established, deterministic equations.

Unfortunately, there is not much Scicomm can do here. Not too many Scicomm people are equipped to deal with such a situation that involves *two* complex systems—climate change and decision making. Basically, it is because Scicomm's focus in the past is on simple systems, and most

Scicomm workers are not educated in the science of complex system. But is it really nothing Scicomm can do? See Sec. 7.2 for an answer.

7 What Happened and What to Do

Here is a historical, cultural, and social outline (or first-order approximation) of what happened since the big bang, with emphasis on the last 100 years. It is followed, accordingly, by recommendations for PS, HS, SS, and Scicomm in the near future.

7.1 The beginning

It all began with the big bang, or earlier, about 13.7 billion years ago (bya). At 10^{-5} seconds later, "long" after the cosmic inflation (which happened about 10^{-37} seconds after the bang), protons and neutrons appeared. They combined to form nucleus at 0.01-300 seconds after the bang, with *atoms* formed 380,000 years later. Stars appeared 13.4 bya; our solar system, of which Earth is a member, 4.7 bya; life on earth, 3.7 bya [Turner 2009]. The important point is that *all living and nonliving material systems on earth are made up of atoms.*

While the origin of life on earth is an unsolved problem, we do know from Darwin's evolutionary theory (1859) that humans were evolved from simpler lives. However, we are not descendants of the monkey, but of fish.[36] In fact, six million years ago, the human and chimpanzee lineages split. We, *Homo sapiens*, appeared 195,000 years ago in Africa, moved out from there 60,000 years (and earlier) ago and spread all over the world [Lam 2011]. That is, all living human beings are relatives from the same family tree. We share the same genes, more or less.

A turning point appeared 10,000 years ago while villages were created and agriculture began. Written history is known for only 5,000 years or so. Long before that, with plenty of free time available, our ancestors began wondering what happened around them—everything, down on earth and up in the sky, including, very likely, the crucial question of why we are here. Without much objective information, they came up with a reasonable answer for every question they asked: God did it. That was perfectly alright except that it was not very enlightening.

7.2 Ancient times

Then about 2,600 years ago, Thales and others like him showed that they could do better by reasoning. And that was the beginning of Philosophy. With slaves doing all the mundane works in ancient Greece, the citizens not just were financially secure once they were born but were able to live a lifestyle with plenty of leisure. They exercised their bodies and gave us the Olympic Games. But others—like Socrates and Plato but especially Aristotle—did more and gave us all kinds of knowledge about nature, including the human system (ethics, political science, logic, etc.) and nonhuman systems (physics, biology, etc.); i.e., they studied both complex and simple systems. As pioneers, they were really ambitious and fearless. They asked all sorts of basic questions such as what truth, reality, etc. are, and demanded absolute knowledge. Soon after, Archimedes made the breakthrough in a simple system by discovering the principle on buoyancy force [Lam 2008a]. The works by Thales, Aristotle, and Archimedes on inanimate systems (the supernatural-free parts) are later recognized as part of "natural science."

Meanwhile, at about the same time as Plato and Aristotle, in the Spring and Autumn period (770-476 BC) in ancient China, Confucius and Laozi chose to work on the complex system of humans, essentially on ethics and political science at the empirical level. Since then the Chinese paid dearly for this "unlucky" choice of ignoring the simple systems; they are still playing catch up in science. The reason is that modern science since Galileo got its breakthrough by focusing on simple (inanimate) systems, while complex systems like human matters are so difficult that it has not advanced much since Aristotle. In other words, modern science did not arise in China because the Chinese picked the wrong topic in their research [Lam 2008a].[37, 38]

7.3 Modern times

Knowledge acquired by the ancient Greeks was preserved by the Arabs and passed to the Europeans in Renaissance times, forming the basis of modern Western civilization and, in fact, a major part of all civilizations today. In particular, Aristotle's *theoretical* cosmos system was enforced by the Christians until it was overthrown by Galileo's *experimental* findings. But it was the publication of Newton's *Principia* in 1687 that was the dividing line that ushered in the rapid development of the "natural science" which are about (mostly simple) nonhuman systems and the gradual decline of the humanities which are about the (complex) human system, as we have witnessed in the last 300 years. Why was that? How did this happen?

It all has to do with a misunderstanding of what Newtonian mechanics is about and the misconception of the human system. The former is less so today but the latter persists to a large extent. Largely encouraged by Newton's success, the Enlightenment started the next year and lasted 101 years, the goal of which was to create a Science of Man, presumably in the fashion of Newtonian mechanics. The intention was good but the timing was wrong. It was wrong on two accounts [Lam 2014]: (1) Newtonian mechanics is about deterministic systems while the human system is probabilistic. The "clockwork" worldview derived from the former, does not apply to the latter. (2) The tool of probability required to handle the human system was not yet ready [Lestienne 1998].

Fortunately, everything about humans is an emerging property and can be handled without knowing the mechanisms involved; they can be studied at the empirical or phenomenological level, say. As it turned out, Adam Smith (1723-1790) succeeded in creating Economics (1776), the first discipline in social science, by staying away from Newton. On the other hand, after the Enlightenment, Auguste Comte created Sociology (1844), which he initially called Social Physics, along the line of Newton—a mistake corrected later by others. Amazingly, as in the case of psychology, a totally new discipline can be created with everything wrong in the beginning; what is needed is a good concept, persistence, and enough number of followers who make it right later.

While all this was going on in the social science, not much progress had been made in the humanities. The reason is that a single human being is much more complicated than a large number of human beings, in the sense that many approximations that could be made in the latter could not be made in the former (see [Lam 2014] for examples). And as breakthroughs in the "natural science" appeared one after the other, the humanities' scientificity remained kind of flat. It appeared that the humanities had been relegated to the backyard. In the midst of this, in 1859, Charles Darwin published *On the Origin of Species*—a publication as relevant to the humanities as Newton's *Principia* is irrelevant, and as controversial as the latter was uncontroversial when it first came out. Out of the many implications coming out of Darwin's evolutionary theory, the one most important to the humanities has been overlooked by almost everybody; i.e., since humans are a kind of animal evolved from the fish and beyond, then *all studies on humans (like in the case of bees)—the humanities in particular, are part of science.*

7.4 Last century

The first three decades of the 20th century was the peak years in physics. Einstein's theories of special and general relativity (1905 and 1915, respectively) and quantum mechanics (1925/1926) appeared, which, however, are irrelevant to the humanities [Lam 2002]. What is relevant is Einstein's proof (indirectly via Brownian motion) that atoms actually exist, published in 1905. The reason is that since humans are made up of atoms which do exist, then, like any other material system in nature, all human-related matters are part of science.

The Vienna Circle, originated in Vienna (1922) and inspired by Mach, tried to pick up where the Enlightenment failed. They tried to constrict science into a jar (see Sec. 3.1) and failed, too, done in by Einstein and Kurt Gödel [Byers 2011]. Out of Vienna came Popper and Feyerabend, teacher and student.

The mistakes made by these two humanists are opposite to each other, at two extremes. Popper, by restricting scientific theories to those that can be falsified, kept science's door open by a slit. Feyerabend, favoring no restrictions, flung the door wide open. Moreover, Feyerabend, by saying science is not strictly rational without mentioning the RC, opened the door for misunderstanding and misinterpretation about science. What he should say carefully is that the scientific *process* may not be "rational" all the time,[39] but the scientific *results* (that pass the RC) are rational results. Well, maybe he was not aware of the difference or maybe he did not care.

Telling people that science is not strictly rational would get him more public attention in the 1960s era of antiwar and anti-establishment, especially in Berkeley where he was teaching [Gitlin 1993]. And he loved attention and theatrical presentations; remember that he was interested in theater and singing as a teenager?

Kuhn's case was different. With his PhD physics training that emphasizes clarity, it would not be too difficult for him to come out and say something like "Newton's gravitational law or its equivalent could be discovered by smart aliens, too," meaning that the law has validity beyond human constructs.[40] Why didn't he do so? My *conjecture* is that Kuhn enjoyed so much his sudden fame coming with his 1962 book of ambiguity that he decided to prolong it by keeping his positions ambiguous.[41] It is like someone "hiding in the corset," with a secret to keep. And that explains why he ended up feeling frustrated, being misunderstood by others—a condition he could remove easily but "couldn't."

An academic book could become a best seller if (1) it is not too technical (no equations, say), (2) not too thick, (3) happens to satisfy the need of a large number of readers, and it helps if (4) the book is full of ambiguities.[42] Kuhn's 1962 book could be the only one that satisfies all these four conditions. Before we talk about who were among the readers and what are their needs let us recall the two "revolt against (natural) science" in the 20th century America [Kevles 1987]. The first revolt occurred in the 1930s after the great depression of 1929. The complains were twofold: (1) "Natural science," technology in fact, was blamed for dehumanizing the society—machines displacing workers, etc., and (2) the simple-minded opinions offered by some "natural scientists" on social matters and the over respect paid to their opinions by the press. The second revolt happened in the 1960s, the anti-Vietnam War and student movement years [Gitlin 1993; Farber 1994], for similar reasons except that this time it was also against the over-blown influence of the physicists who helped to end WWII with atomic bombs. On top of that, postmodernism had been raging in the 1950s, going strong in the 1960s and spreading from France to American campuses, Berkeley included.[43] In short, the 1960s is the agitative era of antiwar, anti-authority, and anti-establishment. Here, a lot of misunderstandings are involved, on the part of the "natural scientists" and also the critics:

1. The training of a "natural scientist" is very narrow. One who excels in superconductors has not much useful to say about black holes, say, not to mention climate change or human matters. Moreover, contrary to the claim by the physics Nobel laureate Robert Millikan (1868-1953) [Kevles 1987: 183] and others, the critical thinking these "scientists" gained through their experiences in (mathematical and physical) simple systems helps but does not automatically make them experts in dealing with human matters. Human problems are complex systems: Each problem is nonlinear and has multiple solutions; each outcome involves multiple (correlated) factors and cannot be precisely predicted. More importantly, every human problem is history dependent while the only history-dependent system most physicists have learned in their trade is the magnet, nothing comparable to a human being in terms of complexity.

2. Many "natural scientists" are not even aware of the basic differences between the (deterministic) simple systems they are familiar with and the (probabilistic) complex systems to which human matters belong, and cross the line inadvertently when making public comments on the latter.[44] Some comments bordered on scientism and enraged the public.

3. The over attention paid by the press to the "natural science" stars, after the triumphs of physics in the early part of the 20th century and again after its demonstrated contributions in ending WWII, was not the fault of the scientists. It had to do with the nature of the press, reflecting the public's keen interest on stars of any kind. They, the press and the public, also misunderstood the differences between simple and complex systems. Fortunately, things have improved. They no longer asked a "science" Nobel laureate for his opinion on fashion trends, say, nor did they ask Lady Gaga for her opinion on gravitational waves.

4. The adverse effects of "natural science" came from human decisions on when and how to use the scientific results, if at all. And

human decisions are in the domain of the humanities. Even though advances from the nonhuman sciences may help, wiser decisions are made by humans and thus could only come from improving the humanities. So *the problem lies in the underdevelopment of the humanities.* The "revolt against science" actually is a "revolt against humanities" prompted by the aggressive growth of physical science.

And so there was a huge number of potential readers in waiting before Kuhn's book came out in 1962. Kuhn, an insider in the eyes of the public but actually a marginer, with the more than proper credentials of a physics PhD from Harvard and professorship from Berkeley,[45] confirmed to them what they suspected all along: That scientists are just humans, working sometimes with selfish motivations, unlike Superman who lives to save the world; the scientific process is like other human endeavors that could be influenced by social forces. All these are in fact trivially true. They now came out with a deeper research, just like a second-order approximation shows more details than a first-order approximation of anything. The book's true impact outside of PS lies not on whether "paradigm shift" or "incommensurability" are correct or not because if wrong, that will affect only one person—the author's academic reputation. The impact lies on what the book hints at but refuses to admit clearly,[46] that some scientific results (like the gravitational law) that passed the RC do have objective validity and are not human constructs.

The sociologists of science picked up the hint and elaborated on the human aspects by doing lab visits, etc. Opposite to the anthropologist Margaret Mead's working style of trusting too much what the natives told her [Hellman 1998], the sociologists did the opposite by not trusting anything the "scientists" told them. This is not good since the scientific process is history dependent. Why a scientist makes a certain judgment depends heavily on what he picked up all the years before doing that particular experiment. And this information can be provided by the scientist if properly asked; otherwise, it has to be dug out through tedious research of the scientist's past.

Working without accumulating enough number of case studies (directly by observation or indirectly from HS) and rushing to draw conclusions from insufficient data characterize the research works of many of these sociologists. And it appears that many were not aware of the intrinsic limitation of sociology's fieldworks and fell into the generalization trap (see Sec. 5.3). As outsiders, some of these researchers had no idea how science actually works, and insisted that a scientific proof to be as rigorous as a mathematical proof. Such a demand bears the footprint of the absolutism of knowledge dating back to Plato, which, in fact, was found to be unsupportable and unnecessary by modern science which is anchored in observations and experiments [Lam 2014; Shapiro 1983].

Furthermore, some sociologists (see, e.g., [Collins 2014]) and others cling to the outdated misconception that social science which studies the collective behavior of humans, a kind of animal, is fundamentally different from "natural science" which they implicitly and mistakenly take it to mean the study of only inanimate systems. In fact, (1) for example, the bees or chimpanzees studied in "natural science" has an "actor's categories" which are distinct from the scientist's "technical categories" but that does not prevent humans from understanding them, and, similarly, for humans to understand other human groups. (2) Common quantitative laws (e.g., power laws) are found in both inanimate systems and human history [Lam 2008c] which are independent of any researcher's technical categories. (3) The same modeling and analyzing methodologies are shared successfully between "natural science" and social science [Lam 2014]. Therefore, these sociologists are wrong.

Sarton had the chance of getting the contents of HS right since HS as a discipline was established by him after 1905, the year the existence of atom was proved by Einstein. But he failed to pick up the message implied by atoms' existence and, before that, Darwin's evolutionary theory, that social science and the humanities are also part of science. Consequently, he led HS into the narrow alley of studying only "natural science" (and medical science) as we see it today. Moreover, Sarton's characterization of "scientists" is so naïve and wrong [Sarton 1962]. It seems that he was not aware of Newton's giving up physics soon after he assumed the Lucasian professorship at Cambridge and his bitter dispute with Leibniz.

Of course, this could be far from the case. We have no idea what happened on this account, a possible project for historians of science.

Scicomm is not in the business of finding out what science is or was. It depends on the other disciplines, PS, HS, and SS, to provide the answers. And so when those disciplines fall short, Scicomm becomes their victim. Yet, there are a few mistakes that are Scicomm's own doing. One is the narrow definition of "scientific literacy" which may not be definable or useful [Bauer 1994; Broks 2014], even though it is a convenient tool for Scicomm administrators.

In *summary*, the shortcomings of PS, HS, and SS are due to three reasons: (1) The content and practice of science change with time. Old writings that pay no attention to the changes are bound to be wrong or soon become obsolete and thus mislead the unknowing readers. (2) Some early aims of PS (e.g., insistence on absolute knowledge) are misguided; SS suffers from intrinsic limitations in their methodology. (3) Overconfidence of and overclaims by some SS researchers who are not or insufficiently trained in science are aplenty.

On the other hand, Scicomm's central problem is (1) the fuzzy/incorrect concept of science held by and (2) the mixed background of its practitioners. As a discipline that deals with the public, a good Scicomm researcher needs knowledge not just in "natural science" but in human-dependent matters provided by the humanities. Yet, researchers trained in both "natural science" and the humanities are rare.

On top of these difficulties, all four disciplines suffer from the common misperception that science is about "natural science" (and medical science) only, with social science and the humanities excluded. In other words, these disciplines presently are dealing with science in its very narrow, outdated sense. This fundamental misconception hurts not just the development of the disciplines but also humanity, jeopardizing the future of humankind [Lam 2014].

7.5 Near future

Confucius said something like this: To get things straight, the first step is to get the names straight. Thus, it would help if people are more careful in their use of words. The word Science should be reserved to mean what it is: the sum of nonhuman-system science and human-system science. It will be misleading if we continue to call nonhuman-system science by natural science, even though it is not straightly wrong if quotation marks are included in the latter (but it would be clumsy to pronounce the quotation marks every time you utter the term). But *it is wrong to call nonhuman-system science by science*, with or without quotation marks (since "science" includes social science, part of human-system science). Similarly, both the humanities and social science (and medical science) should be recognized as components of human-system science. This kind of clarification should start with the curricula in the educational system, from grade 1 on, but this will take time.

Meanwhile, what we can do is to do what we professionals can do immediately. For researchers in PS, HS, and SS, it is time and in fact fruitful to turn their attention to all the human-system sciences, going beyond biology. For example, the history of the humanities from the vantage point of HS would be surely different from those written by the humanists. This change of direction not merely will enlarge the scope of the disciplines but will make them more relevant to the pressing problems faced by our modern societies.

As for Scicomm practitioners, time is long overdue for them to shift their focus from nonhuman-system science to human-system science. The public need and want information on matters close to their lives, loud and clear. Another thing they can do: Apart from communicating the importance of critical thinking, they could change their emphasis on explaining the mechanisms, the why in science, to the appreciation of science. Start something called Science Appreciation. Learn from the art people! Finally, it is not too late for Scicomm to initiate some crash courses on climate change for themselves and the public, and enable a dialogue between the climate-change scientists and the public.

8 Conclusion: An Old but New Frontier

In the last 400 years or so since Galileo, the study of nonhuman systems under the name of "natural science" or modern science did enlighten deeply our understanding of nature (e.g., big bang), make our living easier (cell phone), and help to prolong our lives (for good or bad). But that is not enough as the future of humanity is concerned, as the so-called "revolt against science" tried very hard to remind everybody. *It is the humanities that determine our quality of life* (e.g., to pollute or not to pollute) *and bring us genuine happiness* (human relationships, arts). While the study in "natural science" should be continued, it is time for us to return to the Aristotelian tradition of treating the human system and nonhuman systems as equally important in our search for knowledge. Unfortunately, this tradition was interrupted by the phenomenal success of modern science.

Deepening the humanities' research and taking it to the next level do not require large increase of the research budget. No smashing machines needed to be built. *What is needed is a change of our concept of science and our perception of priority.* (See Scimat in [Lam 2014].) For example, for the four science-related humanities disciplines covered in this chapter, shifting the focus from simple systems to complex systems, from nonhuman systems to the human system, can be started immediately.

The humanities were the frontier for the ancient Greeks and Chinese but are the new frontier for the rest of us. Go complex! Go humanities!

Note added (April 15, 2024)· To understand why many philosophers of science could be so wrong but still attracted numerous followers and exerted out-of-proportion influences, in academia and beyond, we suggest that one should look at their personal life in some detail and examine the societal conditions that allowed it to happen, as first demonstrated in this chapter. This approach as a new discipline is named *Sciphilogy*—the human science of philosophers of science. For more, see my book *Humanities, Science, Scimat* (World Scientific, 2024).

Notes

1. The quotation marks indicate the conventional use of the term, which is not exactly correct. In our terminology, Science = Natural Science = Scimat +

"Natural Science," with Scimat = Humanities + Social Science + Medical Science, and "Natural Science" = nonhuman-system science; "Science" = "Natural Science" + Social Science (see Sec. 2.)

2. It is by this definition that we retroactively call the Godless part of works by Thales (c.624-c.546 BC) and Isaac Newton (1642-1727) science and these two, scientists, irrespective of their belief in God; we further identify Thales as the "father of science." For this reason, Andrew Cunningham and Perry Williams [1993] are wrong to locate the beginning of science at the Age of Revolution (1775-1848) because the secular study of the natural world (the signature of science) has been going on since Thales, first as part of philosophy and later as part of natural philosophy [Lam 2014]. Whether the secularity guideline is made explicit or not and the scientist's personal motivation (to glorify God or not) are immaterial as long as God is left out in the research's reasoning.

3. Harry Collins and Trevor Pinch's metaphor for science is one (imagined) animal, the golem [1998]. It is wrong on two counts: they confuse science with "natural science," and the application of "natural science" (more in the domain of the humanities) with "natural science" itself.

4. *Scientificity* is a quantity that could be but not yet defined. For example, we could define scientificity as a number S ranging from 0 to 10, with 10 the highest in scientific development. And so, for every discipline, we have a S. For instance, S = 8.5 for Physics, S = 6.2 for Economics, S = 0.5 for Philosophy, say. Like the Dow Jones index for stocks, S depends on a number of factors (such as consistency with available data, predictability, etc.) with different weights. Scientificity could be updated from time to time. It would be a useful index for funding agencies and researchers to gauge the direction of their disciplines/works.

5. For beginners in PS we recommend David Oldroyd's *The Arch of Knowledge* [1986] which adequately and critically covers the subject from the ancient Greeks to the present sociologists; for more in the last 100 years, see Peter Godfrey-Smith's *Theory and Reality* [2003]; for a modern, accessible overview of the nature of science from the perspective of a scientist, see Moti Ben-Ari's *Just a Theory* [2005].

6. The biography of Mach and others below, when not specified otherwise, comes from en.wikipedia.org.

7. Sigmund Freud's psychology—Popper's other pseudoscience target, was a different story. Freud (1856-1939) falsified his patient's record and covered it up by burning all his private papers in 1907 [Dewdney 1997]. There is not much point in examining a researcher's proposition if that person is less than honest.

8. George Sarton (1884-1956) and others' one-dimensional historiography of science (see Sec. 4.2) was criticized by Kuhn [1962] without mentioning Sarton's name. (See [Kuhn 1984] for Kuhn on Sarton and you will understand why; see also [Pinto de Oliveira & Oliveira 2013].) It is at this stage Kuhn was disappointed by Sarton's brand of historiography and set out to do differently. Sarton worked

intermittently at Harvard University from 1916 to 1951, appointed tenured HS professor in 1940. He established the HS discipline with the support of Carnegie Institution and Harvard (see Sec. 4).

9. A succinct description of Kuhn's life and career is given by Thomas Nickles [2003: 8-12]; his early years (1940-1962) are detailed by Karl Hufbauer [2012], Kuhn's Berkeley advisee. According to Wes Sharrock and Rupert Read [2002: 1]: "Thomas Kuhn died in 1996, convinced that his lifework had been misunderstood, and failing to complete a categorical restatement of his position before his death."

10. A theory that passes the RC is like a boat got anchored; in contrast, a theory that does not is like a boat not anchored and will drift away and disappear given enough time. Another way to visualize this is that before Galileo, Aristotle's mechanics was thought to be a component of the human-independent part of the Knowscape—the landscape of knowledge [Lam 2014]. But after Galileo, Aristotle's mechanics is shifted to the human-dependent part—the HS part.

11. A collection of informative essays from Feyerabend's friends and students is given in [Preston et al. 2000].

12. en.wikipedia.org/wiki/Herodotus (Mar. 21, 2014).

13. A concise introduction to the historical development of science (including "natural science" and social science) is given in en.wikipedia.org/wiki/ History_of_science (Nov. 1, 2013).

14. Even though in the common view, mathematics is not part of science. See [Livio 2011] for a discussion of whether mathematics is invented or discovered by humans.

15. en.wikipedia.org/wiki/Mount_Everest (Mar. 21, 2014).

16. These approximations kind of parallel those used in building physics models. If HS is compared to a gas system, Sarton's approximation is like the independent-particle approximation used in the ideal gas model; Kuhn's is to include the internal states of the particles; sociologists' is like adding in the interparticle interactions and external fields; feminists' is to take note of the two kinds of particle existing in the gas. What kinds of factors to be included depend on the aim of the model; e.g., whether it is for illustrating the mechanism or explaining a trend, or for detailed comparison with existing data.

17. A Class-F person is one who believes all knowledge is socially constructed; e.g., the gravitational law could not be discovered by smart aliens. Class-F people include self-proclaimed relativists as well as some sociologists and postmodernists, but not any practicing "scientists"—deceased or alive. It is not clear that there are really any truly Class-F persons in practice because no one has claimed that he has passed the book-drop test. Adolf Hitler (1889-1945) is identified as a Class-F person [Ferris 2010:. 245] when he supposed to have said, "Science is a social phenomenon…The slogan of objective science has been

coined by the professorate..." [Rauschning 1939: 221]. But the credibility of Hermann Rauschning's quotes is not established.

18. *The Golem* mislabels 1919 as 1918 [Collins & Pinch 1998: 48].

19. The discarding of discrepant data points in the eclipse experiments was less arbitrary than *The Golem* conveys; it involves careful, technical considerations [Kennefick 2009].

20. The 1919 eclipse experiments did result in newspaper headlines like "Einstein's theory confirmed," and maybe so even for some scientists at that time. But as science goes on, more people learned to be more careful. One experimental proof is just a tentative confirmation while a "conclusive" confirmation needs a number of independent positive experiments. Unfortunately, in the case of the eclipse experiments, one has to wait many years before the same experiments can be repeated. In fact, in the physics community, the validity of the general-relativity theory (or any theory) is not built on that one result, but on a long string of confirming experiments (on different aspects of the theory) that have poured in over the years [Bederson 1999: 74-83] and are still coming in. For the laypeople, the working of the Global Positioning System (GPS) could be the most important confirmation. (See also [Mermin 1996].)

21. For example, we know from Newton's universal gravity that a piece of rock falling from the sky is attracted by all other masses in the universe. But when we calculate how fast the rock will hit the ground, we ignore all other masses except that of the nearby Earth—an approximation.

22. For these sociologists, Steve Mirksy's article, "Physics uncowed" [2012] is recommended. It describes how physicists "prove" that the moon is not made of cheese without going to the moon to sample it, similarly for the impossibility of bending spoons with the mind. For spoon bending, see [Randi 1982; Collins 1992].

23. Some people call computer simulations as computer experiments. That is confusing; it is actually wrong since experiment means you have to get your hands dirty [Lam 2006].

24. I have worked at the Nevis Labs of Columbia University and Bell Labs, Murray Hill (where semiconductor was invented); I have done table-top experiments on pattern formation at San Jose State University. And I have published in all three kinds of theory.

25. This kind of "common" knowledge is available in scientists' memoirs and popular-science (popsci) books. What one needs is to read enough number of them, or just talk to the graduate students.

26. An example: The physics Nobel laureate Chen-Ning Yang was advised by Edward Teller to switch from experiment to theory in his PhD training after he had shown himself to be obviously inept in the former [Chiang 2002].

27. Einstein first submitted a manuscript to *Physical Review* titled "Do gravitational waves exist?" with a negative answer, as inferred, erroneously, from the general-relativity theory he invented. After mistakes were pointed out by an anonymous referee, Einstein reversed his answer, reworked his manuscript, changed the title to "On gravitational waves" and got it published in another journal, without thanking the original referee [Kennefick 2005].

28. In Alan Cholos' report on the gravitational wave discovery, he writes: "The *good* news is that many experiments are poised *not* to repeat the measurements exactly but to add potentially confirmatory or contradictory evidence" (my italics). Except for simple table-top experiments (like Faraday's and those in high-temperature superconductors), discovery experiments were rarely repeated. Sometimes it is because of the difficulty (like Chien-Shiung Wu's parity-nonconservation experiment) but the deeper reason is subtle. To confirm that a person is coming out of a hotel, the same picture from two different cameras is not as strong and informative as more pictures taking from different angles. This is a point that some SS sociologists failed to understand.

29. Apart from confirming theories and bringing in unexpected discoveries, control experiments enable the building of reliable equipment and comparison with data collected from uncontrollable sources (such as the optical spectra originating from the stars).

30. A useful, brief introduction to Scicomm is given in en.wikipedia.org/wiki/Science_communicaton (April 17, 2014).

31. A top non-government Scicomm organization is the Science Squirrels Club (songshuhui.net), which favors biological, health, and medical topics [Chen et al. 2014]—all complex systems. The Club's microblog attracts one million fans vs. government counterparts' thousands or tens of thousands [Wang & Tang 2014].

32. See, e.g., "Copernicus to Kepler" in Johann Sommerville's "History 351: Seventeenth century Europe" (faculty.history.wisc.edu/sommerville/351/351-182.htm, May 25, 2014).

33. According to "Climate change impacts in the United States," the government's National Climate Assessment released in May 2014, global warming is already here (nca2014, globalchange.gov, May 28, 2014).

34. Classical physics, relativity theories, and quantum physics are "neat physics" which are well tested. We have worked with them for over hundred(s) of years and have confidence on them. The equations involved are deterministic (even though the predictions in quantum mechanics are probabilistic). In contrast, weather forecasting is relatively new and is "messy physics." The weather equations are well known but chaotic, and can only be solved by powerful computers. Because of the "butterfly effect," weather forecasting is accurate up to two or three days only. In comparison, the coupled equations in climate change are much larger in number and are less certain; it is "very messy" in content and in practice. Arguing that ten computer models converge to the same result is not

good enough. A computer model is as good as what one puts into the model; garbage in, garbage out (see Freeman Dyson's interview in [Lemonick 2009]). The climate-change predictors owe the public the explanation of why their system is not chaotic, why they can predict for 20 or 100 years while the much simpler weather forecasting is good for only a few days, and how do they ascertain the unimportance of the many factors left out in their models.

35. There is such a possibility: No matter how good a computer model is, the calculations are always done approximately since the numbers in a computer are expressed with a finite number of digits. Or, probability is involved in the model building. In either case, predictions can only be given in terms of probabilities. Contrary to many people's understanding, an event with a small probability, no matter how small, could still happen [Hand 2014]. Therefore, if global warming is predicted to be 99% certain, there is still a 1% chance that it will not happen if nothing is done. And if that 1% prediction is actually realized people would say the prediction is "wrong." That, of course, is not true; people just misunderstand probabilities.

36. See PBS's Nova program, "Your inner fish," and [Shubin 2009].

37. The other ingredient essential to the development of science in the West is "debating with each other" as advocated by Socrates [Lyold 1970]. Debating means "getting to the bottom of things" or "finding out what happens." Unfortunately, debate is not encouraged by Confucianism. Ironically, more than 2,000 years ago, the Chinese already realized, correctly, that "heaven and man are oneness" [Lin 2010]. What it means is that "heaven" (i.e., nature) and humans can influence each other and are organized similarly. The former is trivial now and the latter is borne out by science in the last few decades; i.e., there exist three "universal" organizing principles that cut through living and nonliving systems— chaos, fractals, and active walks [Lam 2008a]. By this "oneness" doctrine, the Chinese could equally choose to study "heaven" (such as buoyancy or ants) in order to understand humans and they would invent "science" but they did not. The above is our answer to the so-called Needham Question [Lam 2008a].

38. Similar experience occurs all the time in physics research. If you pick a difficult research topic you may get nowhere for a long, long time, if not forever.

39. If "rational" means induction and deduction only then science is definitely not always rational since good science involves intuition or (educated) guessing [Lam 2004].

40. And it would not be too difficult for him to realize, after 1962, say, that the difference between Aristotle's mechanics and Newton's mechanics is that the latter passes the RC (within its applicability domain) while the former does not. But if so and if he admitted it openly, he would have to give up his incommensurability that helped to make him famous. Was Kuhn ever aware of the RC difference? This will be an interesting project for historians of science.

41. Feyerabend, Kuhn's Berkeley colleague, has written: "I venture to guess that the ambiguity is *intended* and that Kuhn wants to fully exploit its propagandistic potentialities" (my italics) [Hoyningen-Huene 2000: 109-110]. A detailed and thorough biography of Kuhn could help to settle this conjecture.

42. That is why no physics book ever became a best seller. Stephen Hawking's *A Brief History of Time* (1988) is a different story. It is a popsci book, found incomprehensible by most people; many bought it out of sympathy for the author. The book is incomprehensible not because the author tried to write ambiguously but because he was a novice popsci writer. The publisher should bear the blame because as professionals, they knew that they should find Hawking a coauthor but they did not. Of course, it is easier to come up with a best seller by avoiding scientific details. Examples are Eric Segal's *Love Story* (1970) and Robert Waller's *The Bridges of Madison County* (1992), which are about human relationships, based on the science of love at the observational/empirical level; no dopamine mentioned though. (See [Fisher 2004].)

43. Postmodernism is a late-20th century movement in the humanities. It is anticonvention and anti-authority by playing on words and "nonwords" in the name of semantics. See [Sokal & Bricmont 1998] and en.wikipedia.org/wiki/Postmodernism (May 1, 2014).

44. Yet some very good "natural scientists" were aware of the problem. For instance, Einstein opinioned a lot about matters beyond physics but he was extremely careful in his uttering [Einstein 1982]. When he turned down the offer to be nominated for Israel's presidency, upon the death of Chaim Weizmann (1874-1952), the first president and a chemist, his reason was that since he had devoted his life to objective matters, he lacked "both the natural aptitude and the experience to deal properly with people and to exercise official functions" [Burko 2013]. Enrico Fermi (1901-1954), another physics Nobel laureate, after the completion of the Manhattan Project, turned down an offer to serve on a government committee on the ground that human matters involve multiple solutions while his training and experience equipped him to deal only with problems with unique solutions like those in physics [Fermi 1954]. The same cannot be said about the other two Nobel laureates, Robert Millikan and Niels Bohr (185-1962), and the lesser ones (see [Kevles 1987].)

45. Kuhn's motivation in finishing his PhD in physics while already determined to shift his career plan to HS was that he thought, correctly, the degree will enhance his credentials as a historian in science. Besides, he took some courses in philosophy and found out that the gap was too large for him to cross if he wanted a degree in Philosophy, and so a physics degree was his only option since there was no degree in HS yet [Hufbauer 2012].

46. The game is played somewhat like this. The author writes that when someone's worldview changes, the world changes. The reader asks: When you say "the world changes" do you mean it in the literary sense or do you mean the

physical world changes? And the author refuses to give a straight answer; instead, he replies: It depends on what "physical" means. Etc,, etc.

References

Agin, D. [2006]. *Junk Science: An Overdue Indictment of Government, Industry, and Faith Groups that Twist Science for Their Own Gain.* New York: Thomas Dunne Books.

Bauer, H. H. [1994]. *Scientific Literacy and the Myth of the Scientific Method.* Urbana: University of Illinois Press.

Bederson, B. (ed.) [1999.] *More Things in Heaven and Earth: A Celebration of Physics in the Millennium.* College Park & New York: American Physics Society & Springer.

Ben-Ari, M. [2005]. *Just a Theory: Exploring the Nature of Science.* Amherst, NY: Prometheus Books.

Berg, R. [2012]. Beyond the laws of nature. Philosophy Now, Issue 88, January/February 2012: 24-26.

Bird, A. & Ladyman, J. [2013]. *Arguing about Science.* New York: Routledge.

Bloor, D. [(1976) 1991]. *Knowledge and Social Imagery.* Chicago: University of Chicago Press.

Brakel, J. van [2000]. *Philosophy of Chemistry: Between the Manifest and the Scientific Image.* Leuven, Belgium: Leuven University Press.

Broks, P. [2006]. *Understanding Popular Science.* New York: Open University Press.

Broks, P. [2014]. Science Communication: A history and review. *All About Science: Philosophy, history, sociology & communication*, Burguete, M. & Lam, L. (eds.). Singapore: World Scientific.

Burguete, M. & Lam, L. (eds.) [2008]. *Science Matters: Humanities as Complex Systems.* Singapore: World Scientific.

Burguete, M. & Lam, L. (eds.) [2014]. *All About science: Philosophy, History, Sociology & Communication.* Singapore: World Scientific.

Burko, L. [2013]. Einstein and the presidency of Israel. *APS News*, January 2013: 4.

Byers, W. [2011]. *The Blind Spot: Science and the Crisis of Uncertainty.* Princeton: Princeton University Press.

Chiang, Tsai-Chien (江才健) [2002]. *Biography of Yang Chen-Ning: The Beauty of Gauge and Symmetry.* Taibei: Bookzone.

Chiang, Tsai-Chien [2014]. *Madame Wu Chien-Shiung: The First Lady of Physics Research*, Wong Tang-Fong (trans.). Singapore: World Scientific.

Chen, Mei-Ting, Chen Lu-Yao & Chong Lu [2014]. The bias of disseminating content and audience attention about popular science on network: Taking

Science Squirrels Club's microblog, blog, website for example. Study on Science Popularization **9**(1): 39-45.

Chodos, A. [2014]. Polarization measurement detects primordial gravitational waves. Phys. Today, May 2014: 11-12.

Collins, H. [1992]. *Changing Order: Replication and Induction in Scientific Practice*. Chicago: University of Chicago Press.

Collins, H. [2014]. Three waves of science studies. *All About Science: Philosophy, history, sociology & communication*, Burguete, M. & Lam, L. (eds.). Singapore: World Scientific.

Collins, H. & Pinch, T. [1998]. *The Golem: What You Should Know about Science*. Cambridge: Cambridge University Press.

Cunningham, A. & Williams, P. [1993]. De-centring the "big picture": The *Origins of Modern Science* and the modern origins of science. The British Journal for the History of Science **26**: 407-432.

Dewdney, A. K. [1997]. *Yes, We Have No Neutrons: An Eye-Opening Tour through the Twists and Turns of Bad* Science. New York: Wiley.

Duhem, P. [1906 (1954)]. *The Aim and Structure of Physical Theory*. Princeton: Princeton University Press.

Einstein, A. [1982]. *Ideas and Opinions*. New York: Three Rivers Press.

Farber, D. (ed.) [1994]. *The Sixties: From Memory to History*. Chapel Hill: University of North Carolina Press.

Fermi, L. [1954]. *Atoms in the Family: My Life with Enrico Fermi*. Chicago: University of Chicago Press.

Ferris, T. [2010]. *The Science of Liberty: Democracy, Reason, and the Laws of Nature*. New York: Harper Perennial.

Feyerabend, P. [1995]. *Killing Time*. Chicago: University of Chicago Press.

Feyerabend, P. [2010]. *Against Method*, 4th ed. New York: Verso.

Fisher, H. [2004]. *Why We Love: The Nature and Chemistry of Romantic Love*. New York: Owl Books.

Garfield, E. [1985]. The life and career of Geroge Sarton: The father of the History of Science. Journal of the History of Behavioral Sciences **21**: 107-117.

Gitlin, T. [1993]. *The Sixties: Years of Hope, Days of Rage*. New York: Bantam Books.

Godfrey-Smith, P. [2003]. *Theory and Reality: An Introduction to the Philosophy of Science*. Chicago: University of Chicago Press.

Greenemeier, L. [2013]. Lights, camera, atoms: IBM creates the world's tiniest movie. Sci. Am., July 2013: 26.

Gregory, J. & Miller, S. [1998]. *Science in Public: Communication, Culture, and Credibility*. Cambridge, MA: Perseus.

Hand, D. J. [2014]. *The Improbability Principle: Why Coincidence, Miracles, and Rare Evens Happen Every Day*. New York: Scientific American/Farrar, Strass and Giroux.

Hellman, H. [1998]. *Great Feuds in Science: Ten of the Liveliest Disputes Ever*. New York: Wiley.

Henry, J. [1997]. *The Scientific Revolution and the Origins of Modern Science*. New York: St. Martin's Press.

Holton, G. [1993]. *Science and Anti-Science*. Cambridge, MA: Havard University Press.

Hoyningen-Huene, P. [2002]. Paul Feyerabend and Thomas Kuhn. *The Worst Enemy of Science? Essays in Memory of Paul Feyerabend*, Preston, J., Munévar, G. & Lamb, D. (eds.). Oxford: Oxford University Press.

Hufbauer, K. [2012]. From student of physics to historian of science: T. S. Kuhn's education and early career. Physics in Perspective **14**: 421-470.

Johnstone, J. [1914]. *The Philosophy of Biology*. Cambridge: Cambridge University Press.

Ken, R A. [2001]. Global warming: Rising global temperature, rising uncertainty. Science **292**: 192-194.

Kennedy, R. E. [2012]. *A Student's Guide to Einstein's Major Papers*. Oxford: Oxford University Press.

Kennefick, D. [2005]. Einstein versus the *Physical Review*. Phys. Today, Sept. 2005: 43-48.

Kennefick, D. [2009]. Testing relativity from the 1919 eclipse—a question of bias. Phys. Today, Mar. 2009: 37-42.

Kevles, D. J. [1987]. *The Physicists: The History of a Scientific Community in a Modern America*. Cambridge, MA: Harvard University Press.

Kragh, H. [1987]. *An Introduction to the Historiography of Science*. Cambridge: Cambridge University Press.

Kuhn, T. S. [1962]. *The Structure of Scientific Revolution*. Chicago: University of Chicago Press.

Kuhn, T. S. [1984]. Professionalization recollected in tranquility. Isis **75**(1): 29-32.

Lam, L. [1998]. *Nonlinear Physics for Beginners: Fractals, Chaos, Solitons, Pattern Formation, Cellular Automata and Complex Systems*. Singapore: World Scientific.

Lam, L. [2002]. Histophysics: A new discipline. Modern Physics Letters B **16**: 1163-1176.

Lam, L. [2004]. *This Pale Blue Dot: Science, History, God*. Tamsui: Tamkang University Press.

Lam, L. [2006]. Active walks: The first twelve years (Part II). International Journal of Bifurcation and Chaos **16**: 239-268.

Lam, L. [2008a]. Science Matters: A unified perspective. *Science Matters: Humanities as Complex* Systems, Burguete, M. & Lam, L. (eds.). Singapore: World Scientific. pp 1-38.

Lam, L. [2008b]. SciComm, PopSci and The Real World. *Science Matters: Humanities as Complex* Systems, Burguete, M. & Lam, L. (eds.). Singapore: World Scientific. pp 89-118.

Lam, L. [2008c]. Human history: A science matter. *Science Matters: Humanities as Complex* Systems, Burguete, M. & Lam, L. (eds.). Singapore: World Scientific. pp 234-254.

Lam, L. [2011]. Arts: A science matter. *Arts: A Science Matter*, Burguete, M. & Lam, L. (eds.). Singapore: World Scientific. pp 1-32.

Lam, L. [2014]. About science 1: Basics—knowledge, nature, science and Scimat. *All About Science: Philosophy, history, sociology & communication*, Burguete, M. & Lam, L. (eds.). Singapore: World Scientific. pp 1-49.

Latour, B. & Woolgar, S. [1986]. *Laboratory Life: The Construction of Scientific Facts*. Princeton: Princeton University Press.

Lemonick, M. D. [2009]. Freeman Dyson takes on the climate establishment. *Yale Environment 360* (e360.yale.edu/feature/, May 12, 2014).

Lestienne, R. [1998]. *The Creative Power of Chance*, Neher, E. C. (trans.). Urbana: University of Illinois Press.

Li, Da-Guang (李大光) [2008]. Evolution of the concept of science communication in China. *Science Matters: Humanities as Complex* Systems, Burguete, M. & Lam, L. (eds.). Singapore: World Scientific. pp 165-176.

Lin, Ke-Ji (林可济) [2010]. *Oneness of Heaven and Man and the Subject-Object Dichotomy: An Important Vantage Point in Comparing Chinese and Western Philosophies*. Beijing: Social Sciences Academic Press.

Liu, Bing (刘兵) & Zhang Mei-Fang (章梅芳) [2014]. Scientific culture in contemporary China. *All About Science: Philosophy, history, sociology & communication*, Burguete, M. & Lam, L. (eds.). Singapore: World Scientific. pp 290-303.

Liu, Dun (刘钝) [2008]. History of science in globalizing time. *Science Matters: Humanities as Complex* Systems, Burguete, M. & Lam, L. (eds.). Singapore: World Scientific. pp 177-190.

Livio, M. [2011]. Why math works. Sci. Am., August 2011: 80-83.

Lloyd, G. E. R. [1970]. *Early Greek Science: Thales to Aristotle*. New York: Norton.

Lucibella, M. [2014]. Gaps widen in attitudes toward science. APS News, Mar. 2014: 3. (The source of this article is the biennial *Science and Engineering*

Indicators report issued by the National Science Foundation, which brings together numerous surveys.)

Mason, S. F. [1962]. *A History of the Sciences.* New York: Macmillan.

Mermin, N. D. [1996]. The golemization of relativity. Phys. Today, April 1996: 11-13.

Mirsky, S. [2012]. Physics uncowed: You don't have to say cheese to get the picture. Sci. Am., Jan. 2012: 86.

Morris, R. [2002]. *The Big Questions: Probing the Promise and Limits of* Science. New York: Times Books.

Nickles, T. (ed.) [2003]. *Thomas Kuhn.* Cambridge: Cambridge University Press.

Nickles, T. [2013]. The problem of demarcation: History and future. *Philosophy of Pseudoscience: Reconsidering the Demarcation Problem*, Pigliucci, M. & Boudry, M. (eds.). Chicago: University of Chicago Press.

Oldroyd, D. [1986]. *The Arch of Knowledge: An Introductory Study of the History of the Philosophy and Methodology of* Science. New York: Methuen.

Pinto de Oliveira, J. C. & Oliveria, A. J. [2013]. Kuhn, Sarton, and the history of science. Preprint (phisi-archive.pitt.edu/id/eprint/10078, May 12, 2014).

Popper, K. [1963]. *Conjectures and Refutations: The Growth of Scientific Knowledge.* London: Routledge & Kegan Paul.

Preston, J., Munévar, G. & Lamb, D. (eds.) [2000]. *The Worst Enemy of Science? Essays in Memory of Paul Feyerabend.* Oxford: Oxford University Press.

Sanders, N. [2011]. How to succeed at engaging the public's interest in science. Phys. Today, May 2011.

Sanitt, N. (ed.) [2005]. *Motivating Science: Science Communication from a Philosophical, Educational and Cultural Perspective.* Luton, UK: The Pantaneto Forum.

Sarton, G. [1962]. *The History of Science and the New Humanism.* Bloomington: Indiana University Press.

Schiebinger, L. [1999]. *Has Feminism Changed Science?* Cambridge, MA: Harvard University Press.

Schiele, B., Claessens, M. & Shi Shun-Ke (石 顺 科) [2012]. *Science Communication in the World: Practices, Theories and Trends.* New York: Springer.

Shapiro, B. J. [1983]. *Probability and Certainty in Seventeenth-Century England: A Study of the Relationships between Natural Science, Religion, History, Law, and Literature.* Princeton: Princeton University Press.

Sharrock, W. & Read, R. [2002]. *Kuhn: Philosopher of Scientific Revolution.* Malden, MA: Blackwell.

Shi, Shun-Ke & Zhang Hui-Liang [2012]. Policy perspective on science popularization in China. *Science Communication in the World: Practices,*

Theories and Trends, Schiele, B., Claessens, M. & Shi Shun-Ke. New York: Springer.

Shubin, N. [2009]. *Your Inner Fish: A Journey into the 3.5-Billion-Year History of the Human Body*. London: Pantheon/Allen Lane.

Sismondo, S. [2004]. *An Introduction to Science and Technology Studies*. Malden, MA: Blackwell.

Sokal, A. & Bricmont, J. [1998]. *Fashionable Nonsense: Postmodern Intellectuals' Abuse of* Power. New York: Picador.

Sommerville, R. C. J. & Hassol, S. J. [2011]. Communicating the science of climate change. Phys. Today, Oct. 2011: 48-53.

Thurs, D. P. [2011]. Scientific methods. *Wrestling with Nature: From Omens to Science*, Harrison, P., Numbers, R. L. & Shank, M. H. (eds.). Chicago: University of Chicago Press.

Turner, M.S. [2009]. The universe. Sci. Am., Sept. 2009: 36-43.

Wang, Yu-Hua & Tang Shu-Kun [2014]. A comparative study on microblog science communication between government and civil science popularization organization. Study on Science Popularization 9(1): 32-38.

Weinberg, S. [2001]. *Facing Up: Science and Its Cultural Adversaries*. Cambridge, MA: Harvard University Press.

Worrall, J. [2003]. Normal science and dogmatism, paradigms and progress: Kuhn "versus" Popper and Lakatos. *Thomas Kuhn*, Nickles, T. (ed.). Cambridge: Cambridge University Press. pp 65-100.

Yin, Lin (尹霖) [2014]. Popular-science writings in early modern China. *All About Science: Philosophy, history, sociology & communication*, Burguete, M. & Lam, L. (eds.). Singapore: World Scientific. pp 330-345.

Zhang, Yi-Zhong & Ren Fu-Jun (任福君) [2012]. The development and prospects of legal system construction for science popularization in China since the science and technology popularization law was issued. Science Popularization 7(3): 5-13.

Ziman, J. [2000]. *Real Science: What It Is, and What It Means*. Cambridge: Cambridge University Press.

Zimmer, C. [2001]. *Evolution: The Triumph of an Idea*. New York: HarperCollins.

Published: Lam, L. [2014]. About Science 2: Philosophy, history, sociology and communication. *All About Science: Philosophy, history, sociology & communication*, Burguete, M. & Lam, L. (eds.). Singapore: World Scientific. pp 50-100.

22

Scicomm, Popsci and The Real World

Lui Lam

A physicist's experience in science communication (Scicomm), popular science (Popsci), and the teaching of a Scimat (Science Matters) course *The Real World* is presented and discussed. Recommendations for others are provided.

1 Introduction

Yes, yes, I know. I know that I am not supposed to use abbreviations in a chapter title; I should spell out the whole words. But like the French say: rules are set to be broken. And indeed it happened: Newton (1643-1727) broke the rules set by Aristotle (384-322 BC) in dynamics, and replaced them with his own three laws; Einstein (1879-1955) in turn broke Newton's three laws and replaced them with his theory of special relativity. This is called innovation or, in rare occasions, revolution. Rules could and should be broken when one has a good reason. And I have *two* good reasons.

My background as a scientist is not atypical. I have been working in physics research in the last 40 years. I am now a professor in California, a job involving both research and physics teaching (with an unbelievable teaching load of 12 credits[1] plus office hours per semester). My research was first in condensed matter physics and later in nonlinear physics and complex systems.

My involvement in Science Communication (*Scicomm*) began in 1994, in Mexico City, Mexico. In that year I was invited by Rosalio Rodriquez and gave a public lecture "Nonlinear physics is for everybody!" at Universidad Nacional Autonoma de Mexico. Since then, I have been doing physics research, teaching, and Scicomm simultaneously, trying to synthesize my activities in these three areas and, most importantly, to be creative and have fun in doing that. In recent years, these activities are heavily influenced by my involvement in *Histophysics* (physics of history) [Lam 2002, 2008b] and *Scimat* (Science Matters) [Lam 2008a].[2] What follows is the adventure I went through in the wonderland of Scicomm and my innovation in education on behalf of Scimat, and my recommendation for others.

2 Science Communication

Science communication [Gregory & Miller, 2000] involves four components:

1. Funding and organized effort from the government and learning societies.
2. Engagement of scientists as individuals.
3. Participation of the public.
4. Development of Scicomm as a research discipline, by scholars and students.

Engaging scientists and the science community to participate actively and regularly is a daunting task. What the government can do is to provide funding and encouragement to scientists who are willing and qualified. The other part of the game concerns the scientists themselves, at the individual level. Leon Lederman, a Nobel laureate and physicist, proposed that working scientists should devote 10% of their time to communicating science. This may not be very practical for those professors who do not yet have tenure, because the competition in research is very keen and research requires undivided attention, not to mention that Scicomm is not always appreciated and rewarded by the administrators. But let us say, a scientist—tenured or not—wants to contribute to Scicomm, what can she or he do? This Section addresses this problem, from the perspective of a working physicist.

Six items concerning what science professors or teachers can do in Scicomm are presented here [Lam 2006a].

1. <u>What every science professor/teacher can do: Integrate popular</u>
   <u>science books into science teaching</u>

The quick pace of interdisciplinary development in science and the ever-changing job market demand a broad knowledge base from our students. For five or more years, I integrated popular science (*popsci*)[3] books[4] into my physics classes by giving extra credits to the students who would buy a popsci book,[5] read it and write up a report [Lam 2000a, 2001, 2005a]. The instructor does not actually teach the books, and hence will *not* find the teaching load increased—an important factor in any successful educational reform. It is like a supplementary reading, a practice commonly found in English classes but rarely adopted by science instructors. The aim of this practice [Lam 2000a] is to:

(1) Broaden the knowledge base of students.

(2) Show the students the availability and varieties of popsci books in their local bookstores.

(3) Encourage the students to go on buying and reading at least one popsci book per year for the rest of their life..

(4) Become a science-informed citizen—a voter or perhaps a future science-friendly legislator.

It is about lifetime learning of science matters. Professors in other universities have copied this approach, with equal success. It is equally applicable to high schools. Adopting this practice in the whole country or worldwide in large scale will fundamentally improve the science education of our students—the future average citizens. An immediate side effect is that in a few short months, all the popsci books on the bookshelves of every bookstore will be wiped out. The popsci book market will be drastically improved, attracting more skillful writers into the popsci books profession, benefiting everybody.

2. <u>What every science professor can do (I): Inject popular science talks</u>
   <u>into departmental seminars, or set up a separate popular science</u>
   <u>seminar series in the department</u>

Since 1994, I have been giving public talks on science, history, and religion, starting with a title the audience are interest in and leading them to the topics such as the scientific process that I really want them to learn. The titles include:

- Wu Chien-Shiung: The First Woman President of American Physical Society
- Does God Exist?
- The Real World
- The Birth of a Physics Project: What Happened to My New Book
- Why the World Is So Complex
- How to Model History and Predict the Future

I usually tried them out first in my physics department. In almost all universities around the world, there is a weekly departmental seminar. Recent research results are presented by either outside speakers or the faculty members. These talks are usually boring and quite often poorly attended. The exceptions are popsci talks, because they are easy to understand, even for undergraduates.

What every science professor can do is to insert popsci talks into their departmental seminar series, which can be given by themselves or outsiders. If the department does not allow it, a separate popsci seminar series can be set up within the department, with the help of the student science club if it exists—if not, help the student to set up such a club; your department chair will be thankful. And, of course, these seminars are open to the general public.

3. What every science professor can do (II): Set up a popular science lecture series in the university for general audience

In December 1999, I established a public lecture series "God, Science, Scientist" at San Jose State University (SJSU). The first three speakers (Fig. 1) are:

(1) Michael Shermer, who gave a talk in May 2000 on "How people believe: The search for God in the age of science." Shermer, a monthly columnist for *Scientific American*, is the founding publisher and editor of *Skeptic* magazine. He is the author of many popsci books such as *Why People Believe Weird Things*, *How*

People Believe, Denying History, The Borderlands of Science, The Science of Good and Evil, Why Darwin Matters, and *The Mind of the Market.* He is also a professor of history and science associated with Caltech and the Occidental College at Los Angeles.

(2) Eugenie Scott, the executive director of the National Center for Science Education in El Cerrito, California. Scott is a nationally known authority on creationism and evolution controversy.

(3) Charles Townes, the Nobel laureate in physics and co-inventor of laser.

These talks were attended by a large audience from different walks of life and were well received. I still get letters/emails from the fans who attended the lectures.

Every science professor can set up a popsci lecture series in their university, which will be highly appreciated by the administrators. It is not that difficult to do if you limit yourself to one or two speakers per semester. And don't forget to invite your Dean or President to introduce the distinguish speakers.

Fig. 1. The first three speakers of the "God, Science, Scientist" public lecture series at SJSU. *Left to right*: Michael Shermer, Eugenie Scott, and Charles Townes.

4. What every scientist can do: Give popular science talks in high schools, the community, and other places

For a period of 11 years, I gave invited popsci talks in various high schools,[6] universities, TV interviews (CCTV, Dec. 18-19, 2003), and conferences in Mexico, the United States, Taiwan, Hong Kong, and mainland China.

In November 2000, Shermer [2001] was one of four popsci experts I invited, in my capacity as a member of the International Advisory Committee, as a speaker at the International Forum on Public Understanding of Science, Beijing, organized by China Association for Science and Technology (CAST)—the first *international* popsci conference in China. We became good friends. I wrote an article on active walks for his magazine *Skeptic* [Lam 2000b].

This article led to an unexpected invitation from the Foundation For the Future (in Bellevue, WA), as a keynote speaker in their annual seminar, Humanity 3000, held in Seattle, 2001. The 23 invited "participants" included the famous Edward O. Wilson (from Harvard) and Richard Dawkins (from Oxford); I was the only physicist there. I gave a talk on "How to model history and predict the future" [Lam 2003], and became a Futurology expert, ipso facto.

After that, I was invited by Doug Vakoch of the SETI Institute (Search for Extraterrestrial Intelligence, based in Mountain View, CA), who also attended this Seattle seminar as an "observer," to go to Paris in March 2002 and talk about what science-and-art message to send to the extra-terrestrials (ET), in case they exist. I proposed to beam them digitally the recipes to create the Sierpinski gasket, a fractal.[7] Vakoch liked the idea and included it in his workshop report [Lam 2004a]. And I suddenly found myself an ET expert.

One thing led to another, like in a chain reaction. I met some artists during this Paris workshop, and we have been trying to collaborate on a physics-art-music (PAM) project called "Candle in the Wind."[8]

Another participant in that Seattle seminar was Clement Chang, cofounder of Tamkang University in Taiwan. In December 9-11, 2003, I was invited to give the Tamkang Chair Lectures (Fig. 2). My host was Kuo-Hua Chen, Chair of the Graduate Institute of Futures Studies and Director of the Center for Futures Studies. The result is my first popsci book, *This Pale Blue Dot: Science, History, God* [Lam 2004b] (Fig. 3).[9]

It is not true that every science professor is good at giving popsci talks, but everyone can try and be successful. You just keep practicing, giving the same talk many times and modifying it with the help of PowerPoint. And as shown in my story above, the reward could be significantly large: It gains you many new friends, from all walks of life; it might even take you to Paris.

Fig. 2. The poster of my Tamkang Chair Lectures, titled "Dialogue between science, history, and religion. Pale blue dot—viewing Earth from the universe."

Fig. 3. The covers of my first popsci book *This Pale Blue Dot: Science, History, God* (2004).

5. <u>What some scientists can do but all can try: Contribute to science communication research</u>

Science communication as a discipline is at its very early stage; it is a profession without a formal name[10]—unlike the case in physics, say. A new and short word is needed. My suggestion is to call it Scicomm or Popsci.

It is rare to find a Scicomm course in American universities. In contrast, China has a lead here; there are already degree programs in Scicomm in at least four universities, and a research institute on Popsci (under CAST) in Beijing—the China Research Institute for Science Popularization. Obviously, the contribution of working scientists in making Popsci a mature discipline is much needed; e.g., they can provide different perspectives and help to clarify science issues.

In June 2004, I collaborated with Li Da-Guang (李大光) of CAST (now at Chinese Academy of Sciences, or CAS for short) and Yang Xu-Jie (杨虚杰) of the *Sciencetimes* (科学时报, a Beijing daily published by CAS) and presented a paper at the International Conference on Scientific Knowledge and Cultural Diversity, Barcelona, Spain, June 3-6, 2004, on the absence of *professional* popsci book authors in China [Lam et al. 2005] (see Sec. 4). This was followed by a paper on a new concept for science and technology museums, presented at the International Forum on Scientific Literacy, Beijing, July 29-30, 2004 [Lam 2006b]. The idea is that unified themes governing natural and social sciences [Lam 2008a] should and could be injected into the display in science museums, to avoid the possible misconceptions conveyed to the visitors that the two are completely separated from each other (see Sec. 3). And, reporting for *Sciencetimes*, Yang and I co-wrote an article reporting on the 10th International Conference on the History of Science in China, Harbin, August 4-7, 2004 [Yang & Lam 2004].

6. What some science professors can do: Merging science with the humanities

Science and the humanities are considered by some as "two cultures" [Snow 1998; Lam 2008a]. But in fact, the humanities are about humans, which is a (biological) material system of *Homo sapiens*. Thus, the humanities could and should be part of natural science, which is about *all* material systems. The two can be integrated, but how?

In 1992, two years after I founded the International Liquid Crystal Society [Lam 2005b, 2005c, 2014], I came up with a new paradigm for complex systems. I named it *Active Walk* (AW), reviewed in [Lam 2005b, 2006c]. An active walker is one that changes a landscape—real or mathematical— as it walks; its next step is in turn influenced by the deformed landscape. Active walk is now widely applied in natural and social sciences [Lam 2008a; Han et al. 2008].

By 2000, the year that Shermer and I first met each other, I have been trying to create a new discipline by merging AW with a branch of the humanities/social science. Contact with Shermer, himself a historian [Shermer & Grobman 2000], made me look at history seriously. Two years later, I presented my *first* paper [Lam 2002] on the physics of history, or *Histophysics* [Lam 2008b], at the workshop celebrating the 80th birthday of Chen-Ning Yang, a physics Nobel laureate, at Tsinghua University, Beijing. Histophysics is a successful example of Scimat, the new multidiscipline that treats all human matters as part of science [Lam 2008a]. My work in Histophysics leads us to the discovery of two *quantitative*, historical laws concerning Chinese dynasties (from Qin to Qing) and a new general phenomenon in nature called the *Bilinear Effect* [Lam 2006c, 2008b; Lam et al. 2008]. My knowing of Shermer, made possible through our shared activities in Scicomm, played an important role in the creation of this new multidiscipline, Histophysics [Lam 2005d]. Subsequently I expanded my interest from history to the overall basic situation of the humanities/social science and came up with the idea of Scimat [Lam 2008a, 2008c].

In the summer of 2005, I presented a paper on the history of Histophysics [Lam 2005d] and worked as a reporter for *Sciencetimes* [Lam 2005e, 2005f] at the XXII International Congress of History of Science, Beijing, China, July 24-30. I met Maria Burguete, another participant from Portugal. Upon a cup of Chinese tea, she invited me to visit her. Next year in March, with the award of US$ 1,000 travel money from SJSU, I found myself in Portugal, one of the few countries in Europe that I never visited before, despite my two-year stay in Belgium and West Germany during 1975-1977. It was at the bar of Vila Galé in Ericeira and after a few drinks that we decided to do something together next year, and that was how the First International Conference on Science Matters, Ericeira, Portugal, May 28-30, 2007, co-chaired by Maria Burguete and Lui Lam, came about [Sanitt 2007].

My involvements in Scicomm actually include something more. To help China's fight against pseudoscience, and sometimes "evil religions," I became the Chinese-copyright agent of Michael Shermer and James Randy. I got Shermer's *Why People Believe Weird Things* (Hunan Education Press, 2001) and Randy's five books on magic and anti-pseudoscience works (Hainan Press 2001) published in Chinese.

There are many things scientists can do in Scicomm, as individuals and without funding. Six of these are recommended above, with the first four suitable even for untenured professors. Scicomm is fun and adventurous; it enables one to meet interesting new friends/colleagues beyond their own discipline, or even helps one's research career. Chair Mao (毛泽东) once said: When faced with a daunting task, learn from the ants; mobilize the masses and trust them. It worked for China, and will work for Scicomm.

3 A New Concept for Science Museums

A science museum (or a science and technology museum) is an effective medium in helping the public to understand science. However, in contrast to popsci books [Lam 2001, 2005a; Lam et al. 2005] and TV science programs, museums are limited by their physical locations and large budgets. Yet, when available, these museums allow the public to see the real objects and, apart from admiring the wonders of nature itself, learn the science principles behind some natural phenomena.

In China, new science museums appeared rapidly in the last 20 years. In other parts of the world, for example, in Barcelona, Spain, a brand new science museum is under construction. There is no doubt that the importance of science museums is well recognized.

The first step in making a good science museum is to have good exhibits. The next step is to make it physically interactive, partially or completely. Almost all science museums *stop* here. This could create a problem and is most unfortunate; most unfortunate because the problem is easily removable. What is needed is a new concept.

3.1 Possible misconceptions imparted to the visitors

The exhibits in all science museums are displayed according to their subject matter, in other words, in compartments. For example, the exhibits may be put into four divisions: inanimate matters, life, intelligent matters, and civilizations. This classification is based upon the hierarchic construction of the material world, according to what we know. The world is made of atoms; in increasing size, atoms form molecules, molecules form condensed matter—inorganic matters and organic matters. Organic matters form living matters—plants and animals. Animals consist of cells and organs. In particular, we have human bodies; a group of humans form a society, leading to civilizations. (See Fig. 1.2 in [Lam 2008a].) Consequently, the four divisions of the exhibits are logical and there is nothing wrong with that. However, science museums with these compartmental exhibits could create two misconceptions for the visitors:

1. The visitor may leave with the impression that science is neatly divided into compartments; i.e., there is no unifying themes or principles behind many of those exhibits.

2. Since almost all science museums are limited to natural science only, the visitor may go home thinking that there is a rigid demarcation separating the natural science from the social science.

The fact that social science should and could only be based on natural science [Lam 2002, 2006c; Wilson 1998] is easy to see, but is sometimes overlooked. As explained in Sec. 2, the reasoning goes like this: Social science is about the study of human behaviors and human societies. Humans are (biological) material bodies which, of course, are part of natural science since natural science is about *all* material systems. (See [Lam 2008a] for more discussion.)

3.2 A simple remedy

How can these two misconceptions be avoided and corrected? Very simple! Before the exit of every science museum, there should be a room or a space showing some established principles that are able to unify many different phenomena found in nature, with examples taken from both natural and social sciences. There are three such principles: fractal, chao, and active walk [Lam 1998]. (See [Lam 2008a] for a brief introduction to these three general principles.)

It is gratifying to note that in some science museums in China[11] [Ai 2004][12] and perhaps elsewhere, some (but not all) of the three general principles mentioned above have been included in their exhibits. However, there is still no emphasize on the theme that social science and natural science are an integral whole, and the former is based on the latter, with unifying principles. And we would like to see that this is the case in *all* museums in the world.

Lastly, to have the greatest and lasting impact on the visitors, I still think that putting the unifying themes concerning all natural and social phenomena before the exit of a science museum is the best choice.

4 Science Popularization in China

In China, the term "popular science" or "science popularization" (abbreviated as *kepu* 科普 in Chinese) is favored over "science communication," due mostly to the fact that the former two terms (especially the second one) have been in use for a long period of time. A brief history of science popularization in China, from the time of late Qing dynasty and up to 2006, can be found in [Li 2008].

Essentially, before 1949, the year the People's Republic of China was established, Popsci was advanced by the intellectuals with hands free from the government; many of these people were educated in the West or Japan. After 1949, like everything else in the New China, Popsci was managed from the top by the government. The advantage is that Popsci is financially secure; the disadvantage, as pointed out by Li [2008], is that there were less free discussion and exchange of idea among the practitioners or scholars. As mentioned in item 5 of Sec. 2, in Scicomm, China actually has a lead over many other countries in terms of scales. A summary of the current situation—official policies, programs, activities and studies of Popsci in China is available.[13] Those interested in Popsci research in China could consult the journal *Science Popularization* which is based in Beijing.[14]

Here is an interesting Popsci problem: Why *professional* popular-science book authors do not exist in China? The easy answer to this question would be that, like some other non-English-speaking countries, the sale of popsci books written not in English (and hence no worldwide sales) is not large enough to support the authors full time. But China is a huge country with 1.3 billion people. The story is more complicated than this. The answer to this question and solution of the problem in China's case could be unique.

Before we proceed to the answer, let us first review why this question is important, not merely to China but to the whole world. And, after the answer, recommendations to improve the situation, applicable to China and *beyond*, will be given.

4.1 The importance of popular-science books

Popular-science books have a long history in existence [Gregory & Miller 2000]. Unfortunately, they are a neglected tool in the science education of students and ordinary citizens [Lam 2005a]. Popsci books are unique among the science media: They are available in every bookstore in every town, unlike the technical science books which are available in special bookstores in a university town.

1. Many popsci books are written by the pioneers themselves, Nobel laureates, or very gifted science writers who could be journalists or other scientists.

2. These books are affordable to almost everybody (about 20 Yuan in China, and 15 dollars for a paperback in USA).

3. These books are the place to learn how research was actually done and how discoveries were made in very recent times.

4. These books, at least in the USA and for the majority of them, contain no equations; they, if well written, are easy and entertaining to read.

Obviously, to ensure the continuous supply of new and good popsci books, a large number of competent authors must be available.

4.2 Popular-science book authors in China

In spite of China's large population of 1.3 billion, there is not yet a single *full-time* professional popsci book author in this vast country. This is in contrast to the case in literature, because China does have professional writers who can support themselves by publishing novels. And this is not due to lack of support from the Chinese government. In fact, the Chinese government recognizes science and technology as an important pillar in raising the living standard of its population and the economic well-being of the country as a whole. In 2002, China passed the popsci *law*,[15] the one and only one such law in the world, which protected and encouraged science popularization at every level of government.

In the years from 1949 to about 25 years ago and *before* market economy was introduced, every writer in China was government employed. During this period of time, the government saw the need to support full-time novelists, but not full-time popsci writers. Obviously in China (and everywhere else in the world) popsci books are not deemed to be equally important as literary books.

These days, *after* market economy is in place, quite a number of self-employed literary writers already exist and, as usual, *the government still supports a sizable number of literary writers*. Yet, we still see no full-time popsci book authors in China, self-employed or government employed. Why? To find out what happened, we interviewed a number of popsci book authors and publishers in China [Lam et al. 2005]. We were told that:

1. Science popularization is considered lower in status compared to science research or teaching.

2. Work in science popularization is not counted in job evaluations in many places.

3. Lack of systematic and large-scale government effort or program to train popsci professionals.

4. Insufficient personal income to support free-lance, full-time popsci writers.

Points 1 and 2 are definitely true in almost every other country; some countries are doing something to tackle point 3; point 4 is untrue, for example, in USA.

Point 4 is particularly interesting. With such a huge population in China, how can this happen? In fact, presently, the sale of an average popsci book in China is less than 5,000 copies. There are exceptions: for example, *The Complete Book of Raising Pigs* did sell 3 million copies. What this implies is that a popsci book (not on pig raising) geared to the need of the masses is still waiting to be written.

4.3 Recommendations

To address points 3 and 4 above, here are six recommendations:

1. The government should recognize the importance of popsci books, in line with the popsci law they put into effect in 2002, and support popsci writers the same way they support literary writers.

2. The government could extend the policy of supporting literary book projects to popsci books, too. That is, prospective writers can apply for a grant to write a particular popsci book.

3. In every science funding agency, for example, the Chinese National Natural Science Foundation, a new division of funding should be set up to support popsci activities, including book writing.

4. In major research institutes, such as those in the Chinese Academy of Sciences, one-year visiting positions for prospective writers could be established, enabling them to observe the research in action, learn about recent major research findings, and discuss with the experts or perhaps even collaborate with them to write popsci books.

5. Most importantly, to guarantee that popsci books will be sold in large quantities in the immediate future, all science teachers in high schools and universities should incorporate the use of popsci books in their classes. It is done by offering the students extra credit if they buy a popsci book, read it, and write a brief report. This is a sure way to excite the students in science and to enlarge their knowledge base. (See item 1 in Sec. 2.)

6. Since natural science forms the basis of all social sciences [Lam 2008a; Wilson 1998], and since science and literature are equally important in shaping modern lives, the time has come to include several popsci books—such as James Watson's *The Double Helix* [Watson 2001]—into the list of required readings in the general education of every student in every university.

In points 1-4, the prospective popsci writer should be allowed to come from any place (especially magazines and newspapers) as long as the candidate is qualified. Naturally, points 5 and 6 are equally applicable to other countries. China is a country with a strong central government and these recommendations do not need that much new funding; they can be implemented quickly. What is needed is the willpower to do so. Luckily for China, there is a tremendous amount of willpower, as impressively demonstrated in her organization of the *Olympics 2008*.

5 Educational Reform: A Personal Journey

Educational reforms in universities could involve any of these three components:

1. Contents of the course.
2. The teaching method of the instructor.
3. The learning method of the student.

No matter how it is done, an unavoidable constraint that will crucially affect the success of the reform is usually not mentioned, or ignored completely by the reformers, i.e., *the reform should not increase the teaching load of the instructor*. Also, the quality of the students taking a course—like the quality of a sample in a physical experiment or the raw material used in a factory—is of primary importance; this factor is never emphasized enough. Obviously, with a defective sample, no good experimental result can be expected, no matter how skillful the experimentalist is. This last factor points to the need to start any education reform from grade one on, or even better, from the kindergartens. And I am not kidding.

With the constraints understood and resources limited, I tried to do my best as a teacher. There is not much we can do about item 3 above. It is very hard for the student to change their learning habit after being wrongfully taught for 12 years before they show up in college, and this is not their fault. I therefore concentrated my effort in the first two items.

On item 2, the instructor's teaching method, I have tried something radically different. It is called MultiTeaching MultiLearning (MTML) [Lam 1999]. We note that in a physics class, the instructor usually does not have enough time to cover everything. The attention span of a student is supposed to be about 15 minutes; students in a class have different learning styles; some students are more advanced than others. Active learning and group learning are good for students.

Around 1999, to overcome these problems in the teaching of two sections of a freshmen course in Mechanics, I have tried a zero-budget and low-tech approach. In this course, we covered about one chapter per week, using *Physics* by Resnick, Halliday and Krane as the textbook. In each course, there were three classes per week, each 50 minute long. In the last session of every week, the class was broken up completely. Different "booths" like those in a country fair were set up in several rooms, manned by student volunteers from the class. The rest of the class was free to roam around, like in a real country fair, or *like what professional physicists do in a large conference with multiple sessions*. In this way, we were able to simultaneously offer homework-problem solving, challenging tough problems for advanced students, computer exercises, website visits, peer instruction, and one-to-one tutoring to the students. The students seemed to enjoy themselves and benefited from it. However, this approach was soon discontinued. It did require a little bit of extra preparation from the instructor; but more importantly, it did not seem to significantly raise the grades of the students. The inferior-raw-material factor might be at work here.

The next thing I tried, with better luck this time, is to integrate popsci books into my physics classes, as described in item 1 in Sec. 2. This practice was quite successful; the students liked it very much.[16]

This popsci-book program is not trying to alter the course content per se. My first attempt in this direction, item 1 in education reform above, actually happened earlier. Soon after I started teaching at SJSU in 1987, I created two new graduate courses: Nonlinear Physics and Nonlinear Systems.[17] But these two courses were for physics majors.

In Spring 1997, I established a general-education course called *The Real World*, opened to upper-division (i.e., third and fourth years in college) students of *any* major. It results from my many years of research ranging from nonlinear physics to complex systems [Lam 1998]. The description of this course is given in the flyer in Fig. 4. There were only nine students in class, majoring in physics, music, philosophy and so on, plus two physics professors sitting in. It was fun. The course stopped after one semester due to nonacademic reasons, falling victim to the sociology of science education.

Five years later in Fall 2002, the course was resurrected with the same name but modified to suit incoming freshmen students. It is this general-education freshmen course that will be described in detail in the next section.

6 The Real World

In 2001 we have a new provost in campus. This very energetic and ambitious man, Marshall Goodman, wanted to make SJSU distinctive among the 20 plus campuses of the California State University system. Introducing international programs with a global outlook was his way of doing that. But perhaps more important, with lightning speed as administrative things went, he was able to push through the university senate and actually had 100 brand new freshmen general-education courses set up and running in about half-a-year's time. Each of these courses is limited to no more than 15 incoming *freshmen* students. The program starting in Fall 2002 was called the Metropolitan University Scholar's Experience (MUSE). Here is the official description of the MUSE program:

> University-level study is different from what you experienced in high school. The Metropolitan University Scholar's Experience (MUSE) is designed to help make your transition into college a success by helping you to develop the skills and attitude needed for the intellectual engagement and challenge of in-depth university-level study. Discovery, research, critical thinking, written work, attention to the rich cultural diversity of the campus, and active discussion will be key parts of this MUSE course. Enrollment in MUSE courses is limited to a small number of students because these courses are intended to be highly interactive and allow you to easily interact with your professor and fellow students. MUSE

courses explore topics and issues from an *interdisciplinary* focus to show how interesting and important ideas can be viewed from different perspectives.

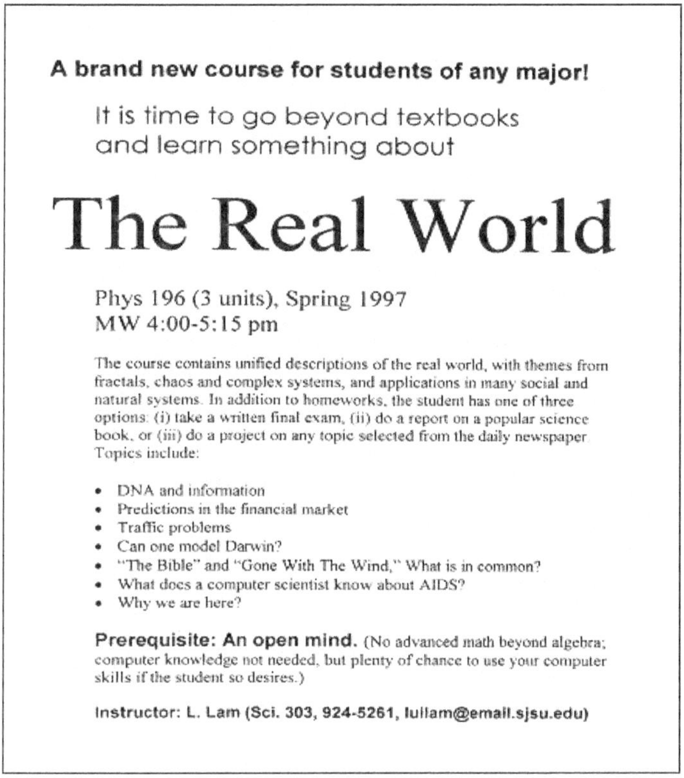

A brand new course for students of any major!

It is time to go beyond textbooks and learn something about

The Real World

Phys 196 (3 units), Spring 1997
MW 4:00-5:15 pm

The course contains unified descriptions of the real world, with themes from fractals, chaos and complex systems, and applications in many social and natural systems. In addition to homeworks, the student has one of three options: (i) take a written final exam, (ii) do a report on a popular science book, or (iii) do a project on any topic selected from the daily newspaper. Topics include:

- DNA and information
- Predictions in the financial market
- Traffic problems
- Can one model Darwin?
- "The Bible" and "Gone With The Wind," What is in common?
- What does a computer scientist know about AIDS?
- Why we are here?

Prerequisite: An open mind. (No advanced math beyond algebra; computer knowledge not needed, but plenty of chance to use your computer skills if the student so desires.)

Instructor: L. Lam (Sci. 303, 924-5261, lullam@email.sjsu.edu)

Fig. 4. The upper-division course, Phys 196: The Real World, offered by Lam in Spring 1997 at SJSU.

6.1 Course description

"MUSE/Phys 10B (Section 3): The Real World," created and taught by me (Fig. 5),[18] was one of the 100 incoming-freshmen MUSE courses.

1. Course description

To *understand how the real world works from the scientific point of view*.[19] The course will consist of two parallel parts. (1) The instructor will introduce some general paradigms governing *complex systems*—fractals, chaos and active walks—with examples taken from the natural and social sciences, and the humanities. (2) Students will be asked to pick *any* topic from the newspapers or their daily life, and investigate what had been done scientifically on that topic, with the help from the web, library, and experts around the world. Outside speakers and field trips are part of this course.

Fig. 5. The plastic card certifying Lui Lam as a MUSE faculty.

2. Student learning objective and goals specific to this course

After successfully completing this course, the student will:

- Realize that there are general paradigms—fractals, chaos, and active walks—governing the functioning of complex systems in the real world, *physical and social systems* alike.
- What nonlinearity is.
- How "dimension" is defined mathematically.
- The meaning of self-similarity and fractals.
- Recognize and able to evaluate data to show that any physical structure or pattern in the real world is a fractal or not.
- What a chaotic system is.

- Able to distinguish a chaotic behavior from a random behavior given the time series of a system.

- To realize that many complex systems in the real world can be described by Active Walk, and be familiar with a few examples.

- Recognize that there are multiple interpretations or points of view on some ongoing, forefront research topics, and that these interpretations can co-exist until the issue is settled when more accurate data and a good theory become available.

- Know the difference between science and pseudoscience, and the real meaning of the so-called "scientific method."

- How scientific research is actually done.

- Able to find out the latest scientific knowledge about any topic of interest in the future.

- Have improved your skills in communicating both orally and in writing.

- Have increased your familiarity with information resources at SJSU and elsewhere.

3 Course material

The following book is required:

Lui Lam, *Nonlinear Physics for Beginners: Fractals, Chaos, Solitons, Pattern Formation, Cellular Automata, and Complex Systems* (World Scientific, 1998), paperback (list price: $28). Reading assignments from this book will be announced in class. Additional material will be provided by the instructor. Other information could be found from the web, magazines, research journals, and books from the library.

4. Grading

The final grade of 100% for each student is split among several items. A *term project* is required. It is a group effort with three to four students in a group. The topic will be chosen by the group, with the help and consent of the instructor. Progress of project will be presented by group members orally in class throughout the semester. A written progress report is to be handed in about the middle of the semester, and a written final report is due at end of semester.

Homework	20%
Tests (3 total, including final; 10 points each)	30%
Term project and presentation	20%
MUSE activities	15%
Field trip	5%
Participation	10%
Total	100%

5. Teaching philosophy

The class is run like a *research* group, with flexibility in content and timing according to the progress and need of the students, and with the injection of other foreseeable and unforeseeable academic activities. The instructor will teach some basic knowledge about complex systems, while each term-project group will be treated like a research group. Each student will be trained to be a scholar, working individually and as a member of a team.

6. Topics covered by the instructor

PART I

1 *The World is Nonlinear*
    1.1 Nonlinearity
    1.2 Exponential growth
    1.3 Gaussian distribution (the bell curve)
    1.4 Power laws
    1.5 Complex systems are nonequilibrium systems

2 *Fractals*
    2.1 Classification of patterns
    2.2 Self-similarity
    2.3 Definition of "dimension"
    2.4 What is a fractal?
    2.5 Fractal growth patterns

3 *Chaos*
   3.1 Sensitive dependence on initial conditions
   3.2 The logistic map
   3.3 A dripping faucet
   3.4 Chaotic vs. randomness

4 *Active Walks*
   4.1 What is an active walk?
   4.2 Examples of active walks

5 *Conclusion*
   5.1 Simplicity can lead to complexity
   5.2 Order can arise from chaos
   5.3 The world can be understood scientifically

PART II

These special topics will be inserted between the chapters in Part I, as time allowed:

- How scientific is the "scientific method"?
- Science vs. pseudoscience.
- How research topics are born.
- Diversity: The first woman president of the American Physical Society.
- Does the world have any meaning?

6.2 The outcome

There were 12 students in the class. In the beginning, every student was asked to buy and read a newspaper, pick out the topics that interested her or him, which could be about international conflicts, movies or television programs, sports, or anything. After class discussion, three topics—Creativity, Prediction, and What Is Life?—were chosen. Three groups with four students each were formed; each group focused on one of the three topics. Each group tried to find out the current status and the frontier in the scientific study of the chosen topic—through books, the web, and interviewing of experts. Each group gave regular progress report in class and, at the end of semester, handed in a written report after orally presenting it. Simultaneously, the instructor gave lectures on nonlinear and complex systems (see item 6 in Sec. 6.1).

At the end, we were all exhausted. The students seemed to have a good time. Did they really get the message that the real world can be understood and is governed by some unifying principles? Only time can tell. But it was a nice try.

My feeling is that this course is better offered to non-freshmen who are more mature and motivated. In fact, this course—with the content and approach intact but the depth of coverage modified—could be taught at any level, for undergraduates or graduate students.

7 Conclusion

Looking back, ever since I published my first paper on nonlinear physics, on propagating solitons in liquid crystals in *Physical Review Letters* in the year 1982 [Lin et al. 1982] while I was working at the Institute of Physics, Chinese Academy of Sciences, I have been doing research on systems of increasing complexity—from solitons to pattern formation to chaos and to complex systems. After the invention of active walks in 1992 [Lam 2005b, 2006c] and after 1998, the year *Nonlinear Physics for Beginners* [Lam 1998] was published, I tried to apply AW to human matters, ending with the creation of Histophysics in 2002 [Lam 2002]. From that point on, it was easy for me to enlarge the vision and come up with the idea of Scimat [2008a], focusing myself on studying the humanities from the perspective of complex systems.

The review of my past activities in Scicomm/Popsci as well as teaching presented in this chapter makes it clear, at least to me, that my research direction is strongly coupled to and influenced by these activities; vice versa. I hope this example will encourage others to try the same. Many of the experiences I went through could be easily borrowed by others, or hopefully, would inspire them to innovate, in the interest of Scicomm, Scimat, and educational reform.

At this point, I hope you have found out and understand my two reasons for breaking the rule in writing the title of this chapter. If not, please go back to read item 5 in Sec. 2.

Appendix: Popular-Science Books Selected in Classes

Sample lists of popsci books selected in my classes are presented in Tables 1 and 2 here. (See also [Von Baeyer & Bowers 2004].)

Table 1. Popular-science books chosen and bought by students themselves in a freshmen calculus-based physics class in Spring 2000.

Title	Author	Year
The Art of Happiness	Dalai Lama/Cutler	1998
Beyond Einstein	Kaku/Thompson	1995
The Big Bang Never Happened	Lerner	1992
Black Holes, Worm Holes, & Time Machines	Al-Khalili	1999
A Brief History of Time	Hawking	1998
Calendar	Duncan	1998
Clones & Clones	Nussbaum/Sunstein	1998
Comets	Levy	1998
Computer	Campbell-Kelly/Aspray	1996
Darwin On Trial	Johnson	1993
The Diamond Makers	Hazen	1999
Faster Than Light	Herbert	1988
Fuzzy Logic	McNeill/Freiberger	1994
Fuzzy Thinking	Kosko	1993
Genesis & the Big Bang	Schroeder	1990
The Hidden Heart of the Cosmos	Swimme	1996
Immortality	Bova	1998
The Little Book of the Big Bang	Hogan	1998
The Meaning of It All	Feynman	1998
The Mind of God	Davies	1992
Night Comes to the Cretaceous	Powell	1998
101 Things You Don't Know About Science and No One Else Does Either	Trefil	1996
The Physics of Star Trek	Krauss	1995
The Real Science Behind the X-files	Simon	1999
Relativity Simply Explained	Gardner	1997
Science, Technology & Society	Bridgstock et al.	1998
Seven Ideas that Shook the Universe	Spielberg/Anderson	1987
Sex & the Origins of Death	Clark	1996
Skeptics & True Believers	Raymon	1998
Skies of Fury	Barnes-Svarney	1999
Steven Hawking's Universe	Filkin/Hawking	1997

There Are No Electrons	Amdahl	1991
To Engineer is Human	Petroski	1992
The Universe and the Teacup	Cole	1998
Why the Earth Quakes	Levy/Salvadon	1995
Why Sex is Fun?	Diamond	1997

Table 2. Popular-science books selected by the instructor for the students to pick, in the upper-division class of Thermodynamics and Statistical Physics in Spring 2000.

Author	Title	Year	Remark
H. C. von Baeyer	Warmth Disperses and Time Passes: The History of Heat	1998	Story of heat and the scientists involved; Maxwell's Demon; time's arrow.
T. Schachtman	Absolute Zero and the Conquest of Cold	1999	Story of how scientists lower the temperature; not that exciting, author not a scientist.
M. Riordan & L. Hoddeson	Crystal Fire: The Invention of the Transistor and the Birth of the Information Age	1997	Very exciting story; shows how good science was done in Bell Labs.; a must read especially if you live in the Silicon Valley.
G. Johnson	Fire in the Mind: Science, Faith, and the Search for Order	1995	Science and religion near Santa Fe, including studies in information and complexity.
A. Guth	The Inflationary Universe: The Quest for a New Theory of Cosmic Origins	1997	Written by the inventor of inflationary universe; unique; exciting physics and story.
T. A. Bass	The Eudaemonic Pie	1985	The story of UC Santa Cruz students, applying what they learn about Newtonian mechanics and chaos to beat the roulette in Las Vegas.
W. Poundstone	The Recursive Universe: Cosmic Complexity and the Limits of Scientific Knowledge	1985	All about cellular automata, with computer program for Game of Life.
J. D. Barrow	The Artful Universe: The Cosmic Source of Human Creativity	1995	Power laws, fractals, music.
M. Schroeder	Fractals, Chaos, Power Laws: Minutes from an Infinite Paradise	1991	Fits our course; highly recommended

Notes

1. At San Jose State University, as the instructor's teaching load is concerned, an undergrad lab of 3 hours is counted as 2 credits (versus 3 credits in the community college of City University of New York, a great city). I end up teaching 2 courses and 3 labs per week.

2. Scimat (Science Mattes) is a new multidiscipline that treats all human matters as part of science.

3. The term Popsci is inspired by Pop Art, advocated by Andy Warhol (1928-1987).

4. See Sec. 4.1 for the reasons of why popsci books are important.

5. See Appendix for a sample of books bought by my students.

6. Such as the Provincial Senior High School, Hsinchu, Taiwan, whose graduates include Yuan-Tseh Lee (Li Yuan-Jie, 李远哲), a Nobel laureate in chemistry.

7. A fractal is a self-similar mathematical or real object with possibly a fractional dimension [Lam 1998].

8. One of the two artists is Aprille Glover (www.aprille.net); she and her husband are two Americans living in Lavardin, France [Glover 2000].

9. "Pale Blue Dot" refers to our dear Earth when observed from far, far away in space; it comes from the title of Carl Sagan's popsci book [Sagan 1994]. My book contains three chapters: Why the world is so complex, How to model history and predict the future, and Does God exist?

10. The absence of a formal name for the Scicomm discipline or profession is due to the fact that the practitioners cannot agree on a single name, partly due to the shifting emphasis or concept in Scicomm. Some favor Popular Science or Science Popularization; others, Public Understanding of Science; etc. In fact, these different terms could be the names of subfields within a single discipline—Scicomm, like atomic physics and condensed matter physics, two subfields in physics.

11. China Science and Technology Museum, Beijing: "Science Tunnel" (http://old.shkp. org.cn/xinxi/suidao/shuidao003.htm).

12. Ai's article is an introduction to the Shandong Science and Technology Museum in Jinan, Shandong Province, China.

13. *2007 Science Popularization Report of China*, published by China Research Institute for Science Popularization, CAST (Popular Science Press, Beijing).

14. This journal is published by the China Research Institute for Science Popularization, CAST. Since its inception in 2006, the author is a member of the

editorial board. (The journal later changed its name to *Studies on Science Popularization*.)

15. *Law of the People's Republic of China on Popularization of Science and Technology*, issued June 29, 2002 (Popular Science Press, Beijing).

16. American students are crazy about extra credits in a course, even though the time they would spend to do the extra-credit work could or should be used in learning the course itself. It is a psychological thing, probably frequently used by teachers from grade one on.

17. These two courses resulted in two textbooks, one for undergraduates [Lam 1998] and the other for graduate students [Lam 1997].

18. I was so enthusiastic about this course that I delayed my sabbatical leave by one semester, from Fall 2002 to Spring 2003, in order to teach it in Fall 2002.

19. Scimat, by design, restricts itself to the scientific study of humans; it is thus part of this course which is about everything in the universe, as indicated by this statement (and the contents of the course). In turn, Histophysics by definition is part of Scimat.

References

Glover, W. [2000]. *Cave Life in France: Eat, Drink, Sleep...* Lincoln, NE: Writer's Showcase.

Gregory, J. & Miller, S. [2000]. *Science in Public: Communication, Culture, and Credibility*. Cambridge, MA: Perseus.

Han, X.-P., Hu, C.-D., Liu, Z.-M. & Wang, B.-H. [2008]. Parameter-tuning networks: Experiments and active-walk model. EPL **83**: 28003.

Lam, L. [1997]. *Introduction to Nonlinear Physics*. New York: Springer.

Lam, L. [1998]. *Nonlinear Physics for Beginners: Fractals, Chaos, Solitons, Pattern Formation, Cellular Automata and Complex Systems*. Singapore: World Scientific.

Lam, L. [1999]. MultiTeaching MultiLearning: A zero-budget low-tech reform in teaching freshmen physics. Bull. Am. Phys. Soc. **44**(1): 642.

Lam, L. [2000a]. Integrating popular science books into college science teaching. Bull. Am. Phys. Soc. **45**(1): 117. (Also reported in *APS News*, March 2000.)

Lam, L. [2000b]. How Nature self-organizes: Active walks in complex systems. Skeptic **8**(3): 71-77.

Lam, L. [2001]. Raising the scientific literacy of the population: A simple tactic and a global strategy. *Public Understanding of Science*, Editorial Committee (ed.). Hefei: Science and Technology University of China Press.

Lam, L. [2002]. Histophysics: A new discipline. Mod. Phys. Lett. B **16**: 1163-1176.

Lam, L. [2003]. Modeling history and predicting the future: The active walk approach. *Humanity 3000, Seminar No. 3 Proceedings*. Bellevue, WA: Foundation For the Future. pp 109-117.

Lam, L. [2004a]. A science-and-art interstellar message: The self-similar Sierpinski gasket. Leonardo **37**(1): 37-38.

Lam, L. [2004b]. *This Pale Blue Dot: Science, History, God*. Tamsui: Tamkang University Press.

Lam, L. [2005a]. Integrating popular science books into college science teaching. The Pantaneto Forum, Issue 19 (2005).

Lam, L. [2005b]. Active walks: The first twelve years (Part I). Int. J. Bifurcation and Chaos **15**: 2317-2348.

Lam, L. [2005c]. The origin of the International Liquid Crystal Society and active walks. Physics (Beijing) **34**: 528-533.

Lam, L. [2005d]. The story of Histophysics: History in the making. Presented at *XXII International Congress of History of Science*, Beijing, China, July 24-30, 2005.

Lam, L. [2005e]. From history of physics to popular-science book writing: The story of Lillian Hoddeson. Sciencetimes, Aug. 4, 2005: B1.

Lam, L. [2005f]. The dialogue between science and religion: What to talk about. Sciencetimes, Aug. 12, 2005: B2.

Lam, L. [2006a]. Science communication: What every scientist can do and a physicist's experience. Science Popularization, No. 2, 2006: 36-41.

Lam, L. [2006b]. A New concept for science museums. The Pantaneto Forum, Issue 21 (2006).

Lam, L. [2006c]. Active walks: The first twelve years (Part II). Int. J. Bifurcation and Chaos **16**: 239-268.

Lam, L. [2008a]. Science Matters: A unified perspective. *Science Matters: Humanities as Complex Systems*, Burguete, M. & Lam, L. (eds.). Singapore: World Scientific.

Lam, L. [2008b]. Human history. A science matter. *Science Matters: Humanities as Complex Systems*, Burguete, M. & Lam, L. (eds.). Singapore: World Scientific.

Lam, L. [2008c]. Science Matters: The newest and biggest interdicipline. *China Interdisciplinary Science*, Vol. 2, Liu Zhong-Lin (刘仲林) (ed.). Beijing: Science Press.

Lam, L. [2014]. The founding of the International Liquid Crystal Society. *All About Science: Philosophy, History, Sociology & Communication*, Burguete, M. & Lam, L. (eds.). Singapore: World Scientific. pp 209-240.

Lam, L., Bellavia, D. C., Han, X.-P., Liu, A., Shu, C.-Q., Wei, Z.-J., Zhu, J.-C. & Zhou, T. [2010]. Bilinear effect in complex systems. EPL **91**: 68004.

Lam, L., Li Da-Guang & Yang Xu-Jie [2005]. Why there are no professional popular-science book authors in China. Presented at *International Conference on Science Knowledge and Cultural Diversity*, Barcelona, Spain, June 3-6, 2004; The Pantaneto Forum, Issue 18 (2005).

Li, Da-Guang [2008]. Evolution of the concept of science communication in China. *Science Matters: Humanities as Complex Systems*, Burguete, M. & Lam, L. (eds.). Singapore: World Scientific.

Lin, L. (Lam, L.), Shu C.-Q., Shen, J.-L., Lam, P. M. & Huang, Y. [1982]. Soliton propagation in liquid crystals. Phys. Rev. Lett. **49**: 1335-1338; **52**: 2190(E) (1984).

Sagan, C. [1994]. *Pale Blue Dot: A Vision of the Human Future in Space*. New York: Random House.

Sanitt, N. [2007]. The First International Conference on SCIENCE MATTERS: A unified perspective, May 28-30, 2007, Ericeira, Portugal. The Pantaneto Forum, Issue 28 (2007).

Shermer, M. & Grobman, A. [2000]. *Denying History: Who Says the Holocaust Never Happened and Why Do They Say It?* Berkeley: University of California Press.

Shermer, M. [2001]. Starbucks in the Forbidden City. Sci. Am., July 2001.

Snow, C. P. [1998]. *The Two Cultures*. Cambridge: Cambridge University Press.

Von Baeyer, H. C. & Bowers, E. V. [2004]. Resource letter PBGP-1: Physics books for the general public. Am. J. Phys. **72**: 135-140.

Watson, J. D. [2001]. *The Double Helix: A Personal Account of the Discovery of the Structure of DNA*. New York: Touchstone.

Wilson, E. O. [1998]. *Consilience: The Unity of Knowledge*. New York: Knopf.

Yang, Xu-Jie & Lam, L. [2004]. Research on the history of science in China: Getting hotter. *Sciencetimes*, Aug. 20, 1998.

Published: Lam, L. [2008]. SciComm, PopSci and The Real World. *Science Matters: Humanities as Complex Systems*, Burguete, M. & Lam, L. (eds.). Singapore: World Scientific. pp 89-118.

23

Solitons and Revolution in China: 1978-1983

Lui Lam

Historically it is rare that one could do scientific research and political revolution at the same time. Such a chance was offered to me in China from 1978 to 1983. Throughout these six years, solitons (i.e., localized waves that travel without, or with slight, change in velocity and shape) were one of my major research topics at the Institute of Physics, Chinese Academy of Sciences. It was a hot topic in the physics community worldwide. In this chapter, the development of soliton research and political revolution in China experienced by the author is reported. The aim is not just to keep a record for those memorable years but also to convey the excitement of the so-called Science Spring of 1978, the year China's reform-and-opening up revolution began.

1 Introduction

It is rare that one can participate in history by doing scientific research and carrying out political revolution simultaneously. A famous example is the case of Condorcet (1743-1794), a French philosopher, mathematician, and political scientist [Baker 1975]. He held many liberal ideas and participated actively in the French Revolution (1789-1799). In 1794, after being branded a traitor and while hiding as a fugitive from French Revolution authorities he finished his masterwork, *Sketch for a Historical Picture of the Progress of the Human Mind* [Lukes & Urbinati 2012]. Soon after that, he was arrested and died in prison, at age 50.

A more recent example is the case of the thousands of scientists working in the 1950s in the new China, trying their best to do science in the midst of a socialist revolution. These include those who returned to China before and after 1949, the year the People's Republic of China was established; a lot of them earned PhD degrees from top universities in Europe and America [Wang & Liu 2012].

However, when I returned to China in 1978 I was unaware of Condorcet's story and the full story of the Chinese scientists did not come out yet.

2 Returning to China

I was born in mainland China but grew up in Hong Kong. After receiving my bachelor's degree at University of Hong Kong, I went for graduate studies first at University of British Columbia, Canada, and then at Columbia University, New York City [Tsui & Lam 2011: 209-211]. It was at Columbia as a physics student that I encountered the anti-Vietnam War student movement (1968)[1] and subsequently the Chinese *Baodiao Movement* (1970). Baodiao means Protect Diaoyutai. Diaoyutai (also called Diaoyudao or Diaoyu Islands) is the group of small islands near Taiwan that was and still is under dispute among the Chinese and the Japanese (Fig. 1) [The Seventies Monthly 1971].[2]

Although I had participated actively in Baodiao and lived in Chinatown to do community work [Lam 2010], my decision to go back and settle in China was not motivated by these experiences. Rather, it was China's The Great Proletarian Cultural Revolution. The Cultural Revolution began in 1966, the year I arrived at Columbia. The ideals put forth by this revolution and the heroic stories coming out from China stimulated the leftists and radicalized many young people all over the world. What could be more exciting than shutting down the whole country and rebuilding its entire infrastructure, leading to a new society not just in China but in every place in the world? How romantic? And it was such a basic solution to all human problems. At Columbia, we were trained and urged to tackle important, basic problems.

While living in Manhattan Chinatown (starting summer, 1971) I joined the editorial board of *China Daily News*, a newspaper established during the Sino-Japan war years by some overseas Chinese in New York who were sympathetic to the Chinese communists [Lai 2010]. One of the founders is Tang Ming-Zhao (唐明照). He went back to New China early on and returned as a member of China's first permanent representatives to the United Nations stationed in New York. On the evening he arrived, he asked his personal driver to deliver a box of Tsingdao beer to the *China Daily News*' office.

Fig. 1. The Protect Diaoyutai movement. *Left*: One of the Diaoyudao Islands. *Right*: Book cover of *Truth Behind the Diaoyutai Incident* [The Seventies Monthly 1971]. Inside the book, the article "Hong Kong, Taiwan, Diaoyutai" (pp 103-105) is a reprint written by the author (under the pen name Huang Shi-Zhi, 黃石之, which is omitted in this book; "shi zhi" means "stone it" in English).

In the summer of 1974, I (and three other persons) was invited to tour China for seven weeks and that was well before China opened its door to tourists. China was obviously underdeveloped but was green and unpolluted. Everywhere we went, including the scenic West Lake in Hangzhou, we were the only guests in town.

After the trip I decided to go back to China to join the revolution.[3] But it was not that easy. Zhou En-Lai (周恩来, 1898-1976), the premier at that time, kept telling visitors that China was not ready to welcome us back. And that was true given that the whole country was practically shut down, a condition that even the premier was not allowed to tell. We kept pressing.

My three-year, postdoctoral appointment at City College came to an end in 1975, and my doctoral mentor, Philip Platzman (1935-2012) at Bell Labs [Hamann & Isaacs 2012], found me a job at Antwerp, Belgium. I postponed my new appointment to October 1, so that I could finish my duty as the stage manager of the Chinatown event celebrating the national day of China—a position that I held for couple of years based on the skills I picked up as a member of the Chinatown Food Co-op [Kuo 1977].

In Belgium, as in New York, I worked with the local Chinese. From time to time, I visited the Chinese embassy in Brussels enquiring about the status of my application and received the same negative answer. After ten months in Antwerp, my job shifted to Saarbrücken, West Germany. The same story, except the Chinese embassy was in Bern, a two-hour trip from my town. Finally, something happened: Chairman Mao (Ze-Dong) (毛泽东主席, 1893-1976) passed away in September 1976. And in 1977, the embassy arranged me to join a Taiwanese-student delegation from Europe to observe the national day celebration in Beijing and to find out about my application status over there.

I was more than excited. To increase my chance of success, I flew back to New York, drove three hours to the Baltimore home of Ren Zhi-Gong (1906-1995; Fig. 2, left)[4] and asked him to write me a referral letter. Ren was in his early seventies at that time. He came out from China in 1946 and had served as department chair at Johns Hopkins University. Because of his seniority he knew all the physicists in China. But more importantly, he and Wang Hao (1921-1995; Fig. 2, right)[5] were the two famous professors who participated actively and whole heartily in Baodiao. Ren was not just a nationalist; he was an anti-imperialist. Ren handed me a sealed envelope, addressed to Qian San-Qiang (1913-1992; Fig. 3).[6]

To save money, I took a train to West Berlin, crossed the border and entered East Berlin, and flew to Beijing via Romania. In Beijing, the weather was perfect. I handed the referral letter to the person welcoming us. After touring the oil fields in Daqing and other places, we returned to Beijing just in time for the National Day celebration. We watched fireworks from the open stalls at the Tiananmen Gate, ending with a lot of firework droppings on our hair. The food of the national banquet at the

Great Hall of the People was a little bit disappointing. But who cared about the food while dining at People's Hall? Being there was a treat by itself.

Fig. 2. Two senior professors participating heavily in Baodiao: Ren Zhi-Gong (任之恭, left) and Wang Hao (王浩, right). Both earned PhD from Harvard University; Ren in 1931 and Wang in 1948.

Fig. 3. Two senior physicists I associated with in Beijing: Yan Ji-Ci (严济慈, left) and Qian San-Qiang (钱三强, right). Qian was the protégé of Yan. Both earned PhD in Paris, France; Yan in 1927 and Qian in 1940.

My date of leaving Beijing was approaching, and suddenly I was informed that Qian San-Qiang invited me to dinner. Qian was an experimentalist in nuclear physics; as the father of China's "two bombs" (atomic and hydrogen bombs), he played the roles of J. Robert Oppenheimer (1904-1961) and Edward Teller (1908-2003) combined. He was the most important Chinese scientist, well-known inside and outside of the country. The dinner was at the Donglaishun (东来顺) restaurant at Dong'an Market (now rebuilt as New Dong'an, 新东安), famous for its mutton hotpot. After all the delicious food was consumed, Qian informed me that my application was approved and I was assigned to work at the Academia

Sinica (called Chinese Academy of Sciences, or CAS, today). I was prepared to assume any job anyway, but out of curiosity I politely asked: Which Institute? Qian replied: You will know when you return.

As I learned later, Qian honestly did not have the answer. At that point, the Institute of Physics and the Institute of Semiconductors, both under CAS, were trying their best to get me assigned to their own institute.

3 Arriving Beijing and the Early History of Institute of Physics

China started sending students abroad in 1872 during the Qing dynasty. Since then, there are a total of 11 generations of overseas students, counting those going out officially or privately. Many graduated from the top universities in Europe and America and came back to contribute significantly in the country's modernization. After 1949, there are three generations of *Haigui* (海归), the nickname for those going abroad privately and subsequently returning to mainland China *voluntarily* after new China was formed in 1949 (see Appendix). I belong to the 9th generation going abroad and the 2nd generation of haigui.

In early January, 1978, my wife and I, together with our eight-month old daughter, stopped in Hong Kong for a few days to buy clothes and some personal items (camera, watch, and two bicycles—which were shipped to Beijing), all made in China. I also bought two foreign-made items: a color TV (upon the advice of the embassy in Bern) and a HP scientific calculator (costing 150 USD).[7] We walked across a bridge at the border and entered Shenzhen; that was one of the happiest moments in my life. We took the train to Guangzhou, stayed overnight and flew to Beijing. I was informed that I was assigned to the Institute of Physics (IoP), CAS.

The history of IoP is closely linked to two names, Cai Yuan-Pei (蔡元培, 1868-1940)[8] [Zhao 1998] and Yan Ji-Ci (1900-1996) [Jin 2000][9] (Fig. 3, left). In 1928, during the Republic of China times, Cai established the government-funded National Academia Sinica (NAS, the forerunner of CAS) in Nanjing *and* the National Academy of Art (now called China Academy of Art) in Hangzhou [Zhao 1998].[10] The NAC's Institute of Physics was located in Shanghai (part of which moved to Nanjing in 1948). Next year, the privately-funded National Academy of Beiping

(NAB) was established in Beiping (Beijing today). Yan headed NAB's Institute of Physics in 1931-1937 and turned it into the most important place in physics research in China, in the first half of last century [Wu 2005: 237-238].

Yan Ji-Ci graduated in 1923 from Southeast University (renamed National Central University in 1928), Nanjing, and went to Paris, France, the same year; he belonged to the 5th generation of overseas students. He was a brilliant experimentalist, obtained his PhD in 1927 from University of Paris, and became the first Chinese earning a PhD in France, which was widely reported in French and Chinese newspapers [Jin 2000: 11]. Yan returned to China with fame the same year, got married, and visited Paris again for two years, 1929-1930; he worked briefly in Madam (Marie) Curie's laboratory.

During his reign at NAB's Institute of Physics, he published 53 physics papers, which, with two exceptions, were all published in foreign journals [Jin 2000: 92]. He trained a large number of young physicists. For example, he took in Qian San-Qiang as his assistant in 1936 and brought him to Paris the next year, to work as a graduate student under Irène Curie and her husband. In 1948, Yan became an academician of National Academia Sinica and the president of the Chinese Physical Society. In short, Yan brought back physics from Europe and developed the physics enterprise in his country, similar to what I. I. Rabi (1898-1988), a Nobel laureate at Columbia University, did for America [Rigden 2000]. As far as research counted, *Yan Ji-Ci is the father of Chinese physics.*

Soon after the establishment of the People's Republic of China [Dietrich, 1986], in 1949, the Academia Sinica was established in Beijing, the capital. Next year, the two physics institutes (of NAS and NAB) were combined to form the Institute of Applied Physics, with Yan Ji-Ci as the director. Shi Ru-Wei (1901-1983) became the new director in 1957 (Fig. 4, left). The institute changed name to Institute of Physics and moved to Zhongguancun, the present location, in 1958. In 1981, Shi stepped down (and passed away in 1983 [Zhao 2005]), and Guan Wei-Yan (1928-2003)[11] [Li 2004] succeeded as director (Fig. 4, right) [Sun 2008; Zhao et al. 2008].

In the first two decades of new China, the top physicists were assigned to the secret "two bombs" project; scientific research was carried out mainly in the numerous institutes of CAS (about 100 in number presently); the universities concentrated in teaching but not in research. Many of the physics-related institutes of CAS such as the Institute of Semiconductors, was split off from IoP. When I arrived in 1978, IoP employed more than 900 people, with about 600 scientists. With its history and scale, the IoP was easily the top research place in the whole country.

Fig. 4. The two directors of Institute of Physics during my stay there: Shi Ru-Wei (施汝为, left) and Guan Wei-Yan (管惟炎, right). Shi earned PhD from Yale University in 1934; Guan studied in USSR (1953-1960) and earned Associate Directorate from Institute of Physical Problems in 1960.

4 Life at Institute of Physics

Life in Zhongguancun and at Institute of Physics is described, including the "Science Spring 1978," my 1979 trip to USA as a visiting scholar, and other things.

4.1 Living in Zhongguancun

I first met Hao Bai-Lin (Fig. 5) when I drove from Saarbrücken, West Germany, to Fribourg, Switzerland, to attend his seminar there. Hao was the one who showed me around in Zhongguancun when I first arrived at IoP. He rode his bike in front of mine and demonstrated the use of hand signals while making turns, while no other bikers in the street did the same.

The theoretical physics group in IoP was disbanded during Cultural Revolution on the ground that basic science was useless. Luckily, a Theory

Group was preserved in the Department of Magnetism. I worked in this group, consisting of ten persons working in theory (see Fig. 5) plus six in computers. I shared an office room with three colleagues (Wang Ding-Sheng, Li Tie-Cheng, and Cai Jun-Dao) on the fifth floor of the main building (Fig. 6).

Fig. 5. Five of my colleagues in the Theory Group, Department of Magnetism, Institute of Physics. *Left to right, top to bottom*: Li Yin-Yuan (李蔭远, 1919-2016), Pu Fu-Ke (蒲富恪, 1930-2001), Hao Bai-Lin (郝柏林, 1934-2018), Yu Lu (于渌), and Wang Ding-Sheng (王鼎盛). All five became academicians of CAS. In 1978, apart from these five and the author, the other four members working in theory are: Shen Jue-Lian (沈觉涟), Li Tie-Cheng (李铁成), Feng Ke-An (冯克安), and Cai Jun-Dao (蔡俊道).

There was only one very old elevator, which was not safe enough to carry people and was used occasionally to transport heavy equipment. Even Shi Ru-Wei, the 77-year old director of IoP, had to climb the stairs to his office on the third floor. Shi received me soon after I arrived; we had a nice chat. Watching Shi's back moving slowly but surely up the stairs, like in a slowed-down video, was the most touching experience I had at IoP. The image conveyed the long history of difficulties China had to overcome to

move forward and her perseverance and success in moving forward, one
step at a time.

Fig. 6. The author in his office, Room 518, main building, Institute of Physics
(1978). The wooden bookcase was made by IoP specially for me before my return.
The books, mostly in English, were brought back by me; note *The Feynman
Lectures on Physics* in red color in second row from top. The newspaper on my
desk is *People's Daily*; my iron rice bowl for lunch is next to bookcase. The
switches mounted on a wooden board at the back wall controlled the lightings.
They "leaked" when touched; getting electrically shocked was a daily experience
for the occupants of this office.

Everyone was poor; the Institute was poor. But even with money, there
was nothing to buy. Everything was in shortage. There was no doufu in
the market, and no white paper in the shop.[12] To write notes and do
calculations, I used the back of thin and semitransparent papers meant for
writing Chinese characters. Months before my arrival, IoP had to find
wood to make all the furniture for my future apartment, including a bed, a
desk, a chair, a wardrobe, and the bookcase shown in Fig. 6.

We stayed at the Overseas Chinese Hotel in downtown for three months while IoP desperately tried to find a family who was willing to exchange their apartment in Zhongguancun for the apartment assigned to me by CAS' headquarters which was more than one hour away by bus from IoP. Eventually, IoP succeeded; our assigned three-room apartment became a two-room apartment, which is two-minute by bike away from IoP.

But a month or so before that happened, I insisted and was allowed to start working at IoP. It was not that simple. It meant that the hotel kitchen had to prepare breakfast for one-person, me, at 5:30 am, six days per week. At 6:00 am, I walked across the street to the China Art Gallery; rode a bus for an hour to the Beijing Zoo; stood for half-an-hour in the freezing cold, with a number of my fellow workers, in an open truck that was sent by the IoP. One could not survive the trip without a "Lei Feng hat" (雷锋帽, which is made of fur and covers the two ears); luckily I brought one in from Hong Kong. But there was nothing to complain about; overcoming "hardship" was quite expected in doing revolution. I joyfully look forward to work every morning, in spite of the four-hour daily trip.

I soon understood why the Chinese had to take a nap after the lunch. The breakfast's calories were able to support you for four hours; after a bare lunch and without a nap, you would not be able to work the afternoon. After half a year, the nutrition stored in my body from Germany waned off and I found myself getting sick from time to time. There were no private doctors. To visit the hospital, I rode a bike for 30 minutes; got a number and waited, and waited. The whole process took half a day.

There was no taxi. What if you were so sick that you could not ride a bike or take a bus to the hospital? It never happened to me, but it happened to Wu Ling-An (吴令安). Wu returned to China with her parents from the United Kingdom; she spoke perfect Oxford English. In one evening, Wu had to be rushed to the hospital to give birth to a child. No problem. Her husband put her on a bike, pushed or rode her to the hospital in time. A healthy baby was born.

Almost everything was rationed, including rice, flour, cooking oil, and cotton cloth; others had quotas, such as bicycles and pregnancies; still others required permission from your "work unit," such as marriages and divorces—the latter was rarely granted. The exception was soy sauce. So after we moved in to Zhongguancun, I went to buy soy sauce. I said: "I want to buy some soy sauce." The saleswoman: "Where is your (empty) bottle?" I replied: "What bottle?" It turned out that the Bern embassy did remind me to bring in a TV and a bicycle but forgot to mention the soy-sauce bottle. No bottle, no soy sauce; that was it. (The problem was solved after my colleague kindly gave me a spare bottle.)

All these difficulties were just inconveniences. What was really handicapping physics research at the Institute, apart from other things, was the six-day work schedule. Formally speaking, we did research Monday to Friday; Saturday was for political studies which no one was allowed to skip. On Sunday, the only free day in the week, one had to wash dirty clothes by hand and attend to family chores. There were no holidays apart from the national holidays of the Labor Day, National Day, and the Chinese New Year (called Spring Festival); no Christmas because it is religious and Western, and no Mid-Autumn Festival because it is feudal. One's body was stretched to the limit; not much energy was left on Monday. Yet, all my fellow members of IoP were more than eager to work after the Cultural Revolution.

4.2 Science Spring 1978

Hua Guo-Feng (华国锋, 1921-2008) succeeded Chairman Mao (1893-1976) in 1976. Deng Xiao-Ping (邓小平, 1904-1997), who studied and worked in France (1920-1925) and studied in Moscow, USSR (1926-1927), belonged to the 5th generation of overseas students; he resumed working in July, 1977. He oversaw post-Mao reforms in science and education. In March 18-31, 1978, the National Conference on Science was held in Beijing, in which Deng gave the speech that recognized intellectuals as part of the working class and science as the first production force [Luo 2008].[13] As an honor and for my education, I was sent by IoP to attend the conference for half-a-day; it was not the opening day and I did not see Deng.

There was a lot to catch up. Feng Kang[13] (Fig. 7) came to IoP and gave an introductory talk on solitons, in the only large auditorium room in the main building. That was the same room used by IoP to deliver important messages to its senior members. On that day, the room with several hundred seats was filled to capacity and every theorist in Beijing seemed to be there. The reason: Soliton was a brand new topic in China, but more importantly, the theorists were ready to roll up their sleeves and work again after theory was discredited or banned in the last ten years [Hao 2012].

Fig. 7. Feng Kang (冯康, 1920-1993) and siblings. *Left to right*: Elder brother Feng Huan (冯焕), Feng Kang, little brother Feng Duan (冯端, 1923-2020), and elder sister Feng Hui (冯慧) [Tang el al. 2010].

Fang Fu-Kang (方福康, 1935-2019) of Beijing Normal University gave a series of lectures on stochastic differential equations. We biked for 30 minutes to attend every lecture. English classes were offered at IoP. A technical member of our group, infected with hepatitis B, was so enthusiastic to improve himself to better serve the country that he joined the class and studied very hard, ignoring the advice that someone of his condition should take a lot of rest. Pretty soon, the hepatitis progressed to liver cancer and he died before the class was finished.

I soon met Yan Ji-Ci in person. When I was a student in Hong Kong I bought his high-school physics textbook, which was superbly written. The book helped me to become the top student in class. So when I first met him it seemed that I already knew him. Shaking his hand was a historical moment for me; it was like finally making connection to the Chinese physics tradition and becoming a part of it. I met him again a few more times. The last time I saw him was in 1982 when he laid sick in Hospital 301; he survived. Hua Guo-Feng was also a "patient" in the hospital; he invited me to his room and we chatted for half-an-hour. Even though Yan was the vice president of CAS from 1978-1981, I was unaware of his appointment; there was not yet webpage to check.[15]

I did meet Qian San-Qiang again several times. There was this time that he conducted a meeting of a few theorists, me included, to discuss the role of theoretical physics. At one point, he emphasized the need of "theory linked to *shiji* (实际, practicality)". I intercepted, "It is also important to emphasize 'theory linked to *shiyan* (实验, experiments)'," He, a trained experimentalist, seemed to find it refreshing since, apparently, no one mentioned this to him before. Actually we were referring to two different aspects of theory: Qian was talking about the application of theory; I, how theory should be conducted in research. Linking theory to experiments is the proven way of good research that I picked up at Bell Labs (see Sec. 5 for an example). This was well recognized by Peng Huan-Wu (1915-2007) (Fig. 8), too, but Peng was not at that meeting [Peng 2001: 82-83].

Peng Huan-Wu graduated from Tsinghua University and spent nine years (1938-1947) in Europe, obtaining his PhD under the Nobel laureate Max Born (1882-1970) in 1940. He was responsible for the theoretical aspects of China's nuclear bomb project [Peng 2001, book-back cover], the counterpart of Hans Bethe (1906-2005) regarding American's atomic bomb. He became the first director of the Institute of Theoretical Physics (ITP), CAS, in May 1978. He and I had several exchanges during my stay in Beijing (see Secs. 5 and 6).

Huang Kun (1919-2005) (Fig. 8) was very kind to me. He studied and worked in Europe from 1945-1951. After earning his PhD (1948) under the Nobel laureate Nevill Mott (1905-1996), he was a postdoc with Max Born [Chen & Yu 2008]. It is the book *Dynamical Theory of Crystal Lattices* (1954) by Born and Huang, a classic in solid state physics, that made Huang the best known Chinese physicist in the Western academic circles. He returned to China in 1951 and married his European girlfriend the next year. Perhaps because of that, Huang was not drafted into the bomb project; he remained in the academia and had the chance to make important contributions in solid state physics [Zhu 2000].

Fig. 8. Two other senior physicists I associated with in Beijing: Peng Huan-Wu (彭桓武, left) and Huang Kun (黄昆, right). Both earned PhD in England; Peng in 1940 and Huang in 1948. The Nobel laureate Max Born was Peng's doctoral thesis advisor and Huang's postdoctoral supervisor.

During the Cultural Revolution, Huang Kun was the only scientist who wrote in *Hongqi* (红旗), the foremost political journal; however, he was deemed politically incorrect after the Revolution. With the approval of Deng Xiao-Ping, Huang was appointed the director of the Institute of Semiconductors, CAS, in 1977. He invited me to visit his institute. The reason was that he learned from my resume that I had worked in semiconductors while in Belgium and he was eager to rebuild the theory group in his institute. I met him again at the Lushan conference (see below) and other meetings. The last time I saw him was in 2002 during Chen-Ning Yang's 80th birthday banquet in Beijing. He was sitting alone at the end of a very long table. I walked up to him and said hello.

In August of 1978, the Chinese Physical Society resumed its "annual" meeting in Lushan after ten years [Wang & Yang 2012]. Lushan is the mountain resort that Chiang Kai-Shek (蒋介石, 1887-1975), president of Republic of China, spent his cool summers but is also the place that an important meeting was held during the Great Leap Forward years in 1959. All the important physicists showed up. I was allowed to go, too. It was a very interesting experience. And, it was at the end of this conference I learned that a new Institute of Theoretical Physics was being added to CAS.

In August, 1978, Chia-Wei Woo (吴家玮), Chair of the physics department at Northwestern University, USA, visited IoP as a guest professor and stayed for a few months. Since his research involved liquid crystals and to prepare for his visit, I was asked to give a series of lectures on liquid crystals to my colleagues. As it happened, liquid crystals became my research field during my six years at IoP [Lam 2014]. In spite of China's large population, the number of physicists was very small and it was easy for one to be counted as an "expert" on a chosen subject. Thus, with only one paper on liquid crystals published before I returned to China, I became an expert in the field of liquid crystal theories. Near the end of his stay, Woo invited the IoP to send a delegation of physicists to his department to work as visiting scholars, with everything paid by his university except for the air tickets. This proposal was quickly approved at the highest level, presumably by Hua Guo-Feng himself [Lam 2010].

Before 1978 ended, I published my first paper in China, "Microscopic theory of first-order phase transitions in liquid crystals" [Lin 1978a]. This paper explains an experimental result from Canada and is the precursor to my 1979 paper in *Physical Review Letters* [Lin 1979] (see Sec. 4.3). In fact, it is also my first paper written in Chinese; before that I had published 22 papers in English and one in French [Lam 1973].

4.3 My 1979 trip to America as a visiting scholar

In December 1978, the most important Third Plenum (of the 11th Central Committee Congress of the Communist Party of China) was held in Beijing, which officially shifted the government's focus from class struggle to economic development. This meeting is regarded as the beginning of China's reform-and-opening up revolution[16] of the last 30 something years [Tang 1998]. A few days after this meeting, on January 1, 1979, China and USA established formal diplomatic relationship. From January 29 to February 4, 1979, Deng Xiao-Ping visited the United States.

The IoP at that time was effectively run by Guan Wei-Yan, the deputy director, due to the old age of the director, Shi Ru-Wei. The IoP picked eight members to form the delegation to Northwestern: Qian Yong-Jia (钱永嘉), Li Tie-Cheng, Zheng Jia-Qi (郑家祺), Shen Jue-Lien, Wang Ding-

Sheng, Cheng Bing-Ying (程丙英), Gu Shi-Jie (顾世杰), and Lin Lei (林磊, aka Lui LAM) [Lam 2010]. The delegation left Beijing on February 9, 1979, five days after Deng's visit. We arrived at the university in Evanston on February 15, via Paris, New York, and Washington, DC. While in New York, *China Daily News* interviewed me (Fig. 9, left).

I spent three months at Northwestern while the rest of the delegation stayed behind for two years. Before leaving, I was invited by quite a number of universities to give physics seminars, and general talks on my experience in China (sponsored by the Chinese students there). The reception was very enthusiastic.

I went back to Beijing, via Hong Kong, in June, 1979. While in Hong Kong, I was interviewed by *Wen Wei Po* (Fig. 9, right) and was invited to write a popular-science article in this newspaper. I also wrote up my work on liquid crystals and submitted it to *Physical Review Letters*, the foremost international (weekly) journal in the physics profession. This paper, "Nematic-isotropic transitions in liquid crystals," turns out to be the first one ever with only mainland-China authors that got published in this top journal [Lin 1979]. (For more details, see [Lam 2010].) The publicity gained through the interviews and campus tours helped to make me a well-known figure among the overseas Chinese, which later played a positive role in helping my research in solitons (see Sec. 5).

4.4 Other things

I was the first returnee assigned to CAS after the Cultural Revolution. The support I received from the higher-ups and my colleagues was fantastic. I was interviewed by the Chinese magazines (see Fig. 10 for an example) and invited to write popular-science articles in the newspapers [Li 1978b].

I soon found myself being useful to China. The first instance occurred when the IoP was showing a foreign film without subtitle to all the members gathered in the canteen. I was called upon to do a translation with my broken Chinese. On another occasion, the official Xinhua News Agency phoned me out of the blue and asked me to translate an English word for them. Academically, there are two physics terms the Chinese translations of which are due to me: *Jiguai Xiyinzi* (奇怪吸引子) for "strange attractor" and *Guzi* (孤子) for "soliton" [Committee on Physics Terms 1996].

Fig. 9. My two interviews by Chinese newspapers at the beginning and end of the 1979 trip to USA. *Left*: "The most welcomed, unexpected guest: An interview with Lin Lei, a researcher from the Chinese Academy of Sciences visiting America," *China Daily News*, New York, first of three installments, March 1, 1979. *Right*: "From University of Hong Kong to Europe and America and to Beijing: Dr. Lin Lei talks about his path of scientific research," *Wen Wei Po*, Hong Kong, June 1979.

Perhaps more importantly, in 1980, I helped to found the Chinese Liquid Crystal Society and served as a vice president and the secretary-general. Before I showed up in Beijing, Xie Yu-Zhang (1915-2011)[17] (Fig. 11, left) was the only physicist working on liquid crystal theory. In the 1970s, when in his sixties Xie started working in liquid crystals to support the liquid crystal display (LCD) group at Tsinghua University [Zhao et al. 1980]. The group received funding even during the Cultural Revolution since Jiang Qing (江青), the wife of Chairman Mao, learned that large-size LCD screens would enhance political propaganda works. Unfortunately, the project never fulfilled its promise [Lam 2022a]. I had published papers with Xie and his student, Ou-Yang Zhong-Can (Fig. 11, middle).

In January 1979, the Theory Group was separated from the Department of Magnetism and became the Theoretical Physics Section directly under IoP, like the other Departments. I was elected the Associate Head of this Section; Pu Fu-Ke,[18] the Head. Somewhat later, the Section consisted of two research scientists (equivalent to full professors), three associate scientists, four assistant scientists, and 11 graduate students.[19]

Fig. 10. My interview in the magazine *China Construct*, "In their primes: Interviews with three young overseas returnees," February 1981.

Fig. 11. Three physicists in liquid crystals. *Left to right*: Xie Yu-Zhang (谢毓章), Ou-Yang Zhong-Can (欧阳仲灿), and Shu Chang-Qing (舒昌清; photo taken in 1984). Ou-Yang became director of the Institute of Theoretical Physics and academician of CAS.

From 1979 on, I had attended international conferences on liquid crystals and on solitons abroad, and stayed as a visiting scholar for a few months once in USA (1979) and two times in Paris (1980 and 1982). The two important occasions are the International Conference on Liquid Crystals (December 3-8, 1979) in Bangalore, organized by the Indians, and The Eighth International Conference on Liquid Crystals (June 30-July 4, 1980) in Kyoto, Japan. As international liquid crystal conferences are concerned, the former was the first one ever attended by a delegation from mainland China and the latter, the first one in the biennial series; both delegations were headed by me. In Kyoto I was nominated (and subsequently elected) to the Planning and Steering Committee that oversaw the biennial series, a crucial event that led to the forming of the International Liquid Crystal Society ten years later [Lam 2014]. (See Sec. 5 and [Lam 2014] for more on liquid crystals in China.)

Liquid crystal is an important member of a large group of materials called soft matter. The Liquid Crystal Group I formed at IoP is in fact the forerunner of IoP's soft-matter laboratory established in 2001.

The signatures of any revolution are the existence of extreme risks and, very likely, the loss of human lives. In this case, there were two. First, the possible downfall of Deng Xiao-Ping accompanied by the appearance of another Cultural Revolution was like a knife hanging over our head daily (especially since Chairman Mao had urged a Cultural Revolution every ten years). Second, one could catch hepatitis B by eating in the canteen and die of liver cancer. The former did not happen, but the latter did. Liao Qiu-Zhong (廖秋忠), a PhD in linguistics from University of California, Berkeley, died of hepatitis B just a few years after returning to China. Some years later, the same happened to Cai Shi-Dong (蔡诗东, 1938-1996), a 1973 returnee at IoP with a PhD in plasma physics from Princeton University.[20]

5 Solitons in China

Feng Kang's 1978 lecture on solitons was very exciting and it was the first lecture on solitons I ever attended. However, at that time I was working on phase-transition problems in liquid crystals and solitons seemed to be irrelevant to my research.

A (rigorous) soliton is a localized wave that travels without change of velocity and shape. It was discovered by John Scott Russell (1808-1882) in 1834. However, in many real situations this definition is relaxed a bit and the velocity and shape are allowed to change a little bit while traveling [Lam 1992]. It was a hot topic in physics from the late 1970s to early 1980s, especially after it was widely applied to condensed matter [Bishop & Schneider 1978].

In the summer of 1981, I went to Tsinghua University to attend a master's degree thesis defense given by a student of Xie Yu-Zhang. After the meeting, Zhu Guo-Zhen (朱国桢), a colleague of Xie, led me to his laboratory and showed me his experimental results of three dark lines that propagate in a liquid crystal cell. Zhu just had his manuscript on this experiment rejected by *Acta Physica Sinica*, the major physics journal of China. The reason was that the sound wave equation, a linear equation, given there was at odd with the experimental observations. In the experiment, the width of each line *narrows* as the line moves while the width of a sound wave always increases as time increases. What he needed now was an interpretation of his experimental results, which was an integral part of any experimental work and could be more challenging than doing the experiments itself.

To me, it was obvious that these lines could not be a sound wave for the reason explained above. In fact, the lines looked like solitary waves and Feng Kang's lecture came to mind.

Knowing that good theory has to go side-by-side with good experiments, better be one's own experiments if possible—a lesson I learned when doing my PhD thesis at Bell Labs—I proposed to Zhu that I would send Shu Chang-Qing (Fig. 11, right) to his laboratory to do more experiments. The condition was that any paper coming out of this joint effort would have Shu as the first author. He agreed. At that point in time, Shu just finished his three-year master's degree with me and was continuing with his doctoral studies.[21]

I started reading soliton literature earnestly. The library of IoP did not have originals of the foreign journals but it did have copies.[22] By this time, IoP had one copying machine, made in China. It broke down several times per day and was managed by a young technician, Chen Jie (陈捷).[23] Every time I went to the copying room at the back of the main building, I passed a cigarette to Chen. After the smoking session, I gave him the journals I wanted to be copied, and I usually got it back the next day. It helped tremendously. Moreover, a few young overseas Chinese visited Beijing and asked to see me (probably because my name was known in those circles). Before departing, they would ask me what they could do to help my work (i.e., helping China). And I got several English soliton books this way.

Zhu's experiment was very simple. He put a thin plate between the two glass plates of a liquid crystal cell (which is nothing but two parallel, thin glass plates coated with indium oxide in the inner surfaces, with liquid crystal in between); the thin plate was placed near one end of the cell and was pushed forward toward the other end for a short distance. Three dark lines were generated and moved forward [Zhu 1982].

Pretty soon, I figured out what happened. When the plate is pushed forward, the molecules at the middle layers move faster than those near the cell's boundaries and a velocity distribution, called shear, is created. The shear induces the rod-like molecules, which are initially vertical, to incline forward at an angle throughout the cell. When the movement of the plate is stopped, opposite shear is generated and the molecules near the plate end will incline backward, resulting in an orientation distribution like this: Two domains with opposite angles at the two ends are separated by a region of (nearly) vertical molecules. In the soliton literature, this distribution is called a "kink." As more and more molecules at the plate end reverse their orientation, the vertical-molecule region shifts forward and we have a propagating kink (Fig. 12). With this new understanding that the shear was what generated the solitons, I directed Shu to do the experiments systematically using pressure difference to create the shear because pressures could be controlled more precisely.

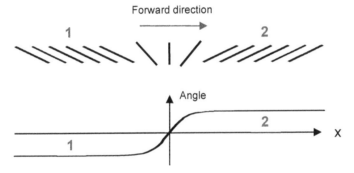

Fig. 12. A soliton in shearing liquid crystals. *Top*: Orientations of the rod-like molecules: A region of (nearly) vertical molecules is sandwiched between two regions of oppositely oriented molecules (labeled 1 and 2, respectively). *Bottom*: The corresponding orientational distribution: a kink soliton.

I very quickly wrote down the equation of motion: the damped sine-Gordon equation. We were excited except that we did not know how to solve it; nothing relevant could be found in the literature. I sent Shu to Feng Kang's home to ask for his advice.

Feng could not solve it either. But Shu came back deeply impressed by Feng's extremely strong memory and enormous knowledge. That no one can solve it is not surprising, even today. The problem is that solitons are solutions of particular nonlinear equations which, unlike linear equations, generally do not succumb to systematic analysis. Furthermore, for Feng and other mathematically inclined soliton researchers, rigorous solitons are what interest them; the damped sine-Gordon equation belongs to the domain of non-rigorous solitons [Lam 1992: 11].

Luckily the soliton solutions of the equation could be figured out qualitatively and we were able to understand the major features of the experiments. This understanding and some initial experimental results obtained by Shu (e.g., a *single* dark line was generated) were briefly summarized within a general review on solitons I gave in Wuhan in 1981 [Lin 1982a].

The review is titled "Solitons in condensed matter" (Fig. 13). It was presented at the First National Conference on Statistical Physics and Condensed Matter Theory, held at Huazhong Institute of Technology. Lin Pei-Wen (林沛文)[24] of ITP and I shared a campus dorm room throughout the conference. Incidentally, two other talks on specific soliton systems were presented at this conference: "Soliton model of polyacetylene and fractional charges" by Yu Lu (ITP), and "Solitons in ferromagnetic chains: A new example of inverse scattering method" by Pu Fu-Ke (IoP).

Pretty soon we found an analytic solution under a special condition which corresponds to the experimental situation. Calculated results from this special solution match very well the experimental observations. To speed up the work, a small group on this project was formed, which consisted of Shu Chang-Qing, Shen Jue-Lian, and me from IoP, Huang Yun (黄芸) from Peking University, and Lin Pei-Wen from ITP. This group met several times at IoP.

In early 1982, we were ready to write a paper for *Physical Review Letters* (PRL). But our initial effort to submit a joint paper with Zhu describing both the theory and experiments failed (because Zhu turned down the invitation). At the end, two papers were written: An experimental paper written by Zhu and a theoretical paper written by our group. Both manuscripts reached the PRL office in May, 1982; our paper arrived May 5, two days earlier than Zhu's. In those years, all outgoing official mails had to go through an appropriate office in one's affiliated institute or university for screening. Apparently, the mailing apparatus of IoP was more efficient than that of Tsinghua.

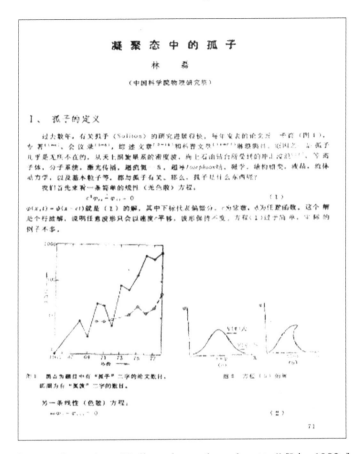

Fig. 13. The review, "Solitons in condensed matter" [Lin, 1982a].

Both papers were rejected. I wrote up a reply to the referees, and did the same for Zhu. Upon that, the editor changed his mind, and the two double-second PRL papers with only mainland-Chinese authors were published in the Nov. 1, 1982 issue of PRL [Lin et al. 1982; Zhu 1982]. Historically, in the liquid crystal literature, "Soliton propagation in liquid crystals" by Lin et al. (Fig. 14) is the first paper with the word "soliton" appearing in the title and the first paper that talks about propagating solitons. This paper from China enabled us to lead the field for ten years, culminating in the review book *Solitons in Liquid Crystals* [Lam & Prost 1992] (Fig. 15, left).[25]

VOLUME 49, NUMBER 18   PHYSICAL REVIEW LETTERS   1 NOVEMBER 1982

Lin *et al.*[8] in which the dark lines are interpreted as solitons.

I am indebted to Yu Hao and Xu Laoli for designing and constructing the mechanical device used to push the exciter, Lin Lei for his interest in this work and his assistance in improving the English writing of this manuscript, Zhao Nanming for discussion at an earlier time, Professor J. L. Ericksen for telling me his recent view about director waves in a personal communication at my request, and finally, my teacher Professor Meng Chaoying for his encouragement.

[1] J. L. Fergason and G. H. Brown, J. Am. Oil Chem. Soc. 45, 120 (1968).
J. L. Ericksen, J. Acoust. Soc. Am. 44, 444 (1968).
[3] F. M. Leslie, in *Advances in Liquid Crystals*, edited by G. H. Brown (Academic, New York, 1979), Vol. 4.
[4] Zhu Guozhen, J. Qinghua Daxue Xuebao 21(4), 83 (1981).
[5] Zhu Guozhen, in *Proceedings of the Chinese Liquid Crystal Conference*, Guilin, China, 20–25 October 1981 (to be published).
[6] Lin Lei and Shen Juelian, in *Proceedings of the Chinese Liquid Crystal Conference*, Guilin, China, 20–25 October 1981 (to be published).
[7] Zhu Guozhen *et al.*, to be published.
[8] M. Born and E. Wolf, *Principles of Optics* (Pergamon, New York, 1975).
[9] Lin Lei *et al.*, following Letter [Phys. Rev. Lett. 49, 1335 (1982)].

Soliton Propagation in Liquid Crystals

Lin Lei,[a] Shu Changqing, and Shen Juelian
Institute of Physics, Chinese Academy of Sciences, Beijing, China

and

P. M. Lam
Institute of Theoretical Physics, Chinese Academy of Sciences, Beijing, China

and

Huang Yun
Department of Physics, Beijing University, Beijing, China
(Received 5 May 1982)

Soliton propagation in nematic liquid crystals under shear is shown to be possible and studied theoretically. Calculations including those pertaining to the modulation of monochromatic or white light passing through such a liquid-crystal cell are presented. Recent experiments are interpreted accordingly and are in good agreement with the theory presented here.

PACS numbers: 61.30.-v, 03.40.Kf, 05.70.Ln, 47.15.-x

Solitons are important and have been found in various objects ranging from celestial bodies to laboratory systems.[1,2] However, unlike the first observation of solitons in shallow water by Scott Russell, many of the recent experimental evidences of solitons in condensed matter are indirect in nature. The experiments[3] on the ordered fluid [3]He are no exception. In this regard, we note that in another type of ordered fluid, viz., liquid crystal, because of the strong coupling of the director with light, it may be possible to observe the motion of the molecules and the solitons rather directly.

Discussions of solitons in liquid crystals[4] was first given by Helfrich[5] and subsequently by de Gennes,[6] Brochard,[6] and Leger.[7] In their work in nematics, the solitons (called "walls") are magnetically generated and are small in width (e.g., a few microns). Experimentally, the observation[7] of these solitons is delicate and a polarizing microscope has to be used. Recently, there has been more but still limited attention[8] paid to the role of solitons in the physics of liquid crystals.

In this Letter, we first point out and discuss a new case in liquid crystals, viz., nematics under uniform shear, in which solitons can exist and propagate. In contrast to the magnetic case[5-7]

1335

Fig. 14. One of the two double-second PRL papers with only mainland-Chinese authors, "Soliton propagation in liquid crystals" [Lin et al, 1982].

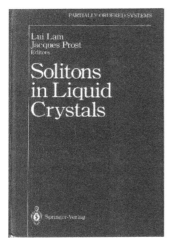 

Fig. 15. *Left*: The world's first book on solitons in liquid crystals [Lam & Prost 1992], which is also the starter in the Springer book series Partially Ordered Systems founded by the author. *Right*: Book cover of *Introduction to Nonlinear Physics* [Lam 1997] which includes two chapters on solitons, based on the review article published in China [Lin 1982a] (see Fig. 13).

Like my first PRL paper, no one at IoP paid any attention. But the reception from outside of China was quite different.[26] When the paper came out in print, I was spending two months at the Université de Paris-Sud, Orsay. I gave a seminar on my PRL soliton paper and solitons in liquid crystals were picked up by a group there [Madhusudana et al. 1992]. At the Chinese University of Hong Kong where I stopped over on my way home to Beijing, Chen-Ning Yang, the Nobel laureate of parity-nonconservation fame, invited me to give a seminar on this topic.

At a certain point after this paper, Shu and I went to talk to Peng Huan-Wu at his home. We wanted to know how instability of a wave solution is studied theoretically. He received us warmly and kindly educated us on this topic.

Overall, I published 23 papers on solitons in liquid crystals in 12 international journals. All, except for the first one, were published after I left China (see next section). Incidentally, in the last conference I attended before leaving, the Second National Conference on Statistical Physics and Condensed Matter Theory, Nanning, November 14-18, 1983, a paper on solitons in polyacetylene was coauthored by Huang Kun. And Shu Chang-Qing did finish his thesis "Propagating solitons in shearing nematic liquid crystals" in April 1984, which, officially, is China's first PhD thesis in liquid crystal physics.[27]

6 Leaving China

Leaving China used to be quite impossible. The government took it personally, reasoning that since the revolution/society was such a splendor thing in the world, anyone chose to leave implied a disagreement with this assessment and should be discouraged. A married couple, Chen Ruo-Xi (陈若曦) and Duan Shi-Yao (段世尧), who returned to China in 1966 and worked in Nanjing, left with permission in 1974, after Zhou En-Lai, the premier, had declared to the Baodiao visitors that the official policy regarding returnees was "Welcome to return; coming and leaving are both free" [Chen 2008]. Apparently, Zhou was already thinking of how to attract talents from abroad to strengthen the country's rebuilding once the Cultural Revolution was finished.

My wife went back to New York for a home visit upon her father's death, taking along our daughter, in the summer of 1980. She then announced that she would not return. It was a total surprise to me; we never discussed this before (or after). In fact, it was hard to discuss anything personal when we were separated in two cities, in two continents. There was no private phone at home, and any form of communication across the border was more likely than not to be monitored. To make a phone call to her, I would ride my bike one hour to the telephone and telegraph building in Chang'an Avenue, downtown. The cost was 10 yuan per minute; my monthly salary would last 20 minutes.

Three years later, her "green card" application was approved and I had to decide. There were a few returnees of my generation who already left, either because they found the societal reality not matching what they envisaged or because their skills were not adequately put to use, or both. (In all cases, none of them left because of low living standards.) My case was different. Since I helped to bring my daughter to the world I had the responsibility to see her grow up properly.

I resigned and left China in December, 1983. Before that, I had a private talk with Guan Wei-Yan, the then director of IoP. After clearing up all work-related business, I asked him to appoint me as a Guest Professor so that there would be no misunderstanding for people abroad that there was bad relationship between China and me. I told him this would be good for China, and he agreed.[28]

I went to say goodbye to Peng Huan-Wu at his home. When I mentioned that I proposed IoP appointing me as a guest professor since it was good for China, he nodded his head and immediately said: "Yes, 'good for China' is very important." Only years later I learned that "Good for China" was his motto in leading his whole life. I last talked to him at Tsinghua University in 2005 during a recession of the workshop organized to celebrate his 90th birthday, which was after he gave a long presentation on some physics topics. He looked healthy and alert; he said, without me reminding him: "You worked on liquid crystals."

Just before leaving, I was invited to have lunch with a vice president of CAS. He asked me for suggestions. The one suggestion I gave was that CAS could set up a stand at the entrance of CAS' main library, a central spot in Zhongguancun, which would post seminars coming up in the different institutes so that people could attend them freely. It was a low-tech, costless way to encourage interdisciplinary research. From what I know, that never materialized.

On the issue of leaving China, the 2012 Nobel laureate in Literature, Mo Yan (莫言), said on his first press conference after the announcement of the award: "One should not think that whoever leaving the country must be unpatriotic and whoever staying behind must be patriotic."[29]

7 Conclusion: The Missing Link

During the ten years of Cultural Revolution, normal research activities in China were practically put on hold. A whole generation of scientists was lost, resulting in a missing link: the absence of mature scientists in their sixties presently (with a few exceptions). That is why most leaders in the scientific enterprises are now in their fifties.[30] An important consequence is the lack of enough number of widely-respected senior scientists who could referee and help to maintain the high level of research in many fields of study. Counting papers is thus the alternative chosen by the administrators in academia, with disastrous results [Lam 2022b]. The reason is that it takes time in writing papers and fighting the referees to get them published. Unnecessary publications thus reduce the amount of valuable time needed for doing high-quality research.

In the 1950s, China's progress in science suffered from embargos from abroad, i.e., due to external factors. Today, it suffers from internal and self-inflicted factors, like counting papers. The first step would be stop running the universities like running an IBM company. Maybe it is time to resume the tenure system in the top six universities, at least [Lam 2024].

Thirty years passed and China changed a lot. For instance, the words "soliton" and "revolution" are no longer heard. Taxis and cars are everywhere in Beijing, perhaps too many.

Appendix: A Brief History of Chinese Students Going Abroad and the Returnees

The following history is updated from [Lam 2010]. Before 1949, there are eight generations of Chinese students going abroad to study, by the Western Returned Scholars Association's counting.[31]

1. *First generation* (1872-1875): The Qing dynasty government sent out 120 children, aged 12-15, to the USA to study (including the famous Rong Hong (容宏, 1828-1912) [Rong 2005]).

2. *Second generation* (1877): Nearly 100 naval students sent to Europe in the early years of Qing dynasty's Emperor Kuang-Xu (光绪).

3. *Third generation*: Students going to Japan in the early 20th century (including Zhou En-Lai [Hong Kong Museum of History 2003]).

4. *Fourth generation*: Students going to the USA under the auspices of the Boxer Indemnity (义和团赔偿).

5. *Fifth generation*: Students going to France to study and work (including Yan Ji-Ci [Jin 2000], Zhou En-Lai [Hong Kong Museum of History 2003], and Deng Xiao-Ping).

6. *Sixth generation*: Students going to USSR during the 1920s.

7. *Seventh generation* (1927-1937): Students going abroad (including Ren Zhi-Gong, Wu Da-You [Qiu 2001], Shi Ru-Wei [Zhao 2005], Chien-Shiung Wu (吴健雄) [Chiang 2013], and Qian San-Qiang).

8. *Eighth generation* (1938-1948): Students going to Europe and USA (including Peng Huan-Wu and Huang Kun; China's first two Nobel laureates, Chen-Ning Yang (杨振宁) [Chiang 2002] and Tsung-Dao Lee (李政道) [Zi 2010]; and Wang Hao [He & Wen 2006]).

Continuing with this counting, after 1949, we have three more generations of Chinese students going abroad, viz.:

9. *Ninth generation* (1949-now): A large number of students from Taiwan and Hong Kong, and a few from Macau, went to USA and Europe [including the four Nobel laureates: Ting Chao-Chung (Samual Ting, 丁肇中) and Lee Yuan-Tseh (李远哲) from Taiwan, Tsui Chee (Daniel Tsui, 崔琦) and Kao Kuen (Charles Kao, 高锟, 1933-2018) from Hong Kong].

10. *Tenth generation* (1950s): Students going to USSR sent by the Chinese government (including Guan Wei-Yan, Pu Fu-Ke, Hao Bai-Lin, and Yu Lu).

11. *Eleventh generation* (1978-now): Students from mainland China, going to USA, Europe, etc., sent officially or going privately [Cao 2009].

The history of students *returning* to China to settle down and work is equally interesting. Before 1949, a large number of these students willingly returned to China (even during the Sino-Japan war years) and contributed to the modernization of their motherland. After 1949, there are three generations who go abroad *privately* and return to mainland China voluntarily, which are collectively and informally called *haigui* (海归; homophonic 海龟, sea turtles; meaning "returnees from overseas"):

1. *First generation* (1950s): Those coming back mainly from USA and Europe, soon after People's Republic of China was established (including Qian Xue-Sen (钱学森) [Chang 1995], Li Yin-Yuan, and Xie Yu-Zhang).

2. *Second generation* (1975-1985): Nearly 100 students of the 9th generation returned to China, mostly after the Cultural Revolu-tion.

3. *Third generation* (after 1980): These are the 11th generation students returning when the reform-and-opening up process in China is picking up speed.

Notes

1. http://en.wikipedia.org/wiki/Columbia_University_protests_of_1968 (Sept. 21, 2013).

2. "China and Japan: Could Asia really go to war over these?" *The Economist*, Sept. 22, 2012.

3. The trend of overseas students returning to China in the 1970s and 1980s resulting from the Baodiao Movement was first predicted in an article in *Zhongwen Yundong* (Chinese Language Movement), a hand-written magazine co-founded by the author, which published a total of three issues, all in 1971. The article, "Facing squarely Hong Kong's political future" [Lam 1971], was reprinted by *Undergrad*, the student newspaper of University of Hong Kong, September 16, 1971, which is reproduced on p 7 of *Impact 100*, a study celebrating HKU's achievements in its first 100 years (daaoweb.hku.hk/UserFiles/Image/publication _book/CNews/Autumn2012/Impact100.pdf, June 19, 2013).

4. Ren Zhi-Gong, graduated from Tsinghua University (1926), MIT (1928), Harvard (1931, PhD in physics); returned to China in 1933; professor at National Southwest Associated University, Kunming; came to US in 1946 and stayed; fellow of National Academy of Sciences (USA) (baike.baidu.com/view/2336180. htm, Oct. 1, 2012).

5. Wang Hao, graduated from National Southwest Associated University, PhD from Harvard University (1948), a logician and professor at Rockefeller University [He & Wen 2006: 220-229] (see also: baike.baidu.com/view/58173. htm#sub6821460, Sept. 30, 2012). He lived in New York City (NYC) and once came to a Baodiao meeting at the Chinatown loft (5th floor, 22 Catherine Street) I lived; like everybody else, he sat on the floor. On the other hand, Ren Zhi-Gong lived in Baltimore and had to drive three hours to attend the Baodiao meetings in NYC. And there were many of these meetings since NYC was the nerve center of Baodiao; I saw them often.

6. Qian San-Qiang, graduated from Tsinghua University (1936), studied in France with Madame (Irène) Curie and her husband (1937-1940, PhD in nuclear physics), returned to China in 1948, father of China's atomic and hydrogen bombs (baike.baidu.com/view/ 3908.htm, Oct. 1, 2012).

7. The calculator turned out to be crucial in helping me to produce China's first *Physical Review Letters* paper with only mainland-Chinese authors [Lin 1979];

the story is given in [Lam 2010]. Even though the bicycles were made in China, the embassy advised us to buy it in Hong Kong because bicycles in China, like many other items, were rationed.

8. Cai Yuan-Pei, a revolutionary and educator, went to Germany at age 40 to study philosophy, psychology, and aesthetics, and visited France at age 46. He was the president of Peking University, 1917-1926, and a strong advocate for liberal education and academic freedom. He resigned briefly in 1919 to protest the government's arrest of students during the May Fourth Movement. He was president of the National Academic Sinica, 1928-1940. Because of the war, he lived in Hong Kong, 1937-1940, and died there at age 72.

9. See also: baike.baidu.com/view/46332.htm (Oct. 1, 2012).

10. Incidentally, in the same year, Mickey Mouse was born.

11. Guan Wei-Yan, studied low temperature physics in the 1950s in USSR (guided by the Nobel laureate Peter Kapitsa), obtained Associate Doctorate (equivalent to PhD in the West) in1960, returned to China same year, director of IoP, president of the University of Science and Technology of China, academician of CAS (1980). He passed away in Taiwan.

12. In contrast, in the war years at National Southwest Associated University, Wu Da-You (Wu Ta-You, 吴大猷, 1907-2000) in Kunming was able to find white papers to type up the manuscript of his book *Vibrational Spectra and Structure of Polyatomic Molecules* (1939) [Wu 2005: 553].

13. Equally important, one month later, Deng sent Xi Zhong-Xun (习仲勋, 1913-2002, father of Xi Jin-Ping, 习近平, current Party Chairman and Head of State) to head the Guangdong Province and create the Special Economic Zone in Shenzhen, signaling the start of socialist market economy. (See: "Editorial: The old path, evil path, and bloody path after the Chinese communists' 18th National Congress," *World Journal*, Nov. 24, 2012: A5.)

14. Feng Kang, BS in physics at National Central University (1944), worked at Tsinghua University and Institute of Mathematics, CAS, and in Moscow (1951-1953). In 1978, he was appointed director of the newly founded Computing Center, CAS, until 1987 when he became the Honorary Director (en.wikipedia.org/wiki/Feng_Kang, Sept. 30, 2012). His brother, Feng Duan, was the only one among the siblings who never studied abroad. He is a physicist in condensed matters at Nanjing University and an academician of CAS, and had served as president of the Chinese Physical Society.

15. The president of CAS was Fang Yi (方毅, 1916-1997), whom I did not meet personally until after I left China (in the Great Hall of the People). I was among a group of high-T_c superconductor physicists from IoP that was received by Fang.

16. The reform-and-opening up is called "a new and great revolution" by Xi Jin-Ping, then Head of the Central Party School, in his talk "Review and thoughts on

party construction at the 30th anniversary of reform-and-opening up," presented on September 1, 2008 (http://news.xinhuanet.com/politics/2008-09/08/content_ 9849759.htm, July 30, 2013).

17. Xie Yu-Zhang, BS and MS from Tsinghua University, associate professor at physics department of National Central University (1945-1948), studied at Vanderbilt University in the United States and earned a PhD (1948-1950) (baike.baidu.com/view/ 2711126.htm, Sept. 30, 2012). He returned to China in 1957, became professor at Tsinghua University, spent four years in jail (1968-1972) during the Cultural Revolution [Wang & Liu 2012], assumed the first presidency of the Chinese Liquid Crystal Society, and retired in 1986.

18. Pu Fu-Ke graduated from Tsinghua University (1952) and studied in USSR (1956-1960), finished with an Associate Doctorate degree; worked all his life at IoP and was an academician of CAS. [See: "Academician Pu Fu-Ke passed away in Beijing," Physics (Wuli) **30**: 382 (2001).]

19. *Biennial Report, 1978-1979, Institute of Physics, Academia Sinica*, Beijing, p 59.

20. *Remembering Cai Shi-Dong*, published by Institute of Physics, Chinese Academy of Sciences; Asian African Association for Plasma Training; and Society for Plasma Research (Beijing, 1997).

21. Incidentally, Shu's master's degree thesis on phase transitions in liquid crystals [Shu & Lin 1984] shows that, among other things, Chia-Wei Woo had misinterpreted his own theoretical results in his first liquid crystal paper which was published in *Physical Review Letters*.

22. China was short in foreign currencies; she bought two copies of every foreign journal and stored them in Beijing and Sichuan Province, respectively, just in case, and made copies for the major libraries. (See also [Xiong 2012].)

23. Chen Jie rode a motorcycle to work (while motorcycles were extremely rare in Beijing). In the 1980s he was sent by IoP to work at the Lawrence Radiation Laboratory, Berkeley. He managed to stay in the US and became the anchor of the daily Mandarin news at KTSF, a local Chinese TV station in San Francisco.

24. Lin Pei-Wen (P. M. Lam), high school in Macau; BS, San Diego State University, CA; PhD, Washington University (Missouri); postdoc, West Germany. He returned to China end of 1980, worked at Institute of Theoretical Physics, CAS; got married and left China for West Germany with his wife in 1985. He is now a physics professor in USA.

25. In the same year, I published the prediction of bowlic liquid crystals [Lin 1982b] which was synthesized three years later in Europe [Lam 1994]. I coined the terms "bowlic" and "bowlic liquid crystal" which are now recognized officially by the IUPAC and appear formally in *Handbook of Liquid Crystals*. Bowlic is one of three existing types of liquid crystals in the world, with important

potential applications. This Chinese invention from its own backyard is not yet recognized in IoP's official showroom.

26. Many years later, when I met Stanley Liu, the editor of PRL who handled my papers, at a physics conference in the United States, he still remembered my two papers, including this one.

27. Unlike other countries, China's PhD thesis advisors have to be approved by the government, and I was the *only* one among a population of 1.1 billion certified in the field of liquid crystal *physics* while I was there.

28. My appointment as Guest Professor of IoP, approved by CAS, was signed by Guan Wei-Yan, dated June 25, 1984.

29. www.ktsf.com/en/nobel-winner-mo-urges-china-dissidents-freedom (Oct. 12, 2012). I think he had Gao Xing-Jian (高行健), the 2000 Nobel laureate in literature, in mind when he made this remark.

30. This was noted by Zhou Guang-Zhao (周光召), the past president of CAS, in his speech at Peng Huan-Wu's 90th birthday workshop, Tsinghua University, 2005.

31. www.coesa.cn/info/categorymore.shtml?Cid=C01 (Mar. 20, 2009). (See also [Hong Kong Museum of History 2003].)

References

Baker, K. M. [1975]. *Condorcet: From Natural Philosophy to Social Mathematics*. Chicago: University of Chicago Press.

Bishop, A. R. & Schneider, T. [1978]. *Solitons and Condensed Matter Physics*. New York: Springer.

Cao, Cong (曹聪) [2009]. "Brain drain," "brain gain," and "brain circulation" in China. Science & Cultural Review **6**(1): 13-32.

Chang, I. [1995]. *Thread of the Silkworm*. New York: BasicBooks.

Chen, Cheng-Jia & Yu Li-Sheng [2008]. *Demeanor of a Great Scholar: Remembering Huang Kun*. Beijing: Peking University Press.

Chen, Ruo-Xi [2008]. *Perseverance, Regretless*, Taibe: Chiuko Press.

Chiang, Tsai-Chien (江才健) [2002]. *Biography of Yang Chen-Ning: The Beauty of Gauge and Symmetry*. Taibei: Bookzone.

Chiang. Tsai-Chien [2013]. *Madame Wu Chien-Shiung: The First Lady of Physics Research*, Wong, Tang-Fong (trans.). Singapore: World Scientific.

Committee on Physics Terms [1996]. *Physics Terms*. Beijing: Science Press.

Dietrich, C. [1986]. *People's China: A Brief History*. Oxford: Oxford University Press.

Hamann, D. R. & Isaacs, E. R. [2012]. Philip Moss Platzman. Phys. Today **65**(5): 64-65.

Hao, Bai-Lin [2012]. My mentor Wang Zhu-Xi. Physics (Wuli) **41**: 455-357.

He, Zhao-Wu (何兆武) & Wen Jing (文靖) [2006]. *Going to School*. Beijing: SDX Joint Publishing.

Hong Kong Museum of History (ed.) [2003]. *Boundless Learning: Foreign-Educated Students of Modern China*. Hong Kong: Hong Kong Museum of History.

Jin, Tao [2000]. *Yan Ji-Ci*. Shenyang: Liaoning Educational Press.

Kuo, Chia-ling [1977]. *Social and Political Change in New York's Chinatown: The Role of Voluntary Associations*. New York: Praeger.

Lai, H. M. [2010]. *Chinese American Transnational Politics*. Urbana: University of Illinois Press.

Lam, L. [1971]. Facing squarely Hong Kong's political future: From supporting to burying the Legalization of Chinese Language Movement. Zhongwen Yundong, Issue No. 2, Aug. 1971: 1.

Lam, L. [1973]. Surfaces de Fermi, profil Compton et effets a N-corps. Phys. Lett. A **45**: 409-410.

Lam, L. [1992]. Solitons and field induced solitons in liquid crystals. *Solitons in Liquid Crystals*, Lam, L. & Prost, J. (eds.). New York: Springer. pp 9-50.

Lam, L. [1994]. Bowlics. *Liquid Crystalline and Mesomorphic Polymers*, Shibaev, V. P. & Lam, L. (eds.). New York: Springer. pp 324-353.

Lam, L. (ed.) [1997]. *Introduction to Nonlinear Physics*. New York: Springer.

Lam, L. [2010]. The first "non-government" visiting-scholar delegation in the United States of America from People's Republic of China, 1979-1981. Science & Culture Review 7(2): 84-94.

Lam, L. [2014]. The founding of the International Liquid Crystal Society. *All About Science: Philosophy, History, Sociology & Communication*, Burguete, M. & Lam, L. (eds.). Singapore: World Scientific. pp 209-240.

Lam, L. [2022a]. Liquid crystal research at Tsinghua University in the 1970s and 1980s. Int. J. Mod. Phys. B (doi.org/10.1142/50217979222300067).

Lam, L. [2022b]. *Research and Innovation*. San Jose: Yingshi Workshop.

Lam, L. [2024]. The invention of Bowlics and its lessons for Chinese innovation. *Scimat Anthology: Histophysics, Art, Philosophy, Science*. Singapore: World Scientific

Lam, L. & Prost, J. (eds.) [1992]. *Solitons in Liquid Crystals*. New York: Springer.

Li, Ya-Ming (ed.) [2004] *Guan Wei-Yan's Memoir: An Oral History*. Hsinchu: National Tsinghua University.

Lin, Lei (Lam, L) [1978a]. Microscopic theory of first-order phase transitions in liquid crystals. Kexue Tongbao **23**: 715-718.

Lin Lei [1978b]. Liquid crystals lead us in walking. Guangming Daily, Sept. 29, 1978.

Lin Lei [1979]. Nematic-isotropic transitions in liquid crystals. Phys. Rev. Lett. **43**: 1604-1607.

Lin, Lei [1982a]. Solitons in condensed matter. *Recent Developments in Statistical Mechanics and Condensed Matter Theory*. Wuhan: Huazhong Institute of Technology Press. pp 71-86.

Lin, Lei [1982b]. Liquid crystal phases and the "dimensionality" of molecules. Wuli (Physics) **11**: 171-178.

Lin, Lei, Shu Changqing, Shen Juelian, Lam, P. M. & Huang Yun [1982]. Soliton propagation in liquid crystals. Phys. Rev. Lett. **49**: 1335-1338.

Lukes, S. & Urbinati, N. (eds.) [2012]. *Condorcet: Political Writings*. Cambridge: Cambridge University Press.

Luo, Ping-Han (罗平汉) [2008]. *Spring: Chinese Intellectuals in 1978*. Beijing: People's Press.

Madhusudana, N. V., Palierne, J. F., Martinot-Lagarde, Ph. & Gurand, G. [1992]. Charged twist walls in nematic liquid crystals. *Solitons in Liquid Crystals*, Lam, L. & Prost, J. (eds.). New York: Springer. pp 253-263.

Peng, Huan-Wu [2001]. *Poems and Essays by Peng Huan-Wu*. Beijing: Peking University Press.

Qiu, Hong-Yi (丘宏义) [2001]. *Wu Da-You: The Father of Chinese Physics*. Taibei: Triumph.

Rigden, J. S. [2000]. *Rabi: Scientist & Citizen*. Cambridge, MA: Harvard University Press.

Rong, Hong [2005]. *My Life in China and America*. Beijing: Unity Press.

Shu, Chang-Qing & Lin Lei [1984]. Theory of homologous liquid crystals. I. Phase diagrams and the even-odd effect. Mol. Cryst. Liq. Cryst. **112**: 213-231.

Sun, Mu (孙牧) (ed.) [2008]. *IPCAS 80th Anniversary*. Beijing: Institute of Physics.

Tang, Tao; Yao Nan & Yang Lei [2010]. Feng Kang: The story of an outstanding mathematician. Mathematical Culture **1**(1): 24-38.

Tang, Ying-Wu [1998]. *Choices: The Road of Chinese Reform since 1978*. Beijing: Economic Daily Press.

The Seventies Monthly (ed.) [1971]. *Truth Behind the Diaoyutai Incident*. Hong Kong: The Seventies Monthly.

Tsui, Hark (徐克) & Lam, L. [2011]. Making movies and making physics. *Arts: A Science Matter*, Burguete, M. & Lam, L. (eds.). Singapore: World Scientific. pp 204-221.

Wang, De-Lu & Liu Zhi-Guang [2012]. The home-bound journeys in the 1950s and later experience of some American-trained Chinese scientists. Science & Cultural Review **9**(1): 68-87.

Wang, Shi-Ping (王士平) & Yang Guo-Zhen (杨国桢) [2012]. Eighty years of the Chinese Physical Society. Physics (Wuli) **41**: 506-512.

Wu, Da-You [2005]. Reminiscences of early development of Chinese physics. Physics (Wuli) **34**(3): 165-170; **34**(4): 233-239; **34**(6): 399-404; **34**(8): 551-554.

Xiong, Wei-Min (熊卫民) [2012]. Resumption of China's foreign scientific exchange: An interview with Hu Ya-Dong. Science & Culture Review **9**(5): 106-115.

Zhao, Jian-Gao (赵见高) [2005]. A pioneer in China's modern magnetism enterprise: Academician Shi Ru-Wei. Physics (Wuli) **34**(10): 758-764.

Zhao, Jing-An (赵静安); Tong Shou-Sheng (童寿生) & Ruan Liang (阮亮) [1980]. Research on the electro-optic effect due to cholesteric-nematic phase transition used in television displays. Wuli (Physics) **9**: 10-13.

Zhao, Qing-Yuan [1998]. *Cai Yuan-Pei*. Hefei: Anhui People's Press.

Zhao, Yan (赵岩); Chen Wei (陈伟), Wang Yu-Peng (王玉鹏) & Sun Mu (孙牧) [2008]. Searching knowledge for eighty years, glorious physicists: Eighty years anniversary of Institute of Physics. Physics (Wuli) **37**(6): 363-371.

Zhu, Bang-Fen (ed.) [2000]. *Selected Papers of Kun Huang with Commentary*. Singapore: World Scientific.

Zhu, Guo-Zhen [1982]. Experiments on director waves in nematic liquid crystals. Phys. Rev. Lett. **49**: 1332-1335.

Zi, Cheng (季承) [2010] *Lee Tsung-Dao*. Beijing: International Cultural Press.

Published: Lam, L. [2014]. Solitons and revolution in China: 1978-1983. *All About Science: Philosophy, history, sociology & communication*, Burguete, M. & Lam, L. (eds.). Singapore: World Scientific. pp 253-289.

Liquid Crystal Research at Tsinghua University in the 1970s and 1980s

Lui Lam

Liquid crystal research in China started during the Cultural Revolution years in the late 1960s. It was motivated partly by the potential application in liquid-crystal-display (LCD) digital television (TV) and movies, deemed good for the government's political propaganda agenda. The large-screen LCD TV/movie project was initiated and done at Tsinghua University, Beijing, which became successful in 1976 with prototypes built. Though never reached the commercial stage the research achievement was world-leading then. Here, the history of this project and other liquid crystal research at Tsinghua University in the 1970s and 1980s are described for the first time, with appropriate historical backgrounds provided. Included is also Tsinghua's significant contributions in basic LC research, and a description of the founding and workings of the Chinese Liquid Crystal Society. Finally, lessons drawn from this piece of science history are presented.

1 Introduction

Liquid crystals (LCs) were discovered in Europe in 1888. There are several important events in LC developments afterwards. In 1970 the twisted-nematic liquid crystal display was invented. In 1988 LCD televisions appeared. In 2017 LCD became a 100 billion US dollars industry and China became the top LCD producer in the world [Lam 2014a; Kawamoto 2002].

Academically, from the late 1960s to the late 1980s, with the joining of mainstream theoretical physicists such as Pierre-Gilles de Gennes (1932-

2007) of the Université de Paris-Sud in France and Paul Martin (1931-2016) of Harvard University in the United States, LC became a prominent topic in physics research. De Gennes' influential text *The Physics of Liquid Crystals* was published in 1974 and he won the Nobel Prize in 1991 partly due to his pioneering works in LC [Gennes, de 1974].

China's LC research began in the late 1960s during the Cultural Revolution, almost in sync with the international community [Lin 1983]. In 1969 Tsinghua University in Beijing organized an interdisciplinary and inter-institute LCD digital TV/movie research team, which saw its success in 1976. The screen-display speed reached the standard required for practical applications, leading the world in this field then.

Before China's reform-and-opening up in 1978 [Luo 2001] and the resumption of the college entrance examination in 1977, Tsinghua's basic-education department created the LC Physics Research Group, which became a national leader in LC teaching and personnel training. Also, Tsinghua made significant contributions to LC's basic research. For example, Tsinghua University published in 1982 China's second article (one of two) in *Physical Review Letters* (with only mainland-Chinese authors) [Zhu 1982]; synthesized a new kind of bowlic liquid crystals [Wang & Pei 1988], invented by Lin Lei[1] in 1982 [Lin 1982]; published the only and well-received Chinese textbook on LC physics [Xie 1988], and educated a future academician of the Chinese Academy of Sciences (CAS). Additionally, Tsinghua, in collaboration with the Institute of Physics (IoP) of the CAS, jointly raised China's LC research to a world-class level in the 1980s, resulting in the establishment of a "first-class discipline" in its campus.

In terms of professional organizations, Tsinghua led the establishment of the Chinese Liquid Crystal Society (CLCS) in 1980 (the world's first national LC society), which influenced the formation of the International Liquid Crystal Society (ILCS) in 1990 [Lam 2014a, 2017].

The story of this piece of science history, focusing on the LC research at Tsinghua University in the 1970s and 1980s, is told here for the first time. The relevance to China's present ambition of constructing so-called "double first-class" universities and recommendations for China's science historians will be discussed in the final section.

2 Tsinghua's LCD Digital Television/Movie Project (1969-1978)

China's Great Cultural Revolution run from 1966 to 1976, officially speaking [Dikötte 2016]. In 1967, with the direct approval of the Chinese Communist Party Chairman, Mao Ze-Dong (毛泽东), China started its national research program of finding a cure for malaria, which resulted in the invention of artemisinin. This success led to the award of the Nobel Prize in Physiology or Medicine to Tu You-You (屠呦呦) in 2015 [Tu 2015]. It was also during this Revolution, in 1969, scientists at Tsinghua University, Beijing, organized an interdisciplinary program to research and construct an LCD digital TV/movie system, the latter of which using large-screen projection. The project, headed by Ruan Liang (阮亮) and Zhao Jing-An (赵静安) (Fig. 1), was approved by Li Zhuo-Bao (李卓宝), head of Tsinghua's department of basic research [Ruan 2017].

Fig. 1. Two LC scientists from Tsinghua University: Ruan Liang (left) and Zhao Jing-An (right).

Ruan Liang was born in Shanghai in 1939. He finished undergraduate studies in physics at Tsinghua in 1962; his undergraduate thesis mentor was Wang Ming-Zhen (王明贞).[2] After graduation, Ruan worked immediately at Tsinghua's physics department, doing teaching and research in both theory and experiments. He retired from there in 1999. Zhao Jing-An, born in 1929, did his undergraduate studies in physics at Fu Jen Catholic University (辅仁大学) and started working at Tsinghua in 1952.

The project in fact involved the collaboration of scientists from several universities and research institutes: in Beijing, Tsinghua's physics department (Ruan Liang, Zhao Jing-An, Tong Shou-Sheng 童寿生), chemistry department (Wang Liang-Yu 王良御), electronic department, and the language department; in Shanghai, the Institute of Organic Chemistry, CAS (Liu Zhu-Jin 刘铸晋, Hong Xi-Jun 洪熙君), and Fudan University (Huang Jia-Hua 黄嘉华). Figure 2 shows the first page of Ruan Liang's recollection of this 1969 project, written in August 2009.

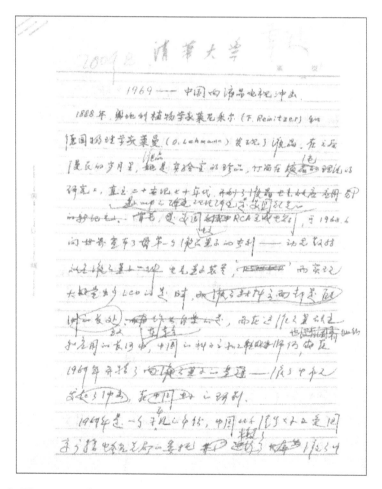

Fig. 2. First page of Ruan Liang's hand-written manuscript: 1969—Impact of China's LC television (Aug. 2009).

LCD TV first appeared in 1988 while digital movie, 1999.[3] Before that, big-screen movie projection uses high-power light shining through a running film in front of it. In contrast, instead of the running film, digital movies use a stationary plate consisting of a matrix of LC cells—the LCD. Each cell is controlled individually to allow light to pass through or not. In this way, the LCD modifies the light beam to create an image.[4] If the movie's big screen is replaced by a smaller transparent screen close to the LC matrix, then one has a LCD TV of some kind.[5]

In Tsinghua's 1969 project, the LC molecules in the LCD are of the cholesteric type. The molecular orientation can be influenced by an external electric field E. When $E = 0$, the molecules arranged in helixes; the cell blocks the light; the LC cell appears black. When the vertical E field is on ($E \neq 0$) the molecules are oriented vertically; the cell allows the white light (vertical arrows in Fig. 3) to go through; the LC cell appears white in color (Fig. 3, left). The basic mechanism is explained in a 1978 paper by Zhao Jing-An et al. [1978] (Fig. 4).

The completed projector is shown in Fig. 5, with the cathode tube shown in Fig. 6, left, and one of the screen images shown in Fig. 6, right.

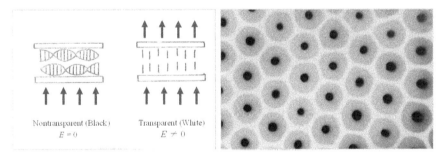

Fig. 3. *Left*: Sketch of the basic mechanism of the LC cell switch (see text). *Right*: Arrangement of the LC cells (dots) on a plane.

3 Success and awards

In 1975 the project achieved switch speed in the LCD appropriate for TV and movie projection. In 1976 it reached a response time of ~1ms and a 3×4 m^2 size in the projected image.[6] At this point in time these achievements were at the top of the world, which was included in the liquid crystal book by Japan's Okano Koji (冈野光治). Finally, in 1978 the project received China's Science Conference Award [Ruan 2017].

Later, apart from the digital TV/movie, a self-storage large-size LCD board for Chinese characters was invented (Fig. 7). It received a Chinese patent in 1985, one of the first batch of patents awarded by the country (Fig. 8).

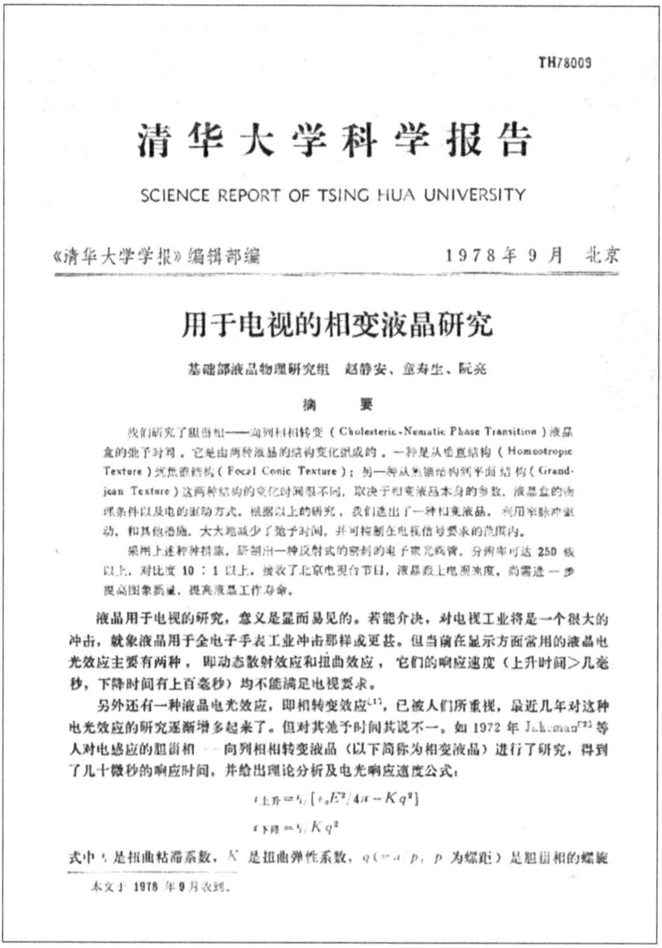

TH78009

清华大学科学报告

SCIENCE REPORT OF TSING HUA UNIVERSITY

《清华大学学报》编辑部编　　　　1978年9月　北京

用于电视的相变液晶研究

基础部液晶物理研究组　赵静安、童寿生、阮亮

摘　要

我们研究了胆甾相——向列相相转变（Cholesteric-Nematic Phase Transition）液晶盒的弛予时间。它是由两种液晶的结构变化组成的。一种是从垂直结构（Homeotropic Texture）到焦锥结构（Focal Conic Texture）；另一种从焦锥结构到平面结构（Grandjean Texture）这两种结构的变化时间很不同，取决于相变液晶本身的参数，液晶盒的物理条件以及电的驱动方式。根据以上的研究，我们选出了一种相变液晶，利用窄脉冲驱动，和其他措施，大大地减少了弛予时间，并可控制在电视信号要求的范围内。

采用上述种种措施，研制出一种反射式的密封的电子束充电管，分辨率可达250线以上，对比度10：1以上，接收了北京电视台节目，液晶盘上电视速度，尚需进一步提高图象质量，提高液晶工作寿命。

液晶用于电视的研究，意义是显而易见的。若能介决，对电视工业将是一个很大的冲击，就象液晶用于全电子手表工业冲击那样或更甚。但当前在显示方面常用的液晶电光效应主要有两种，即动态散射效应和扭曲效应，它们的响应速度（上升时间>几毫秒，下降时间有上百毫秒）均不能满足电视要求。

另外还有一种液晶电光效应，即相转变效应[1]，已被人们所重视，最近几年对这种电光效应的研究逐渐增多起来了。但对其弛予时间其说不一。如1972年Jakeman[2]等人对电感应的胆甾相——向列相相转变液晶（以下简称为相变液晶）进行了研究，得到了几十微秒的响应时间，并给出理论分析及电光响应速度公式：

$$t_{上升} = \gamma_1/[\epsilon_a E^2/4\pi - Kq^2]$$

$$t_{下降} = \gamma_1/Kq^2$$

式中 γ_1 是扭曲粘滞系数，K 是扭曲弹性系数，$q(=2\pi/p$，p 为螺距）是胆甾相的螺旋

本文于1978年9月收到。

Fig. 4. First page of the Tsinghua University paper explaining the mechanism of the LCD part of the projection TV [Zhao et al, 1978].

Fig. 5. The completed projector of the LCD TV/movie project.

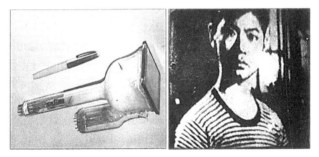

Fig. 6. *Left*: A component of the projector. *Right*: one of the projected images.

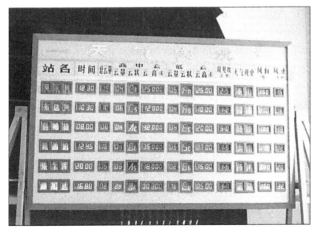

Fig. 7. The self-storage large-size LCD board for Chinese characters.

Fig. 8. Patent awarded to the self-storage large-table Chinese-character LCD device (1985).

4 Tsinghua's Basic Research in Liquid Crystals (1978-1988)

After the Cultural Revolution, national university-entrant examination was resumed in 1977. In December the next year the open-and-opening up revolution (or movement) officially started. In the same year Tsinghua's basic-education department (later renamed physics department) established its own "liquid crystal physics teaching and research group," which was then the first liquid-crystal educational entity in the country. It admitted immediately four master's degree students, including Ou-Yang Zhong-Can (欧阳钟灿), Wang Xin-Jiu (王新久), and Ruan Li-Zhen (阮丽真) (Fig. 9). Apart from Zhao Jian-An and Ruan Liang, the teachers included a senior fellow, Xie Yu-Zhang (谢毓章, 1915-2011) (Fig. 10, left), whose life story is quite different from the other two's.[7] Xie's major work consists of rederiving in detail the static theory of LCs which culminated in a textbook *Physics of Liquid Crystals*, published in 1988, and is well received by Chinese readers (Fig. 10, middle).

One of Xie's graduate students is Ou-Yang Zhong-Can. He was born in 1946, in Quanzhou, Fujian Province; graduated from Tsinghua's department of automatic control in 1968; reentered Tsinghua as a graduate student in 1978; received MS in 1981, PhD in 1984 in the field of optics[8] (Fig. 11). His two PhD mentors are Xu Yi-Zhuang (徐亦庄) and Xie Yu-Zhang; PhD thesis is titled "Theory of second-harmonic-generation effect in nematic liquid crystals." In 1986 Ou-Yang went to the Freie Universität Berlin with a Homburg Scholarship for two years, to work on membrane-shape theory with Wolfgang Helfrich. He became an academician of the Chinese Academy of Sciences in 1997 and was the director of the Institute of Theoretical Physics, CAS, in 1998-2007.

Fig. 9. Students and teachers of Tsinghua University's physics teaching-and-research group (Jan. 16, 1981).

It turned out that Lui Lam was the only non-Tsinghua person in Ou-Yang's PhD thesis committee, for a good reason (see below). Lam was born in Guangdong Province, China, educated in Hong Kong (from grade one to receiving a BS from University of Hong Kong in 1965), earned a MS from University of British Columbia, Canada, and a PhD in physics from Columbia University (1973). He did his postdoc at City College of New York, CUNY (1972-1975), worked a year at Universitaire Instelling Antwerpen, Belgium (1975-1976), and another year at Universität des Saarlandes, West Germany (1976-1977). In January 1978, Lam settled down in Beijing, China, and worked at the Institute of Physics, CAS (1978-1983) [Li 2003; Lam 2014b].[9] In short, for this period Lam's life trajectory runs a full circle from China to China, via Guangdong, Hong Kong, Canada, USA, Belgium, West Germany, and Beijing. He was a second-generation *haigui* while Xie belongs to the first generation [Lam 2014b].

Fig. 10. *Left*: Xie Yu-Zhang. *Middle*: *Physics of Liquid Crystals*, a textbook by Xie Yu-Zhang, published by Science Press, Beijing (1988). *Right*: Inscription from Xie to Lui Lam in the latter's copy, a gift from Xie (July 1988).

Fig. 11. After Ou-Yang Zhong-Can's PhD thesis defense, Tsinghua University, July 2, 1984. From the third person on, *left to right*: Ou-Yang Zhong-Can, Xu Yi-Zhuang, Tong Shou-Sheng, Xie Yu-Zhang, Ruan Liang, Meng Zhao-Ying (孟昭英), Zhao Jing-An, and Lui Lam.

During Lam's six-year stay at the IoP he worked on LCs, on both theory and experiments. Historically, he was the one who published the *first* paper with purely mainland-China authors in the prestigious journal *Physical Review Letters* (PRL) [Lin 1979], and was the one and only one PhD mentor on LC *physics* in China, officially speaking. In 1982 Lam invented *Bowlics*, the third type of LCs in the world (Fig. 12) [Lin 1982; Wang et al. 2017], and pioneered the study of propagating solitons in (shearing) LCs [Lin et al. 1982; Lam & Prost 1992].

Fig. 12. Three types of liquid crystals and their discoverers.

The propagating soliton paper by Lin et al. (1982) has coauthors from three institutes: IoP, CAS; Institute of Theoretical Physics, CAS; and Peking University (called Beijing University then), which was one of the two *second* papers with purely mainland-China authors published in PRL. However, it was the *first* PRL paper ever from the latter two places (while only the *second* PRL paper from IoP). This is kind of important[10] as reflected by the fact that (for a long period of time since the late 1980s) any PRL paper was rewarded with a large sum of money by many respectful institutes when China's academic world went into the frenzy of counting (foreign journal) papers, to boost the pride and prestige of the country.[11]

This 1982 PRL paper by Lin et al. predicts the existence of solitons in shearing nematic LCs, which was confirmed by the experiments of Zhu

Guo-Zhen (诸国桢) [1982] and subsequent research carried out by Lam's LC group at the IoP [Lin & Shu 1984]. Zhu worked on sound (a *linear* phenomenon) at Tsinghua's department of fundamental courses. Yet, soliton is a *nonlinear* phenomenon [Lam 1997, 1998] which was then a new field in science and especially so within China [Lam 2014b]. China's works on LC solitons are subsequently summarized in the book *Solitons in Liquid Crystals* [Lam & Prost 1992]. In total, in the field of solitons, 23 papers in 12 mostly foreign journals were published by Lam and his coauthors, apart from 1 PhD thesis and 1 MS thesis by his students at IoP (Fig. 13).

During Lam's six-year stay at IoP (1978-1983) he mentored five graduate students who completed five MS theses and one PhD thesis. One of the students is Shu Chang-Qing (舒昌清) who did both his MS and PhD thesis with Lam. The MS thesis is on phase transition in LCs [Shu & Lin 1984], which overthrows someone's theory published in PRL. Title of the PhD thesis is "Propagating solitons in shearing nematic liquid crystals," which was officially China's first PhD thesis in the field of LC *physics* since Lam was then the *only* certified PhD mentor in this field (see footnote 7). From 1983 to 1988, Lam published five articles with Tsinghua's LC scientists [Xie & Lin 1983; Xie et al. 1983; Lin & Zhao 1987; Zhao & Lin 1987; Lam et al. 1988].

Coauthored with IoP students		Total (including papers coauthored with SJSU students and reviews)
1982	1	
1984	1	
1985	5	23 papers (in 12 journals)
1986	2	· Physical Review Letters
1987	1	· Acta Physica Sinica
Coauthored with NNU people		· Chinese Physics
1986	1	· Physics Letters A
1987	4	· Journal of Mathematical Physics
1988	2	· Journal of Statistical Physics
		· Molecular Crystals and Liquid Crystals
* 1 PhD thesis (Shu Chang-Qing, 1984)		· Molecular Crystals and Liquid Crystals Letters
		· Nanjing Normal University Bulletin (Natl. Sci.)
		· Physical Review A
* 1 MS thesis (Xu Gang, 1986)		· Liquid Crystals
		· Chaos Solitons Fractals

Fig. 13. Soliton papers published by Lam and his coauthors. IoP = Institute of Physics, CAS; NNU = Nanjing Normal University; SJSU = San Jose State University.

Lastly, Tsinghua's chemist Wang Liang-Yu (王良御) made important contributions in basic LC research. He was one of the first few who synthesized a new type of bowlics [Wang & Pei 1988; Lam 1994]. In 1988 Wang and his colleague published a book on LC chemistry [Wang & Liao 1988].

5 The Chinese Liquid Crystal Society

The LC conference in Bangalore, India, December 3-8, 1979, was the *first* international LC conference ever attended by mainland China's LC scientists. Lui Lam was invited by the organizer, Sivaramakrishna Chandrasekhar (1930-2004), to give a talk [Lin 1980]. He led a group of three to attend, with two of them from the IoP and one from Peking University. Apart from Chandrasekhar, the top LC people present there include Patricia Cladis, Dietrich Demus, Christian Destrade, Elisabeth Dubois-Violette, Charles Frank (1911-1998), Wolfgang Helfrich, Jerzy Janik, Shunshuke Kobayashi, Sven Lagerwall, Alan Leadbettter, Anne Levelut, Jacques Prost, Alfred Saupe (1925-2008), and Adrian de Vries. Incidentally, it was at this conference that Lam got his idea of the Bowlics [Lam 1994; Lam 2014a].

Six months later, the 8th International Liquid Crystal Conference (ILCC) in Kyoto, Japan, June 30-July 4, 1980, was the *first* ILCC conference ever attended by mainland Chinese. A delegation of several Chinese LC scientists attended, led by Lam again.

At the Kyoto conference, Lagerwall and Chandrasekhar expressed their wish to visit their counterparts in Beijing. The visa was rapidly granted, and they arrived in time to join the founding ceremony of the Chinese Liquid Crystal Society (CLCS). Lagerwall brought along a LC switch device he just invented, and he kindly let the Chinese bought it from him.

The CLCS was cofounded by Zhao Jing-An, Rong Liang, Xie Yu-Zhang and Lui Lam in Beijing. The founding ceremony was held at Tsinghua University on July 18, 1980 (Fig. 14). However, organizationally speaking, even today, the CLCS is actually the Liquid Crystals Branch of the Chinese Physical Society. Lam wrote the constitution and served as the Vice President and Secretary General from 1980 to 1983 (Fig. 15) while Xie was its founding President, all by election.

Fig. 14. Founding photo of the Chinese Liquid Crystal Society (July 18, 1980).

Afterward, Lam wrote to De Vries at the Liquid Crystal Institute of Kent State University, about the existence of the CLCS. De Vries replied with these words:

> Congratulations with the formation of the Chinese Liquid Crystal Society, and with your election as Secretary General and Vice President. To my knowledge, too, this is the first such society in the world. Maybe more countries will follow now, and maybe the International Planning and Steering Committee will then also be reorganized in a more formal way. I think it is time for that.

De Vries wrote this on September 4, 1980, long before anyone had the idea of the International Liquid Crystal Society (ILCS) in mind (see below) [Lam 2014a].

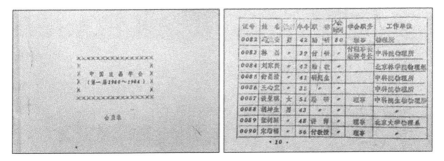

Fig. 15. Membership book of the Chinese Liquid Crystal Society: First Term, 1980-1984.

From June 29 to July 3, 1979, a major national LC conference was held in Shanghai, in which the LC research in Japan and the USA was reviewed (Fig. 16). Two years later, an important LC conference organized by the CLCS was held in the scenic city Guilin, Guangxi Province (Fig. 17).

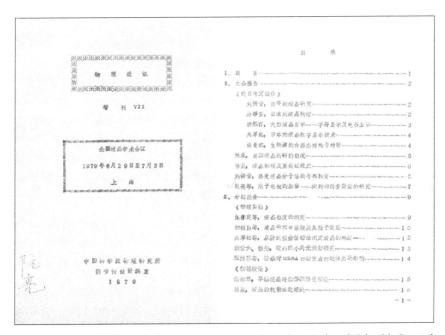

Fig. 16. A partial list of the papers presented at the National Liquid Crystal Conference held in Shanghai, June 29-July 3, 1979.

Fig. 17. Participants at the LC conference in Guilin, Guangxi Province, organized by the CLCS (1981).

Seven years later, from June 27 to July 1, 1988, a special *international* LC conference titled "Centenary Conference of Liquid Crystal Discovery: 1888-1988" was organized by the CLCS, held in Huilongguan (迴龙观), Beijing (Fig. 18). This is probably the *only* LC centenary conference organized by anybody in the world then. Many prominent LC scientists from around the world attended, including G. Barero (Italy), Lajos Bata (Hungary), Sivaramakrishna Chandrasekhar (India), Georges Durand (France), Shunshuke Kobayashi (Japan), R. Lierau (Switzerland), Martin Schadt (Switzerland), and Hideo Takezoe (Japan) (Figs. 18-21).

Pierre-Gilles de Gennes, a pioneer in LC physics and Nobel laureate in 1991, was also invited. He could not attend though but did send in a photo of himself holding up a picture he drew for this special occasion (Fig. 22).

The ILCS was in fact the *first* national LC society in the world, founded five years ahead of the British LC Society. Exactly ten years later, on July 27, 1990, the International Liquid Crystal Society (ILCS) was established in Vancouver, Canada, during the 13th ILCC. It was not by accident that

Lam, who cofounded the ILCS, was the one who initiated and orchestrated the founding of the ILCS [Lam 2014a, 2017].

Fig. 18. The international LC conference, Centenary Conference of Liquid Crystal Discovery: 1888-1988, organized by the CLCS and held in Huilongguan, Beijing, June 27-July 1, 1988.

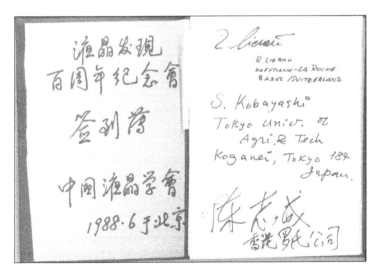

Fig. 19. The sign-in book of the Centenary Conference of Liquid Crystal Discovery: 1888-1988, Huilongguan, Beijing, June 27-July 1, 1988.

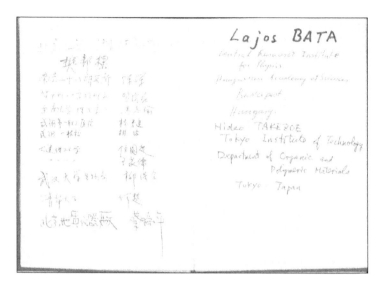

Fig. 20. The sign-in book of the Centenary Conference of Liquid Crystal Discovery: 1888-1988, Huilongguan, Beijing, June 27-July 1, 1988 (cont.).

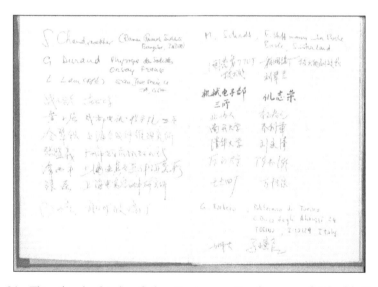

Fig. 21. The sign-in book of the Centenary Conference of Liquid Crystal Discovery: 1888-1988, Huilongguan, Beijing, June 27-July 1, 1988 (cont.).

Fig. 22. Pierre-Gilles de Gennes (left) holding up a picture he drew (right) for the Centenary Conference of Liquid Crystal Discovery: 1888-1988, to which he was invited but could not attend. De Gennes did not explain the meaning of the picture. But a plausible interpretation is that the woman represents the French people while the buffalo, the Chinese people, indicating the close friendship between peoples of the two countries.

6 Conclusion

During the ten years of Cultural Revolution (1966-1976), there were three important scientific achievements in China with international impacts: (1) Analysis of insulin's crystal structure. (2) Invention of artemisinin. (3) Tsinghua's LCD digital television/movie project. The first one was well-known due to government's advertising [Wang & Gu 2010]. The second one was hardly known until recent years which, through the efforts of a few, eventually won the Nobel Prize in 2015 [Tu 2015]. The last one, the topic of this article, was rarely known, if at all, outside of China.

The achievements of Tsinghua University's LC research in the 1970s to 1980s are summarized in Sec. 1. Here are the lessons one can draw from this piece of science history.

1. Under *very* poor conditions one can still do first-class, innovative research. In contrast, the history of the National Southwest Associated University, Kunming, during the Sino-Japanese war years [Liu 2013], shows that under *extremely* poor conditions one can still produce two future physics Nobel laureates (Tsung-Dao Lee 李政道 and Chen-Ning Yang 杨振宁) and a first-rate mathematical logician (Wang Hao 王浩).

2. The key to innovative works in science is the picking of important research topics [Lam 2022], which requires dare thinking and dare doing,

and *not* counting papers in the universities and research institutes (see footnote 10).

3. China's LC research in the 1970s and 1980s shows that, with first-class scientific management, harmonious and fruitful collaborations between indigenous scientists and the first-/second-generation returnees could happen and indeed happened. Their works helped to raise the country's level in research and industrial developments.

4. The invention of *Bowlics* (one of three types of LC molecules) in 1982 at the IoP remains a benchmark in China's long pursuit of innovative research [Lam 2024a], not quite surpassed yet in many Chinese institutes/universities despite the more-than-sufficient funding in the last thirty years. The crux of the problem is that scientists spend too much time in writing research proposals, the split of salary into three parts (with the basic salary counting only about one third of the total), and the abandonment of the tenure system in the 1980s [Lam 2022, 2024b]. Innovative work requires freedom to think, plenty of leisure time, and concentration in work (not worrying about money, say).

5. Tsinghua University in the 1980s already established a *first-class* discipline in its campus: LC research (which, unfortunately, has since declined). We hope China's science historians would pay more attention to recent, contemporary, and significant science history in their own country. Better still, someone will do a master's thesis on Tsinghua's glorious LC research in the early years. And, the newly established Science Museum at Tsinghua would include and preserve the historical relics of its own, in basic and applied LC research.

Acknowledgment

This article is based on my talk[12] given on July 1, 2017, at Tsinghua University's Inaugural Meeting of the Department of History of Science and the 3rd National Forum on the Historiography of Science and Technology, June 30-July 1, 2017. We thank Wu Guo-Sheng for invitation to give the talk and Ruan Liang for private communications as well as providing Figs. 1, 2, 3 (right), 5-9, 14, and 16-22.

Notes

1. In this article, Chinese names are written with family name first (except for Lui Lam whereas Lam is the family name).

2. There was no degree system in China then, which started later in 1980 [Nofri 2015; Ou-Yang 2004].

3. https://cinepedia.com/history/an-early-history-of-digital-cinema/ (April 22, 2022).

4. https://entertainment.howstuffworks.com/digital-cinema5.htm (April 24, 2022).

5. In comparison, the flatscreen LCD TV we see today combines the light source, the LCD matrix, and the screen into one slab, enabling the TV to be so thin.

6. Wang Liang-Yu's assessment in 1988: The Tsinghua TV project "preliminarily solved the response time problem but the images' resolution is poor, and the LC's lifetime is short" [Wang 1988].

7. Xie Yu-Zhang is a first-generation *haiqui* (returnees to China voluntarily after 1949) [Lam 2014b]. He was born 1915 in Suzhou, Jiangsu Province; got his bachelor's and master's degree in science from Tsinghua University, in 1936 and 1942, respectively; became a teacher of a special course on telecommunications at the National Southwest Associated University, Kunming, in 1942; was associate professor and secretary of the president's office at National Central University, 1945-1948. He went to the United States to further his study at Vanderbilt University, 1948-1950, and obtained a PhD in physics there. In 1955-1957 he was an associate professor at Wichita State University in USA. Xie returned to China in 1957. In 1957-1968 and again in 1975-1986 Xie worked at Tsinghua, first in the physics teaching-and-research group of fundamental courses committee and later as professor in the modern applied physics department, respectively. In between these two periods during the Cultural Revolution, he spent four years (1968-1972) in *prison* and then was forced to undergo re-education [Wang & Liu 2012]. After the Cultural Revolution his research interest turned to LC theory. In 1980-1986, he was elected president of the CLCS (see Sec. 5). Xie retired from Tsinghua in 1986 and passed away in Beijing in 2011.

8. In China, since the late 1970s when the PhD degree system was introduced all PhD mentors must be approved officially by the government. Xie was approved to be a PhD mentor in *optics* but not in LC physics. Lui Lam in fact was the first official PhD mentor in LC *physics* in the country.

9. During Lui Lam's six years working in China his name in English is spelled Lin Lei (林磊), whereas Lin is his family name in pinyin; Lam, in Cantonese. In the References, Lam, L. = Lam, Lui.

10. According to the Chinese version of Tsinghua's physics department official webpage: "In the years after the end of the Cultural Revolution, the scientific

research work of the Department of Engineering and Physics and the Physics Teaching and Research Group has steadily improved. The research results of the Optics, Accelerator and Nuclear Physics Teaching and Research Groups have won a series of scientific research awards. In 1982, Zhu Guo-Zhen's research work of the Ultrasonic Research Group was published in *Physical Review Letters*, becoming the first paper published in the journal since the establishment of the Department of Physics of Tsinghua University." (https://www.phys.tsinghua. edu.cn/gk/wlxls.htm, June 4, 2022).

11. Counting papers published in foreign journals in considering promotions is a practice initiated by the physics department of Nanjing University, in 1986. In the early 1990s, the practice already spread to the whole country, including institutes of the CAS. One important consequence (that affects only China) of this ill-conceived practice is that it makes truly innovative works impossible since the number of papers is considered more important than their quality [Lam 2022, 2024b]. Another consequence (that affects the whole world) is the rapid and inevitable appearance of many fraudulent papers in various international research journals [Qin 2017; Stigbrand 2017].

12. PPT of this talk is available: https://www.sjsu.edu/people/lui.lam/scimat/ 170630-Tsinghua%20LC-70701.pdf

References

Dikötte, F. [2016]. *The Cultural Revolution: A People's History, 1962-1976*. New York: Bloomsburg Publishing.

Gennes, de, P. G. [1974]. *The Physics of Liquid Crystals*. Oxford: Clarendon Press.

Kawamoto, H. [2002]. The history of liquid-crystal displays. Proceedings of the IEEE **90**: 460-500.

Lam, L. [1994]. Bowlics. *Liquid Crystalline and Mesomorphic Polymers*, Shibaev, V. P. & Lam, L. (eds.). New York: Springer. pp 324-353.

Lam, L. [1997]. *Introduction to Nonlinear Physics*. New York: Springer.

Lam, L. [1998]. *Nonlinear Physics for Beginners*. Singapore: World Scientific.

Lam, L. [2014a]. The founding of the International Liquid Crystal Society. *All About Science: Philosophy, History, Sociology & Communication*, Burguete, M. & Lam, L. (eds.). Singapore: World Scientific. pp 209-240.

Lam, L. [2014b]. Solitons and revolution in China: 1978-1983. *All About Science: Philosophy, History, Sociology & Communication*, Burguete, M. & Lam, L. (eds.). Singapore: World Scientific. pp 253-289.

Lam, L. [2017]. Prehistory of the International Liquid Crystal Society, 1978-1990. Mol. Cryst. Liq. Cryst. **647**: 351-372.

Lam, L. [2022]. *Research and Innovation*. San Jose: Yingshi Workshop.

Lam, L. [2024a]. The invention of Bowlics and its lessons for Chinese innovation. *Scimat Anthology: Histophysics, Art, Philosophy, Science.* Singapore: World Scientific.

Lam, L. [2024b]. *Humanities, Science, Scimat.* Singapore: World Scientific.

Lam, L. Ou-Yang, Z.-C. & Lax, M. [1988]. Ab initio theory of linear and nonlinear optics of liquid crystals. Phys. Rev. A **37**: 3469-3474.

Lam, L. & Prost, J. [1992]. *Solitons in Liquid Crystals.* New York: Springer.

Li, Yuan-Yi (李元逸) [2003]. The unbroken China complex: The life story of a Chinese scientist. Sciencetimes, Aug. 8, 2003.

Lin, Lei (Lam, L.) [1979]. Nematic-isotropic transitions in liquid crystals. Phys. Rev. Lett. **43**: 1604-1607.

Lin, Lei [1980]. Critical properties of nematic-isotropic transitions in liquid crystals. *Liquid Crystals*, Chandrasekhar, S. (ed.). London: Heyden.

Lin, Lei [1982]. Liquid crystal phases and the "dimensionality" of molecules. Wuli (Physics) **11**: 171-178.

Lin, Lei [1983]. Liquid crystal research in China: 1970-1982. Mol. Cryst. Liq. Cryst. **91**: 77-91.

Lin, Lei & Shu Chang-Qing [1984]. Soliton propagation in shearing liquid crystals. Acta Phys. Sin. **33**: 165. [English translation in Chin. Phys. **4**: 598 (1984).]

Lin, Lei, Shu Chang-Qing, Shen Jue-Lian, Lam, P.M. & Huang Yun [1982]. Soliton propagation in liquid crystals. Phys. Rev. Lett. **49**: 1335-1338.

Lin, Lei & Zhao Jing-An [1987]. Liquid Crystals. *Chinese Encyclopedia: Physics.* Beijing: Chinese Encyclopedia Press.

Liu, Yi-Qing [2013]. *Great Masters: National Southwest Associated University and Spirit of Scholars.* Nanjing: Jiangsu Literature and Art Publishing House.

Luo, Qi [2001]. *Chin's Industrial Reform and Open-door Policy 1980-1997.* New York: Routledge.

Nofri, E. M. [2015]. The university system in China. http://www.albertoforchielli.com/the-university-system-in-china/ (June 4, 2022).

Ou-Yang, Kang [2004]. Higher education reform in China today. Policy Futures in Education **2**(1): 141-149.

Qin, A. [2017]. Fraud scandals sap China's dream of becoming a science superpower. New York: New York Times. (https://www.nytimes.com/2017/10/13/world/asia/china-science-fraud-scandals.html, June 4, 2022)

Ruan, Liang [2017]. Private communication.

Shu, Chang-Qing & Lin Lei (1984). Theory of homologous liquid crystals. I. Phase diagrams and the even-odd effect. Mol. Cryst. Liq. Cryst. **112**: 213-231.

Stigbrand, T. [2017]. Retraction note to multiple articles in *Tumor Biology.* Tumor Biol. (https://doi.org/10.1007/s13277-017-5487-6)

Tu, You-You [2015]. Discovery of artemisinin: A gift from traditional Chinese medicine to the world. Stockholm: The Nobel Foundation.

Wang, Da-Cheng & Gu Xiao-Cheng [2010]. A brief account on the study of the insulin crystal structure. In retrospect: Forty years after the determination of insulin's crystal. Science China (Life Sciences) **53**(1): 13–15.

Wang, De-Lu & Liu Zhi-Guang [2012]. The home-bound journeys in the 1950s and later experience of some American-trained Chinese scientists. Science & Cultural Review **9**(1): 68-87.

Wang, Liang-Yu (1988). Progress in liquid crystal television. Wuli (Physics), **17**(9): 521-524, 527.

Wang, Liang Yu & Liao Song-Sheng (廖松生) [1988]. *Chemistry of Liquid Crystals*. Beijing: Science Press.

Wang, Liang-Yu & Pei Xue-Feng (裴学锋) [1988]. Bowlic liquid crystals. J. Tsinghua University (Natural Science) **28** (Supplement 4); 80-84.

Wang, L., Huang, D., Lam, L. & Cheng, Z. [2017]. Bowlics: history, advances and applications. Liq. Cryst. Today **26**(4): 85-111.

Xie, Yu-Zhang [1988]. *Physics of Liquid Crystals*. Beijing: Science Press.

Xie ,Yu-Zhang & Lin Lei [1983]. Brief report on the 1981 Chinese Liquid Crystal Conference. Mol. Cryst. Liq. Cryst. **91**: 93-96.

Xie, Yu-Zhang, Ou-Yang Zhong-Can & Lin Lei [1983]. Electric field effects on the elastic constants of nematics. I. Mol. Cryst. Liq. Cryst. **101**: 19-33.

Zhao, Jing-An & Lin Lei [1987]. Theory of liquid crystals. *Chinese Encyclopedia: Physics*. Beijing: Chinese Encyclopedia Press.

Zhao, Jing-An, Tong Shou-Sheng & Ruan Liang [1978]. Research on liquid crystal phase transitions used in television. Science Report of Tsing Hua University, Sept. issue, 1978.

Zhu, Guo-Zhen [1982]. Experiments on director waves in nematic liquid crystals. Phys. Rev. Lett. **49**: 1332-1335.

Published: Lam, L. [2022]. Liquid crystal research at Tsinghua University in the 1970s and 1980s. International Journal of Modern Physics B (doi.org/10.1142/S0217979222300067).

The Invention of Bowlics and its Lessons for Chinese Innovation

Lui Lam

1 Introduction

For more than 2,000 years in history, according to the management system, there have been three successful models of scientific innovation: Greek, Chinese, and Italian [Lam 2022: 81-86]. Bowlic liquid crystals (called Bowlics) is one of the three types of liquid crystals in the world, invented by Lam in China, in 1982 (while working at the Institute of Physics of the Chinese Academy of Sciences), with the Italian model in place [Lin 1982; Lam 1994]. Since then, many Chinese scientists have participated in it and made important contributions, and bowlics has been formally recognized internationally.

The invention of bowlics is a successful case of Chinese innovation. This article reviews the invention and development process of bowlics through first-hand information, and discusses the enlightenment of this case for current independent innovation in China.

2 On Scientific Innovation

There are two categories of innovation in terms of its *importance*: $1 \rightarrow N$ or $0 \rightarrow 1$. The first type, $1 \rightarrow N$, means major improvements on something already invented by others; the second type, $0 \rightarrow 1$, ground-breaking inventions (Fig. 1). Obviously, $0 \rightarrow 1$ is less common than $1 \rightarrow N$ and is valued much more, even though, in some cases, it is easier to make (see below).

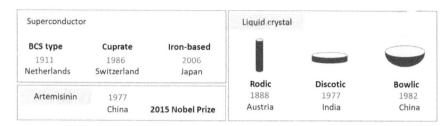

Fig. 1. *Left top*: As an example of 1 → N innovation, Chinese scientists had significantly raised the T_c of high-temperature superconductors twice, once for the cuprates and then for the iron-based, and won the top State Natural Science Award (2013) and the State Science and Technology Award (2016). The cuprates was invented by the Swiss and iron-based by the Japanese. *Left bottom and right*: Since 1949, there are two cases of 0 → 1 scientific innovations in China: artemisinin in 1977 and bowlics in 1982. The former won Tu You-You (屠呦呦) the 2015 Nobel Prize in physiology or medicine; the latter was invented by Lui Lam (林磊) at the Institute of Physics, Chinese Academy of Sciences [Lin 1982].

There are also two types of innovation in terms of its *speed*: slow or quick. Examples of the former are Einstein's general theory of relativity (takes a decade) and the LIGO gravitational wave detection (two decades). Example of the latter is the single-atomic layer graphene (one day), done with everyday adhesive tape. The last two works won the Nobel Prize; the first one didn't but worth more than a Nobel Prize. Both types require a *bold* idea, but the slow ones require additionally difficult mathematics or complicated apparatus while the quick ones do not.

There are three successful models of scientific innovation in terms of the *administrative* style: Greek, Chinese, and Italian [Lam 2022].

1. The Greek model: Ask questions, constantly asking why (Socratic method), do it for curiosity and fun, and maximum freedom of exploration. The model was invented by ancient Greeks and is commonly adopted in Western countries.

2. The Chinese model: Set a goal with cooperation from the whole country, with central coordination from the top. This model was used in the invention of artemisinin during 1967 to 1971 in China (ending with the 2015 Nobel Prize), which is difficult to replicate, especially when the target is unknown.

3. The Italian model: Stright control in the humanities but lose control in doing natural science. The model was practiced by the Vatican 400 years ago (with Galileo inventing modern science), the Soviet Union in the last century, and new China at the Institute of Physics, CAS, around 1980. This model is abandoned in today's China.

3 Bowlics

The Bowlics was invented by Lam in 1982 at the Institute of Physics, CAS, which belongs to the 0 → 1 category, a quickie, and was done within the Italian model. It was first published in the magazine *Wuli* (Physics), an official publication of the Chinese Physical Society [Lin 1982]. Since then, the invention of Bowlics has been recognized by the peers (Fig. 2).

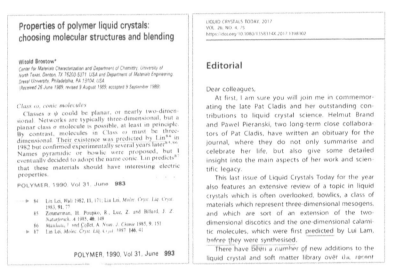

Fig. 2. *Left*: A paper by Witold Brostow in *Polymer* (1990), *Right*: The Editorial by editor Ingo Dierking in Vol. 26, No. 4 of *Liquid Crystals Today* (2017).

Subsequently, many Chinese scientists and others have contributed to it [Wang & Pei 1988; Xu & Swager 1993; Zeng 2001; Wang et al. 2017] (see Sec. 3.3). Presently, the word Bowlic has appeared frequently in the title of liquid crystal papers (such as [Miyajima et al. 2009; Dong et al. 2009; Xu & Swager 1993; Wang & Pei 1988] and more in Fig. 3), has been officially recognized by the IUPAC (International Union of Pure and

Applied Chemistry) [Barón & Stepto 2002], and has been included in the *Handbook of Liquid Crystals* [Demus et al. 2008].

Fig. 3. Sample of liquid crystal papers with the word Bowlic or Bowl in the title, published in 1993, 2008, 2009, 2018 and 2020, respectively.

Bowlics belongs to the new field of strategic materials, and has very important application potential in ultra-high-speed liquid crystal displays, room-temperature superconductors [Lam 1988a, 1988b], etc. And there are already two filed patents based on it (Fig. 4).

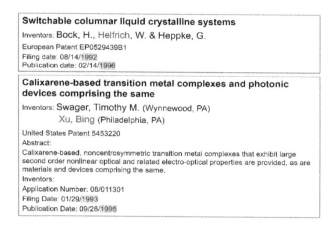

Fig. 4. Two patents based on bowlics, filed in Europe and the USA, respectively. Note that Wolfgang Helfrich is the co-inventor of the ubiquitous liquid-crystal display using twisted nematic cells.

3.1 Invention

The idea of Bowlics was conceived in India in 1979 (see Sec. 3.3). Publication of the invention spreads over three papers: The first one announces the prediction of bowlic monomers and is written in Chinese, published in *Wuli* (1982), which contains a few hundreds of words and two diagrams only (Fig. 5). The second paper in *Molecular Crystals and Liquid Crystals* (1987) is a review, providing more details (Fig. 6, left), and mentions bowlic polymers near the end. The third one extends it to bowlic polymers in detail, also published in MCLC (1988) (Fig. 6, right).

Fig 5. The *Wuli* paper announcing the prediction of bowlic monomers [Lin 1982].

3.2 Promotion

In 1982, after the publication of the Chinese article in *Wuli*, the part of the article on bowlics was translated into English and sent to major foreign colleagues. The next year, in the article summarizing the LC research in China (1970-1982), a paragraph on the prediction of bowlics is included [Lin 1983].

Subsequently, two review articles were published, respectively, in 1994 (Fig. 7) and 2017 (Fig. 8, left) [Lam 1994; Wang et al. 2017].

Fig. 6. *Left*: The review on bowlic monomers [Lin 1987]. *Right*: The paper predicting bowlic polymers and room-temperature bowlic superconductors [Lam 1988a]. Both papers are published in *Molecular Crystals and Liquid Crystals*.

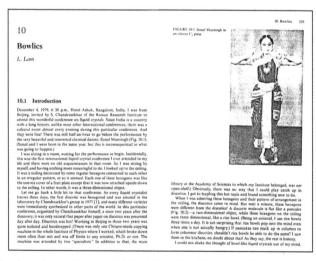

Fig. 7. The bowlic review published in *Liquid Crystalline and Mesomorphic Polymers* [Lam 1994]. The Indian dancer depicted is Sonal Mansingh.

Fig. 8. *Left*: The bowlic review published in *Liquid Crystals Today* [Wang et al. 2017]. *Right*: The new bowlic liquid crystals invented at Tsinghua University [Wang & Pei 1988].

3.3 Contribution of Chinese scientists

There are three Chinese scientists who contributed to the early development of bowlics.

1. Wang Liang-Yu

In December 1979, I was invited by S. Chandrasekhar to give a talk at the liquid crystal conference he organized at the Raman Research Institute, Bangalore, India. On December 4, 1979, 6:30 pm, at Hotel Ashok, I was sitting alone in a big room waiting for Sonal Mansingh to perform the Indian traditional dance next room (Fig. 7). Incidentally, I looked up to the ceiling and saw a number of three-dimensional protruded hexagons connected to each other as a decoration. I thought to myself, these hexagons could easily be stacked to each other to form columns, a form of liquid crystals. And that is how bowlics was born [Lam 1994].

After I went back to Beijing, I talked to Wang Liang-Yu (王良御), a chemistry professor at Tsinghua University and urged him to make the bowlics. He didn't since no one had done it before; instead, he worked on discotics, invented by Chandrasekhar and his team in 1977, which was a hot topic then. Wang waited until 1988, three years after two European teams synthesized the bowlics (monomers) in 1985, not a complicated process [Malthête & Collet 1985; Zimmermann et al. 1985], and published the first bowlic paper in China [Wang & Pei 1988]. In other words, Wang missed the chance to be *the* pioneer in bowlic chemistry because he was not interested or dared to be a 0 → 1 innovator.

Wang did subsequently publish more papers on bowlics, and had coauthored a book called *Chemistry of Liquid Crystals* (1988).

2. Bing Xu

I didn't know Bing Xu personally, and not his Chinese name (Xu is the last name). From what I can gather, Xu received his BS (1987) and MS (1990) from Nanjing University. In 1993, together with his doctoral supervisor Timothy Swager, he applied for a patent related to bowlics (Fig. 4), and obtained his PhD in 1996 from University of Pennsylvania.

In 1996, at the 16th International Liquid Crystal Conference, he won the Glenn Brown Award (for best doctoral dissertation) of the International Liquid Crystal Society for his works on bowlics.

Subsequently, Xu did his postdoc at MIT (1996-1997) and Harvard University (1997-2000); worked from assistant professor to full professor at chemistry department, Hong Kong University of Science and Technology (2000-2010); and became full professor at Brandeis University, USA, since 2009.

3. Zeng Er-Man

Zeng Er-Man (曾尔曼) received his BA from Xiamen University (1992), MS (1998) and PhD (2001) from Georgia Institute of Technology, USA.

While doing his doctoral work in 1999, Zeng, inspired by my review article [Lam 1994], succeeded in synthesizing bowlic polymers for the first time in the world [Zeng 2001]. He emailed me and, with his girlfriend, came to meet me at San Jose State University. He told me he intended to return to China to work, and he did. But before that, he had worked as a postdoc at MIT (2001).

Subsequently, in Xiamen, he led a group in synthesizing more bowlic polymers [Zeng et al. 2009]. But he also ventured into other areas such as innovation management, Marxist economics, and molecular nano-technology. Zeng is the coauthor of *Introduction to Marxist Economics of Productive Forces* (2013).

4 Lessons for Chinese Innovation

It seems that $0 \to 1$ scientific innovation appeared only two times in new China (Fig. 1), and somehow is absent after 1982. Why? To answer this question, one should first note that there are *three* preconditions for it to happen: (1) At the individual level, there has to be someone who is ambitious enough from the beginning, someone who is determined to innovate and will not be satisfied in merely publishing papers. (2) At the vicinity level, there has to be successful $0 \to 1$ innovators nearby who will inspire, if not guide, the candidates. (3) At the external environmental level, the candidates need tenure (called "iron rice bowl"), high enough single salary, and no assessments.

Unfortunately, all these three conditions are absent since the mid-1980s or a little bit later. On top of that, starting in 1986 with the physics department at Nanjing University, counting papers—only those published in foreign journals—swept the academic circles throughout the country [Lam 2022: 91-92]. The iron-rice bowl was abandoned around 1990 for all professions, including the academics, and replaced by the three-part wage system (basic salary + administrative bonus + merit bonus). And because of the merit part, annual and frequent assessments are unavoidable. Counting papers is like nailing the innovation dragon to the ground, and the new wage system is like cutting the dragon into three sections. Can the dragon fly? Of course not.

How did these things happen? China's reform-and-opening up movement started in December 1978. Since then, many policies were inspired or copied from the West, except the counting-paper policy which was self-invented. To address the shortage problem of research and innovation talents in science, in 1986, borrowing from the National Science Foundation of the United States, the National Natural Science Foundation of China was established. It looked good until it was infiltrated by the paper-counting practice and the traditional culture of nepotism, and innovation became impossible.

When copying from the West, the inexperienced policy makers could overlook some details. For example, in capitalist America, only private businesses are driven by money considerations; the universities are not. In other words, universities while care about balancing budgets, are not money-driven; associate and full professors are tenured with no assessments; i.e., they are provided with a carefree and secure environment, to explore, to advance knowledge, and to innovate.

Thus, the first step in encouraging $0 \rightarrow 1$ innovations is to retore immediately the tenure system and stop paper-counting [Lin et al. 2018]; piling more money into the research enterprise won't do, as witnessed by the lack of progress in the last 30 years or so.

Lastly, note that tenure and no paper-counting are only the necessary but not sufficient conditions in innovation, as demonstrated by the cases in Hong Kong and Taiwan and elsewhere.

5 Conclusion

1. Bowlics is a $0 \rightarrow 1$ independent innovation, which is one of only two times since the founding of new China more than 70 years ago.

2. Important papers should be published first in domestic journals.

3. The road to innovation has been blocked, since 1986, by the new rule of counting papers.

4. To resume innovation, it is necessary to return to the Italian model (humanities tight, science loose) practiced around 1980, and immediately restore the tenure system, single salary, and no assessment in the scientific research circle.

5. Unfinished business: Synthesis of room-temperature super-conducting bowlic polymers (Fig. 9).

Fig. 9. My paper on bowlic high-T_c superconductors [Lam 1988b], first predicted in [Lam 1988a]. For updates, see Sec. 10.4.3 in [Lam 1994].

References

Barón, M. & Stepto, R. F. T. [2002]. Definitions of basic terms relating to polymer liquid crystals. Pure Appl. Chem. **74**: 493-509

Demus, D., Goodby, J. W., Gray, G. W., Spiess, H. W. & Vill, V. [2008]. *Handbook of Liquid Crystals*. New York: Wiley.

Dong Yan-Ming, Chen Dan-Mei, Zeng Er-Man, Hu Xiao-Lan & Zeng Zhi-Qun [2009]. Disclination and molecular director studies on bowlic columnar nematic phase using mosaic-like morphology decoration method. Science in China Series B: Chemistry **52**: 986-999.

Lam, L. [1988a]. Bowlic and polar liquid crystal polymers. Mol. Cryst. Liq. Cryst. **155**: 531-538.

Lam, L. [1988b]. Possible liquid crystalline high T$_c$ superconductors. *3rd Asia Pacific Physics Conference*, Chan, Y. W., Leung, A. F., Yang, C. N. & Young, K. (eds.). Singapore: World Scientific.

Lam, L. [1994]. Bowlics. *Liquid Crystalline and Mesomorphic Polymers*, Shibaev, V. P. & Lam, L. (eds.). New York: Springer. pp 324-353.

Lam, L. [2022]. *Research and Innovation*. San Jose: Yingshi Workshop.

Lin, Lei (Lam, L.) [1982]. Liquid crystal phases and the "dimensionality" of molecules. Wuli (Physics) **11**: 171-178.

Lin, Lei [1983]. Liquid crystal research in China: 1970-1982. Mol. Cryst. Liq. Cryst. **91**: 77-91.

Lin, Lei [1987]. Bowlic liquid crystals. Mol. Cryst. Liq. Cryst. **146**: 41-54.

Lin, Lei (林磊), Liu Li (刘立) & Sun Nan (孙楠) [2018]. Make full use of domestic journals to obtain priority in publications. Science and Technology China (科技中国), July 2018, Issue 7: 48-50.

Malthête, J. & Collet, A. [1985]. Nouv. J. Chemie **99**: 151.

Miyajima, D., Araoka, F., Takezoe, H. & Aida, T. [2009]. Synthesis of new bowlic liquid crystals responsive to electric field. 12th International Conference on Ferroelectric Liquid Crystals (FLC'09).

Wang, Liang-Yu (王良御) & Pei Xue-Feng (裴学锋) [1988]. Bowlic liquid crystals. J. Tsinghua University (Natural Science) **28**, Supplement 4: 80-84.

Wang, L., Huang, D., Lam, L. & Cheng, Z. [2017]. Bowlics: history, advances and applications. Liq. Cryst. Today **26**(4): 85-111.

Xu, B. & Swager, T. M. [1993]. Rigid bowlic liquid crystals based on tungsten-oxo CaliM41arenes: Host-Guest effects and head-to-tail organization. J. Am. Chem. Soc. **115**: 1159-1160.

Zeng, Er-Man (曾尔曼) [2001]. Design, synthesis and characterization of columnar discotic and bowlic liquid crystals. PhD thesis, Georgia Institute of Technology, Atlanta, USA.

Zeng, Er-Man, Chen Dan-Mei (陈丹梅), Dong Yan-Ming (董炎明) & Hu Xiao-Lan (胡晓兰) [2009]. Synthesis of two new bowlic molecules and their study on fibrous crystals. Chem. Bull. 2009, Issue 3: 258-264

Zimmermann, H., Roupko, R., Luz, Z. & Billard, J. [1985]. Naturforsch. **40a**: 149.

Unpublished: Lam, L. [2024]. Bowlic liquid crystals: A Chinese innovation story. Based on my talk given on December 26, 2023 at the Annual Meeting of Chinese Society of Science and Technology, December 26-27, 2023, Beijing.

The Founding of the International Liquid Crystal Society

Lui Lam

The story of the founding of the International Liquid Crystal Society in 1990 is told here for the first time. The founding process lasted three years starting 1987 and is quite different from the usual case concerning other learned societies. A personal account of the why and how as well as the background and crucial events is given. It is written for those working or interested in science, liquid crystals in particular, and for science historians.

1 Introduction

Liquid crystal is a state of matter intermediate between liquid and crystal. The molecules of the organic compounds that exhibit liquid crystal phases may be rodic, discotic, or bowlic in shape [Lin 1982, 1987]. Rodic liquid crystals are the ones used in liquid crystal display (LCD) today and were discovered in 1888 by the Austro-Hungarian botanist Frederick Reinitzer (1857-1927). Since the industrial application of liquid crystals as display was proposed in the 1960s, there has been a resurrection of intense interest in these materials [Kawamoto 2002]. The explosive commercialization of LCD televisions since 2007,[1] a $100 billion industry, makes the study of liquid crystals as a research field more important than ever.

An important landmark in the history of liquid crystals (LC) is the establishment of the Liquid Crystal Institute (LCI) in 1965 by Glenn Brown (1915-1995), an American chemist, at Kent State University, Ohio.[2] The Institute has been and remains an important driving force in LC's study and application. In August the same year, the first official

International Liquid Crystal Conference (ILCC) was held at LCI, with 50 scientists worldwide and 42 papers presented.[3] Three years later in 1968, the 2nd ILCC was hosted by Kent, too, and since then, it has become a biennial series.[4]

An international "Planning and Steering Committee" (PSC) was formed to oversee this series. More precisely, it is the "PSC for ILCCs" because its only function was to decide, at each ILCC, who would organize the next ILCC. The PSC was disbanded in 1990 in Vancouver, Canada, when the International Liquid Crystal Society (ILCS) was established during the 13th ILCC held there. It was the Conference Committee (chaired by the author, Fig. 1) within the ILCS that took over the function of the PSC.

The ILCS differs from the PSC in many ways; the most important difference is that the ILCS is an *open* organization that is owned by the whole LC community while the PSC was essentially a "gentlemen's club." In any case, 23 years later, the ILCS remains the only game in town. According to the ILCS' official website, "Since 1990, the ILCS has attracted nearly 900 members in 43 countries and territories on six continents. The Society also serves as an umbrella organization for regional societies established in recent years."[5] By all standards, it is a success story.

But then something odd happens. Apart from the implied statement that the ILCS first came into being in 1990, there is no mention of its pre-history on the ILCS's website. For example: (1) Where did the ILCS come from? (2) Who initiated it? (3) Why the initiation? (4) Who saw it through? (5) How was it established? (5) Why it was 1990 and not sooner or later? Answers to these questions could neither be found in the first issue of *Liquid Crystals Today* (LCT), ILCS' official publication (Fig. 2). Seven years later, the then President of ILCS, Atsuo Fukuda, put in what he knew: "The International Liquid Crystal Society (ILCS) replaced the Planning and Steering Committee (PSC) for ILCCs in 1990" [Fukuda 1998], which was a (small) step forward but did not help in answering the questions above.

International Liquid Crystal Society
Conference Committee

Chairman: *Lui Lam*
Department of Physics, San Jose State University
San Jose, CA 95192, USA

Phone: (408)924-5261 Fax: (408)9244815
Bitnet: luilam@calstate Telex: 171 171 UD

May 2, 1991

Prof., Dr. Sci. Viktor V. Titov
NIOPIK,
Ul. B. Sadovaya, 1/4.
103787 Moscow, GSP-3
USSR

Dear Viktor:

Thank you very much for your Christmas card which I received in Jan. 1991. I am very glad to learn that Dr. Yuri Molchanov of Leningrad State Univ. in ready to organize the 15th Int. Liquid Crystal Conf. in 1994. Sorry I was not able to write to you earlier, but I just finished running a conf. here and now have more time to attend to this matter.

Enclosed here is a copy of the letter that I send to Dr. Molchanov. Hope you will continue to help in organizing the conf. and keep in touch.

Best regards.

Sincerely,

Lui Lam

Lui Lam
Professor

Fig. 1. Letter from Lui Lam, Chair of Conference Committee, ILCS, to Viktor Titov (1935-1966), May 2, 1991. Seven months later, on December 26, 1991, the USSR was dissolved; five years later, Titov's life was abruptly terminated [Dunmur & Sluckin 2011: 231-234]. Up to now, neither USSR nor Russia has hosted an ILCC.

More strangely, in the recent book by David Dunmur and Tim Sluckin [2011] on the detailed history of LC, the term of International Liquid Crystal Society (or ILCS) is completely absent from the Timeline (1854-1997), the Index, and the rest of the book, even though the LCI at Kent and ILCC are prominently mentioned. Note that Dunmur was the Secretary of ILCS and the Editor of LCT in the early years starting from the beginning.

What happened? How did the ILCS come about? Obviously, it is a story never told.[6] The first step in clearing the mystery is by providing an answer to Question 2: Lui Lam is the one who initiated the ILCS. But to answer the other questions and provide the context of the happenings, we have to go deeper and reach back earlier and ask: Why Lam? Where did he come from?

Liquid Crystals Today **Vol 1**, 1, (1991) 1

LIQUID CRYSTALS

Vol. 1   No. 1
January 1991

Today

Newsletter   of   the   International   Liquid   Crystal   Society

A Message from the President

Liquid crystal research has burgeoned in recent years to become an exciting interdisciplinary field with important practical applications, and of course the international community of liquid crystal scientists has also grown accordingly. The formation of an International Liquid Crystal Society (ILCS), announced a few months ago, is therefore most timely and fulfils the need for a forum for the exchange of information between scientists and for promoting greater interaction between basic and applied research.

Amongst the various objectives of ILCS (which include the sponsorship of conferences and workshops, and the preparation of a world directory of liquid crystal scientists), probably the most important one is the publication of a journal. Its purpose is to provide up-to-date reports on recent advances in the field, new materials and devices, forthcoming events, job opportunities, etc. A beginning has been made, and we are happy to present the first issue of *Liquid Crystals Today*.

Greetings for the New Year.

S CHANDRASEKHAR

WELCOME TO READERS

Two contributions in this first issue of *Liquid Crystals Today* give an industrial perspective on the development of Liquid Crystal science.

Although the potential of liquid crystals as optical devices was appreciated in the 1930's, it was not until stable materials became commercially available in the 1970's, that the liquid crystal industry really took off. As the article from Dr Castellano shows, the future remains bright for continued growth in the range of applications and the value of business. However these developments will rely on having a secure research base for liquid crystal science in Universities and Research Institutes, and a steady supply of top-grade trained liquid crystal scientists to provide the back-up for industrial development and production. The responsibility for securing the liquid crystal science base necessary into the next century rests with professional liquid crystal scientists world-wide, and it is hoped that the International Liquid Crystal Society, through its various activities will play an important role in the guardianship and development of liquid crystal science.

Liquid Crystals Today is intended to provide a forum for liquid crystal scientists world-wide, through which information and ideas can be exchanged. It is a magazine for members of the ILCS, and the contents and style will be determined by the membership. Thus contributions for the next issue are invited and suggested topics might be:

> News items; notices of meetings; conference reports; short review articles; letters; new product information; new books information; vacant positions; experimental tips; historical items and reminiscences.

A number of countries now have National Groups of liquid crystal scientists, and it would be particularly interesting to receive reports on liquid crystal activities at national level. It is hoped to feature a different national report in each issue.

Comments and Contributions should be sent to:

Dr D A Dunmur,
Secretary to the ILCS
Department of Chemistry,
University of Sheffield,
Sheffield S3 7HF, UK

Fax: (742) 738673
E-mail: CH1DAD@UK.AC.
SHEFFIELD.PRIMEA

COPY DATE for next issue:
31st MARCH 1991

In This Issue:

LOGO COMPETITION

We need a LOGO!!
Please submit your designs for a LOGO for the International Liquid Crystal Society, to be judged by members of the Board of Directors. A prize (to be decided) will be awarded to the person sending a design suitable for adoption by the ILCS, to be used on its stationery and incorporated into the masthead of this newsletter. Closing date: 31 March, entries to D A Dunmur.

Fig. 2. First issue of *Liquid Crystals Today*, Vol. 1, No. 1, January 1991.

2 My Involvement in Liquid Crystals

I was born in mainland China. I lived in Hong Kong (1949-1965) where I went through grade schools and earned my BS, with first-class honors, from the University of Hong Kong.[7] I received my MS from the University of British Columbia (UBC), Vancouver, Canada; and PhD from Columbia University, New York City, with thesis done at Bell Labs.[8] I had never worked on LC before I became the postdoc of Melvin Lax (1922-2002) at City College of City University of New York (CCNY), in 1972.[9]

2.1 In the West (1972-1977)

Neither he nor I knew anything about the subject when Melvin Lax asked me to work on LC. Mel was one of the founders of quantum optics and was known to be a very thorough researcher on any topic he set his eyes on. He just published a 37-page long paper on a new formulation of the electrodynamics of crystals starting from a Lagrangian [Lax & Nelson 1971; Nelson 1979]. He saw a recent paper from Harvard's Paul Martin[10] that claims that liquid crystals are crystals [Martin et al. 1970] (which turned out to be wrong) and so Mel thought that he, actually I, could do better using his superior Lagrangian theory.

And so I started reading LC papers and that was two years before the LC book of Pierre-Gilles de Gennes (1932-2007) [1974] came out. I had no clue to proceed. In summer of 1973, I attended the Les Houches Summer School on molecular fluids. De Gennes gave a series of lectures on nematodynamics based on his book manuscript. Unfortunately, it was delivered in French and I learned nothing. But I did get something out of this school: I got to know Roland Ribotta from Orsay because we shared the same room, and convinced Michel Mirkovitch to translate a short manuscript of mine on Compton profile, from English to French, and got it published in *Physics Letters A* (which was never quoted by anyone even though it was a good paper) [Lam 1973]. Before we left, we were given a few chapters from De Gennes' book, which, luckily, were written in English.

Back to New York, I checked every formula in the chapters and found a serious mistake in De Gennes' derivation of the Frank free energy. I wrote to De Gennes and he sent back a postcard, acknowledging the mistake but saying that the mistake had been found by someone else ahead of me. And so there was this footnote in the book acknowledging the help of someone (not me) in correcting a mistake in the manuscript.

And I still had no clue to proceed. The next summer, I went away to tour China for seven weeks [Lam 2014a] with full pay from Mel. The breakthrough came one day when, by chance, I came across an article in CUNY's library that uses a dissipation function to formulate irreversible thermodynamics in continuum mechanics.[11] I ended up using Lax's Lagrangian to take care of the reversible processes and a dissipation function for the irreversible (dissipative) processes. It preserves the beauty and simplicity of Lax's theory in that everything, including the necessary space-time symmetries and the material symmetries, are built into the Lagrangian and the dissipation function from the very beginning. The conservations laws and reciprocal relations emerge naturally and consistently from the equations of motion.

I left CUNY for Universitaire Instelling Antwerpen, Belgium, in 1975 without writing any paper for or with Mel. He was not displeased at all and even offered to find me a job at Los Alamos National Laboratory, which I declined, thinking that, wrongly, the job required a US citizenship. I worked on semiconductors and structural phase transitions in Belgium and then on superionic conductors at Universität des Saarlandes, West Germany, the following year. I was anticipating returning to China any time soon and so in 1977, I wrote up four papers summing up my dissipation-function formulation of the thermo- and hydro-dynamics of molecular liquids and solids [Lam 1977a, 1977b, 1977c, Lam & Lax 1978]. One of these four papers is specifically on LCs [Lam 1977c],[12] which is my one and only one LC paper before I returned to settle down in China [Lam 2014a].

2.2 In China (1978-1983)

In January of 1978, I returned to China to join the revolution. I was assigned to do physics at the Institute of Physics (IoP), Chinese Academy of Sciences, in Beijing. I could work on any topic. The fact that I ended up doing LC in China was purely incidental, which was related to the stay of

Chia-Wei Woo (吴家玮) at the IoP in the last few months of my first year there [Lam 2014a].

1. The USA visit (1979)

From February to May, 1979, I was a visiting scholar at Chia-Wei Woo's Physics Department, Northwestern University, Evanston. The work I did there was published in *Physical Review Letters* (PRL), my fourth LC paper and my second one while working in China [Lin 1979]. Historically, this is the first LC paper and the first one with only mainland-Chinese author(s) in PRL ever came out from China [Lam 2010]. By going beyond the mean-field approximation, among other things, it shows that the gap exponent in the nematic-isotropic transition is temperature dependent, clarifying the experimental results of Keyes and Shane related to tricritical points.

2. The Bangalore conference (1979)

Late in 1979, I received suddenly an invitation from Sivaramakrishna Chandrasekhar (1930-2004) to his LC conference in Bangalore, December 3-8, 1979.[14] I, together with a colleague from IoP and another one from Peking University, attended this conference [Lin et al. 1980]; I was housed in the guest house at Raman Research Institute. Like every visitor who visits India for the first time, I was culturally awakened. This conference was held less than two years after the discovery of discotics by Chandra's group. It was attended by a Who's Who list of liquid crystalists[15] (except Pierre-Gilles de Gennes who, a speaker at the previous conference in 1973, was no longer active in LC). Apart from Chandra, the people I met include Patricia Cladis,[16] Dietrich Demus, Christian Destrade, Adrian de Vries, Elisabeth Dubois-Violette, Charles Frank (1911-1998), Wolfgang Helfrich, Jerzy Janik, Shunshuke Kobayashi, Sven Lagerwall, Alan Leadbettter, Anne Levelut, Jacques Prost, and Alfred Saupe (1925-2008).

I presented a summary of my work on the critical properties of nematic-isotropic transitions [Lin 1980].[17] It was at this conference, while looking up to the ceiling, that I got the idea for the three-dimensional bowlic LCs [Lam 1994].

3. My first Orsay visit and the Kyoto conference (1980)

In next year's summer, upon the invitation of Roland Ribotta, I spent a month at CNRS' Laboratoire de Physique des Solidides at Université de Paris-Sud, Orsay. The lab was established by André Guinier (1911-2000), Jacques Friedel, and Raimond Castaing (1921-1998) in 1959; De Gennes spent 10 years (1961-1971) there and did his LC works. I was received warmly by Friedel in his office. Before I left, I got Guinier's permission to translate his popular-science book *Structure of Materials* into Chinese [Lin et al. 1985], and finalized the arrangement for an Orsay LC group to visit China.[18]

Right after I returned to Beijing, I led a delegation of several scientists to attend the 8th ILCC in Kyoto, Japan, June 30-July 4, 1980—the first ILCC conference ever attended by mainland Chinese. As it was the case, scientists from Taiwan were listed as coming from Republic of China in the program. By my understanding, we were prohibited to show up in any conferences under such circumstances, a rule self-imposed by the Chinese government [Xiong 2013]. The deadlock was resolved after I told Shunshuke Kobayashi, general secretary of the conference, that he only had to make an announcement at the beginning of the conference that there was a "mistake" in the program.[19]

It was at this Kyoto conference that the PSC announced that they decided to open its door a little bit on two accounts: (1) Invite the LC community to nominate candidates to the PSC; and (2) each PSC member will be allowed to serve a maximum of eight years, starting from 1980. At the end of the conference, Chia-Wei Woo told me that he had nominated me to the PSC.

4. Founding the Chinese Liquid Crystal Society (1980)

China's LC research began in 1970 amidst the Cultural Revolution [Lin 1983]. On July 18, 1980, the Chinese Liquid Crystal Society (CLCS) was founded by Zhao Jian-An, Ruan Liang, Xie Yu-Zhang, and me. I served as the Vice President and Secretary General from 1980 to 1983.[20]

At the Kyoto conference, Lagerwall and Chandra expressed their wish to visit us in Beijing. The visa was rapidly granted and they arrived in time to join the founding ceremony of the CLCS. Lagerwall brought along a LC switch device he just invented and he kindly let us buy it from him.

Afterward, I wrote to De Vries at Kent about the existence of the CLCS. He replied with these words:

> Congratulations with the formation of the Chinese Liquid Crystal Society, and with your election as Secretary General and Vice President. To my knowledge, too, this is the first such society in the world. Maybe more countries will follow now, and maybe the International Planning and Steering Committee will then also be reorganized in a more formal way. I think it is time for that.

De Vries wrote this on September 4, 1980, long before anyone had the idea of ILCS in mind.

5. Martin Gordon's visit (1981)

Martin Gordon, Chairman of Gordon & Breach that published *Molecular Crystals and Liquid Crystals* (MCLC), ventured into China early, in 1981. He invited me to serve as an associate editor of the journal (which I did, 1981-1993) and asked me how he could help. I asked for a copy machine which he promised but never delivered. But I did write up a review on China's LC research for his journal, in which my 1982 prediction for Bowlics is mentioned [Lin 1983].

6. My second Orsay visit and Bangalore's ILCC (1982)

In August 1982, I gave a review on solitons in liquid crystals at the "Solitons '82" conference at the Riccarton campus of Heriot-Watt University outside of Edinburgh, UK, marking Scott Russell's centenary discovery of the solitons.[21] On November 1, 1982, my LC soliton paper came out in PRL [Lin et al. 1982; Lam 2014a][22] while I was spending two months at Orsay, working with Ribotta on LC's electroconvection pattern formation [Ribotta et al. 1986]. On my way back to Beijing, I attended the 9th ILCC, December 6-10, in Bangalore.[23] Kobayashi greeted me at the Raman Research Institute and congratulated me for being nominated to the PSC. However, my formal admission to the PSC had to wait until 1984 (Table 1).

Table 1. Members of the Planning and Steering Committee (PSC), 1980-1990. Data collected from the program book of each ILLC; the Chair of PSC is marked in grey cell. Since 1980, no member served more than a maximum of eight years (marked by five x)—a rule set by PSC in 1980, with two exceptions: (1) Glenn Brown, the founder of Kent's LCI and the ILCC series, after serving as PSC's Chair for many years, was elevated to honorary chair after Kyoto. (2) The other exception, Chandra, resulted from PSC breaking its own rule (see Sec. 3, Act II).

	PSC member	Country	1980	1982	1984	1986	1988	1990
		ILLC	8th	9th	10th	11th	12th	13th
	Number of PSC members		19	20	20	20	21	20
1	Ambrose, E. J.	UK	x	x				
2	Baur, G.	FRG						x
3	Blinov, Lev	USSR			x	x	x	x
4	Bouligand, Y.	France	x	x	x			
5	Brown, Glenn	USA	x	x	x	x	x	x
6	Chandrasekhar, S.	India	x	x	x	x	x	x
7	Chistyakov, I.	USSR	x	x				
8	Clark, Noel	USA					x	x
9	Demus, Dietrich	DR				x	x	x
10	Doane, William	USA					x	x
11	Durand, Georges	France			x	x	x	x
12	Figueiredo-Neto, Antonio	Brazil				x	x	x
13	Friberg, S. H.	USA		x	x			
14	Fukuda, Atsuo	Japan					x	x
15	Gennes, Pierre-Gilles de	France	x	x				
16	Gray, George	UK	x	x	x	x	x	
17	Gerristsma, C. J.	Netherlands	x	x	x			
18	Hosemann, R.	FRG	x	x				
19	Janik, Jerzy	Poland			x	x	x	x
20	Jeu, Wim de	Netherlands					x	x
21	Kahn, Fredic	USA	x	x	x	x		
22	Kelker, H.	FRG	x	x	x			
23	Kobayashi, Shunshuke	Japan	x	x	x	x		

24	Lagerwall, Sven	Sweden			x	x	x	x
25	Lam, Lui (Lin, Lei)	USA (China)				x	x	x
				x				
26	Leadbettter, Alan	UK			x	x	x	x
27	Lister, J. David	USA			x	x	x	x
28	Meier, Gerhard	FRG	x	x	x	x	x	
29	Okano, K.	Japan					x	x
30	Onogi, S.	Japan	x	x	x			
31	Porter, R. S.	USA	x	x				
32	Prost, Jacques	France				x	x	x
33	Rustichelli, Franco	Italy			x	x	x	x
34	Sackmann, Horst	DDR/ GDR	x	x	x			
35	Saupe, Alfred	USA	x	x				
36	Skoulios, A.	France	x	x				
37	Stegemeyer, Horst	FRG				x	x	x
38	Vries, Adrian de	USA	x	x	x	x		

2.3 In New York (1984-1987)

I left China at end of 1983 for family reasons [Lam 2014a]. From 1984-1987, I continued to work in LC at City University of New York. I visited Beijing every summer and set up a LC group at Nanjing Normal University, resulting in ten papers on two-dimensional solitons and viscous fingering in LC disc cells (1986-1989).

In 1985, I attended the 6th Liquid Crystal Conference of Socialist Countries, August 26-30, Halle (Saale), German Democratic Republic, which was chaired by the legendary Horst Sackmann (1921-1993) [Dunmur & Sluckin 2011: 167-172]. In Halle, I learned from Christian Destrade that bowlics had been synthesized in Europe. I stopped over at Orsay to meet Anne Levelut who just finished doing an x-ray study of these materials [Lam 1994: 327]. After I informed her that I had predicted bowlics while in China, she kindly mentioned my 1982 work in her paper [Levelut et al. 1986].

In summer of 1986, the 11th ILCC was held in Berkeley, California. I gave a review on bowlics [Lin 1987] and arranged for Xie Yu-Zhang (谢毓章, 1915-2011), president of CLCS and my coauthor of two papers in MCLC

(1983), to give a talk there. This review is the one often quoted in the bowlic literature (see, e.g., [Palffy-Muhoray ct al. 1988; Brostow 1990; Golubović & Wang 1992; Xu & Swager 1993]).

The second LC journal, *Liquid Crystals*, started in 1986; I served in the editorial board from 1986-1990. More importantly, in December 1986 I started planning the first book in the new Springer book series *Partially Ordered Systems*, which I initiated.[24] Historically, the series' first book *Solitons in Liquid Crystals*, edited by Lam and Prost [1992], is the first LC book that concentrates on a single topic. The contributors include liquid crystalists from China, France, Germany, India, Japan, USA, and USSR.

In the summer of 1987, I was preparing to move to San Jose State University (SJSU) in California. But there was this LC polymer conference in Bordeaux, July 20-24, which I planned to go, partly because I wanted to sit down with Prost and write the Preface and Chapter 1 for the soliton book together.[25] As was often the case, I picked a conference (or city) I wanted to go before I have an idea for an appropriate paper.[26] I never worked on polymers before; for this conference, I forced myself to come up with a presentation on bowlic and polar LC polymers and wrote it up later [Lam 1988a].[27]

Afterward, I visited Lajos Bata[28] in Budapest before traveling by car to Pardubice, Czechoslovakia, to attend the 7th Liquid Crystal Conference of Socialist Countries, August 31-September 4, 1987. At the conference I gave a review talk on bowlics and was presented with a red plastic bowl as a souvenir (Fig. 3, left).

At this point, I had worked in LC for 15 years, published 56 LC papers, co-founded the CLCS, invented a new kind of LC called Bowlic, pioneered the study of propagating LC solitons in shear flow, set up a Springer series that includes LC books, sit on the editorial board of the only two LC journals, and was a current member of the PSC—the power center of the LC profession. And I knew nearly everybody in LC, East and West. Following the example of De Gennes, my time to phase out of LC research was overdue. Besides, the job at San Jose required me to work and publish with undergraduates and that would be in nonlinear physics [Lam 1998: 328-331].

Fig. 3. Two presents I received in 1987. *Left*: A red plastic bowl from the LC socialists in Pardubice (with the characters "OPP ROCKYCANY 0,5" shown on the bottom). *Right*: A liquid crystal thermometer from Kent's LCI. I am responsible for the imperfection at the left upper corner, resulted from the thermometer slipping from my fingers and falling to the floor in San Jose.

3 Founding the International Liquid Crystal Society (1987-1990)

On my way to the LC polymer conference in Bordeaux, I stopped over at Orsay. I was shocked to learn that Mireille Delaye (1951-1987) had passed away. I visited Mireille's lab five years ago; she was obviously a young and bright liquid crystalist. After shock, came sadness and disappointment. The liquid crystal community was relatively small; it was almost like a family. How come I, a colleague in the same field, did not know of Mireille's death before I came to Orsay, and has to find out about it almost by accident? It occurred to me that the liquid crystal community should be organized better and should have a publication like *Physics Today* which includes an obituary in every issue. And so, I decided to establish an international liquid crystal society.

1. Prologue (July 1987-March 1988)

I asked several people, including Chandra, at the Bordeaux conference and they *all* agreed to this idea of an ILCS. In between the Bordeaux and the Pardubice conferences, I spent 12 days driving my Honda Accord from New York to San Jose. I stopped over at Kent to meet Glenn Brown, Alfred Saupe, Bill Doane, and others; and at Boulder to meet Noel Clark. They all supported an ILCS, too. At Kent, the LCI presented me with a gift, an

LCD thermometer (Fig. 3, right). At SJSU, I received a support letter from Doane, written in his capacity as LCI's director and dated August 10, 1987, which includes these words: "Your idea of a Society of Liquid Crystal Researchers is excellent. If we at Kent can be of help in this endeavor, please let us know." The LCers in Budapest and in Pardubice also gave their support.

To ensure the success of this project, I adopted a strategy based on Chairman Mao's two important teachings: (1) Mobilize the masses; trust the masses. (2) Encircle the city from rural areas (i.e., secure the surroundings before overcoming a stronghold). Being a ballet fan, I labeled each stage of the strategy an Act. And being employed as an untenured full professor, with a teaching load of 15 hours per week, I did not find time to act until March, 1988.

2. Act I (March 2, 1988)

In February of 1988 I gave a seminar on bowlics and refreshed my memory of the maple leaves at UBC, Vancouver, where I spent a year in the graduate school studying physics. My host, Birger Bergersen, helped to solicit support for the ILCS among the Canadians. In March and April of 1988, I contacted a selected group of liquid crystalists, including all current PSC members, by mail.[29] Each mail contains four pages: (1) an open letter explaining why an ILCS should be formed; (2) a page called Act I laying down my vision of the ILCS (Fig. 4); (3) a copy of Doane's support letter; (4) a questionnaire asking the recipient for comments and her/his choice of support or not support in writing, to be returned to me no later than April 25, 1988.

In fact, the need for an ILCS was lying bare there for many years, as noted by De Vries in 1980 (see Sec. 2). The overall discontent from both developing and developed countries was twofold: (1) The PSC was run like a private club without outside overseeing; (2) the club, with all the smart people in the field, did only one thing, i.e., selecting a site for a conference every two years—and absolutely nothing else, while the LC profession in many countries were not getting its fair share of attention from their respective scientific community, and jobs for new PhDs in LC declined very fast in number.

As shown in Fig. 4, the plan was quite in detail. All the suggestions proposed here, including the name of the society, the organizational structure, the publication of an official "newsletter," and option 3 in connection with the existing PSC were indeed incorporated into the ILCS. The item "Membership Due" was written with developing countries in mind, derived from my China experience. It was motivated to give people from developing countries an equal voice and would not be discriminated simply because they could not afford to go to the biennial ILCC.

The one exception that did not go as planned was the "Date of Founding"; instead of being suggested here to be in 1988, the centenary of LC's discovery, the ILCS was founded two years later. The reason for the delay is explained in the next subsection.

3. Act II (May 25, 1988)

In May of 1988, the responses to Act I were summarized in Act II (Fig. 5). It was based on 61 replies from 14 countries that I received; they were all positive, with nine coming from current members of PSC and two from past members. Act II was mailed to all 61 repliers and to others. As shown in Act II, more details on the future ILCS were provided; the planned founding date remained to be in August, 1988, in Freiburg, during the 12th ILCC. The plan was to hold an organizing meeting open to all participants of the conference; with enough consent, the PSC would declare itself transformed into an ILCS; and people sign up to join. Like the founding of a new country, the bylaws (the constitution) could be drafted later. Naturally, this assumed that the PSC members were among the consenters.

At this point, the missing piece was the absence of Chandra and George Gray (1926-2013), the current and previous PSC chair, respectively (see Table 1), among the repliers to Act I. I waited until July 15, the deadline to respond to Act II, and then August 15, the beginning of the Freiburg conference, but there were still no words from Chandra and Gray (see Fig. 6). Were they opposing to the idea of an ILCS? If not, what was holding them up? The answer came at the PSC meeting in Freiburg.

3/2/88

Founding the International Liquid Crystal Society - Act I

(The ideas suggested below are meant to induce discussion. There is nothing final about it - L.L.)

Name: International Liquid Crystal Society

Aim: 1. To serve the interest of liquid crystalists as individuals.
      2. To provide a forum for people and organizations to exchange information and ideas.
      3. To serve the need of the liquid crystal community, scientists and industries, in general.

Membership: Individuals join as members; lab or group may join as a unit; company and national liquid crystal group may join as corporation member.

Membership Due: Keep it as low as possible. Members from countries with hard currency pay in US$ (or equivalent); those from other countries pay in their own currency. Reduced rate for students.

Function: Publish a newsletter (something like the Physics Today but less formal, incorporating some features of newspapers) once every two months (or may be 3 months in the first year). Organize meetings, summer or winter schools, short courses. The newsletter may be sent to each country or district in bulk and distributed by the local chapters using the local currency collected from the membership dues.

Organization: Local chapters may be formed in each country (if there is enough membership), groups of countries, or different geographical sections within a country, upon the approval of the governing board.

Relationship with the existing International Steering and Planning Committee: There are three possibilities:
      1. completely independent of each other.
      2. incorporate the Committee under the Society.
      3. form the Society from the Committee by expanding it, and then reorganize or keep the Committee intact and put it under the Society.
      (It seems to me that 3. is most practical.)

Date of founding: August 1988 during the 12th International Liquid Crystal Conference at Freiburg, West Germany. It is the centenary discovery of liquid crystal and is a good occasion; time frame is right too.

Fig. 4. Act I (March 2, 1988): Initial formulation of the ILCS.

5/25/88

Founding the International Liquid Crystal Society - Act II

The response to Act I (3/2/88) questionnaire is very positive. At this point there are 61 replies from 14 countries, including those from 9 present members and 2 past members of the International Planning and Steering Committee for International Liquid Crystal Conferences. Summing up the comments expressed by those who responded,

o The International Liquid Crystal Society should be founded at the 12th Int. Liq. Cryst. Conf. at Freiburg, Aug. 15-19, 1988.

o It should be organized with the existing International Planning and Steering Committee as the basis.

o There should be open discussion by those attending the Freiburg conference before the founding.

o The publication of Newsletter by the Society is desired. One suggested to have it monthly; another one for quarterly; the rest seems to content to have it bimonthly.

o The membership fee should be as low as possible. (For clarification, the membership dues from members in countries without hard currency can be collected as long as it is not transferred outside. This will be used within those countries to mail the newsletter and for functions sponsored or endorsed by the Society. For countries in which paper is a scarce material printed newsletter has to be sent in in bulk; for others a good copy can be sent and reprinted there.)

o Concern has been expressed (by one) on the autonomous of local chapters. If there is no financial support from the central office to local chapter, the local chapter should be autonomous, in contrast to the practice in the Society of Information Display, it was suggested. Please keep this issue in mind and we will discuss it at Freiburg.

o In view of the positive and extensive responses outlined above I have written to S. Chandrasekhar, Chairman of the Int. Planning and Steering Committee, and to the organizers of the Freiburg Conference requesting that the International P&S Committee to start discussion on these suggestions, and to make arrangements (time and place) during the Conference for the participants to hold meetings, respectively.

o A name list of all those who sent back their questionnaire to me as of today is enclosed here for your reference. All support the founding of the Society.

● A brief statement as an open letter to the participants of the Freiburg Conf. is prepared here and will be distributed to all during the registration period (assuming the consent of the organizers, of course). If you want to remove or add your name to the list please mail the form back to me before July 15, 1988 (the date the form should reach me). If you agree to have your name there, do not take any action. Of course, your comments are always welcome. Please copy these pages and help distribute to your colleagues.

Lui Lam

Lui Lam

For Lui Lam, Phys. Dept., SJSU, Telephone: (408)924-5261, Fax: (408)924-1018, Telex: 171171 UD (Note new fax number)

Fig. 5. Act II (May 25, 1988): A summary of responses to Act I.

On August 16, 1988, PSC held its business meeting in a room without windows. After the meeting was declared open by Chandra and before anyone had the chance to raise the question of electing the next PSC chair, as required by the rules since Chandra had reached his eight-year limit (see item 3 of Sec. 2), Gray said, "Let Chandra continue to be the next Chair." Everyone was silent. I was struck dumb; I looked at Doane who sat next to me at my right and saw his jaw dropped, or so it seemed. And so Chandra began his third term as PSC chair, violating the rules he and Gray helped to write and pass just eight years ago.

I presented the Petition signed by 82 people (Fig. 6) and proposed that we transformed the PSC to an ILCS. Gray asked immediately: "What will happen to the PSC members?" I answered without thinking: "They will be the Board members of this new ILCS." And so, the resolution was passed unanimously. Someone proposed a subcommittee (of five members) to formulate the details of the transformation.[30] The decisions were announced by Chandra to the conference participants the next day. On August 18, I was asked to draft the bylaws because no one else in the subcommittee wanted to do it (see Fig. 7).

4. Act III (September 20, 1988)

A month later, I sent out Act III (Fig. 7) to all the 82 signees whose name appeared on the final petition (Fig. 6), thanking them for their fruitful support.

5. Act IV: Writing the Bylaws (1988-1990)

Before the summer of 1990 I was untenured. Needless to say, in my first three years at SJSU, I had to work like crazy. I created and taught two new graduate courses in nonlinear physics [Lam 1997, 1998]; at one point, I supervised nine students in research and wrote papers with them; I attended several conferences and organized one or two conferences every year.[31] My research focus was no longer in LC but in nonlinear and complex systems. During this period, my LC publications are merely extensions of my previous works.

Being overloaded, I asked Hiap-Liew Ong (IBM) and then Flonnie Dowell (Los Alamos) to write the bylaws; both turned me down. Finally, I asked Ong to send me a copy of the bylaws of the Society for Information Display (SID). After reading the (1987) Bylaws of SID and the Constitution and Bylaws of the American Physical Society, I wrote ILCS' bylaws myself in July 1989 (Fig. 8).

The spirit and suggestions in Acts I and II were incorporated into the bylaws. In particular, I named the official magazine *Liquid Crystals Today* and stipulated that the society will *not* publish any research journals (in deferment to the two existing journals, MCLC and *Liquid Crystals*). More importantly, the President is limited to two consecutive terms of two years each term, and has to wait two years before reelected to the same position; after consecutive four years of having one (or two) President(s) from the same country, the next President has to be elected from another country. In other words, no same person or country could have the presidency for more than four years in a row.

Finally, in the summer of 1990, everything was ready for the ILCS to be formed at the 13th ILCC. And so I had to go to Vancouver. But there was a problem. I was scheduled to present a LC paper in that conference but I did not yet have enough material to write this paper.

6. Finale: Birth of the ILCS (July 22-27, 1990)

It turned out that 1990 was my very busy year. In January I organized a "Winter School in Nonlinear Physics" at SJSU. In March, in Anaheim, California, I chaired the American Physical Society Symposium on "Instabilities and Propagating Patterns in Soft Matter Physics," which I proposed. In June I was the Director of the NATO Advanced Research Workshop on "Nonlinear Dynamical Structures in Simple and Complex Liquids," in Los Alamos. And in July I attended briefly the "Nonlinear and Chaotic Phenomena" conference in Edmonton, Canada [Lam 1991] before flying directly to Vancouver to attend the 13th ILCC, July 22-27, at the UBC campus—my Alma Mater and the first place I lived after Hong Kong 35 years ago.

It was a few days before this Edmonton conference that we rushed through a simple experiment in our Nonlinear Physics Laboratory at SJSU, in Room 55 in the basement of the Natural Science Building. The experiment was so simple that it could be finished in less than one second. What we did was taking a LC cell, putting in LC or oil, and applying a high enough voltage across the cell. After a flash of light, the experiment was finished and we found a complicated filamentary pattern left on the inner surfaces of the coated glass plates. I wrote up the paper [Lam et al. 1991] right before I rushed to the airport and presented it in Vancouver, which led to my invention of *Active Walk*, a new paradigm in complex systems [Lam 2005a].

In Vancouver, before the ILCC began, Chandra carefully sounded me out by telling me his wish. Here was the exchange:

> Chandra: "Lui, I want to be the President of the ILCS."
>
> Lam: "OK."

Being the President never crossed my mind nor was it my wish when I started this campaign to found the ILCS. I did it for Mireille and for the LC community.

Now we needed a Vice President. Chandra and I could not come up with a candidate. And so we stood at the entrance of the conference canteen watching people going in and out, and still did not see a suitable candidate. After five minutes, the conversation went like this:

> Chandra: "Lui, why don't you be the Vice President?"
>
> Lam: "That wouldn't be appropriate because then both the President and Vice President are Asians."
>
> Chandra: "But I never see you as an Asian."

That was probably true; I acted and talked more like a New Yorker than a Chinese. Luckily, at this point, we saw Martin Schadt approaching the canteen. That was it; we asked and Schadt kindly agreed to be the Vice President.

Call for Establishing an

INTERNATIONAL LIQUID CRYSTAL SOCIETY

Colleagues:

The 12th International Liquid Crystal Conference Marks the centenary of the discovery of liquid crystals. The number of people working in this field, the interdiscipliniary character of the research on these unique and fascinating materials, the productive collaboration between industry and university and the importance of liquid crystals in new industries require a permanent organization dedicated to serving the international liquid crystal community. We invite all participants at this Conference to attend an organizational meeting to plan the formation of an INTERNATIONAL LIQUID CRYSTAL SOCIETY.

Signed:

BRAZIL
* Figueiredo Neto, Antonio
Fujiwara, Fred Yukio
Santos, Marcus B. L.

BULGARIA
Derzhanski, Alexander I.
Petrov, Alexander G.

CANADA
Bergersen, Birger
Burnell, E. Elliott
Dong, Ronald Y.
Gilson, Denis
Leigh, William James
Tracey, Alan S.

CHINA
Dong, Chuchuan
Li, Guozhen
Liang, Ruan
Liang, Zhong Cheng
Liu, Han-Ming
Shao, Ren-Fan
Wang, Bin
Wang, Liang Yu
Wang, X. J.
Xi, Guangeng
Xi, Hua
Xie, Yu-Zhang
Yang, Shong Ling
Zhao, Jing An
Zheng, Shu
Zhong, Guofu
Zhengmin, Sun

CZECHOSLOVAKIA
Pirkl, Slavomir

FRANCE
Charvolin, Jean
* Durand, Georges
Gasparoux, Henry
Hardouin, Francis
Noel, Claudine
* Prost, Jacques

GERMAN DEMOCRATIC REPUBLIC
* Demus, Dietrich

HUNGARY
Bata, Lajos
Buka, Agnes

ITALY
Rustichelli, Franco

JAPAN
Kawamura, Yasuaki
+ Kobayashi, Shunsuke

NETHERLANDS
* De Jeu, Wim H.

POLAND
Janik, Jerzy A.

SWITZERLAND
Lierau, Rolf R.
Schadt, Martin

UNITED KINGDOM
Clark, Michael George

UNITED STATES
Acree, William E., Jr.
Armitage, David
+ Brown, Glenn
* Clark, Noel A.
Davis, Frederick
+ De Vries, Adriaan
* Doane, William J.

Dowell, Flonnie
Drzaic, Paul S.
Fishel, Derry L.
Gelbart, William M.
Gelerinter, Edward
+ Kahn, Frederick J.
Keast, Sandy S.
Kumar, Satyendra
* Lam, Lui
* Litster, James David
Mahmood, Rizwan
McAdams, Larry R.
McRuer Robert N.
Meyer, Robert B.
Neubert, Mary E.
Ong, Hiap Liew
Petschek, Rolfe George
Rosenblatt, Charles
Shen, Y.R.
Ukleja, Paul
Vargas-Aburto, Carlos
Vaz, Nuno A.
Vora, Rasiklal A.
Westerman, Philip W.
Wu, Shin-Tson

USSR
* Blinov, Lev M.
Shibaev, Valery Petrovich

YUGOSLAVIA
Blinc, Robert
Zeks, Bostjan

———
* Present member.) of the International Planning and Steering
) Committee for International Liquid Crystal
+ Past member.) Conferences.

Fig. 6. The Petition: Flyer distributed before August 16, 1988, the date of the PSC meeting, at the 12th ILCC, Freiburg. There are 82 signees (1 from UK and 0 from India).

The Founding of an International Liquid Crystal Society - Act III

    In response to the suggestion of you and many other colleagues (see Act II), the Planning & Steering Committee for International Liquid Crystal Conferences agreed to transform itself into an International Liquid Crystal Society during their business meeting in the afternoon of August 16, 1988, at Freiburg, FR Germany. A subcommittee consists of

        S. Chandrasekhar (India)
        W. Doane (USA)
        A. Fukuda (Japan)
        S. Lagerwall (Sweden)
        L. Lam (USA)

was approved by the Planning & Steering Committee during the meeting. The subcommittee was charged to formulate the details of the transformation and make proposals to the P&S Committee. All the above were announced by the Chairman of the P&S Committee, Dr. S. Chandrasekhar, at the end of the awarding ceremony of the 2nd Glenn Brown Award on August 17, 1988 at the 12th International Liquid Crystal Conference at Freiburg.

    In the subsequent meeting of the Subcommittee on August 18, 1988 at Freiburg, L. Lam was asked by the subcommittee to write the first draft of the bylaws of the upcoming International Liquid Crystal Society.

    In short, our effort to found the I.L.C.S. has been fruitful. From this point on, any questions or comments concerning the future I.L.C.S. should be directed to members of the Subcommittee. In matters relating to the drafting of the bylaws please send your suggestions (or a copy of your writing) directly to L. Lam [Dept. of Physics, San Jose State University, San Jose, CA 95192, USA. Tel.: (408)924-5261]. This will shorten the time of communication and enables the International Liquid Crystal Society to be born earlier.

    Thank you very much for your support in the past and in the future.

                                     Lui Lam

Fig. 7. Act III (September 20, 1988): Announcing PSC's transformation to ILCS, and the subcommittee in charge of formulating the transformation.

With the President and Vice President in place and the Bylaws approved, the PSC dissolved itself. All PSC members automatically became Board-of-Directors members. The new ILCS was announced on *July 27, 1990*, at the closing ceremony and people were invited to sign up to join (Figs. 9 and 10).[32] I assumed the Chair of the new Conference Committee, with all Board members as its members. In other words, the Conference

Committee was the old PSC under a new name. And that was how the ILCS was born!

Tel: (408) 924-5261   Fax: (408) 924-1018
Bitnet: LUILAM@CALSTATE (after 8/25)

July 15, 1989

To: Subcommittee to establish Int. Liq. Cryst. Soc. (ILCS)
(S. Chandrasekhar, W. Doane, G. Durand, A. Fukuda, J.A. Janik, S. Lagerwall, L. Lam, A.J. Leadbetter)

From: Lui Lam   *Lui Lam*

Re: Drafting the bylaws of ILCS

o   I will not be able to attend the 8th LC Conf. of Socialist Countries in Krakow, Poland, Aug. 28 - Sept. 1, 1989. I cannot get away from my teaching duties at that time. Fortunately, Chandra may be able to attend that, and I hope many of you will be there. I agree with Chandra that the occasion could be used for the members to discuss the bylaws and the possibility of merging the "Western" and the Socialist Countries Conferences into a single series. I am all for this idea.

o   Enclosed please find a copy of the SID Bylaws, and the **Constitution and Bylaws of the American Physical Society** for your references.

o   Here is some skeletal form of Bylaws for the ILCS written by me. It is for discussion and as a basis of modification by the Subcommittee. I intend it to be brief and general (as compared to those of SID and APS) since we are doing it for the first time.

Bylaws of the International Liquid Crystal Society

Article 1 - NAME

The Society shall be called The International Liquid Crystal Society, hereafter called the ILCS.

Article 2 - OBJECT AND SCOPE

1. The object of this Society shall be:
   (a) To encourage the scientific, literary and eduactional advancement of liquid crystals.
   (b) To provide a forum for individuals and organizations to exchange information and ideas relating to liquid crystals.
   (c) To serve the need of the liquid crystal community, including both individual scientists and the industry.

2. The scope of the ILCS is non-national.

Article 3 - MEMBERSHIP

1. Grades and Qualifications

   (a) Student Member - An individual pursuiting an undergraduate or graduate degree.
   (b) Associate Member - An individual interested in furthering the object of the ILCS.

Fig. 8. Act IV (July 15, 1989): Draft of ILCS' Bylaws. Shown here is the first page.

4 After 1990

My term as ILCS' Board member ended in 1994. I stopped going to ILCC in 1996 because my research, since 1995, had completely shifted to nonlinear and complex systems [Lam 1997, 1998].[33] But I did show up in Sendai in 2000 on my way home from Beijing to have dinner with Helfrich, Cladis, and others who were at the 18th ILCC there.

I was in Beijing on June 4, 1989 [Morrison 1989]. After that I stopped going to Beijing for eight years. Instead, I frequented Taiwan and worked with the liquid crystalists I met at the Kyoto conference [Pan et al. 1995]. I urged them to join the ILCS. With much delay, the LC society in Taiwan was formed in 1995 and affiliated to the ILCS as the "ROC Taiwan Liquid Crystal Society" [Fukuda 1998]; the CLCS chose to stay out. Interestingly and historically, it was not until 2004 that a petition signer (Fig. 6) got elected to the presidency of ILCS. He was Satyendra Kumar, from Kent.[34]

In short, the International Liquid Crystal Society is a French-inspired, Chinese-initiated, and truly international mass organization. Enjoy!

Notes

1. In 2007, LCD TVs overtook cathode-ray-tube TVs in sales worldwide for the first time (en.wikipedia.org/wiki/LCD_television, April 10, 2013).

2. www.lcinet.kent.edu/index.php (April 17, 2013).

3. The number 50 comes from J. William Doane's "Glenn H. Brown, Honorary Chairman, 11th International Liquid Crystal Conference," in Scientific Sessions and Abstracts Book of the 11th ILCC, June 30-July 4, 1986, Berkeley, California. The number 42 comes from the same Abstracts Book.

4. Up to now, ILCC has been held in Kent five times (1965, 1968, 1972, 1976, and 1996).

5. www.ilcsoc.org/ILCS/aboutILCC.html (April 22, 2013).

6. This is not strictly true. The fact that the ILCS was initiated by Lui Lam is reported briefly in [Lam 2005a: 2318, 2005b: 529].

7. I was absolutely a normal kid. I run away from school only once. One day in Shantou, Guangdong Province, I eluded the guard at the gate of my kindergarten during a recess, run 15 minutes home and hided myself by a sofa. I was three years old then and my mom convinced me that it was not such a good idea.

ANNOUNCEMENT

International Liquid Crystal Society (ILCS)

ILCS is pleased to announce its existence, and to invite applications for membership.

ILCS is a voluntary, non-profit international organization. The objects are (i) to encourage the sicientific and educational advancement of liquid crystals; (ii) to provide a forum for individuals and organizations to exchange information and ideas relating to liquid crystals; and (iii) to serve the need of the liquid crystal community, including both individual scientists and the industry. Membership consists of student member, assocaite member, member, sustaining member and affiliate Society member. (See the bylaws for more details.)

The Board of Directors of ILCS currently consists of G. Baur (Germany), L.M. Blinov (USSR), G.H. Brown (USA), S. Chandrasekhar (India), N.A. Clark (USA), W.H. de Jeu (The Netherlands), D. Demus (Germany), J.W. Doane (USA), G. Durand (France), A.M. Figueiredo Neto (Brasil), A. Fukuda (Japan), J.A. Janik (Poland), S.T. Lagerwall (Sweden), L. Lam (USA), A.J. Leadbetter (United Kingdom), J.D. Litster (USA), K. Okano (Japan), J. Prost (France), F. Rustichelli (Italy) and H. Stegemeyer (Germany).

For further information on the ILCS please contact:

Professor S. Chandrasekhar
Raman Research Institute
Bangalore 560080   INDIA

FAX: 91 812 340492; Tel: 91-812 340122; Telex: 845 2671 RRI IN Grams; RAMANINST, or any member of the Board of Directors.

Fig. 9. Finale (July 27, 1990): Announcement of the *existence* of the ILCS distributed at the 13th ILLC, Vancouver (prepared by the author in June, 1990).

PRELIMINARY APPLICATION FORM FOR MEMBERSHIP

International Liquid Crystal Society (ILCS)

Name: _____
           First   Middle initial   Last

Affiliation (if any): _____

Address: _____

            _____

            _____

Phone: _____   **Fax**: _____

Telex: _____   **E-mail**: _____

Highest Degree obtained: (B.S./M.S./Ph.D.)_____ year_____

                            _____Univ.

Current Interests in Liquid Crystals: (Physics, Chemistry, Biology, Polymers, Applications, etc.)

_____

_____

Type of Membership Applied: (Student member, associate member, member, sustaining member, or affiliate Society member)

_____

Signature: _____   **Date**: _____

P.S. o The **membership dues** will be announced by the ILCS shortly.
          o Please **return this form** to the reception desk of the 13th International Liquid Crystal Conference during the conference, or to Prof. S. Chandrasekhar, Raman Research Institute, Bangalore 560080, India.
          o **Suggestions and Comments**:

Fig. 10. Finale (July 27, 1990): The sign-up form (together with the Announcement and the Bylaws) distributed at the 13th ILLC (prepared by the author in June, 1990).

8. My doctoral work on Compton profile resulted in the Lam-Platzman Correction [Callaway & March 1984; Papanicolaou et al. 1991; Blass et al. 1995] and the Lam-Platzman Theorem [Bauer 1983].

9. While at Columbia and CCNY, I helped to found and build up the Chinatown Food Co-op, a mass organization with a political agenda [Kuo 1977; Lam 2010]. It was at the Food Co-op that I developed my organizational skills which later became handy in founding the ILCS. The Co-op members included Jean Quan (now Mayor of Oakland, CA) and Peter Kwong (now Professor of Asian American Studies at Hunter College, CUNY).

10. In the late 1960s and early 1970s two theorists, Pierre-Gilles de Gennes at Orsay and Paul Martin at Harvard, helped to bring liquid crystals to the mainstream of physics research. Both worked closely with their experimental colleagues. The Orsay team beat the Harvard team by a huge margin. De Gennes was awarded the Nobel Prize in 1991 partly for his LC works.

11. Personal computers have been available since 1976 (the year Apple was founded) but before 1990—the year the World Wide Web was born, the way to do a literature search, if you do not know what you are looking for, is to go to a good library to flip through Chemical Abstracts and walk semi-randomly between the long rows of shelves housing books and journals. It takes time, like treasure hunting without a map. But it is also a serene and spiritual experience, like immersing yourself in a quiet church.

12. In [Lam 1977c], in one stroke, we rederive the Frank free energy and the Ericksen-Leslie equation for both nematics and cholesterics plus the Parodi and other relations among the transport coefficients (which are nothing but the Onsager reciprocal relations). The dissipation-function approach is later used for biaxial nematics [Das & Schwartz 1994]. It remains, in fact, the best and simplest approach in handling any new thermoviscous solids with microstructures and the hydrodynamics of any new molecular fluids (including liquid crystals).

13. LIN Lei (林磊) is my Chinese name in pinyin; LAM Lui represent the same Chinese characters in Cantonese which I picked up in Hong Kong.

14. I thought that invitation came from the attention of my PRL paper; Chandra, the name Chandrasekhar asked me to call him, told me years later that I was recommended by Ron Shen of Berkeley. The historical significance of this Bangalore conference is that it is the first LC conference outside of China ever attended by a delegation from mainland China.

15. In this chapter, a person working in LC is referred as a *liquid crystalist*, or a *LCer*, pronounced LC-er.

16. Patricia Cladis, born in Shanghai, joined Bell Labs in 1972, the year I left. We met each other for the first time in Bangalore.

17. I opened my talk with a remark from Chairman Mao (Ze-Dong, 毛泽东主席) (1893-1976) on dialectics that says every positive thing when left too long, becomes its opposite, referring to the need to go beyond the Landau-de Gennes mean-field model. The talk was closed with a quote from Rabindranath Tagore (1861-1941), India's Nobel laureate in literature: "Truth seems to come with the final word; and the final word gives birth to its next" [Lin 1980].

18. In September 1980, the Orsay's "gang of four" (Jean Charvolin, Georges Durand, Maurice Kléman, and Roland Ribotta), led by Ribotta and accompanied by me, toured Beijing, Shanghai, and Wuhan. Liquid crystal research in China was so far behind then, that these four world-renowned LCers ended up giving popular-science talks everywhere we went. But the tour was very exciting.

19. I learned firsthand the Japanese culture through the protracted negotiations with Kobayashi, who would say "yes" whether he agreed with me or not.

20. See Membership List of Chinese Liquid Crystal Society (First Term, 1980-1984), p 10.

21. I was in the small boat in the Union Canal that tried to recreate the solitary wave John Scott Russell (1808-1882) first observed in 1834. The failed "experiment" was reported in *The Scotsman*, August 26, 1982, and by Ellbeck and Scott [1982].

22. Historically, this is the first paper in LC literature that contains the word soliton in the paper's title *and* the first one that talk about *propagating* solitons in LCs.

23. The 9th ILCC was originally scheduled to be held in Kraków, Poland. But in December 1981, Poland's communist government declared martial law and delegalized the opposition Solidarity and interned its key members (http://en.wikipedia.org/wiki/1982_ demonstrations_in_Poland, Sept. 1, 2013). My French LC colleagues led the petition to move the ILCC out of Poland. Bangalore became the replacement in the last minute and that was why the 9th ILCC was held in winter instead of summer. Subsequently, Lech Wałęsa, cofounder of Solidarity, won the Nobel Peace Prize (1983) and served as President of Poland (1990-1995). The 9th ILCC is my second visit to Bangalore, or for that matter, India. Later, personally, to my regret, my third chance to visit Bangalore in 1986, as an invited speaker at the Second Asia Pacific Physics Conference, did not materialize due to insufficient time in applying for a visa.

24. The Editorial Board consisted of Jean Charvolin, Wolfgang Helfrich, and Lui Lam; the Advisory Board: David Lister, David Nelson, and Martin Schadt. With the consent of Springer, I elevated myself to the series' editor-in-chief in 1999.

25. As it turned out, Prost drove us to a nude beach outside of Bordeaux and he wrote his part lying down nude on the sand while I was absorbing the scenery and the fresh air.

26. I first met De Gennes at the Les Houches Summer School in 1973. I met De Gennes the second time, 14 years later, at this conference. The third and last time I saw him was in Varenna, Italy, during the Enrico Fermi Summer School on The Physics of Complex Systems, July 9-19, 1996, organized by Francesco Mallamace and Eugene Stanley.

27. Bowlic polymers predicted in this paper were synthesized later in USA by Zeng Er-Man (曾尔曼) [2001]. It took 13 years, while the prediction of bowlic monomers took only three years to be confirmed in Europe [Lam 1994]. Eventually, the word "bowl" or "bowlic" appears in the title of LC papers (such as [Cometti et al. 1990; Xu & Swager 1993; Arcioni et al. 1995; Mehta & Uma 1998; Imamura et al. 1999; Dong et al. 2009]) and is recognized officially by the IUPAC [Barón & Stepto 2002: 499] and formally in *Handbook of Liquid Crystals* [Demus et al. 2008]. The paper also contains a prediction of ultrahigh-T_c bowlic (or discotic) superconductors (see also [Lam 1988b]) which was reported by *Superconductor Week* (October 19, 1987: 4), a newsletter newly created to report on high-T_c superconductors.

28. Bata, Durand, and Schadt were among the speakers that I, member of the Preparatory and Advisory Committee, helped to invite to the Centenary Conference of Liquid Crystal Discovery, Beijing, June 27-July 1, 1988, which was organized by the CLCS [Lam 2022].

29. Email was not yet available. Communications between scientists were by phone, fax, or mail; mailing was the cheapest.

30. For this reason, the open organizational meeting envisioned in the Petition never took place.

31. The conferences include two local series: Woodward Conference [Lam & Morris 1989] and Liquid Crystals West. James Fergason (1934-2008) and Ron Shen participated in the latter series.

32. ILCS' first Officers Meeting was held July 27, 1990, at UBC; present: Chandra (President), Shadt (Vice-President), Lam, Doane (Treasurer), and Dunmur (Secretary).

33. The fact that the Chair of the 16th ILCC at Kent refused to waive my registration fee did not help. After *Active Walk* (1992) [Lam 2005a, 2006], I created and worked on *Histophysics* (2002) [Lam 2002] and *Scimat* (Science Matters, 2007) [Lam 2008]. The latter is a new multidiscipline that treats all human matters as part of science [Lam 2008, 2014b, 2024]. See also www.sjsu.edu/ people/lui.lam/scimat.

34. The first few Presidents of ILCS are: Sivaramakrishna Chandrasekhar (India, 1990-1992), Geoffrey Luckhurst (UK, 1992-1996), Atsuo Fukuda (Japan, 1996-2000), and John Goodby (UK, 2000-2004).

References

Arcioni, A., Tarroni, R., Zannoni, C., Dalcanale, E. & Du vosel, A. [1995]. Microscopic heterogeneity in a bowlic columnar mesophase as probed with fluorescence depolarization measurements. J. Phys. Chem. **99**: 15981-15986.

Barón, M. & Stepto, R. F. T. [2002]. Definitions of basic terms relating to polymer liquid crystals. Pure Appl. Chem. **74**: 493-509.

Bauer, G. E. W. [1983]. General operator ground-state expectation values in the Hohenberg-Kohn-Sham density functional formalism. Phys. Rev. B **27**: 5912-5918.

Blass, C., Redinger, J., Manninen, S., Honkimäki, V., Hämäläinen & Suortti, P. [1995]. High resolution Compton scattering in Fermi surface studies: Application to FeAl. Phys. Rev. Lett. **75**: 1984-1987.

Brostow, W. [1990]. Properties of polymer liquid crystals: Choosing molecular structures and blending. Polymer **31**: 979-995.

Callaway, J. & March, N. H. [1984]. Density functional methods: Theory and applications. *Solid State Physics* **38**: 135-220.

Cometti, G., Dalcanale, E., Du vodel. A. & Levelut, A.-M. [1990]. New bowl-shaped columnar liquid crystals. J. Chem. Soc., Chem. Commun., p 163.

Das, P & Schwartz, W. H. [1994]. Continuum and molecular theories of biaxial nematics: Calculation of the 2-Director viscosity coefficients. Mol. Cryst. Liq. Cryst. **239**: 27-54.

Demus, D., Goodby, J. W., Gray, G. W., Spiess, H. W. & Vill, V. [2008]. *Handbook of Liquid Crystals*. New York: Wiley.

Dong Yan-Ming, Chen Dan-Mei, Zeng Er-Man, Hu Xiao-Lan & Zeng Zhi-Qun [2009]. Disclination and molecular director studies on bowlic columnar nematic phase using mosaic-like morphology decoration method. Science in China Series B: Chemistry **52**: 986-999.

Dunmur, D. & Sluckin, T. [2011]. *Soap, Science, and Flat-Screen TVs*. New York: Oxford University Press.

Ellbeck, J. C. & Scott, A. C. [1982]. Solitons galore. Phys. Bull. **33**: 426-427.

Fukuda, A. [1998]. Message from the President, Professor Atsuo Fukuda. Liquid Crystals Today **8**(4): 9. (All issues of *Liquid Crystals Today* could be downloaded free from www.tandfonline.com/loi/tlcy20#.UbqVAOc3uSo.)

Gennes, P. G. de [1974]. *The Physics of Liquid Crystals*. Oxford: Clarendon Press.

Golubović, L. & Wang, Z.-G. [1992]. Anharmoic elasticity of smectic A and the Kardar-Parasi-Zhang model. Phys. Rev. Lett. **69**: 2535-2538.

Imamura, K., Takimiya, K., Aso, Y. & Otsubo, T. [1999]. Triphenyleno[1,12-bcd:4,5-b'c'd':8.9-b''c''d'']trithiophene: The first bowl-shaped heteroaromatic. Chem. Commun., pp 1859-1860.

Kawamoto, H. [2002]. The history of liquid-crystal displays. *Proceedings of the IEEE* **90**: 460-500.

Kuo, Chia-ling [1977]. *Social and Political Change in New York's Chinatown: The Role of Voluntary Associations*. New York: Praeger.

Lam, L. [1973]. Surfaces de Fermi, profil Compton et effets a N-corps. Phys. Lett. A **45**: 409-410.

Lam, L. [1977a]. Dissipation functions and conservation laws of molecular liquids and solids. Z. Physik B **27**: 101-110.

Lam, L. [1977b]. Reciprocal relations of transport coefficients in simple materials. Z. Physik B **27**: 273-280.

Lam, L. [1977c]. Constraints, dissipation functions and cholesteric liquid crystals. Z. Physik B **27**: 349-356.

Lam, L. [1988a]. Bowlic and polar liquid crystal polymers: Mol. Cryst. Liq. Cryst. **155**: 531-538.

Lam, L. [1988b]. Possible liquid crystalline high T_c superconductors. *3rd Asia Pacific Physics Conference*, Chan, Y. W., Leung, A. F., Yang, C. N. & Young, K. (eds.). Singapore: World Scientific.

Lam, L. [1991]. Unsolved nonlinear problems in liquid crystals. *Nonlinear and Chaotic Phenomena*, W. Rozmus, W. & Tuszynski, J. A. (eds.). Singapore: World Scientific.

Lam, L. [1994]. Bowlics. *Liquid Crystalline and Mesomorphic Polymers*, Shibaev, V. P. & Lam, L. (eds.). New York: Springer. pp 324-353.

Lam, L. (ed.) [1997]. *Introduction to Nonlinear Physics*. New York: Springer.

Lam, L. [1998]. *Nonlinear Physics for Beginners: Fractals, Chaos, Solitons, Pattern Formation, Cellular Automata and Complex Systems*. Singapore: World Scientific.

Lam, L. [2002]. Histophysics: A new discipline. Mod. Phys. Lett. B **16**: 1163-1176.

Lam, L. [2005a]. Active Walks: The first twelve years (Part I). Int. J. Bifurcation and Chaos **15**: 2317-2348.

Lam, L. [2005b]. The origin of the International Liquid Crystal Society and Active Walks. Physics (Wuli) **34**: 528-533.

Lam, L. [2006]. Active Walks: The first twelve years (Part II). Int. J. Bifurcation and Chaos **16**: 239-268.

Lam, L. [2008]. Science Matters: A unified perspective. *Science Matters: Humanities as Complex Systems*, Burguete, M. & Lam, L. (eds.). Singapore: World Scientific. pp 1-38.

Lam, L. [2010]. The first "non-government" visiting-scholar delegation in the United States of America from People's Republic of China, 1979-1981. Science & Culture Review 7(2): 84-94.

Lam, L. [2014a]. Solitons and revolution in China: 1978-1983. *All About Science: Philosophy, History, Sociology and Communication*, eds. Burguete, M. & Lam, L. (eds.). Singapore: World Scientific. pp 253-289.

Lam, L. [2014b]. About science 1: Basics—knowledge, nature, science and Scimat. *All About Science: Philosophy, history, sociology & communication*, Burguete, M. & Lam, L. (eds.). Singapore: World Scientific. pp 1-49.

Lam, L. [2022]. Liquid crystal research at Tsinghua University in the 1970s and 1980s. Int. J. Mod. Phys. B (doi.org/10.1142/50217979222300067).

Lam, L. [2024]. *Humanities, Science, Scimat*. Singapore: World Scientific.

Lam, L., Freimuth, R. D. & Lakkaraju, H. S. [1991]. Fractal patterns in burned Hele-Shaw cells of liquid crystals and oils. Mol. Cryst. Liq. Cryst. **199**: 249-255.

Lam, L. & Lax, M. [1978]. Irreversible thermodynamics of thermoviscous solids with microstructures. Phys. Fluids **21**: 9-17.

Lam, L. & Morris, H. C. (eds.) [1989]. *Wave Phenomena*. New York: Springer.

Lam, L. & Prost, J. [1992]. *Solitons in Liquid Crystals*. New York: Springer.

Lam, L. & Shu, C. Q. [1992]. Solitons in shearing liquid crystals. *Solitons in Liquid Crystals*, Lam, L. & Prost, J. (eds.). New York: Springer. pp 51-109.

Lax, M. & Nelson, D. F. [1971]. Linear and nonlinear electrodynamics in elastic anisotropic dielectrics. Phys. Rev. B **4**: 3694-3731.

Levelut, A. M., Malthête, J. & Collet, A. [1986]. X-ray structural study of the mesophases of some cone-shaped molecules. J. Phys. (Paris) **47**: 351-357.

Lin, Lei (Lam, L) [1978]. Microscopic theory of first-order phase transitions in liquid crystals. Kexue Tongbao **23**: 715-718.

Lin, Lei [1979]. Nematic-isotropic transitions in liquid crystals. Phys. Rev. Lett. **43**: 1604-1607.

Lin, Lei [1980]. Critical properties of nematic-isotropic transition in liquid crystals. *Liquid Crystals*, Chandrasekhar, S. (ed.). London: Heyden. pp 355-360.

Lin, Lei [1982]. Liquid crystal phases and the "dimensionality" of molecules. Wuli (Physics) **11**: 171-178.

Lin, Lei [1983]. Liquid crystal research in China: 1970-1982. Mol. Cryst. Liq. Cryst. **91**: 77-91.

Lin, Lei [1987]. Bowlic liquid crystals. Mol. Cryst. Liq. Cryst. **146**: 41-54.

Lin, Lei, Liu Lin & Qiu Ju-Liang (trans.) [1985]. *Structure of Materials: From Blue Sky to Plastics*. Beijing: Science Press.

Lin, Lei, Shu Changqing, Shen Juelian, Lam, P. M. & Huang Yun [1982]. Soliton propagation in liquid crystals. Phys. Rev. Lett. **49**: 1335-1338; **52**: 2190(E).

Lin, Lei, Ye Pei-Xian & Zhou He-Tian [1980]. A brief note on the International Liquid Crystal Conference, 1979, Bangalore, India. Wuli Tongxun, No. 2, Supplement, 1980 (Institute of Physics, CAS, Beijing): 45-46.

Martin, P. C., Pershan, P. S. & Swift, J. [1970]. New elastic-hydrodynamic theory of liquid crystals. Phys. Rev. Lett. **25**: 844-848.

Mehta, G. & Uma, R. [1998]. Oxa-bowls: Formation of exceptionally stable diozonides with novel, $C-H\cdots O$ hydrogen bond directed, solid state architecture. Chem. Commun. (1998), pp 1735-1736.

Morrison, D. (ed.) [1989]. *Massacre in Beijing: China's Struggle for Democracy*. New York: Warner Books.

Nelson, D. F. [1979]. *Electrical, Optic, and Acoustic Interactions in Dielectrics*. New York: Wiley.

Palffy-Muhoray, P., Lee, M. A. & Petschek, R. G. [1988]. Ferroelectric nematic liquid crystals: Realizability and molecular constraints. Phys. Rev. Lett. **60**: 2303-2306.

Pan, Ru-Pin, Sheu Chia-Rong & Lam, L. [1995] Dielectric breakdown patterns in thin layers of oils. Chaos Solitons Fractals **6**: 495-509.

Papanicolaou, N. I., Bacalis, N. C. & Papaconstantopoulos, D. A. [1991], *Handbook of Calculated Electron Momentum Distributions, Compton Profile, and X-ray Form Factors of Elemental Solids*. Boston: CRC Press.

Ribotta, R., Joets, A. & Lin Lei [1986]. Oblique roll instability in an electroconvective anisotropic fluid. Phys. Rev. Lett. **56**: 1595-1597; **56**: 2335 (E).

Shu, Chang-Qing & Lin Lei [1984a]. Theory of homologous liquid crystals. I. Phase diagrams and the even-odd effect. Mol. Cryst. Liq. Cryst. **112**: 213-231.

Shu Chang-Qing & Lin Lei [1984b]. Theory of homologous liquid crystals. II. Orientation correlation functions. Mol. Cryst. Liq. Cryst. **112**: 233-264.

Xiong, Wei-Min [2013]. China's participation in international science: The case of biochemistry, 1949-1982. Science & Culture Review **10**(2): 50-72.

Xu, B. & Swager, T. M. [1993]. Rigid bowlic liquid crystals based on tungsten-oxo CaliM4larenes: Host-guest effects and head-to-tail organization. J. Am. Chem. Soc. **115**: 1159-1160.

Zeng, Er-Man [2001]. *Design, Synthesis and Characterization of Columnar Discotic and Bowlic Liquid Crystals*, PhD thesis. Atlanta: Georgia Institute of Technology.

Published: Lam, L. [2014]. The Founding of the International Liquid Crystal Society. *All About Science: Philosophy, history, sociology & communication*, Burguete, M. & Lam, L. (eds.). Singapore: World Scientific. pp 209-240.

Hawking and His Legacy

Lui Lam

This is a two-hour long lecture I gave at the Department of the History of Science, Tsinghua University, Beijing, May 25, 2018—72 days after the death of Stephen Hawking on March 14, 2018. Stephen Hawking (1942-2018), chair professor at Cambridge University, astrophysicist, film and television actor, and popular-science writer. This lecture will review, analyze, and comment on Hawking's professional achievements and personal life, explain the personal, organizational, and social factors behind the so-called "Hawking phenomenon," and raise some basic questions related to the history of science and science communication. Hype is a central issue to be discussed in this lecture. This lecture consists of nine parts. 1: Introduction; 2: Academic; 3: Personal Life; 4: Science Popularization; 5: Film and Television; 6: Science Idol; 7: After Death; 8: History of Science and Science Communication; 9: Conclusion. This lecture breaks the convention of only focusing on the research achievements of celebrities by regarding Hawking as an ordinary person, which makes Hawking as a character more three-dimensional and comprehensive. Three years after this talk, the book *Hawking Hawking: The Selling of a Scientific Celebrity* (2021) was published, which agrees with our conclusion that Hawking is a master of self-promotion and hype.

1 Introduction

There are two kinds of legacy: One is abstract; the other is material. For example, one of the things Cai Yuan-Pei (1868-1940) left behind is the spirit of Peking University (Beida for short). The Beida spirit is an abstract legacy. To this day, the Beida spirit can still be seen in some Peking University students.

The material legacies are concrete and visible to the eyes. For example, before 1997, before the British government withdrew from Hong Kong, it built a new airport for Hong Kong—of course, with Hong Kong people's money. This new airport is also a legacy, which is concrete and visible. We'll talk about these two aspects concerning Hawking's legacy.

It's quite easy to find information about Hawking. You can easily find two timelines of Hawking by searching online. They differ slightly from each other; the differences can be ignored for our purpose here.

2 Academic

When we talk about academics, we don't care about the person's physical condition. There is only one standard, which is whether you did well or poorly. So while we're talking academics, please forget about Hawking's wheelchair. We just regard him as a scientist, and this is how we talk about his academic work (Table 1).

Table 1. Summary of Hawking's academic life: 1942-2018.

Year	Age	Event
1942	0⁺	Born
1960	18	BA, physics, Oxford Univ.
1963	21	Diagnosed with motor neuron disorder ALS
1966	24	PhD, Cambridge U., thesis "Properties of Expanding Universes"
1970	28	Published Penrose-Hawking singularity theorems
1973	31	Published *The Large Scale Structure of Space-Time* (with George Ellis)
1974	32	Published Bekenstein-Hawking radiation; started using wheelchair fulltime
1979	37	Lucasian Professor at Cambridge U.
1983	41	Published Hartle-Hawking State model
1985	43	Lost ability to speak; used machine to talk
1988	46	Published *A Brief History of Time*
1994	52	Published *Black Holes and Baby Universes and Other Essays*
2013	71	Published *My Brief History*
2018	76	Died

Hawking was born in 1942. He received a bachelor's degree from Oxford University at the age of 20 and a PhD from Cambridge University at the age of 24. His important scientific works were respectively published in 1970, 1974, and 1983. (We will discuss each of them in detail later.) At the age of 37, he became the Lucas Chair Professor at the University of Cambridge. This is just an honor; it is not very important. I will introduce this to you later. He just passed away at the age of 76.

Hawking's grades in middle school were not very good—just above average, and his grades at Oxford University was also not very good. When he graduated, his grades were between second-class honors and first-class honors. Why was this important to him? Because at that time he already knew that he wanted to study PhD in physics at Cambridge University. Cambridge University only accepts first-class honors graduates.

In fact, his first-class honor was the result of "begging." Before graduation, his grades were not very high, so he was banking on interviews. He went to the examination room and told the professors— there were only about 2 or 3 professors. He said: "If you give me second-class honors, I will stay at Oxford University to study for a doctorate; if you give me first-class honors, I will go to Cambridge to study for a PhD. Please give me first-class honors." In the end, they gave him first-class honors.

As we all know, in terms of physics, Cambridge is much higher than Oxford. This story is very important. This person is always on the edge between second class and first class.

Let's go a little further down the Table and look at the so-called important works: the first one done at 28 years old, the second one at 32 years old, and the third one at 41 years old. But when he became known to the world, it was because of this book—*A Brief History of Time*, at 46 years old. So these three works were completed when he was relatively young.

He couldn't speak when he was 43 years old, so he started using the machine. In the previous period, he had lived a relatively normal life.

2.1 Hawking's three major works

1. <u>Penrose-Hawking singularity theorems</u>

His first work was named after two people: the Penrose-Hawking singularity theorems. But these theorem are useless—he said it himself.

That history is like this. In 1939, Oppenheimer (1904-1967) and his students proved, based on Einstein's general theory of relativity—not a proof, but a deduction, saying that there are black holes. In 1965, Roger Penrose used mathematics to prove that black holes are singularities. Well, five years later, Hawking and Penrose jointly proved that big bang is also a singularity.

Well, it sounds great, but it's useless. Because it's like if you see a horse lying on the road and you are quite sure the horse is dead, this guy comes over and says: Well, I will prove to you that the horse is dead. You can't use a dead horse, right? Whether you can prove it to me or not, I know it is a dead horse.

That's what the singularity theorems are like—it's useless. Hawking later said in his book *A Brief History of Time*: These theorems has no practical significance. Why they're useless?

He said: Because of quantum mechanics, there is actually no singularity at the beginning of the universe. Different from classical mechanics, quantum mechanics believes that the world is basically fuzzy, and there is an uncertainty theorem and so on. Therefore, in this universe governed by quantum mechanics, there cannot be a clear (singular) point.

So, Hawking's first "important" work was *useless*. However, Penrose is a very good researcher.[1] He is from Oxford University.

2. Bekenstein-Hawking radiation

This second work is what made him famous and should be the most important to him. Nowadays, most people call it "Hawking radiation," but that's not right. There should be two people's names, and it should be called "Bekenstein-Hawking radiation." If you go to Russia, the first name will be changed to another person's name. This makes sense. In any case, it definitely cannot be called "Hawking radiation." Let me first tell you what this radiation is.

For a black hole, nothing can come out if you throw things in, including light. If you accidentally fall in, you won't be able to listen to this lecture.

At the edge of this black hole is a quantum world. According to physics and quantum mechanics, this edge is a vacuum, and it looks like there is nothing there. But in fact, this vacuum can spontaneously produce a pair of particles, such as a pair of electrons and positrons. Anyway, one pair. One pair is produced and then combines, and another pair produced and combines again. It happens so fast that you can't detect it at all. Everyone agrees on this, even those who study quantum mechanics agree: When a pair of particles is produced and one of the particles is sucked in by the gravitational force of the black hole, the other will run out.

Why did it run outside? Because it has to obey a conservation law, called conservation of momentum. For example, if there is nothing originally, and two particles appear at once, they will inevitably go in opposite directions, because the total momentum added up must remain zero. So, if one is sucked in, the other will inevitably go out. What goes outside is the so-called Bekenstein-Hawking radiation of this black hole. Therefore, the energy of this black hole will become smaller and smaller, and eventually it may completely evaporate. Okay, so where did this idea come from?

First there is Jacob Bekenstein (1947-2015). This man was born in Mexico, and his family immigrated to the United States and lived in Texas. After that, he went to Princeton University to study with a famous teacher named John Wheeler (1911-2008), the doctoral supervisor of Richard Feynman (1918-1988). The doctoral work done by Bekenstein with Wheeler was to give the black hole a so-called entropy—an entropy similar to that in the second law of thermodynamics.

Anyway, in 1972, Bekenstein defined an entropy for a black hole and proved that this entropy is finite (and proportional to the "surface area" of the black hole, Fig. 1). This step is very important. Because of this work, he created a branch of physics called Black Hole Thermodynamics.

The next year (1973), Hawking went to Moscow to meet Zel'dovich and Starobinsky. Yakov Zel'dovich is important; he has many famous works. It was these two people who told Hawking that a rotating black hole would produce pairs of particles, according to quantum mechanics. After Hawking received this idea, he went home, worked on it, and wrote an article. That's why in Russia they call this radiation Zel'dovich-Hawking radiation.

$$S_{BH} = \frac{k_B \, \text{Area} \, c^3}{4G_N \hbar}$$

S_{BH} = entropy of black hole (thermodynamics)
Area = area of black hole
k_B = Boltzmann constant (thermodynamics)
c = velocity of light
G_N = gravitational constant (general relativity)
$\hbar$ = Planck constant/2π (quantum mechanics)

Fig. 1. The entropy formula of Bekenstein-Hawking radiation. "Stephen wanted this equation inscribed on his gravestone," Andrew Strominger (*Physics Today*, Mar. 14, 2018.)

Therefore, none of Hawking's ideas were original. Is there anything original about him? Yes, I'll tell you what's original about him in a moment; wait a minute, very soon.

Finally, his result (equation in Fig. 1) is based on Bekenstein's work, plus Zel'dovich's idea. The interesting thing about this equation is: This k_B is called Boltzmann constant, which is something in thermodynamics; the entropy S_{BH} is something in thermodynamics (the subscript BH stands for Bekenstein-Hawking); this $\hbar$ (Planck's constant/2π) is something from quantum mechanics; the G_N is the universal gravitational constant, something from general relativity. So this equation contains three aspects: thermodynamics, quantum mechanics, and general relativity.

Hawking is very proud of this equation. According to Hawking's student Andrew Strominger's writing (written on the day of Hawking's death), Hawking wanted this equation to be engraved on his tombstone. I don't know if there is such an engraving because I haven't seen it yet.[2]

In fact, although the S_{BH} equation represents Hawking's main work, the contribution of Jacob Bekenstein is as important as his.

3. Hartle-Hawking state model for universe

What about the third work? No one believed in it.

The third work was in 1983, which he did in cooperation with James Hartle (1939-2023). They proposed a no-boundary wavefunction of the entire universe; the work is called "no-boundary universe": Our universe has no borders.

After that, from that 1983 year until his death—this year, he continued to revise this theory. One moment he said there was a boundary, and the next he said there was no boundary. Anyway, no one believed in his third work.

Even if it is believed, it does not mean it is correct; it just means a "belief" (like belief in faith). If you want to be correct in the physics world, you need experimental support and data. If you don't have data, it's all just talk.

2.2 Academic competency

Let's take a look at his academic competency. What is Hawking's competency as a scientist?

As I said just now, every work he had done was initiated by someone else first. But in terms of *methodology*, he had a pioneering work: the radiation theory just mentioned. He put two theories together—general relativity and quantum mechanics. So why don't others do the same?

Because, if you were in your right mind, you wouldn't do it. The reason is that everyone knows that general relativity is *not* consistent with quantum mechanics. This situation is like building a house: If the house is built on a foundation, and half of the foundation is made of stone and the other half is made of wood, will the house be stable? Will it be durable? Unstable and not durable, right? Or try placing a wooden board on two pillars of different heights, and then place something on the wooden board—a glass of water, say. Will this thing be stable? Doesn't it slide? No way, right? Constructing the black hole theory on two inconsistent theories, just like building a house on an unreasonable foundation or placing things on a sloping board as mentioned above, the result will be unbelievable, ineffective, and may even be completely wrong. So, no one in their right mind would do it; Einstein wouldn't do it; Pauli wouldn't do it.

Einstein would first find a way to integrate these two things before doing so. Hawking did not; he knew that these two things are not consistent, but he still did it. What you build there may turn out to be completely wrong. People will know in the future; because there is no experimental verification, we don't know now and will only know in the future.

So let's take a look at Hawking's (academic) competency. Hawking often did not wait for the theory to be completed and made bets with others before the decisive results were obtained. This habit or hobby of his is well known to everyone because he or his friends propagated it.

He advocated certain things (in physics) and made bets (on future results) with others (Table 2). I don't know how long it took for the 1st one to be reversed; for the 2nd one, it took 28 years. Sometimes it was shorter; this 3rd one was even shorter, about 4 or 5 years. After a while, when the conclusion came out, he gave up and admitted that he had made a mistake.

Table 2. Three cases of Hawking's bet that he lost.

Year proposed	Proposal	Year reversed
?	About naked singularities (bet with John Preskill)	1997
1976/1981?	Black holes lose information (bet with John Preskill in 1997)	2004
2002/2008	Higgs boson would never be found (bet with Peter Higgs)	2012

It's a bet. If you lose, you have to pay something—a bottle of wine, etc. This 2nd one took a long time from 1976 to 2004. It took 28 years for him to admit defeat. It was about whether black holes would lose information. I will tell you the details later soon. In short, this guy is often wrong.

Was Einstein ever wrong? We know Einstein was wrong (at least) once, not publicly. He submitted a manuscript to *Physical Review*, and the referee found that he was wrong. Originally only the persons involved knew about this, but later the information was made public and we learned that he was wrong once.

Therefore, any scientist may make mistakes, but the one who has made mistakes so many times and still remains untroubled or unashamed is probably Hawking. Let's look at this example, one of his mistakes.

We know this example because there is a book, *The Black Hole War: My Battle with Stephen Hawking to Make the World Safe for Quantum Mechanics* (2009), written by Leonard Susskind. This man is very famous; he is a physics professor at Stanford University. He has written several popsci books in succession. This book tells a story about the debate between him and Hawking about black holes. In the end, he won and Hawking lost, thus saving the world from being harmed by quantum mechanics. That's it.

How did that come about? Here is the story according to Don Page in his review of *The Black Hole War* published in *Physics Today* (May 2009). Note that Page was Hawking's student, and the student usually will not accuse his teacher unfairly.

According to Page, in 1974, Hawking proposed his black hole radiation theory, and then he went on to say that if something is thrown into a black hole, and everything carries information, and this thing cannot come out after being thrown in, then this information will be gone—permanently lost in the universe. If it really disappears, then it violates quantum mechanics, which says that the sum of all information is unchanged, and information will not disappear.

After that, many people argued about this because it involves the basics of quantum mechanics. Finally, Don Page raised his objections in 1980. He said that his teacher's mistake was to treat black holes as a classical thing—(one of) the inevitable consequences of mixing two inconsistent things, general relativity and quantum theory, as I mentioned earlier. As it turned out, it took 28 years for his teacher to admit that he was wrong—Hawking publicly admitted his mistake.

This is an example, and the entire book is about this matter.

I'll tell you later, why Hawking is like this, why he often bets with others, and why he often loses.

2.3 Academic integrity

When we talk about a scholar, the first thing to look at is his academic competency. Is he highly competent? The second is academic integrity. Is his academic integrity high or low?

Hawking had some (academic) mistakes that he was *unwilling* to admit. This is related to academic integrity. Below is an example related by someone directly involved, Alan Guth, in his popsci book *The Inflationary Universe* (1998).

Guth is qualified to win the Nobel Prize and is now at MIT. He started out doing high energy physics—elementary particles, but couldn't find a good, regular job. He was working as a postdoctoral fellow at Stanford University and was worried about his future. One day, someone went to Stanford to give a talk on Cosmology. This was not Guth's field, but he went to listen. After listening, Guth compared what he knew with what was said in the talk, and came up with his *Inflationary Universe* theory: Not long after the beginning of the universe, it suddenly expanded. That is called inflation, inflationary. As a result, it didn't take long for him to find a job and eventually became a chair professor at MIT.

The cosmic-background-radiation data supported his theory. Although his theory was later improved by others, he eventually became a famous professor at MIT. Guth is such an important person.

The first time he met Hawking was in 1982. The first time they met was because Hawking was holding a conference in Cambridge called "Very Early Universe." In his popsci book, he tells the whole story.

So he went to this meeting. But a month before leaving, Guth and his friend Paul Steinhardt carefully checked an article by Hawking—probably a preprint—and found an important error. He said that Hawking miscalculated one of the quantities, and the error was 10 to the fourth power (10^4); i.e., the number calculated by Hawking is 10,000 different from what it should be.

They arrived in Cambridge. Before the meeting, the two of them went to Stephen, to Hawking, and told him that you were wrong. Hawking refused to admit it and said I was right. Well, in the second week of the conference, it was Hawking's turn to speak at the conference. As it turned out, when talking about this part, Hawking pretended to be nonchalant and took the exact result that Guth told him as his own result. He did not say what result I got before but just told this number. He didn't say thank you so and so for telling me. I miscalculated. Hawking is a famous person. This is not someone who has no name, nor is he someone like you or me. He should have said, thanks to these two people for telling me, I have now revised it to this number, but he did not mention it. So, this is academic *misconduct*.

If no one notices and you correct your mistake without telling them, that's fine. But no, his is not this case. He didn't admit it! Well, this is one of the things that Hawking did not admit. There are many things that he did not admit.

But this is not the only occasion that Guth mentioned this story. On the day of Hawking's death (March 14, 2018), *Physics Today* organized eight people to write an article each. Seven people praised Hawking throughout their entire article. But Guth's article had six paragraphs and he spent the first five paragraphs talking about the Cambridge conference and the above story in detail, before he politely praised Hawking in the last, sixth paragraph. It can be seen that this incident is very important to him.

According to my guess, what he originally wanted to say, he should have said: Hawking is actually a *scumbag*. But that would make it difficult for him to write or he didn't dare to do so. In any case, maybe because he spoke more harshly in the first four paragraphs, he wanted to soften it in the fifth paragraph and "relieve" Hawking. He said: "It showed his willingness to change his opinions when he saw convincing reasons to do so." Of course, anyone who understands will know as soon as he reads this fifth paragraph that what Guth says is an understatement.

In short, for Guth, it is a very serious, very serious matter for Hawking to deny that he was wrong, both privately and publicly. This matter bothered him because it was important to him, as can be clearly seen that 20 years after he wrote about it in his book, he still mentioned it on the day of Hawking's death.

For me, not admitting mistakes is also a very important thing, and it is related to academic integrity. In our profession, if you are right, you are right, and if you are wrong, you are wrong.

It's wrong for you, Hawking, to *cheat*.

2.4 Academic style

So what about his academic style? I mean the style of study in doing research.

Philip Ball had commented on this. This Philip Ball is very famous. He was the editor of the great—the Chinese think it is great—magazine/journal, *Nature*. He served for a long time and later wrote a popular popsci book. After becoming famous, he left *Nature*. So he is now a famous author of popsci books, but he soon goes back and writes for *Nature* or other magazines.

This is his comment written two years ago, in 2016. He said that Hawking's style of study is: brave, smart, but not rigorous. He seemed to speak out based on his own intuition. In the end, it was proved to be not entirely correct.

So why did Hawking do things like this? You haven't seen Einstein did this, Feynman did this, or anyone—a scientist in any profession—did this. Why did Hawking do what he did? Answer: This is related to personal character.

A person's character can be seen from an early age. It is said that "a person's personality is determined at the age of three." Of course we don't know what Hawking was doing when he was three years old, but we know what he was doing at Oxford University. At Oxford University, he didn't have many friends when he first entered, and so later he joined the rowing team in the university. In the rowing team, his job was to steer the boat; i.e., he did not row (the boat) but steered it sitting at the stern of the boat.

His style (at the helm) at that time was reckless. Some people also scolded him and called him "brave" like the devil. For example, there were two pillars in the water ahead of the boat. He thought that the boat could pass through the slit in the middle, so he steered the boat towards those two pillars—the helmsman decides the direction of the boat. As a result, the boat passed between the two pillars, but the oars on both sides of the boat were damaged. This is his life-long decision-making style: *reckless*.

So, based on what I just said, my evaluation or summary is: Hawking is a "Trump-type" scientist. What is the Trump type? (1) It's just capricious: one moment I want to sanction this, the next moment I don't sanction it; one moment I want to talk to that fat man, and the next moment I won't talk to him again. (2) Not thoughtful. (3) Don't necessarily admit that you are wrong.

Of course, Trump is worse than him. Trump has made 10 mistakes, say, but he doesn't admit 9 and a half of them. As for Hawking, he has made 10 mistakes, and he admits about 8 of them. This is their difference—a quantitative difference. But when it comes to academic style, Hawking is totally Trumpian.

Later, due to health problems, Hawking did not know how many years or months he could live. He felt that life was short, so he became even more hasty and less thoughtful in learning and doing things. But it is essentially driven by personal character. As he showed when he was at the helm of a rowing boat as an undergraduate at Oxford, he was not a calm person to begin with. When these two factors—short life and reckless personality—are superimposed, we see that he frequently guessed the answer before he knew how to complete the theory or completed it, and bet with others based on the unfinished theory. After all, if you guess correctly, you will be remembered in history; if you guess wrong, you only lose a bottle of red wine.

So, why can a Trump-type scientist like him come to Cambridge University and become a chair professor there? There should be a reason for that. Right? OK, I'll tell you right away.

2.5 Lucasian Chair of Mathematics

Let's first analyze his Chair—Lucas' mathematics chair.

Mathematics is a term that has existed since ancient times. In Newton's times, the term theoretical physics had not yet appeared, but this does not mean that there was no work that was later called theoretical physics. Newton's three laws of mechanics and universal gravitation fall into this category.

The Lucasian Chair of Mathematics at the University of Cambridge was established in 1663 with a donation from Henry Lucas. The word mathematics in the Chair actually represents two aspects: (pure) mathematics and theoretical physics. This Chair is not something that ordinary professors can sit on. So far, only 19 people have sat on that Chair (Table 3).

Table 3. A partial list of occupants of the Lucasian Chair of Mathematics at Cambridge University.

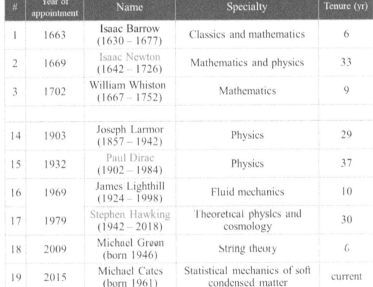

#	Year of appointment	Name	Specialty	Tenure (yr)
1	1663	Isaac Barrow (1630 – 1677)	Classics and mathematics	6
2	1669	Isaac Newton (1642 – 1726)	Mathematics and physics	33
3	1702	William Whiston (1667 – 1752)	Mathematics	9
14	1903	Joseph Larmor (1857 – 1942)	Physics	29
15	1932	Paul Dirac (1902 – 1984)	Physics	37
16	1969	James Lighthill (1924 – 1998)	Fluid mechanics	10
17	1979	Stephen Hawking (1942 – 2018)	Theoretical physics and cosmology	30
18	2009	Michael Green (born 1946)	String theory	6
19	2015	Michael Cates (born 1961)	Statistical mechanics of soft condensed matter	current

You probably haven't heard of the first person. The second one is Newton, the one we all know. You haven't heard of these other people; we physics people know this one (Paul Dirac).

After 1901, there was only one person sitting there who was truly famous—Paul Dirac (1902-1984, Nobel Prize in Physics 1933). He won the Nobel Prize very early and very soon. There is an equation in quantum mechanics called the Dirac Equation. Others (in Table 3), if you are doing physics, you may have heard of these names, but if you are not doing physics, you will not know these names at all.

Therefore, if you read the newspaper, it will mislead you. The newspaper will say: The Lucasian Chair in Cambridge is very important; then it will say: Newton did it, Dirac did it. I used to think that these other people were also exceptionally outstanding. When I checked, only these two people were known to everyone.

By the way, this Chair also has a stipulation that you must retire at the age of 67; that's why Hawking did step down at the age of 67. Hawking has occupied it for 30 years, and Newton has occupied it for 33 years. Dirac has occupied it for a long time. Why is Dirac so long? Because he became famous early. He became famous in his 20s and was about 30 when he won the Nobel Prize.

Therefore, this Lucas' Chair is *not* as valuable as everyone imagined.

Moreover, Table 3 also reveals a result that some people have not noticed: the decline of British physics.

2.6 Decline of British physics

To illustrate this, let's look at Table 4.

This Table contains people who have won the Nobel Prize. I use this Table because the physics Nobel Prize information is easy to look up. In fact, everyone in the Table is British, at least with a British passport, which includes Rayleigh and so on.

Pay attention to the awards of these few people—Rayleigh, Thompson, and so on—in the early 1900s, soon after the Nobel Prize started in 1901. Well, all the names in black are those who do experiments, and those in red are those who do theory, but Thompson did both experiments and theory. Dirac did theory, purely theoretical.

Table 4. List of Nobel laureates in physics from the United Kingdom: 1904-2016.

Year	Nobel Physics Recipient
1904	Lord Rayleigh
1906	Joseph John Thompson
1915	William Henry Bragg & William Lawrence Bragg
1917	Charles Glover Barkla
1927	Charles Thomson Rees Wilson
1928	Owen Williams Richardson
1933	Paul Dirac
1935	James Chadwick
1937	George Paget Thomson
1947	Edward Victor Appleton
1948	Patrick Maynard Stuart Blackett
1950	Cecil Frank Powell
1951	John Douglas Cockcroft
1971	Denis Gabor
1974	Martin Ryle & Antony Hewish
1977	Nevill Francis Mott (1905-1996)
2003	Anthony James Leggett
2009	Charles K. Kao
2010	Andre Geim & Konstantin Novoselov
2013	Peter Higgs
2016	David Thouless, Duncan Haldane & Michael Kosterlitz

Red = theorist

1979 Hawking appointed to Lucasian Chair

Dirac is a very interesting person. If you are into history of science, go buy his biography, which was published a few years ago (*The Strangest Man: The Hidden Life of Paul Dirac, Mystic of the Atom*, 2011). Contrary to Hawking, Dirac is a very quiet person and very thoughtful. He didn't take in (supervise) students. Isn't that awesome?

Some of the experimental people on the list are very important. These others all won the Nobel Prize in the 1940s and 1930s, and up to the 1950s. After that, there are *no* Nobel Prize in the 1960s. By 1977, this guy (Mott) won (the Nobel Prize). In fact, Mott is very important in condensed matter physics. He was a theorist at University of Bristol. He was born in 1905 and was already 72 years old when he won the Nobel Prize. At the age of 72, he cannot be placed in the Lucas (chair professor position). He was too old (over 67).

Then, there was no Nobel Prize for 26 years, not until 2003; this man (Leggett) had a British passport, these two (Thouless and Kosterlitz) also had British passports, and this one (Higgs) is still in the UK. So why aren't these four people assumed the Lucasian Chair? Because when Hawking took that position in 1979, these people had not yet won the Nobel Prize. Later, the first three people had immigrated to the United States.

I received Leggett when I was working in Beijing. At that time he had not won the Nobel Prize or immigrated to the United States. I was in Beijing from 1978 to 1983. Probably in 1979 or 1980, he came to visit the Institute of Physics, and I participated in the reception. I went to Xiangshan (Fragrant Mountain) with him. A year later, I went to England and stayed at his house for one night. Not long after that, he went to University of Illinois in the United States, as an immigrant and stayed. Later he won the Nobel Prize there, and so UK couldn't invite him back —couldn't get him back.

Thouless and Kosterlitz are also in the United States, but they are not in very top schools. Higgs has a fortunate life; i.e., he lives a long time. In 2013 the Higgs boson was confirmed in the laboratory (when Higgs was aged 84). Geim and Novoselov are experimentalists and came from Russia. They are now at the University of Manchester, the city where the football team is said to be very good. Charles Kao is my compatriot from Hong Kong and was trained in the UK. He holds a British passport, so I include him in the Table.

So, in 1979 (ten years after Paul Dirac stepped down), Cambridge University had to find someone to fill the Chair again. They probably wanted a theorist and couldn't find one. The Nobel laureates are either experimentalists, too old, or had immigrated away. Well, Hawking was chosen to sit on it. I didn't say there was a problem with him sitting on it. I'm just saying that this Chair is not that valuable.

2.7 Comparison with Newton and Einstein

Finally, I will tell you what's my evaluation of his academic achievements—an overall review. How does he compare with Newton and Einstein?

Sometimes, when you see him giving talks, he often claimed (academically speaking) to be the heir to Einstein. Ten years after his book *A Brief History of Time* was published in 1988, the president in the White

House was Bill Clinton, and his wife was Hillary. They invited him to give a lecture; there was a lecture series at the White House. You can find that video online. (On the Internet, you search for White House Lecture, Hawking.)

White House Lecture, 1998, is a video. At the beginning of that lecture video—it was a short video, I don't know if it was taken from one of his movies, but it was played by actors: One played Newton, the other played Einstein, and he was there in a wheelchair.

One of his lectures was even more exaggerated. I can't find it yet. I saw it on American TV. The first ppt of his lecture: God, Newton, Hawking (not even Einstein). Can he compare with Newton and Einstein?

Newton, graduated as an undergraduate from Cambridge University, three years older than Hawking did (Table 5). Within two years of graduating with an undergraduate degree, he did everything: mathematics (calculus); physics (three laws of motion, gravity, optics).

Table 5. Comparing Hawking with Newton and Einstein.

Einstein	Newton	Age	Hawking
		20	BA, physics, Oxford U.
Academic Diploma at ETH		21	
	BA, Cambridge U.	23	
	Developed Calculus, Laws of Motion, Gravitational Law, Optics	23-25	
		24	PhD, Cambridge U.
PhD, University of Zurich; 4 breakthrough papers (Brownian motion, Photoelectric Effect, Special Relativity, $E = mc^2$)		26	
	Lucasian Professor at Cambridge U.	27	
		28	Published the Penrose-Hawking singularity theorems
		32	Published "Bekenstin-Hawking radiation"
General Relativity		36	
		37	Lucasian Professor at Cambridge U.
		41	Published the Hartle-Hawking State model
Nobel Prize		42	

In comparison, Hawking got his doctorate at the age of 24 while Newton didn't even need a PhD. Newton already assumed the position of Lucas chair professor at the age of 27; Hawking assumed that at the age of 37.

As for Einstein, he got his undergraduate degree at the age of 21 and his PhD at the age of 26. At the age of 26, he published four important articles (Brownian motion, photoelectric effect, special relativity, $E = mc^2$), each of which could win a Nobel Prize. Ten years later, the general theory of relativity can also win the Nobel Prize. It's a pity that he only got it once, at age 42; i.e., his works deserve five Nobel Prizes. Newton also has no problem; he can get at least two Nobel Prizes. But Hawking couldn't even get one, because his works couldn't be confirmed; or, not confirmed so far.

Well, Hawking's first significant work—the singularity theorems—has to wait until he was 28, and they are useless, as he said.

The Bekenstein-Hawking radiation is the only work worth mentioning about him; he was 32 years old then. There is no way to confirm it now, and it may be impossible to self-consistently establish it because it is based on two inconsistent theories. So, how could he get a Nobel Prize?

But that is not important. This is important: If you say you are the heir to Einstein while there are so many others who are more qualify than you, that's an overclaim or self-boasting. He is ranked 101, in my opinion, and he shouldn't even mention it.

I didn't say that Hawking was not good at physics. But his physics is not rigorous and he often reversed himself; if any university is willing to give him a Chair professorship, this is probably not a problem. He in fact is a genius in some aspects, such as hype.

3 Personal Life

Let's talk about his personal life. This part is *not* about gossip, and personal life is very important. If you want to do history of science about a scientist, you need to understand the persons involved with him, because history of science, like any other human activity, is made by people. You need to understand how he chose his topic, who he collaborated with, and why he took 40 or 20 years to publish it. It's all about people.

3.1 Life history

Let's take a look at his personal life (Table 6). I'll tell you what's important.

Table 6. Hawking's personal life—a summary.

Year	Age	Event
1942	0⁺	Born January 8
1960	18	BA, physics, Oxford Univ.
1963	21	Diagnosed with motor neuron disorder ALS
1965	23	1st marriage (Jane Wilde)
1966	24	PhD, Cambridge University
1967	25	First son born
1970	28	Daughter born
1974	32	Started using wheelchair fulltime
1979	37	Second son born; Lucasian Professor at Cambridge University
1985	43	Lost ability to speak; used machine to talk
1988	46	Published *A Brief History of Time* (became rich)
1995	53	1st divorce; 2nd marriage (Elaine Mason)
2006	64	2nd divorce
2018	76	Died March 14

Marginal notes:
1989 Hire Elaine as nurse
1990 Hawking departed home
1992 Shing-Tung Yau visited Hawking 2nd time

Hawking was born in 1942. He was diagnosed with ALS at the age of 21. This disease basically means that you gradually lose control of your muscles. Generally speaking, you can only live for 2 years. His girlfriend Jane Wilde was very brave. After waiting for 2 years, she found out that he was not dead and married him. She was not taboo about this. Even if she knew about his conditions, she would still follow him. Both sides understood this very well.

Hawking got married at the age of 23, and after that he got his PhD. One year after postdoc, his first son was born; three years later, his only daughter was born. After that, he started to use a wheelchair at the age of 32. Before that, the doctor said that he did not need to use a wheelchair, so he used crutches to support himself. Later, why did he need this wheelchair? Because his condition was getting more and more serious. But later, he (37 years old) had a second son. In the same year, he became the Lucasian professor.

Since the age of 43, he had been unable to speak and had to use a machine to speak. The reason why he couldn't speak was because during a visit to Geneva, he developed inflammation and was sent to the hospital where a hole was cut and he needed to be intubated. He would not be able to speak after that, but he could still speak before. Later, Caltech shipped him a computer with a synthesized voice capacity (when he keyed in the words). From then on, he started speaking with a machine. He was 43 years old then.

Note that by the time he was 43, his three important works had been published. He also became famous because of the Bekenstein-Hawking radiation (at age 32), and he already sat on that Lucasian Chair (at age 37). What happened after that is a tragedy, a British tragedy.

Do you think this can happen in China? A so-called "genius" professor, who is in a wheelchair, needs a nurse. And his university will not provide him with a nurse? Don't say just one, they'll give him three nurses. It's not like Cambridge didn't have money, but Cambridge didn't give him a nurse. He didn't have a nurse. In the early days, he had no nurse, so he relied on his wife to take care of him. The wife already had three children to take care of; in addition, he relied on her to take care of him. I'll tell you more, slowly.

Let me mention this important thing first: In 1988, he published that brief book *A Brief History of Time* and suddenly became rich. As soon as he became rich, problems began. Why did he publish this book? Because he needed money. He suddenly became famous because of the book, just like that (universe) inflation. If a man has money, he will make the same mistakes that other men often make—having mistresses.

Hawking was confined to a wheelchair. If he wanted to play with women, he could only play with the ones next to him. Who's next to him? That's the nurse. He started looking for a nurse. He became rich in 1988 and so, immediately the next year, Elaine Mason was hired as a nurse, not the only nurse, one of them. A year later, he had a fight with his wife and left home. Where did he go after leaving home? He probably went to Elaine's house.

Well, he married his first wife, Jane, in 1965, divorced 30 years later in 1995, and immediately married again to this nurse, Elaine, which lasted 11 years, and divorced (the second time).

During this period of time, in 1992, three years before the first divorce, the famous Harvard mathematician, Shing-Tung Yau (丘成桐, pinyin: Qiu Cheng-Tong), went to meet him. I'll tell you later, the story told by Yau.

3.2 Wife abuse

"Stephen Hawking is a misogynist; and also, quite possibly, a narcissist." This sentence, or judgment, does not come from me; it comes from Tana Gold.

First, Hawking wrote a popsci science book about cosmology called *The Theory of Everything* (2002), and then there was a biographical movie about him with the same title (2014). After the movie came out, Tana Gold commented on the movie and uttered those words (spectator.co.uk, Jan. 10, 2015).

After the first marriage in 1965, Hawking moved forward diligently and got a doctorate within one year (1966). His (first) wife, Jane, received her PhD from the University of London in 1981. She studied medieval Spanish poetry. The divorce followed in 1995. The same year later, Hawking married nurse Elaine. After his second marriage, his second wife Elaine stopped letting the first wife to see him. Therefore, Hawking and Jane are basically separated. The children lived with Jane; they can go see their father.

Four years after the (first) divorce, his first wife Jane wrote an autobiography: *Music to Move the Stars*. In the introduction to this book from amazon.com, it says: "The collapse of the high profile Hawking marriage, provoked by Stephen's affair with a nurse, is related in honest detail."

Hawking divorced nurse Elaine in 2006, his second divorce. A year later, Jane revised her book and changed its title to *Travelling to Infinity*. Gold said in her film review that Infinity means Divorce. So the first version (or edition) was published a little earlier (1999). She was angry because she was divorced just four years ago.

1. Tana Gold's film review

Here is Tana Gold's film review:

What *The Theory of Everything* doesn't tell you about Stephen Hawking

Based on his first wife's memoir, the film refuses to tell her complicated and disturbing story

… Jane knew Hawking might not live long when they married in 1965. The original prognosis was two years. Even so, they made a home, they travelled to conferences abroad, they had three children. She abandoned her scholarly ambitions — the medieval lyric poetry of the Iberian peninsula, if you care, and he didn't — to support his.

Her sacrifice deserves thanks, but no thanks came; when he became the youngest fellow of the Royal Society at 32, he made a speech, but he did not mention his wife. And why would he? She had become "chauffeur, nurse, valet, cup-bearer, and interpreter, as well as companion wife"; that common ghost that haunts university cities—"a physics widow." (Jane notes that Albert Einstein's first wife, Mileva, named "physics" as the co-respondent in her divorce proceedings.) …

The cruellest thing was his refusal to discuss his illness. "It was," she writes, 'the very lack of communication that was hardest to bear." He insisted on "a facade of normality"; yet if he could not acknowledge his own suffering—he "never" talked about the illness—how could he acknowledge hers? …

A genius Professor Hawking may be—what do I know of physics?—but he was, if you believe his wife, and I do—a very bad husband indeed. [spectator.co.uk, Jan. 10, 2015]

Therefore, the conclusion of Gold's review is: Hawking is a very bad husband and *abuses* his wife.

2. Shing-Tung Yau remembers Hawking

Shing-Tung Yau met Hawking twice, in 1978 and 1992. When Yau went to his home in 1992, he saw *six* nurses taking turns to take care of him, and they were jealous of each other—Yau said that. Why? Because Hawking did get rich in 1988. Hawking was no longer a man in a wheelchair, but a *rich* man in a wheelchair. Here is Yau's description of the two meetings:

> I invited Hawking to visit China twice in 2002 and 2006. I first met him in 1978. At that time, I was doing an important work on general relativity, solving the proof of the "Calabi Conjecture." At that time, scholars of general relativity did not believe that mathematicians had the ability to solve this problem. After learning about it, Hawking wrote a letter inviting me to explain the research. After hearing my ideas, he thought it was possible. When I went to see him, I saw that he was very happy and smiling. After chatting for eight hours, he said he wanted to treat me to some good food. He loved eating good, but he ate in a mess because he couldn't swallow it.

> When I went to Cambridge to see him for the second time in 1992, six nurses took turns taking care of him, and they were fighting for his favor. I invited Hawking to eat at his favorite restaurant. Within a quarter of an hour, he drove his electric wheelchair away and used the machine to make calls. His wife became more and more unhappy and ran over to start a quarrel. It turned out that Hawking was chatting with the nurse on the phone. His wife shed tears, but he was still smiling. [*Southern People Weekly*, Mar. 1, 2018]

So, according to Yau, recalled one week after Hawking's death, Hawking was chatting "secretly" with the nurse on the phone; his wife shed tears, but he was still smiling. Isn't this *wife abuse*?

Do all people who do physics abuse their wives? No, let me give you a counterexample.

Feynman, everyone knows. He lived for 70 years, and had cancer in the last 10 years. He fought cancer for 10 years and finally died. He had married three times, and the wife for the first time, Arline Greenbaum (1919-1945), was his first love—his high school classmate. He received his bachelor's degree from MIT and his PhD from Princeton University. He was born in 1918. At age 24 in 1942, he married Arline. Why not earlier? Because the scholarship given to him by Princeton stipulated that he could not get married before graduation.

Therefore, as soon as he graduated with his doctorate in 1942, he immediately married Arline. At that time, he already knew that the woman had lung disease and could not be cured. He knew she was going to die but he still married her. It was said that she could not live for more than 2 years due to the disease. Later, Arline lived for 3 years.

On the wedding day, as soon as the ceremony was over, he immediately took her to a hospital and went to the hospital to see her every weekend. It's not like seeing her every day. He had to work and can only see her on weekends. This story is the reverse of Hawking's story, with the man and woman reversed.

When he graduated with his PhD, he was asked to go to Los Alamos to work on the atomic bomb. He placed Arline in a sanatorium down the mountain because the place where the atomic bomb was made on the mountain was not accessible to families. The city where the sanatorium is located is called Albuquerque, and it takes about two hours to drive there.

So, he was now working on the atomic bomb—the Manhattan Project. Three years later, in 1945, he received the call that she was dying. He borrowed a colleague's car, drove two hours to Albuquerque, and stayed with her for a few hours until she died. That's love and loyalty! Feynman's story, this part of the story, was made into a movie in 1996 called *Infinity*. This word happens to be the same as the word in Jane's book (*Travelling to Infinity*).

When Feynman's wife Arline died, Feynman was 27 years old and Arline was 25. Later, about a year and four months after Arline passed away, Feynman wrote a letter which he sealed and said it could not be opened until after his death. He is dead now and the letter has been opened. That letter is a very intimate one (https://lettersofnote.com/ 2012/02/15/i-love-my-wife-my-wife-is-dead/).

3.3 Abused by wife

Hawking divorced first wife Jane in 1995 and immediately married Elaine, his nurse for six years. In 2006, after 11 years of marriage, Hawking and Elaine went to court for divorce. Why go to court? Fight for property. He was already rich at that time, and when divorcing, she must check how much money he had and how much should be divided between him and her. What was the marriage like before divorce? Here is a report[3] from *Daily Mail*:

> **Hawking's nurse reveals why she is not surprised his marriage is over**
>
> …[F]or years there have been shocking rumours of violence and abuse against the vulnerable scientist—mental as well as physical—supported by his own children no less…In 2000, detectives launched an inquiry after Prof Hawking made a number of visits to Addenbrooke's Hospital, Cambridge, suffering from cuts and bruises, and another inquiry was opened in 2003 after his daughter Lucy rang police. Prof Hawking declined to explain how his injuries had come about. A number of his former nurses… alleged that over the years his wife inflicted a catalogue of injuries on the vulnerable scientist: fractured his wrist by slamming it on to his wheelchair; humiliated him by refusing him access to a urine bottle, leaving him to wet himself; gashed his cheek with a razor, allowed him to slip beneath the water while in the bath, ensuring water entered the tracheotomy site in his throat; and left him alone in his garden during the hottest day of the year so long that he suffered heatstroke and severe sunburn…[dailymail.co.uk, Oct. 20, 2006]

So, Hawking was abused by his second wife after he abused his first wife. This is called retribution, or poetic justice.

3.4 Violent behavior

Hawking not only has violent tendencies but also violent behavior. The wheelchair was his weapon, which got bigger and heavier as time progressed.

As remarked by his old partner Roger Penrose: "He had been known to run his wheelchair over the foot of a student who caused him irritation." [theguardian.com, Mar. 14, 2018]

Therefore, never wear sports shoes to meet him; wear leather shoes. The evidence of Hawking's violent behavior is conclusive.

3.5 Children

Hawking had three children: an older boy, a girl, and a younger boy (Table 7). These three children are lucky. In fact, it is their mother who is a good mom: sentient, normal, useful—useful means the children are satisfied with her.

Table 7. Hawking's three children and their careers.

Year	Hawking's age	Event	Children's profession
1965	23	Married Jane Wilde	
1967	25	Timothy (son) born	Software engineer, Microsoft, Seattle
1970	28	Lucy (daughter) born	Journalist; children book author
1979	37	Robert (son) born	Account manager: Loyalty executive, Lego Group

The elder son Timothy was born when Hawking was 25 years old. When he was a child, he wanted to learn physics from his dad. Later, when he entered college, he changed his mind and studied software. Finally, he worked for Microsoft and lived in Seattle.

The second child is daughter Lucy. She is a journalist who writes, and she eventually co-wrote 5 children's books with her father (see Sec. 4).

The third one, a son named Robert, was born when Hawking was 37 years old. This Robert is the most low-key. He is an account manager and works for the Lego Group.

So, you may ask, why can people in wheelchairs give birth to children?

If you want to know this secret, watch the movie *The Sessions* (2012), which is based on a true story of a paralyzed student who earns a MA in poetry from University of California at Berkeley—a pretty good movie. This movie is like this: This person can't move from the neck down and usually have to live in an iron cabinet within which the air pressure can be controlled so that he can breathe. He could breathe after leaving the iron cabinet, but probably not for too long.

When he was in his 30s, he discovered that he was still a virgin. As a result, he hired a woman. She was not a prostitute, but a first-class professional who specialized in solving such problems for people. She did it to him (on top), and he got what he wanted. Later, he wrote a poem to her, and the two of them found that they fell in love with each other, which seemed to be the case. If you don't watch this, you don't know that this kind of thing exists, and you don't know why Hawking can have his own children, right?

This master's-degree poet is not as great as Hawking, nor did he live as long as Hawking. He may have died in his 30s or 40s. Then, what's so great about Hawking? I'll tell you later.

4 Science Popularization

Let's talk about popular science (popsci). Hawking has a lot of popsci books (Fig. 2). The first one was published in 1988, *A Brief History of Time*. As for the later ones, you've probably never heard of them.

4.1 *A Brief History of Time*

A Brief History of Time was bought by many people. Why? According to Chris Impey, in his review of another Hawking's book *The Universe in a Nutshell*:

> Thirteen years ago, Stephen Hawking turned the publishing world on its head with *A Brief History of Time*. Written in part to help pay for his round-the-clock nursing care, the book sold more than 10 million copies and has been translated into 35 languages. Despite

its phenomenal success, *A Brief History of Time* is an uncompromising book, filled with difficult concepts, uninterrupted by diagrams or pictures, and probably bought by more aunts and uncles (and unread by more nephews and nieces) than any other book in history. Hawking himself has acknowledged that many people probably did not finish or understand it.

Beyond his reputation as a theoretical physicist, Hawking has a second component to his success. *A Brief History of Time* marked his elevation into the public consciousness as an icon of science. Heir to Newton and Einstein, and afflicted by a degenerative disease, Hawking represents the struggle of a brilliant mind trapped in a wasting body. His personal tragedy sharpens the metaphor of science in which humans transcend their ephemeral status by trying to comprehend a vast and ancient universe. [*Physics Today*, April 2002]

Fig. 2. List of popular-science books by Hawking.

So, Hawking wrote this book because he needed money and the book is poorly written. In fact, the publishing house he found immediately lent him a sum of money before writing. After three years, he finally wrote up the first draft and submitted it to the publisher, which is Bantom. At first glance, it seemed that it couldn't be used and needed to be rewritten. Because he was already in a wheelchair, it was very difficult for him to do that, and so in the end they found someone to help him revise, revise, and revise.

The reviewer said that most people have not finished reading this book, or have not read it, and cannot understand it after reading it. This happened to him, too; he said: In order to review this book, I spent a week reading it carefully but did not understand it. But this book actually received some good reviews, especially from reviewers in the physics community, because it is an introduction and summary written by a front-line expert for physics people—others were confused (of course).

The book review says: This book has two parts. The first part introduces quantum mechanics and superstrings to the general public, and the second part talks about his own works, etc. So if you know these things very well, this book is very thin and summarizes it very clearly for you. It tells you what he did, roughly the results obtained, and what other problems there are to be explored. If you don't understand, you won't understand the first sentence, that's it. But why do so many people buy this book? Hey, it's because they support this scientist in a wheelchair.

As a result, the book made Hawking a science idol in the public eye, which I'll get to in a moment. But this assessment is a bit exaggerated: He represents a brilliant mind struggling and striving. The book review reads: "His personal tragedy sharpens the metaphor of science in which humans transcend their ephemeral status by trying to comprehend a vast and ancient universe." These few sentences are an analysis of why *A Brief History of Time* is such a hit.

Well, what is the concept of 10 million copies? No popsci books or academic books can surpass it (Fig. 3, left). And there are only these two books that sold more than 10 million copies: (1) The great *Quotations from Chairman Mao Tse-Tung* (middle) is an order of magnitude more than it. This is merely the statistics from 1968. The counting continued until 1976. The number should definitely be higher. (2) The *Bible* (right) is even more formidable: It is an order of magnitude higher than the previous one (middle) but it is mostly given out as a gift.

10 million	740 million (by 1968)	6000 million

Fig. 3. Three books that sold more than 10 million copies: *A Brief History of Time*, *Quotations from Chairman Mao Tse-Tung*, and the *Bible*.

4.2 *The Grand Design*

What's the quality of Hawking's other popsci books? Let's take *The Grand Design*, published by Hawking and Leonard Mlodinow in 2010, as an example. The main ideas of this book: Philosophy is dead and cannot answer, for example, this profound question: Where did this world come from? Only M theory—i.e., superstring theory—*can* answer this question.

If it can, very good. But not yet.

Here is a book review by Angela Olinto: "The book's assertion that physics has all the answers may be especially provocative in a time of growing intolerance toward science, but certainly it is not accurate." [*Physics Today*, Jan. 2011]

What does it mean? It means: In our era where more and more people are antiscience, impatient and intolerant of science, this book affirms an unproven physical theory—M theory in superstring theory. It is said that it can provide the answer to the big question of where the universe came from. This statement is certainly not accurate and may even be wrong. For you to say this in this era is "especially provocative."

That is to say, by hyping the success of *A Brief History of Time*, Hawking became a *spokesperson* for science—it was not claimed by him, but by someone else, but in the end he became a spokesperson for *antiscience*. The result of being antiscience is obviously detrimental to science because he is too uncritical.

Why was he able to do these things, create this book (*The Grand Design*)? To understand this, one needs to know the background of his coauthor.

The coauthor is a PhD in physics from the University of California, Berkeley, majoring in theory. He went to Caltech (California Institute of Technology) to do a postdoc, but in the end he couldn't find a job. He then went to Hollywood to write some science-related scripts; Caltech is in Los Angeles, next to Hollywood. He also wrote popsci books and became famous after writing a biography of Feynman (*Feynman's Rainbow*, 2003), so he kept writing popsci books.

After that, he wrote a popsci book with Hawking, *A Briefer History of Time* (2005), by rewriting *A Brief History of Time* and adding color pictures, and later cowrote *The Grand Design* (2010).

Therefore, these two people have the same background, both in theoretical physics. They know nothing about human society, the human animal, or the complex world of humans. It was because they knew nothing that they were able to make the claims and write this book (*The Grand Design*).

4.3 Children books

Hawking has written five children books with his daughter, Lucy Hawking (Fig. 4).

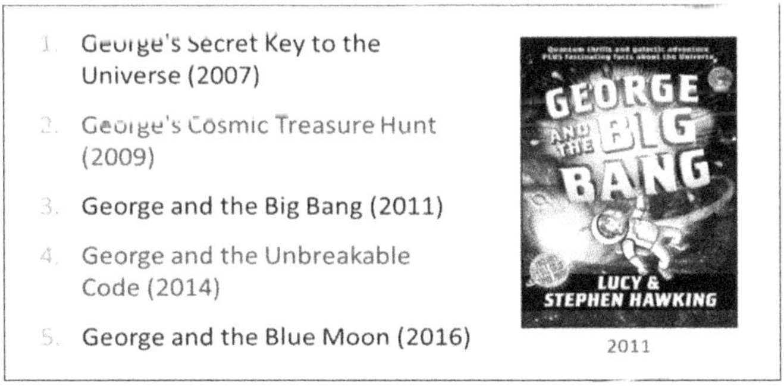
1. George's Secret Key to the Universe (2007)
2. George's Cosmic Treasure Hunt (2009)
3. George and the Big Bang (2011)
4. George and the Unbreakable Code (2014)
5. George and the Blue Moon (2016)
GEORGE AND THE BIG BANG
LUCY & STEPHEN HAWKING
2011

Fig. 4. Five children books coauthored by Lucy and Stephen Hawking.

4.4 Popsci films and series

Well, Hawking has many, many popsci movies and TV shows (Fig. 5). *Stephen Hawking's Grand Design* (2012) is based on his book. *Genius by Stephen Hawking* (2016) is based on the "Genius" competition he hosted. *Stem Cell Universe with Stephen Hawking* (2014 documentary) is interesting, because stem cell is biology, not his field. He did everything.

1. A Brief History of Time (1992)
2. Stephen Hawking's Universe (1997)
3. Hawking – BBC television film (2004), starring Benedict Cumberbatch
4. Horizon: The Hawking Paradox (2005)
5. Masters of Science Fiction (2007)
6. Stephen Hawking and the Theory of Everything (2007)
7. Stephen Hawking: Master of the Universe (2008)
8. Into the Universe with Stephen Hawking (2010)
9. Brave New World with Stephen Hawking (2011)
10. Stephen Hawking's Grand Design (2012)
11. The Big Bang Theory (2012, 2014–2015, 2017)
12. Stephen Hawking: A Brief History of Mine (2013)
13. The Theory of Everything – Feature film (2014), starring Eddie Redmayne
14. Genius by Stephen Hawking (2016)

Fig. 5. Examples of Hawking's popsci films and TV series.

4.5 Sagan and Hawking: two popsci celebrities

Although Hawking has a large number of popsci books and films, even more than Carl Sagan (1934-1996) has, but they should be assessed in quality; quantity is not important. So, let's compare the two of them—in quality (Fig. 6).

Anyone who works in science communication knows Sagan's name. So, how important is this person? When Hawking's *A Brief History of Time* was published in 1988, it was Sagan who wrote the preface to the book. He was just that important. But why then is Sagan's reputation not as great as Hawking's? It's because Sagan is an astronomer and couldn't get a permanent job at Harvard. He later went to Cornell and started doing popsci stuff on his website early on. So, he is just mediocre academically.

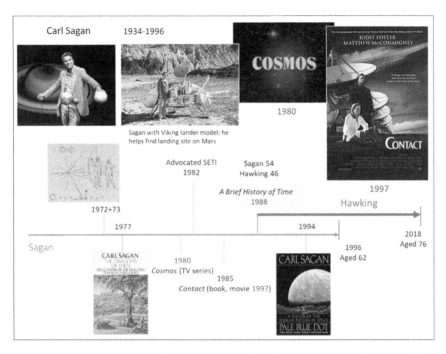

Fig. 6. Comparing the popsci achievements of Carl Sagan and Stephen Hawking.

But he did a great job in popularizing science. Therefore, Hawking stepped on his shoulders and did science popularization on his shoulders. In fact, in popsci, Hawking is not as successful as him. How successful is he?

In 1972 and 1973, he designed this metal plate (engraved with a man and a woman figure; Fig. 6, middle left) which was sent up to space for aliens to see. He was probably one of the two designers, or he may have participated in the design. Why did it take two years? Because a rocket was sent up only once a year, so no hurry.

Sagan's popsci books are not as many as Hawking's. All his popsci things are not as many as Hawking's, but they are all more important than Hawking's!

The Dragons of Eden (1977; Fig. 6, bottom left) is a book about the evolution of human intelligence. *Cosmos* (1980) is a very famous TV series, very famous.

Four years later, in 1984, SETI (Search for Extraterrestrial Research Institute) was formed. At that time, no one supported it. It was established only after a rich man gave one million US dollars. Sagan came out and organized more than 20 famous scientists, including several Nobel laureates, to sign an open letter in support of this institute. Therefore, he advocated finding aliens although he was not the first to propose it.

The next one is his 1985 book *Contact* and the 1997 movie (top right), which is a must-see. Not because of him, but because of her—starring Jodie Foster. She is already a two-time Oscar-winning actress (*The Accused*, 1988; *The Silence of the Lambs*, 1992), and she did not become the best actress through this movie. Sagan's help to find the landing site on Mars for the Viking vehicle (top second left) is also important. I don't know what year it was.

Therefore, adding up the few things he did, everything is quite important; especially in the field of science popularization, he is very influential. Later, when Hawking came out with *A Brief History of Time* in 1988, Hawking surpassed him—surpassed him in terms of public recognition.

Eight years later, in 1996, Sagan passed away at age 62—a short life. (In contrast, Hawking died 76 years old.) And the occupant of the "throne" of science popularization was "officially" replaced. In general, Hawking inherited Sagan's science popularization career, but he did not surpass Sagan.

4.6 Hawking in China

Hawking visited China three times. The first time was in 1985 (Fig. 7, left).

During his last visit in 2006, Hawking said: "I am interested in Chinese culture and food, but I am most interested in Chinese women. They are all beautiful." These are not polite words; he really thinks so, and I agree.

Recently, there is this big telescope called *Tianyan* (天眼, sky eye; Fig. 7, right). Where did this Sky Eye come from? If you want to build something big overseas, you have to write a proposal and persuade people at all levels to tell them why you want to build this and what problems you hope to solve after it is built. In mainland China, this is not the case. A mainland Chinese scientist went to Europe and worked in this field for more than ten years. Finally, he returned to China and said he hoped China would have such a thing to fill the vacuum. He finally succeeded and built this thing. After building this thing, they don't quite know what it is used for.

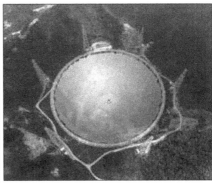

Fig. 7. *Left*: Hawking's first visit to China; photo taken in Hefei (1985). Pay attention to this person in the back row, second from left; his name is Fang Li-Zhi (方励之). *Right*: China's Sky Eye radio telescope in Guizhou Province, called Five-Hundred-Meter Aperture Spherical Radio Telescope (FAST) and started operating in 2016.

So when they promoted it, they said that ours, the largest in the world, can hear messages from the farthest galaxies and find aliens. That's it. So after the construction was completed, they tested it and found the signal of a known neutron star out there. Not long after, another signal was found, and everyone thought it was a signal from aliens.

As soon as he heard about this plan to build the Sky Eye, Hawking said: Don't do it, don't do it. If you want to find aliens, don't do it. Later, in 2017, a year before Hawking's death, when he heard the news that the signal had been received, Hawking immediately said: Don't respond. Why not respond? I agree with what he said, because if aliens really come, the result may be the same as when Columbus discovered the American continent: They will deal with the indigenous people fiercely. And the "indigenous" people are us.

In short, in addition to caring about Chinese beauties, Hawking also cared about aliens attacking China.

5 Film and Television

Hawking appeared in about 20 movies and TV shows, including biographies (*A Brief History of Time*, 1992; *Hawking*, 2013), pop culture productions (*The Big Bang Theory*, since 2007; *Star Trek: The Next Generation*, 1987-1994; *Futurama: The Beast with a Billion Backs*, 2008 Animation; *Simpsons*, appeared multiple times), and other productions (*Quantum Is Calling*, 2016 comedy; *London 2012 Paralympic*, opening ceremony).

Well, these are not important.

6 Science Idol

6.1 Einstein and Hawking: two science idols

It seems that there were no scientific idols before 1919. The 1919 solar eclipse experiments confirmed the general theory of relativity and gave birth to the science idol Einstein. Einstein died in 1955, leaving this science-idol "throne" empty for 33 years before being filled by Hawking (Fig. 8). Hawking occupied it for just 30 years (1988-2018).

You are a scientist. If you want to popularize science, you must do astronomy. Feynman cannot become a great science popularizer because the things Feynman did are not what everyone cares about most.

Why are people most concerned about astronomical things? Because there are three important topics that can be sold in popsci, three origin issues: One is the origin of the universe, one is the origin of life, and the other is the origin of our consciousness. Therefore, many popsci books discuss these three issues. You know that Sagan is an astronomical person, so is Einstein (pioneered modern cosmology), and so is Hawking, which is a bit of a clique.

In short, the science-idol throne had been empty for more than 30 years (1955-1988). Everyone was looking for a new idol. There was no other idol left, so Hawking took the place. This is related to science communication.

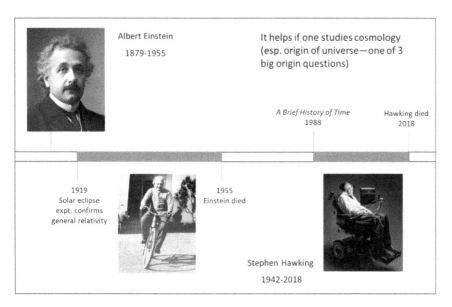

Fig. 8. The vacuum of science idols. Two prominent science idols, Albert Einstein and Stephen Hawking, with a gap of 33 years in between (1955-1988).

6.2 British hype

What is British hype? In 1215, England had the Magna Carta. Later, an emperor with British characteristics was created. What is an emperor with British characteristics? She is a real emperor, she looks like an emperor, or a queen, but she has no real power—from generation to generation.

In *hype*, this principle is copied. However, of course they cannot do this better than us in China. Hype is one of our inherent abilities, and we have been doing it for 2,600 years.

The hyping of Hawking, by himself or others, is a kind of British hype.

6.3 No publicity, no one knows

Without publicity, no one would know. This is what happened to my PhD supervisor, Philip Platzman (1935-2012), who passed away six years ago

(Fig. 9). He has a bachelor's degree from MIT and a doctorate from California Institute of Technology. He has two doctoral supervisors: Gell-Mann (1929-2019) and Feynman (1918-1988), two Nobel laureates.

Fig. 9. Philip Platzman (1935-2012).

Platzman's works with Feynman established his reputation, which is on polaron. After that, he went to Bell Labs as soon as he graduated and stayed there until he retired. Winning the Arthur H. Compton Award (1997) is not important, but in the last ten years of his life, because he fell down the stairs from the second floor of his home, he could not move from the neck down, similar to Hawking's condition. After that, for the next ten years, he went to his office at Bell Labs every day, Monday to Friday, every day. Within ten years, he published 20 articles, of course in collaboration with others.

No one knew—there was no publicity. Why not publicized? Because he didn't have many students, and the most "established" one is me. There was no publicity. You may not find him in Wiki. Go to *Physics Today* (June 2012), look for the period when he died and you'll find his obituary.

So how did he survive? He hired a male nurse, who lived in his home full-time, 24 hours a day. So, this male nurse took care of him, took care of everything. His family didn't have to worry about him.

That Hawking's tragedy is like this; not tragedy, his difficulty, is like this. In the United States, this difficulty does not occur because of its health insurance system. In the case of my PhD supervisor, medical insurance may pay in full or 80%, even though he has retired. Besides, the salary at Bell Labs was very high, and he would have some savings.

This kind of situation definitely calls for one male nurse, not six female nurses. Why do you need a female nurse? You need to hold this big man up, not with two nurses, otherwise the body of the patient in the middle will be broken and injured. A strong man can hold him up with two arms. My PhD supervisor is very strong and used to play tennis a lot. As a result of the male nurse who helped to exercise his body from time to time, during those ten years, his muscles did not deteriorate. We have never seen anyone in such poor condition as Hawking's in his later years.

So there are some problems with the medical insurance system in the UK, as everyone knows. Besides, it was the University of Cambridge that was wrong because it had the resources to assist Hawking.

It would be no problem if it were now. As long as someone goes online and starts a crowdfunding campaign and talks about my old classmate Hawking and how much money is needed, the money will be raised immediately, in one day perhaps, but there was no crowdfunding at that time. However, at that time, it shouldn't have happened. He was a famous university professor and it was easy to take special care of him. Cambridge University has a lot of alumni, doesn't it? I mean, Cambridge didn't take good care of him.

7 After Death

Some people rumored that Hawking in his last 33 years was actually an actor playing him. [*World Journal*, Mar. 20, 2018] Conspiracy theories.

7.1 Words from his colleagues

If you want to read about peers' opinions about him, read *Physics Today*. On the day he died, eight people (including three Hawking's students) wrote something to remember him, with one or several paragraphs each (Fig. 10). Alan Guth is the person I told you who says Hawking never admitted his mistakes. Also, be aware that being his students was not easy; the teacher had violent behaviors.

Words from colleagues

Physics Today (DOI:10.1063/PT.6.4.20180314a 14 Mar 2018 in People & History)

Stephen Hawking (1942–2018)

Colleagues remember the leading cosmologist, whose influence expanded beyond the physics community.

Andrew Grant

1. Don Page (*University of Alberta, Hawking's student*)
2. Marika Taylor (*University of Southampton, Hawking's student*)
3. John Preskill (*Caltech*)
4. Thomas Hertog (*KU Leuven, Hawking's student*)
5. George Ellis (*University of Cape Town*)
6. Alan Guth (*MIT*)
7. William Unruh (*University of British Columbia*)
8. Andrew Strominger (*Harvard University*)

Fig. 10. Colleagues and students invited by *Physics Today* to remember Hawking, published Mar. 14, 2018, the day Hawking died.

7.2 Words from the crowd

What do ordinary people say about him? This one (Fig. 11, top right) represents the positive side; needless to say, everyone knows this: Hawking is the most outstanding scientist, a scientific giant.

Well, look at the rest, from the people out there. They say that many famous people become famous without proving anything, which is shameful. This one says that without good marketing and devoted followers, that is, without publicity and hype, the value of theory is not enough to buy a cup of coffee. This last one says that Hawking had some interesting ideas but didn't actually accomplish anything.

However, Hawking did achieve one thing. I'll tell you later.

Words from the crowd	What kind of person is Hawking? He is a myth, the most outstanding physicist of our time, a scientific giant, and a warrior who challenges destiny.

- **Christopher Johnston** August 9, 2017
  Everything these scientific people do are mostly THEORY. Meaning it is not proven. It is a thought. I fail to see why these people are so famous without ever proving anything.

- **jim m** March 14, 2018
  These arrogant scientists getting awards for coming up with way out ideas that may or may not be true... and then acting like they're brilliant for their perhaps never to be proven opinion. I've seen how these people come to convenient conclusions based on things no honest person would accept. — Shameful — the people that support this nonsense are just as dishonest and egotistical.

- **blade** March 15, 2018
  The value of a theory without good marketing and devoted followers would not be enough to buy a cup of coffee.

- **Lisa Gilmer** March 16, 2018
  So basically he came up with some interesting ideas, but actually didn't accomplish anything.

Fig. 11. Appraisal of Hawking from ordinary people.

7.3 Hawking's funeral

Well, finally his funeral. Funeral is not a burial; it is just a memorial service. A normal burial: It may be transported to the crematorium and burned, or it may be buried. It turned out he was cremated and his ashes were finally buried in Westminster Abbey.[4] Who had been buried there before? Newton, 1727; Darwin, 1882. So there's a conversation between them. Newton: Who's the new guy? Darwin: I heard it's a guy named Hawking.

So, has Hawking done any harm to the world so far? It's not obvious now, but in another 10 or 20 years, it may be seen, if anything. Note that I didn't say there must be. However, there are already two losers. How great these two men were, one Newton and one Darwin. Suddenly, there was a Hawking, and any of his theories were not proven and could be wrong in the future. He got into Westminster Abbey through the use of a wheelchair. Well, actually, Hawking shouldn't be put there (Westminster Abbey).

Presumably, I am the first one to belittle him, but I didn't belittle him 100 percent. Belittle him, yes, but not intentionally—just want to restore a real Hawking.

7.4 The fitting resting place

Actually, it's better to say this to (Elon) Musk: Hey, this scientist, we want to send him up there (using your rockets), because his favorite thing is to go to space. Musk would agree.

By the way, Musk in Chinese is 马斯克; the last two characters when reversed is 马克斯, which sounds the same as 马克思, (Karl) Marx's name in Chinese.

7.5 How much money Hawking left behind

The Chinese title of this lecture is "霍金及其遗产" (遗产 includes money), but the English title is "Hawking and His Legacy." If I had asked the question in English, I wouldn't have to tell you how much money he had left behind. But we use Chinese, so I'll tell you.

Hawking's net worth: $20 million (US dollar). And that's a pretty reliable number.

Here is why. His book *A Brief History of Time* sold 10 million copies. In the United States, paperbacks are generally $15 a book. For simplicity, I assume that the paperback of his book sells for $20, and his author's royalty is 10%, so he gets $2 for every copy sold, and 10 million copies is $20 million. So that's a reliable number! Note that, in addition, he won the Special Fundamental Physics Prize in 2012 and was awarded $3 million, which is a credible number.

So, that's why Elaine, the nurse, wanted to force other nurse competitors away so she could marry this wheelchair-bound man.

8 History of Science and Science Communication

8.1 History of science

History of science is the most important thing in this talk. If you look at the history department of a university, there are three aspects: ancient history, modern history, and contemporary history. Look at any of the literature departments, there is ancient literature, there is modern literature, there is contemporary literature. Exactly the same, right?

So, as a department of the history of science, in addition to these two parts—ancient and modern, shouldn't you have the history of *contemporary* science? Of course we know that the science history department at Tsinghua University is only one year old, and that's why we have a suggestion for it.

Why do you want to do contemporary? I'm not saying you don't do anything else; I'm just saying that in addition to those two, there is contemporary. There are two reasons why: The *first* is to preserve the material for *future* historians. You may not have thought about it, but when we do physics, we often attach great importance to the first step, and the first step is to collect information and collect data. This data can be foreign, natural, or an experiment designed by me. The first step is to gather evidence. So the first step for us to do science history is also to collect evidence: ancient documents, something from the Republic of China, and modern documents. You have to think about the science history in the future. Just imagine: We are sitting here today, if there is basic information about 1,000 years ago, or even 500 years ago: how the Chinese, how the Romans lived, so on and so forth, all the details are here, how helpful it is for us!

So in our generation, we can do this now—collect information. Not necessarily for myself, I collected it and kept it there, and after a hundred years, it will be useful to future historians. This thing can be done, but if you don't do it, who will do it?

The *second* reason, which is actually very important at the moment, is to innovate. You have to promote innovation, and if our department of science history is also willing to promote innovation, not only artistic innovation, but also to promote scientific and technological innovation, then we have to race to the frontline to promote it, right? Especially Tsinghua University, such a great school.

Can your science history department promote innovation? You can tell people how past innovations were done through the history of science. If you give examples of 1,000 years ago, 500 years ago, they are related to the present, but they are not so close, and the stories of innovation in the last 20 or 30 years are closely related to the present. How do they choose the topic? Why does he do this and not that? How does he find information? Who does she/he work with? How does he compete or fight with his peers? And so on. It's the latest story, and it's called Contemporary Science History.

This department (at Tsinghua) is actually specialized in the study of Western history of science. This is no problem, in the most developed place in the country, in China.

We study the history of science in the West—the history of science outside of China. We Chinese do not have the advantage of studying the science history that happened outside China; what is lacking is first-hand information. However, there is a gap that can be filled by contemporary people. This blank is the overseas-Chinese scientists. Westerners don't do it for them; they can't do it because they don't understand Chinese. They don't understand their culture; they can't do it.

So only the Chinese can do it. There are two examples that were done by the same person. In 1996, the book *Wu Chien-Shiung* (1996) came out. How did he do it? He went to Chen-Ning Yang (杨振宁, pinyin: Yang Zhen-Ning) and said I wanted to do this because she was so famous, but no one had written a biography for her and so I really wanted to write it. Yang said, you have to learn the way of writing popsci books in the West; this is a popsci book, you have to interview people who know her, interview her colleagues, friends, and herself.

The author (Chiang Tsai-Chien, 江才健), a reporter from the *China Times* (in Taiwan), took a year off, got a grant, went to the United States to look for information, interviewed many people, and finally wrote this book *Wu Chien-Shiung*. A few years later, he wrote *Biography of Yang Chen-Ning*.

So, someone did it, but it seems that he is the only one who did it.

This book *Wu Chien-Shiung* has been translated into English by a friend of mine, and is currently the *only* English biography of a Chinese physicist in the world (*Madame Wu Chien-Shiung*, World Scientific, 2014). There are several English books for Chinese mathematicians, but for Chinese physicists, you can look it up, and there is only this one.

Chinese scientists, some are great, some are not so great, but no matter how great they are, no one does an English biography for them. So, as a science history department, you study the West, etc., isn't that what you can do and should do?

You don't have to do it yourself; you get money from the government and find someone else to do it. The government will agree to it immediately. It's a work of the united front; isn't it? It's like before that you want to build a science museum first (as planned), and you don't have to do it yourself. With the money, let someone else apply, we support this person to do it, and so on. I didn't say you must do the biography of overseas-Chinese scientists; I mean it's a good topic.

There are many ways to do a biography, including oral history, keeping their information, writing books for them, and so on. In fact, oral history is already being done vigorously in China, at UCAS (University of Chinese Academy of Sciences), but for domestic scientists. The history of science for overseas-Chinese scientists has not yet been done seriously and systematically.

In fact, the content of my lecture today is a contemporary history of science, about Hawking's contemporary history of science.

8.2 Science communication

Isn't science communication, also known as science popularization (*kepu*, 科普 in Chinese), just to convey the spirit of science to the masses? No, but this is the main thing, and of course there are others.

What is the most important thing in the spirit of science? Seeking truth from facts; i.e., telling the masses the truth about what has happened. So, if what a science idol says is true, that's fine, but if he secretly participates in the fake, what would you, a science popularizer, do?

Should a science popularizer be a microphone? Is it just a microphone? A scientist gives you something, or the BBC gives you something and asks you to pass it on to the public, and if you take it all and pass it on directly, you are just a simple microphone.

Or, are you a subjective entity? Are you a person who can think independently? People who do science popularization are not scientists, not necessarily scientists, and cannot make independent judgments. But you can find a way to collect opinions from various sides and interview some other people, and so on. We know this.

Finding out the truth and passing it to the public is the most important job of a science popularizer or science communicator.

9 Conclusion

Hawking's life is absolutely wonderful, and his greatest contribution is to live—have a "terminal illness" but do not give up.

The physics he does is between second-rate and first-class, or barely first-class, but definitely not super-first-class; he definitely not a physics genius. He's a genius for hype, though.[5, 6]

With powerful propaganda, a "first-rate" scientist in a wheelchair can be hyped up to genius.[7] So, do science historians and science popularizers have the responsibility to expose the truth?

Finally, this one is the most important. A female marries a male PhD of physics, think clearly. Have you ever heard of a physics widow? If your husband does physics and does not do well and is looked down upon by his colleagues, he will get no respect; if he does a good job, he has to think about physics 24 hours a day, and he doesn't have time to think of you. So don't marry a physicist unless you run into someone like Feynman. A male marries a female PhD of physics, similar.

The end. Thank you!

Notes

1. In 2020, the mathematician Penrose actually won the Nobel Prize in Physics "for the discovery that black hole formation is a robust prediction of the general

theory of relativity." It seems that the academic standards of the Nobel Prize in Physics have declined in recent years. [Note added in 2021]

2. The S_{BH} equation was not engraved on Hawking's tombstone in the end. What was engraved on his tombstone was another related equation, and there was no trace of Bekenstein on the equation (see Note 4). [Note added in 2019]

3. The *Daily Mail* story is backed up by other reports. For example, *People and the Telegraph* have printed similar startling revelations. In the June 2004 issue, *Vanity Fair* printed an article with the title "A beautiful mind, an ugly possibility: Forty-one years with Lou Gehrig's disease has left famous physicist Stephen Hawking's body utterly vulnerable. After two police investigations, his family and friends reveal why they fear for his life" (https://www.vanityfair.com/ news/2004/06/ hawking200406). On March 14, 2018, the day of Hawking's death, the respectable *Arizona Daily Star* run an article titled "Hawking's second wife was investigated for abuse, but no charges were ever filed," which says: "Cambridge police investigated allegations that his second wife abused him, and his former assistant told *Vanity Fair* in 2004 Hawking only suffered injuries when he was alone with Elaine. But Hawking shot down the allegations, and the investigation was eventually dropped with no charges filed against his wife" (https://tucson.com/hawkings-second-wife-was-investigated-for-abuse-but-no-charges-were-ever-filed/article_c8c33227-1645-5601-830c-bcc947d9e41f.html). [Note added in 2023]

4. This lecture was given on May 25, 2018. It was not until June 15, 2018 that Hawking's ashes were officially buried at Westminster Abbey. It turned out that apart from these words, "Here lies what was mortal of Stephen Hawking 1942-2018," the equation engraved on the tombstone (lying flat on the floor) is $T = \hbar c^3/8\pi GMk$, not the original one (the S_{BH} equation) that Hawking presumably wanted it to be. [Note added in 2019]

5. Stephen Hawking passed away on March 14, 2018, and 72 days later, I gave this Hawking lecture on May 25, 2018 at the Department of the History of Science of Tsinghua University. In addition to introducing and summarizing Hawking's academic work and personal life, I talked about his academic style and the way he deals with people, focusing on his grandstanding and hyped attitude in life. Three years later, *Hawking Hawking: The Selling of a Scientific Celebrity* (2021) by Charles Seife was published. This book argues that Stephen Hawking, a master of self-promotion who declares victory before the problem is solved, is as brilliant in physics as he is in the talent for constructing his own myths. [Postscript added June 10, 2021]

6. We agree with the 2nd part of Seife's assessment of Hawking that he has talent in hype but not the 1st part that he is brilliant in physics. [Note added Jan. 31, 2024]

7. This lecture was given later on April 23, 2019 at Guangxi University for Nationalities in Nanning. After the lecture, the concluding remarks by moderator

Yung Zhi-Yi (容志毅): "Professor Lam broke the convention of only focusing on the academic research achievements of celebrities in previous lectures (by others), regarded Hawking as an ordinary person, and showed his ordinary life to everyone, so that the students' understanding of Hawking as a character is more three-dimensional and comprehensive." [Addendum added Nov. 15, 2021]

Published: Lam, L. [2021]. Hawking and His Legacy. *Lam Lectures: New Humanities, Science, Hawking*, Lui Lam & Jian Xiao-Qing. San Jose: Yingshi Workshop. pp 208-301. Original in Chinese; here is English translation with most of 65 ppt's dropped and words condensed.

Lui Lam's Academic Life

Lui Lam

1949-1955	All Saints Primary School (诸圣小学, Hong Kong)
1955-1961	Clementi Middle School (金文泰中学, Hong Kong)
1961	Ranked second in All Hong Kong Chinese High School Examination
1961-1962	King's College (英皇书院, Hong Kong), Upper Six
1962-1965	Undergrad student in physics and math, University of Hong Kong
1965	BS (First Class Honor), HKU
1965-1966	PhD student in physics, University of British Columbia, Canada (MS, 1968)
1966-1972	PhD student in physics, Columbia University, USA
1972	Finished PhD thesis (after publication in 1974,[1] there are Lam-Platzman Theorem [2] and Lam-Platzman Correction [3] in the physics literature)
1972-1975	Postdoc at Physics Department, City College, City University of New York (created theory of dissipation function for complex materials [4])
1973	PhD, Columbia University
1975-1976	Research associate, Universitaire Instelling Antwerpen, Belgium
1976-1977	Research associate, Universitat des Saarlandes, West Germany
1978-1983	Associate professor/researcher, Institute of Physics, Chinese Academy of Sciences, Beijing
1979	Published first paper with only mainland-Chinese authors in *Physical Review Letters* [5]
1980	Cofounded Chinese Liquid Crystal Society

1982	Invented Bowlic liquid crystals (published in *Wuli* [6]); pioneered study of solitons in (shearing) liquid crystals [7]
1984-	Guest professor, Institute of Physics, CAS, Beijing
1984-1987	Associate professor, Queens Community College, CUNY; adjunct professor, City College, CUNY
1987-1990	Associate professor, San Jose State University, California
1988	Invented bowlic polymers and predicted bowlic room-temperature superconductors [8]
1990	Initiated and established International Liquid Crystal Society [9]
1990-2018	Professor, San Jose State University, California
1992	Invented Active Walk [10]; founder and editor-in-chief, Partially Ordered Systems book series (Springer)
2001-	Guest professor, China Research Institute of Science Popularization, China Association for Science and Technology
2002	Established the new discipline Histophysics (physics of history) [11]
2006	Discovered Bilinear Effect in complex systems [12]
2007	Established the new multidiscipline Scimat [13]; proposed the third definition of Science [13,14]
2008	Founder and editor-in-chief, Science Matters book series (World Scientific); published first Scimat book *Science Matters: Humanities as Complex Systems*
2011	Published a new theory on origin and nature of art [15]
2013	*Renke* (人科), Chinese translation of *Science Matters*, published by China Renmin University Press
2017	Distinguished Service Award, San Jose State University, CA [16]
2018	Published Chinese paper on innovation [17]
2019-	Professor emeritus, San Jose State University, CA
2021	Published 2 Chinese books: *Wenliren* (文理人), *Lam Lectures: New Humanities, Science, Hawking* (林磊演讲录：新文科、科学、霍金)
2022	Published 6 Chinese books: *New Humanities* (新文科), *Science and Scientist* (科学与科学家), *Research and Innovation* (研究与创新), *Being*

	Human (做人做事), *This Pale Blue Dot* (这淡蓝一点), *China Complex* (中国情结)
2024	Published 2 English books: *Humanities, Science, Scimat*; *Scimat Anthology: Histophysics, Art, Philosophy, Science*

Notes and References

1. Lam, L. & Platzman, P.M. [1974]. Momentum density and Compton profile of the inhomogeneous interacting electronic system. I. Formalism. Phys. Rev. B **9**: 5122.

2. Bauer, G. E. W. [1983]. General operator ground-state expectation values in the Hohenberg-Kohn-Sham density functional formalism. Phys. Rev. B **27**: 5912-5918.

3. Callaway, J. & March, N. H. [1984]. Density functional methods: Theory and applications. Solid State Physics **38**: 135-220; Papanicolaou, N. I., Bacalis, N. C. & Papaconstantopoulos, D. A. [1991]. *Handbook of Calculated Electron Momentum Distributions, Compton Profile, and X-ray Form Factors of Elemental Solids*. Boston: CRC Press; Blass, C., Redinger, J., Manninen, S., Honkimäki, V., Hämäläinen & Suortti, P. [1995]. High resolution Compton scattering in Fermi surface studies: Application to FeAl. Phys. Rev. Lett. **75**: 1984-1987.

4. Lam, L. [1977]. Dissipation function and conservation laws of molecular liquids and solids. Z. Physik B **27**: 101; Reciprocal relations of transport coefficients in simple materials. Z. Physik B **27**: 273; Constraints, dissipation functions and cholesteric liquid crystal. Z. Physik B **27**: 349. This dissipative function theory was later used for biaxial nematic phase liquid crystals [Das, P. & Schwartz, W. H., Mol. Cryst. Liq. Cryst. **239**: 27-54 (1994)]. In fact, it remains the best and easiest way to deal with the fluid dynamics of any new thermoviscous solid or molecular fluids with microscopic structures, including liquid crystals.

5. Lin, Lei (Lam, L.) [1979]. Nematic-isotropic transitions in liquid crystals. Phys. Rev. Lett. **43**: 1604-1607.

6. Lin, Lei [1982]. Liquid crystals phases and "dimensionality" of molecules. Wuli (Physics) **11**: 171-178.

7. Lin, Lei, Shu Changqing, Shen Juelian, Lam, P.M. & Huang Yun [1982]. Soliton propagation in liquid crystals. Phys. Rev. Lett. **49**: 1335-1338; Lam, L. & Prost, J. [1992]. *Solitons in Liquid Crystals*. New York: Springer.

8. Lam, L. [1988]. Bowlic and polar liquid crystal polymers. Mol. Cryst. Liq. Cryst. **155**: 531-538.

9. Lam, L. [2017]. Prehistory of International Liquid Crystal Society, 1978-1990: A personal account. Mol. Cryst. Liq. Cryst. **647**: 351-372.

10. Lam, L. [1995]. Active walker models for complex systems. Chaos Solitons Fractals **6**: 267.

11. Lam, L. [2002]. Histophysics: A new discipline. Mod. Phys. Lett. B **16**: 1163-1176.

12. Lam, L. [2006] Active Walks: The first twelve years (Part II). Int. J. Bifurcation and Chaos **16**: 239-268; Lam, L., Bellavia, D. C., Han, X.-P. Liu, C.-H. A. , Shu, C.-Q., Wei, Z., Zhou, T. & Zhu, J. [2010]. Bilinear effect in complex systems. EPL (Europhys. Lett.) **91**: 68004.

13. Lin, Lei [2008]. Science Matters: The newest and biggest interdiscipline. China Interdisciplinary Science, Vol. 2, Liu Zhong-Lin (刘仲林) (ed.). Beijing: Science Press. pp 1-7; Lam, L. [2008]. Science matters: A unified perspective. *Science Matters: Humanities as Complex Systems*, Burguete, M. and Lam, L. (eds.). Singapore: World Scientific. pp 1-38.

14. Lam, L. [2014]. About science 1: Basics—knowledge, nature, science and Scimat. *All About Science: Philosophy, History, Sociology & Communication*, Burguete, M. and Lam, L. (eds.). Singapore: World Scientific. pp 1-49.

15. Lam, L. [2011]. Arts: A science matter. *Arts: A Science Matter*, Burguete, M. and Lam, L. (eds). Singapore: World Scientific. pp 1-32.

16. Jackson, J. H. [2017]. Lui Lam: 2017 Distinguished Service Award. *SJSU Washington Square* https://blogs.sjsu.edu/wsq/2017/02/21/lui-lam-2017-distinguished-service-award/.

17. Lin, Lei (林磊), Liu Li (刘立) & Sun Nan (孙楠) [2018]. Make full use of domestic journals to obtain priority in publications. Science and Technology China (科技中国), July 2018, Issue 7: 48-50.

Index

Song, 355
Tang, 355
Dynasties of the World, 152, 153

E

Eat Art, 349
Earth, 21, 41, 66, 84, 328, 367, 409, 473, 477
East, 297, 311, 371, 381, 388, 396, 616
Economic and Philosophical Manuscripts of 1844, 376
Economics, 174, 192
  birth of, 409
  computer modeling, 216
Eddington, Arthur, 448
Edmonton, 623
Edinburgh, 613
Education, 25, 27, 42, 45, 66, 82, 88
  reform, 27-29, 516
Edmonton, Canada, 623, 624
'85 Art New Wave, 360
Einstein, Albert, 398, 451, 500, 674
Electroconvection, 613
Emergent property, 51
Emperor Kuang-Xu, 560
England, 19, 40, 138, 140, 141, 452, 545, 653, 675
English learning, 129
Enlightenment, 77, 86, 407, 409
Epigene, 251
Ericeira, Portugal, 9-11, 66, 509
Eros, 403
ETH, Zürich, 457
Ethics, 38, 63, 320, 334, 335, 339, 345, 350, 383, 386, 389, 396, 406, 428, 434, 478
Europe, 561, 569
Evanston, 547
Evolution, 192, 367, 398, 405, 409, 477
Experimental philosophy, 390
Experimental Philosophy, 390

F

Falsificability, 453, 457
Fan, Ke-An, 539
Fang, Li-Zhi, 673
Fang, Fu-Kang, 543

Feng, Duan, 543
Feng, Huan, 543
Feng, Hui, 543
Feng, Kang, 542, 543, 551-553
Fermi, Enrico, 493
Feyerabend, Paul, 448, 452, 457, 480
Feynman, Richard, 270, 642, 661, 674
Feynman Lectures on Physics, 540
Fire, 251
Fish, 95, 125, 172, 398, 477, 480
  descendants of, 84, 94, 328
  Microbrachius, 84
Foundation For the Future, 151, 154, 505
Fountain, 256, 259, 325, 327, 331, 342
Fractal, 25, 52, 167
Fragrant Hill Hotel, Beijing, 358
France, 542, 570, 616
Frank, Charles, 581, 611
Freiburg, 619
Freie Universität Berlin, 577
Freud, Sigmund, 323
Fribourg, 538
Friedel, Jacques, 612
Fukuda, Atsuo, 606, 614
Fuzzy logic, 64
Fuzzyism, 392

G

Galileo Galilei, 248, 396, 407, 415, 416
Gang of Four, Orsay, 632
Gates, Bill, 186
Geim, Andre, 652, 654
Gell-Mann, Murray, 158, 159, 294
Gene, 252
General education, 29, 90, 91
General Education and the Plight of Modern Man, 91
Gengzi Epidemic, The, 345, 350
Genius, 308, 426, 656, 658, 660, 670, 684
Genius of Physics, The, 168
Gennes, Pierre-Gilles de, 569, 584, 587, 609, 612, 614, 616
  hand-drawn picture, 587
Georgia Institute of Technology, 600
Germany, 177, 616
  West, 578

Printed in the USA
CPSIA information can be obtained
at www.ICGtesting.com
JSHW061816260824
68693JS00003B/8